A Companion to Political Geography

Edited by

John Agnew
University of California, Los Angeles
Katharyne Mitchell
University of Washington
and
Gerard Toal (Gearóid Ó Tuathail)
Virginia Tech

BLACKWELL PUBLISHING
350 Main Street, Malden, MA 02148-5020, USA
9600 Garsington Road, Oxford OX4 2DQ, UK
550 Swanston Street, Carlton, Victoria 3053, Australia

First published 2003 by Blackwell Publishing Ltd
First published in paperback 2008 by Blackwell Publishing Ltd

1 2008

Library of Congress Cataloging-in-Publication Data

A companion to political geography / edited by John Agnew, Katharyne Mitchell, and Gerard Toal.
p. cm. — (Blackwell companions to geography ; 3)
Includes bibliographical references and index.
ISBN 978-0-631-22031-2 (hardback) — ISBN 978-1-4051-7564-7 (paperback)
1. Political geography. I. Agnew, John A. II. Mitchell, Katharyne. III. Ó Tuathail, Gearóid
IV. Series.

JC319 .C645 2003
320.1′2—dc21

2002003789

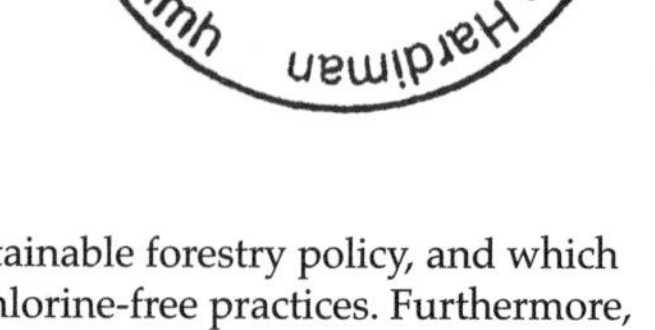

A catalogue record for this title is available from the British Library.

Set in 10 on 12pt Sabon
by Kolam Information Services Pvt Ltd, Pondicherry, India
Printed and bound in Singapore
by Utopia Press Pte Ltd

The publisher's policy is to use permanent paper from mills that operate a sustainable forestry policy, and which has been manufactured from pulp processed using acid-free and elementary chlorine-free practices. Furthermore, the publisher ensures that the text paper and cover board used have met acceptable environmental accreditation standards.

For further information on
Blackwell Publishing, visit our website:
www.blackwellpublishing.com

Contents

Contributors

John Allen is Professor and Head of Geography in the Faculty of Social Sciences at the Open University. His recent publications include *Rethinking the Region: Spaces of Neoliberalism* (Routledge, 1998) with Doreen Massey and Alan Cochrane, and *Lost Geographies of Power* (Blackwell, 2002).

R. Scott Appleby is Professor of History at the University of Notre Dame, where he also serves as the John M. Regan, Jr. Director of the Joan B. Kroc Institute for International Peace Studies. He is the author, most recently, of *The Ambivalence of the Sacred: Religion, Violence and Reconciliation* (Rowman & Littlefield, 2000), and a co-author of *Strong Religion: The Rise of Fundamentalisms in the Modern World* (University of Chicago Press, 2002).

Mark Bassin is Reader in Political and Cultural Geography at University College London. He is the author of *Imperial Visions: Nationalist Imagination and Geographical Expansion in the Russian Far East 1840–1865* (Cambridge University Press, 1999). He has been a visiting professor at UCLA, Chicago, Copenhagen, and Pau (France), and has received research grants from bodies including the American Academy in Berlin, the Institut für Europäische Geschichte (Mainz), and the Fulbright Foundation.

Noel Castree is a Reader (Associate Professor) in Human Geography at the University of Manchester. His interests are in the political economy of environmental change, with a specific focus on Marxian theories. Co-editor (with Bruce Braun) of *Remaking Reality: Nature at the Millennium* (Routledge, 1998) and *Social Nature* (Blackwell, 2001), he is currently researching how economic and cultural value are constructed in the "new" human genetics.

Simon Dalby is Professor of Geography, Environmental Studies and Political Economy at Carleton University in Ottawa where he teaches courses on geopolitics and environment. He is co-editor of *The Geopolitics Reader* (Routledge, 1998) and

Rethinking Geopolitics (Routledge, 1998), and is the author of *Environmental Security* (University of Minnesota, 2002).

Klaus Dodds is Senior Lecturer in Geography at Royal Holloway, University of London. He is author of *Geopolitics in a Changing World* (Pearson Education, 2000) and *Pink Ice: Britain and the South Atlantic Empire* (I B Tauris, 2002). He also joint edited, with David Atkinson, a collection of essays called *Geopolitical Traditions* (Routledge, 2000).

Brendan Gleeson is currently Deputy Director and Senior Research Fellow at the Urban Frontiers Program, University of Western Sydney, Australia. He has authored and co-authored several books in the fields of urban planning, geography, and environmental theory. His most recent book, with N.P. Low, *Governing for the Environment*, was published in 2001. He has undertaken research and teaching in a range of countries, including Britain, Germany, the USA, Australia, and New Zealand.

Richie Howitt is Associate Professor of Human Geography, Macquarie University, Sydney, Australia, where he teaches in the Resource and Environmental Management and Aboriginal Studies programs. His professional work has involved applied research in social impact assessment, native title negotiations, and community development in remote Australia. He has previously published papers on theoretical issues of geographical scale, indigenous rights, and resource management.

Ron Johnston is a Professor in the School of Geographical Sciences at the University of Bristol. He has collaborated with Charles Pattie (see entry below) in a wide range of work on electoral geography since the mid-1980s, including the following books: *A Nation Dividing?* (with G. Allsopp); *The Boundary Commissions* (with D. J. Rossiter); and *From Votes to Seats* (with D. Dorling and D. J. Rossiter).

Gerry Kearns is a Lecturer in Geography at the University of Cambridge and a Fellow of Jesus College. He works on nineteenth-century urban public health, Irish nationalism, and the history and philosophy of geography.

Eleonore Kofman is Professor of Human Geography at Nottingham Trent University, UK. Her research focuses on gender, citizenship, and international migration in Europe, including skilled and family migration, and feminist political geography. She has co-edited *Globalization: Theory and Practice* (Pinter, 1996), and co-authored *Gender and International Migration in Europe: Employment, Welfare and Politics* (Routledge, 2000).

Vladimir Kolossov is Head of the Center of Geopolitical Studies at the Institute of Geography of the Russian Academy of Sciences in Moscow, Professor at the University of Toulouse-Le Mirail (France) and Chair of the International Geographical Union Commission on Political Geography. Recent books include *The World in the Eyes of Russian Citizens: Public Opinion and Foreign Policy* (FOM, 2002, in Russian), and (as co-author) *La Russie (la construction de l'identité nationale)* (Flammarion, 1999).

Sankaran Krishna is Associate Professor and Chairman, Department of Political Science, at the University of Hawai'i at Manoa in Honolulu. He is the author of *Postcolonial Insecurities: India, Sri Lanka and the Question of Nationhood* (Minnesota, 1999).

Victoria Lawson is Professor of Geography and the Thomas and Margo Wyckoff Endowed Faculty Fellow at the University of Washington. Her research and teaching is concerned with the social and economic processes of global restructuring in the Americas with a particular focus on migration, identity formation, and the feminization of poverty. Her most recent work has appeared in journals such as the *Annals of the Association of American Geographers*, *Progress in Human Geography* and *Economic Geography.*

Karen T. Litfin is Associate Professor of Political Science at the University of Washington in Seattle. She teaches and writes primarily on global environmental politics. Her publications include *Ozone Discourses: Science and Politics in Global Environmental Cooperation* (Columbia University Press, 1994) and *The Greening of Sovereignty in World Politics* (MIT Press, 1998).

Nicholas Low is Associate Professor in Environmental Planning at the Faculty of Architecture, Building and Planning at University of Melbourne. His interests include urban planning, politics and state theory, environmental justice, participation, decision making and problem solving, and land markets. Recent books, with B. Gleeson, include *Justice, Society and Nature* (Routledge, 1998) and *Governing for the Environment* (Palgrave, 2001).

Timothy W. Luke is University Distinguished Professor of Political Science at Virginia Polytechnic Institute and State University in Blacksburg, Virginia. He is author of recent books including *Museum Politics: Power Plays at the Exhibition* (University of Minnesota Press, 2002) and *Capitalism, Democracy, and Ecology: Departing from Marx* (University of Illinois Press, 1999), and co-editor, with Chris Toulouse, of *The Politics of Cyberspace* (Routledge, 1998).

Wolfgang Natter is Associate Professor of Geography and Co-Founder/Director of the Social Theory Program at the University of Kentucky. His research has explored the ramifications of various poststructuralisms for understandings of space, aesthetics, nationalism, cultural memory, identity politics, democratic theory, and film, particularly in German and US contexts. He is currently pursuing research on Friedrich Ratzel and the disciplinary history of geography in Germany and the USA prior to the World War II.

David Newman teaches political geography in the Department of Politics and Government at Ben Gurion University of the Negev in Israel. He received his BA from the University of London in 1978, and his PhD from the University of Durham in 1981. He is currently co-editor of *Geopolitics*. He has written widely on territorial aspects of the Arab–Israeli conflict, with a particular focus on boundary and settlement issues and, more recently, has become engaged in the debate over deterritorialization and the "borderless" world.

John O'Loughlin is Professor of Geography and Director of the Graduate Training Program on "Globalization and Democracy" at the University of Colorado at Boulder. He is editor of *Political Geography.* His research interests are in spatial modeling of political processes, the democratic transitions in the former Soviet Union, and Russian geopolitics.

Anssi Paasi is Professor of Geography at the University of Oulu in Finland. He has published extensively on the history of geographical thought, on "new regional geography," region/territory building, and the sociocultural construction of boundaries and spatial identities. His books include *Territories, Boundaries and Consciousness: The Changing Geographies of the Finnish-Russian Border* (Wiley, 1996) and *J.G.Granö: Pure Geography* (co-edited with Olavi Granö) (Johns Hopkins University Press, 1997).

Charles Pattie is a Professor of Geography at the University of Sheffield. His research interests include redistricting, political parties and campaigning, and citizenship and participation. He has published widely in numerous journals and books. Since the mid-1980s he has collaborated with Ron Johnston (see entry above) in a wide range of work on electoral geography.

Paul Routledge is a Reader in Geography at the University of Glasgow. His principal interests concern geographies of resistance movements, geopolitics, South Asia, and the cultural politics of development. He is author of *Terrains of Resistance* (Praeger, 1993), and co-editor (with Gearóid Ó Tuathail and Simon Dalby) of *The Geopolitical Reader* (Routledge, 1998), and (with Joanne Sharp, Chris Philo, and Ronan Paddison) of *Entanglements of Power: Geographies of Domination/Resistance* (Routledge, 2000).

Michael J. Shapiro is Professor of Political Science at the University of Hawai'i. Among his publications are: *Violent Cartographies: Mapping Cultures of War* (University of Minnesota Press, 1997), *Cinematic Political Thought: Narrating Race, Nation and Gender* (NYU Press, 1999), *For Moral Ambiguity: National Culture and the Politics of the Family* (University of Minnesota Press, 2001), and *Reading "Adam Smith": Desire, History and Value* (2nd edition with new Preface; Rowman and Littlefield, 2002).

Joanne P. Sharp is a lecturer in Geography at the University of Glasgow. Her research interests are in political, cultural, and feminist geography with a particular interest in popular geopolitics. She recently published a monograph on the role of the media in the construction of US political culture as *Condensing the Cold War: Reader's Digest and American Identity* (University of Minnesota Press, 2000).

David Slater is Professor of Social and Political Geography in the Department of Geography at Loughborough University. He is author of *Territory and State Power in Latin America* (Macmillan, 1989) and co-editor of *The American Century* (Blackwell, 1999). He is also an editor of the journal *Political Geography.*

Matthew Sparke is an Associate Professor with appointments in both Geography and the Jackson School of International Studies at the University of Washington. He has published in numerous journals including *Society and Space*, *Geopolitics*, and *Gender, Place and Culture*, and is the author of *Hyphen-Nation-States: Critical Geographies of Displacement and Disjuncture* (University of Minnesota Press, forthcoming). He is currently working on a National Science Foundation CAREER project integrating his research on the transnationalization of civil society with educational outreach initiatives in poorer neighborhoods of Seattle.

Lynn A. Staeheli is Associate Professor of Geography and a Research Associate in the Institute of Behavioral Science at the University of Colorado. Her research interests include citizenship, democratization, political activism, immigration, and gender.

Peter Taylor is Professor of Geography at Loughborough University and Associate Director of the Metropolitan Institute at Virginia Tech. Over the last two decades he has developed a world-systems political geography including a textbook (*Political Geography: World-Economy, Nation-State and Locality*, 4th edition with C. Flint; Prentice Hall, 2000), monographs on world hegemony (*The Way the Modern World Works*; Wiley, 1995), and ordinary modernity (*Modernities: a Geohistorical Perspective*; Polity, 1999). Current work focuses on quantitative measures of the world–city network and he is Founder and Co-Director of the Globalization and World Cities (GaWC) Study Group and Network (see www.lboro.ac.uk/gawc).

Karen E. Till is an Assistant Professor of Geography and Co-Director of the Space and Place Research Group of the Humanities Center at the University of Minnesota. Her research has focused on the cultural politics of place and social memory, national identity, and urban public landscapes in the USA and Germany. Her recent publications include a co-edited volume, *Textures of Place: Rethinking Humanist Geographies* (University of Minnesota Press, 2000), and *The New Berlin: Memory, Politics, Place* (University of Minnesota Press, forthcoming), based upon 10 years of ethnographic research.

Gill Valentine is a Professor of Geography at the University of Sheffield where she teaches social and cultural geography, and qualitative methods. She has published widely on a range of topics including geographies of sexuality, consumption and children, youth and parenting. Gill is co-author/co-editor of eight books including: *Children in the Information Age* (Falmer Routledge, 2002), *Social Geographies* (Longman, 2001), *Children's Geographies* (Routledge, 2000), and *Mapping Desire* (Routledge, 1995).

Colin H. Williams is Research Professor in the Department of Welsh, Cardiff University and an Adjunct Professor of Geography at the Department of Geography, University of Western Ontario. He also serves as a Member of the Welsh Language Board and on European government agencies concerned with multiculturalism and multilingualism. He is well known for his scholarly and practical work encouraging the rights of ethnic and religious minorities worldwide.

We dedicate this book to the memory of a colleague who most certainly would have had a chapter in it if he was still with us: Dr. Graham Smith, Cambridge University (1953–99).

Chapter 1

Introduction

John Agnew, Katharyne Mitchell, and Gerard Toal (Gearóid Ó Tuathail)

In a photograph that won a prize in the *Overcoming the Wall by Painting the Wall* exhibition mounted by the museum at Checkpoint Charlie in West Berlin in 1989, Ziegfried Rischar has superimposed a hand breaking through the Berlin Wall that had divided the city from 1961 to 1989 to offer a white rose to an outstretched hand on the other side. It was poster art such as this that carried the messages of many of the protagonists of the "velvet revolutions" that swept through Eastern Europe and into the Soviet Union in the years between 1980 and 1992. The Cold War division of Europe, symbolized most graphically by the Berlin Wall, had to be overcome and replaced by a new, nonantagonistic relationship between "East" and "West." This particular poster is also representative of the sense – wildly popular at the time in Eastern Europe – that old barriers were breaking down and a new world order was about to dawn. Many such hopes have been dashed. Certainly, most of the old barriers have come down. But new ones, such as restricted entry into the European Union, Russia's exclusion from the European "club," and gated communities protecting the affluent from the impoverished, have replaced them.

Human history has rarely seen such a crystalline moment of change as November 9, 1989, when thousands of cheering people climbed upon, dismantled, and overcame the Berlin Wall by passing through it unimpeded. The revolution of ordinary citizens breaking through a geopolitical division in the heart of Europe was the culmination of a long struggle by new social movements to create a cultural space that challenged and moved beyond the geopolitics of the Cold War. With the mass media in the hands of authoritarian Communists until the very end in Eastern Europe, these social movements gave expression to their principles and aspirations in artistic creations and urban street activities. "1989," one commentator noted, "was the springtime of societies aspiring to be civil" (Ash, 1990, p. 147). Vaclav Havel, later president of Czechoslovakia and the Czech Republic, noted: "In November 1989, when thousands of printed and hand drawn posters expressing the real will of the citizens were hanging on the walls of our towns, we recognized what power is hidden in their art" (quoted in Smithsonian Institution, 1992, p. 25).

At least two lessons seem to emerge from the events captured by Rischar's image. One is that the last decade of the twentieth century was one of the most dramatic periods in the reordering of the world's political geography. Between 1945 and 1989, most political leaders and commentators around the world thought that the Cold War geopolitical divisions were more or less permanent. We now know better. In fact, with hindsight we can see that geopolitical order and the relative barrier to movement and interaction posed by national boundaries have never been fixed but always historically contingent (Agnew and Corbridge, 1995). We can also see that power is not simply concentrated in the hands of states and other organizations (such as transnational corporations and the mass media), but is also a capacity available to people when they mobilize collectively to realise their aspirations (as social movements and new group identities) and pursue their material and symbolic interests. One of the great surprises of 1989 was how the commitment of vast masses of people overcame the coercive apparatus of the states arrayed against them. Of course, we should not be naïve enough to think that coercion could not have worked if external conditions (such as the absence of Soviet military intervention) and internal changes (such as the demoralization of police forces) had not been favorable. "Resistance" does not in itself guarantee political success (Sharp et al., 2000).

What is Political Geography?

We begin this *Companion to Political Geography* with the theme of divisions and power because of the centrality of orders and borders to contemporary "political geography." As an area of study, "political geography" has changed historically but the themes of borders and orders, power, and resistance are always central to its operation. For us, political geography is about how barriers between people and their political communities are put up and come down; how world orders based on different geographic organizing principles (such as empires, state systems, and ideological-material relationships) arise and collapse; and how material processes and political movements are re-making how we inhabit and imagine the "world political map." Barriers are not only global or international, but also operate between regions within countries, and between neighborhoods within cities. They are conceptual and ideological as well as economic and physical. Politics is likewise not simply state-oriented, but includes the collective organization of social groups to oppose this or that activity (such as land-use changes they do not like) or to pursue objectives that transcend political boundaries (such as environmental or developmental goals). Political movements can be open and inclusive, asking critical questions of power structures and always pushing at the limits of human freedom of expression and how humans can live. Alternatively, they can be exclusive and closed to change, radically seeking a return to an idealized past or simplified moral universe, containing and corralling the possibilities of human freedom.

Reflecting on the historical evolution of "political geography" is instructive in situating what we have gathered in this volume to represent contemporary political geography. The use of the term "political geography" dates only from 1750 when the French *philosophe* Turgot coined it to refer to his attempts to show the relationship between geographic "facts," from soils and agriculture to settlement and ethnic distributions, to political organization. Political geography, in other words, was conceived as a branch of knowledge for government and administration – as state knowledge. As a self-confessed sub-area of academic Geography, the term is even more recent, dating from the 1890s. As reinvented at that time, the field was particularly oriented to justifying and providing advice about the colonial ventures in which the Great Powers were then engaged (Godlewska and Smith, 1994). The word "geopolitics" was also invented in the 1890s, by the Swedish political scientist Kjellen, to refer to the so-called geographic basis of world politics. In the 1920s this word was expropriated by a group of right-wing Germans to offer justification for German territorial expansion. Thereafter disavowed for many years by professional geographers, the word has undergone a recent revival both in the hands of politicians and among political geographers. The former use it to refer to "hard headed" global strategies, whereas the latter are typically interested in how geography figures in the making of foreign policies (Parker, 1998).

But with respect to the political organization of earthly space and the links between places and politics, political geography pre-dates use of the term as such. From this viewpoint, it is an ancient enterprise with such venerable practitioners as Aristotle, Thucydides, Sun Tzu, and Livy and more recent exponents as Machiavelli, Hobbes, Montesquieu, Madison, Rousseau, and Hegel. Thucydides' (Strassler, 1996) idea of the fundamental opposition between sea- and land-powers – exemplified for him, respectively, by Athens and Sparta – has repeatedly been recycled as a key idea in modern geopolitics. A book published as recently as 1999 is organized around it but without citing the great man himself (Padfield, 1999). Far-right geopoliticians from South America to Russia and the United States still evoke variants on such radically simplifying deterministic categories (on Russia, see Smith, 1999). Jean Gottmann, possibly the greatest political geographer of the twentieth century, saw each of the historic figures in political thought wrestling intellectually, among other things, with how space is and should be organized politically. He was rightly critical of much of what had been made of them by later generations (Gottmann, 1952, 1973).

Early twentieth century political geography was largely in thrall to the great nation-states of the time, reflecting the common thinking of the era across most fields in the social sciences. A tendency to read geography in largely physical terms was combined with a reductionist understanding of politics as the activities of states and their elites. Thus, successful states were explained in terms of their relative location on a world scale and the resource bases they could exploit. Much effort was taken up with exhaustive accounting of state assets and with boundary disputes of one sort or another (see Kasperson and Minghi, 1969). Little or no attention was paid to politics outside the purview of states or to normative and ethical questions about the nature of rule or the "best" type of political organization for this or that problem. There were exceptions, such as Gottmann (1952) and Wilkinson (1951). But they are the exceptions that help to prove the rule.

Since the 1960s, the field has gone through a long period of reinvention using very different theories and methods than those that characterized political geography in the first part of the twentieth century. Although still focused broadly around questions of political territoriality and boundary-making, the old interest in global geopolitics has been revitalized in various types of "critical geopolitics" which problematize powerful geopolitical discourses (Ó Tuathail 1996), and new research areas, such as place and political identities and geographies of ethnic conflict, have been engaged (e.g. Miller 2000). This revitalization has produced a veritable explosion of research and publication, including new journals and new research organizations.

Currently, three broad currents of thought run across the field. One adopts a spatial-analytic perspective to examine geographic patterns of election results or international conflicts and relate these to place differences, the spread of democratic practices, or the global pattern of interstate hostility (see, e.g., O' Loughlin, 1986 and chapter 3, this volume). A second takes a political–economic approach to understanding the historical structures of global political dominance, hegemonic competition between Great Powers, the development of a new geopolitical order based around major world cities (such as New York, Tokyo, and London), and the political economy of "law and order" [see, e.g., Glassman, 1999; Helleiner, 1999; Herbert 1997; chapters 4 (Taylor) and 29 (Gleeson and Low), this volume]. A third sees power as always mediated by modes of representation or ways of talking about and seeing the world [e.g. Hyndman, 2000; Ó Tuathail, 1996; chapters 6 (Slater), 18 (Shapiro), 19 (Till), and 20 (Krishna), this volume]. In this postmodern approach, international conflicts are understood in terms of the competing narratives or stories each side tells about itself and the other, nationalist identities are seen as constructed around popular memories that need repeated commemoration and celebration at sites of ritual or "places of memory," and groups invent or maintain identities by associating with particular places and the images such places communicate to larger audiences (see, e.g., Sharp, 2000). These currents are hardly sealed off from one another and innovative thinking frequently works across them. But as a rough and ready way of characterizing the theoretical structure of contemporary political geography the threefold division has considerable merit.

We would argue that three influences have helped to raise the profile of political geography around the world after a long period of intellectual stagnation following World War II (particularly during the early Cold War). The first was the slow erosion of the intellectual grip of the Cold War mentality beginning with the Vietnam War and ending with the Soviet collapse. In a wide range of fields the Cold War had intellectually stultifying effects (see, e.g., Siebers, 1993). Not surprisingly, given its subject matter, political geography was especially affected. Cold-War thinking led to a refusal on both sides to consider the historical character of geopolitical arrangements, a tendency to see each side as concentrated entirely in the capital cities of the two major (non)combatants, a freezing of international boundaries around the world to diminish the chances of military escalation if local conflicts brought in the two Superpowers, and national security states that were put beyond question for domestic criticism or proposals for alternative security arrangements. The final collapse of the Soviet Union was the icing on the cake, so to speak.

The second has been the recruitment into the social sciences in general and political geography more specifically of people from a wider range of geographic and social backgrounds. At one time, political geographers were overwhelmingly European and American males from upper and middle class backgrounds in the various Great Powers. Today, this is much less the case. This diversification of backgrounds has undoubtedly encouraged perspectives less oriented to the central political importance of states – particularly the Great Powers – and research interests that focus on the problems and prospects of subordinated social groups and identities.

The third is the synergy with a number of powerful intellectual influences originating both within Geography and in other fields. Good examples would be the influence of that political–economic thinking which originated with radical economic geography in the 1970s and the infusion of feminist approaches over the past twenty years. More recently, the variety of intellectual movements and trends grouped (often crudely) under the labels "postmodernism," "poststructuralism," and "postcolonialism" have underscored the significance of the issues political geographers struggle to engage: de-territorialization and re-territorialization, the macro- and micro-geopolitics of states and systems of control, space, power, and place. These influences are examined in several chapters of this book.

Together these trends have produced a contemporary political geography that is dynamic and diverse, an intellectual enterprise open to geographers and non-geographers that is distinguished by the critical nature of the questions it asks and the themes it pursues. We have no doubt that the themes and questions that distinguish contemporary political geography will change over the coming decade. Just as the collapse of the Berlin Wall was one of the most important events at the close of the twentieth century, the destruction of the World Trade Center in Manhattan after terrorist attacks (9.11) is one of the defining events of the opening of the twenty-first. The attacks were shocking reminders of the still active legacies of the wars of the late twentieth century, wars that left Afghanistan destroyed and then ignored after its utility as a Cold War pawn ended, and Saudi Arabia as an explicit American protectorate after Iraq's ill-fated invasion of Kuwait in 1991. The "blowback" from these geopolitical wars of world ordering took the form of a transnational network of radicalized Islamic militarists, *Al-Qaeda*, that declared a *jihad* against the perceived oppressive and corrupt empire of the United States [see chapter 15 (Luke) in this volume]. Networks are organizational systems that do not rely on sharply hierarchical arrangements, but rather, work through embedded, relational linkages. In contrast with slower and more inventory-intensive organizational hierarchies, networks allow fast and flexible movement in response to a rapidly changing environment. Celebrated as the organizational future of capitalism by Wall Street in the 1990s, networks were suddenly powerful because advances in information technology allowed them to function in such dynamic and flexible ways. Informational system networks have also transformed the practice of geopolitics since the end of the Cold War. Many of the same principles of relational, nonhierarchical linkages, and flexibility are evident in the rising power of non-state networked organizations, including transnational criminal and terrorist networks (e.g. Castells, 1996). While the borders of states remain vitally important and legal, and legitimate networks must negotiate with the political geographic order established by states, illicit and covert transnational networks such as *Al-Qaeda*, coordinate activities *through* and *around* state territories in a manner that eludes border controls and challenges territorial sovereignty in a novel way.

The geopolitical questions and moral dilemmas posed by events like 9.11, bioterrorism, and the open-ended war on global terrorism that followed are reminders of the continuing relevance of political geographic themes of (b)ordering in contemporary global affairs. This volume is the first *Companion to Political Geography* but it will certainly not be the last collection covering the best that political geography has to offer.

Approach and Organization of the Volume

This book is not a survey of the history of political geography or of its "great thinkers." Neither is it a dictionary nor an encyclopedia. A dictionary is a compilation of technical concepts. An encyclopedia is an official record of a field. This is a "companion." As such it is designed to both guide a reader through the main concepts and controversies of the field, and offer fresh and stimulating perspectives on the range of topics covered in contemporary political geography. The purpose is to introduce you to the energy and vitality of research and writing that characterizes today's political geography. Many of the authors are geographers, because in Anglo-American universities most of what goes for political geography is undertaken by geographers. Yet there are also many chapters by those working outside of Geography in other disciplines and domains of knowledge. Political geography has always been interdisciplinary, so it is both limiting and disingenuous to limit authorship to geographers. We have tried to recruit authors who are active contributors to the contemporary field rather than simply senior figures or professional commentators.

The overall purpose of the volume is to provide advanced undergraduate students and graduate students, and faculty both inside and outside political geography, with a substantive overview of contemporary political geography. Our interest is not so much in empirical findings as in the ideas, concepts, and theories that are most debated in the field today. We hope that the essays convey a sense of the intellectual dynamism and diversity that presently characterize political geography. The chapters collected herein differ not simply by the topics they address but by the heterogeneity of perspectives, positions, and analytical frameworks they articulate. Yet while there are many "voices" in the volume – and undoubtedly some "silences" too – the conversation they make possible is political geography at its best.

The book is organized into six sections. The first, *Modes of Thinking*, provides an overview of the philosophical diversity of the field. This is necessarily selective. But it does cover what we consider the most significant modes of thinking in past and contemporary political geography. As our orientation is primarily to the present we cannot possibly provide a survey of all modes of thinking that have affected political geography. Following the first essay, which examines the content and impact of environmental determinism, subsequent essays explore in turn the spatial analysis tradition, Marxism, feminism, and postmodern approaches to political geographic questions and themes. These perspectives differ considerably in terms of their assumptions, theories, and methodological emphases. Essays in later sections cannot avoid taking positions in relation to these modes of thinking. Whether oriented to conceptual analysis or substantive themes, they cannot but situate themselves in relation to one or more of the modes of thinking. It is important to bear this in mind as you read the essays in the other sections.

The second section addresses what are arguably the most important concepts in political geography. These *Essentially Contested Concepts* are power, territory, boundary, scale, and place. The purpose is to survey the range of meanings associated with these concepts and show how they figure in different theoretical frameworks and substantive studies. The point about calling these concepts "essentially contested," a phrase drawn from Gallie (1956), is not to suggest that there are such profound disagreements about their meanings that they cannot be communicated to "non-believers." Rather, the purpose is a "rhetorical stratagem" to "call attention to a persistent and recurring feature of political discourse – namely, the perpetual possibility of disagreement" (Ball, 1993, p. 556). Indeed, this disagreement is to be valued as a resource for making present and future conversations restlessly critical and self-reflexive.

One of the motifs that connects contemporary political geography to its past is that of "geopolitics." In its most recent manifestation, geopolitics has reappeared in political geography as *Critical Geopolitics*: the study of the ways in which geopolitical thinking has entered into the practical reasoning of politicians and mass publics and how formal geopolitical analysis both represents and communicates essential features of the "modern geopolitical imagination." The essays in this section cover the competing imperial geopolitical visions at the beginning of the twentieth century, Nazi geopolitics, Cold War geopolitics, "postmodern" geopolitics, and the century-long tradition of resistance to geopolitical discourse which forms an "anti-geopolitics."

Another historic focus of political geography has been on *States, Territory, and Identity*. If in the past the relationship between the three elements was often taken for granted, today it is the subject of intensive investigation. Four of the most important substantive foci of contemporary research are opened up in this section: nation-states, places of memory, boundaries in question, and transnational regions. The intent is to provide a sense of how these phenomena are examined from political–geographic perspectives.

More recently, much energy has also gone into exploring *Geographies of Political and Social Movements*. Here attention is directed to the geographic formation and mobilization of groups directed towards affecting, disrupting, undermining, and supporting various policy goals and institutional frameworks. The classic focus on political parties and elections is the subject of the first essay. The following essays consider nationalism, religious movements, civil rights and citizenship, and sexual politics. Reflecting the politics of the day, these are all "hot" topics in contemporary political geography.

Last, but by no means least, political geography has begun to engage once more with questions of the physical environment. As part of Geography this might appear appropriate and unsurprising. But if in the past a causal arrow was seen as running from the physical environment to political outcomes (as in local geology causes predictable electoral outcomes!), today the interest is in how the natural environment is (mis)managed politically and how this generates political activities of one sort or another. *Geographies of Environmental Politics* addresses this emerging area of political geography with essays on the geopolitics of nature and resources, green geopolitics, environmental justice movements, and the appearance of planetary environmental politics.

The essays in the later sections can be read without having read the first two sections. It is our conviction, however, that a more informed reading of the more substantive essays would result from some familiarity with the modes of thinking and concepts examined at the outset. The hope is that you will come away from this book with a well-versed sense of the wide range of topics and approaches in contemporary political geography. We also hope that you will identify gaps and openings for your own research and writing – "silences" that need to be articulated.

In the final analysis, and in spite of the diversity, we hope that you see a common objective at work: to understand the ways in which people divide themselves up geographically and use these divisions for political ends. This is no small task in a world still stratified by barriers and walls of many kinds.

BIBLIOGRAPHY

Agnew, J. A. and Corbridge, S. 1995. *Mastering Space: Hegemony, Territory, and International Political Economy*. London: Routledge.

Ash, T. G. 1990 *We the People: The Revolution of 89 Witnessed in Warsaw, Budapest, Berlin and Prague*. Cambridge: Granta.

Ball, T. 1993. Power. In R. E. Goodin and P. Pettit (eds.) *A Companion to Political Philosophy*. Oxford: Blackwell.

Castells, M. 1996. *The Rise of the Network Society*. Oxford: Blackwell.

Gallie, W. B. 1956. Essentially contested concepts. *Proceedings of the Aristotelian Society*, 56, 167–98.

Glassman, J. 1999. State power beyond the territorial trap: the internationalization of the state. *Political Geography*, 18, 669–96.

Godlewska, A. and Smith, N. (eds.). 1994. *Geography and Empire*. Oxford: Blackwell.

Gottmann, J. 1952. *La politique des états et leur géographie*. Paris: Armand Colin.

Gottmann, J. 1973. *The Significance of Territory*. Charlottesville, VA: University Press of Virginia.

Helleiner, E. 1999. Historicizing territorial currencies: monetary space and the nation-state in North America. *Political Geography*, 18, 309–39.

Herbert, S. 1997. *Policing Space: Territoriality and the Los Angeles Police Department*. Minneapolis, MN: University of Minnesota Press.

Hyndman, J. 2000. *Managing Displacement: Refugees and the Politics of Humanitarianism*. Minneapolis, MN: University of Minnesota Press.

Kasperson, R. E. and Minghi, J. V. (eds.). 1969. *The Structure of Political Geography*. Chicago: Aldine.

Miller, B. A. 2000. *Geography and Social Movements: Comparing Antinuclear Activism in the Boston Area*. Minneapolis, MN: University of Minnesota Press.

O' Loughlin, J. 1986. Spatial models of international conflicts: extending current theories of war behavior. *Annals of the Association of American Geographers*, 76, 63–80.

Ó Tuathail, G. 1996. *Critical Geopolitics: The Politics of Writing Global Space*. London: Routledge.

Padfield, P. 1999. *Maritime Supremacy and the Opening of the Western Mind: Naval Campaigns that Shaped the Modern World, 1588–1782*. London: Pimlico.

Parker, G. 1998. *Geopolitics: Past, Present and Future*. London: Pinter.

Sharp, J. 2000. *Condensing the Cold War: Reader's Digest and American Identity, 1922–1994*. Minneapolis, MN: University of Minnesota Press.

Sharp, J., Routledge, P., Philo, C. and Paddison, R. (eds.). 2000. *Entanglements of Power: Geographies of Domination and Resistance*. London: Routledge.

Siebers, T. 1993. *Cold War Criticism and the Politics of Skepticism*. New York: Oxford University Press.

Smith, G. 1999. The masks of Proteus: Russia, geopolitical shift and the new Eurasianism. *Transactions, Institute of British Geographers, N. S.*, 24, 481–500.

Smithsonian Institution. 1992. *Art as Activist: Revolutionary Posters from Central and Eastern Europe*. New York: Universe.

Strassler, R. B. (ed.). 1996. *The Landmark Thucydides: A Comprehensive Guide to the Peloponnesian War.* New York: Free Press.

Wilkinson, H. R. 1951. *Maps and Politics: A Review of the Ethnic Cartography of Macedonia.* Liverpool: Liverpool University Press.

Part I Modes of Thinking

Chapter 2

Politics from Nature
Environment, Ideology, and the Determinist Tradition

Mark Bassin

Introduction

In 1997, the Harvard economist Jeffrey Sachs published a lengthy thinkpiece in the *Economist* under the rather unlikely title "Nature, Nurture, and Growth." The title was unlikely insofar as Sachs – whose international fame (or notereity) came from his work as the number-crunching patron saint of the "shock therapy" approach to economic reform in post-communist Eastern Europe – never seemed very preoccupied with environmental or ecological concerns. Yet as the essay makes clear, these latter have now moved to the very center of his analytical interests. In his essay, Sachs considers the current prospects for economic convergence and equalization between the various regions of the globe, now that communism no longer operates as a divisive factor and thus, "for the first time in history," almost all of humanity is bound together in a single network of global capitalism. Yet despite this circumstance – which Sachs obviously believes is a very good thing – his conclusions are not positive, and he speaks rather about the "limits of convergence;" that is to say the eventuality that despite capitalism's new universality, many developing countries are going to be left behind nonetheless. The reasons for this, he argues, are not only or even primarily political or ideological. Rather, they relate to the objective environmental or geographic conditions within which less-developed countries find themselves. An entire range of countries, Sachs argues, are "geographically disadvantaged," indeed "cursed" with what he variously terms a "geographical penalty," a "geographical deficit," or "poorer geographical endowments." This is particularly true of countries in the tropics, where endemically poor soils together with climatic conditions favorable to the proliferation of debilitating diseases act as "fundamental geographical barriers" to economic development and prosperity. The great geographical contrast, unsurprisingly, is offered by the countries of the "temperate zone," that is to say Europe and North America. Quite unlike the blighted tropics, these regions are geographically "blessed" with moderate conditions favoring industry and the expansion of agricultural production. And while Sachs is at pains to "guard against a kind of geographical determinism" that he apparently feels the manner in which he marshals his facts might suggest, he nonetheless concludes that in the short and medium terms, "for much of the world bad climates, poor soils and physical isolation are likely to hinder growth whatever happens to policy." Indeed, for the tropics in particular prosperity can only be assured through a sort of tenuous symbiosis with the developed world, through which the former will be fed chiefly by "temperate-zone exports" (Sachs, 1997).

Despite his protestations, Sachs is in fact offering a distinctly geo-deterministic argument, which he has further elaborated in a series of highly visible articles (Sachs, 2001; Sachs et al., 2001). It is, moreover, an argument which broadly resonates with the views of other scholars. A sort of corresponding historical scenario has been presented, for example, by Sachs' Harvard colleague David Landes, whose much-praised overview of the history of global economic development is premised upon the "unpleasant truth" that "nature like life is unfair, unequal in its favors, [and] further that nature's unfairness is not easily remedied" (Landes, 1999, pp. 4–5; see also Diamond, 1998). In a similar spirit, a belief in the critical salience of physical–geographic conditions to political affairs is fundamental also to the international renaissance of geopolitics, as betrayed in Zbiginew Brzezinski's succinct observation that "geographical location still tends to determine the immediate priorities of a state" (Brzezinski, 1997, p. 38).

Exactly why this preoccupation with environmental influences should be gaining popularity at this particular moment is a complex question, but at least one contextual factor already mentioned would seem to be fairly significant. This is the collapse of the communist system, the existence of which served to bifurcate global relations into two exclusive and opposing networks whose political and ideological oppositions could themselves be taken as the ultimate source of variation and difference between societies across the globe. As we have seen, Sachs in principle happily heralds the burgeoning universality of triumphant capitalism, but importantly refuses to draw Francis Fukuyama's comforting "end of history" conclusion about the universalization and standardization of social life that should ensue (Fukuyama, 1992). Quite to the contrary, Sachs makes it clear that divisions between societies and regions are going to persist, and that economic–material – and thus human – conditions will most decidedly not converge.

Such scepticism does not sit entirely easily with capitalism's own distinctly more optimistic vision of the universal well-being that it can bring to the world if provided full freedom of operation, and insofar as communism is no longer available for convenient fingering as the culprit obstructing capitalism from realizing its universal mission, then something else has to be found. And the physical conditions of the natural world, which can be plausibly invested with a virtually endless variety of meanings and implications, prove in this regard to be very useful.

As a substantial literature already makes quite clear, what we may call the "argument from nature" has a rich and controversial history (Bassin, 1993, 1996; Bergevin, 1992; Glacken, 1967; Lewthwaite, 1966; Martin, 1951; Montefiore and Williams, 1955; Peet, 1985; Tatham 1951). The aim of this chapter is to provide some insight into the tradition of geo-determinist thinking, as developed in the work of three very influential scholars of the late nineteenth and early twentieth centuries:

the German geographer Friedrich Ratzel, the American historian Frederick Jackson Turner, and the Russian revolutionary Marxist Georgii Valentinovich Plekhanov. The point is not to find in their writings antecedents in a strict sense to the sorts of theories advanced in our own day by Sachs, Landes, Brzezinski, and others, for while there are indeed some striking parallels, the political and intellectual worlds that they operated in were entirely different. To try and read them in terms of contemporary concerns and preoccupations would not help us very much in appreciating what they were in fact attempting to do. Rather, the affinity across the centuries is to be sought on a more general and structural level. My argument is that environmental or geographic determinism must be understood as an ideological phenomenon, at least in certain dimensions. This is by no means a dismissal of the complex issue of environmental influences on human societies as a legitimate scientific problem, but simply an affirmation that theoretical discussions from the social sciences as to what such influences might mean for the evolution and constitution of human civilization invariably take place in political–historical contexts which themselves have an articulated influence on the nature of the arguments made and conclusions drawn. It is in this confluence of theory, ideology, and politics that the meaningful and real continuity with the determinist thinking of today is to be found, and not in the nature of the ideas involved or how they are applied. The following discussion will seek to elaborate the various contexts of a century ago and to explore how determinist argumentation was variously formulated in terms of them.

Determinism, *Politische Geographie*, and Political Expansion

Without any question, the best-appreciated deployment of determinist arguments in the nineteenth century was as part of the ideology of imperialism. The so-called "Age of Imperialism," which gave *fin-de-siècle* European politics its distinctive stamp, was not limited to the practicalities of diplomatic rivalry, colonial acquisition, and imperial administration. Much more than this, it was a state of mind – a political mentality founded on the unshakeable conviction that the healthy development of an advanced state in the modern world was conditional upon the ever-greater physical extension of its territorial base. Failure to expand or grow, it was piously believed, could have existential consequences for the state's future welfare. The new preoccupation with expansion was a pan-European passion, to be sure – Cecil Rhodes, for one, mused dreamily that he "would annex the whole world" if only he could – but the particular implications for each of the nations involved were distinct. This was certainly the case for Germany, whose situation differed from that of its rivals Britain, France, and Russia in two fundamental respects. Unlike its European neighbors, Germany as late as 1880 still possessed no extra-European colonial domains whatsoever, and it thus entered into the newest round of imperial competition at a distinct disadvantage. Yet more significant was the circumstance that, for Germany, territorial expansion was not only a question of colonial annexations outside of Europe in Asia or Africa. The German population had achieved national consolidation of sorts through the establishment of the Second Reich in 1871. Bismarck's intention had been to create what would effectively serve as a nation-state, but its success was undermined from the outset by the fact that significant concentrations of German population in Central and Eastern Europe remained well outside its political boundaries. The imperative of territorial expansion thus represented a challenge on two different geographical levels – "domestic" European as well as global – and was impelled by a rationale that flowed from two distinct sources. These conditions lent this imperative a specific urgency, and it is in terms of this urgency that the political–geographic system of Friedrich Ratzel must be understood.

Ratzel's *Politische Geographie* represents an attempt to develop a theory of political expansionism in which the need for more or less constant physical growth was explained, as it were, "scientifically" in the manner popular for the age: by direct analogy with the plant and animal world. Ratzel was heavily influenced by Darwin's teachings on natural evolution, and while by no means a Darwinist in a strict sense (e.g. 1905, p. 399), his logic and argumentation derived a great deal of their inspiration from them. Throughout all of his work, he vociferously advocated the essential unity of all organic life on Earth, as part of which he very much included the anthropological realm. This meant, among other things, that the nature and operation of human societies were to be understood in terms of precisely the same laws that govern the natural world (Ratzel, 1869, pp. 478–9, 482; 1901–02, II, p. 554). This premise then supplied the fundamental supposition of his *Anthropogeographie*, or human geography; namely, that human populations are as dependent as all other forms of organic life upon the conditions of the external natural environment. As one of Ratzel's most gifted disciples, the American geographer Ellen Churchill Semple put it in an inspired passage:

> Man is a product of the earth's surface. This means not merely that he is a child of the earth, dust of her dust; but that the earth has mothered him, fed him, set him tasks, directed his thoughts, confronted him with difficulties that have strengthened his body and sharpened his wits, given him his problems of navigation or irrigation, and at the same time whispered hints for their solution. She has entered into his bone and tissue, into his mind and soul.... Man can no more be scientifically studied apart from the ground which he tills, or the lands over which he travels, or the seas over which he trades, than polar bear or desert cactus can be understood apart from its habitat.... Man has been so noisy about the way he has "conquered Nature," and Nature has been so quiet about her persistent influence over man, that the geographic factor in the equation of human development has been overlooked (Semple, 1911, pp. 1–2).

For his political geography, Ratzel's goal was to create a "science" which would parallel that of physical geography and would carry the full explanatory authority and conviction of natural science (Ratzel, 1885, pp. 248–9; Overbeck, 1965, pp. 63–4). He derived the central element of this political geography – a theory of expansionism based on the notion of *Lebensraum* – from a biogeographic consideration of the nonhuman organic world. Ratzel argued that every living organism required a specific amount of territory from which to draw sustenance and labeled this territory the respective *Lebensraum*, or living space, of the organism in question. He continually emphasized the elemental significance of the *Lebensraum* concept, to the extent indeed that the idea of life itself could not for him be separated from its attendant space-need: "[Every] new form of life needs space in order to come into existence," he argued, "and yet more space to establish and pass on its characteristics." (Ratzel 1899–1912, I, p. 231; 1901, p. 146). Importantly, Ratzel conceived of an organism not only as an individual entity, such as single trees or elephants, but also applied the concept to entire, homogenous, and spatially coalesced populations of these individuals, such as forests or herds. These Ratzel termed aggregate-organisms, and as such they had their own independent *Lebensraum* requirements. And because the laws of nature and organic reproduction dictated that the size of the populations which comprised such aggregate organisms would steadily increase, so too would the attendant space-need of the latter, leaving them with the inescapable alternative of expansion or decline.

In order to apply this biogeographic scheme to human society, it was necessary only to locate in society the organism on to which the space-need concept could be transferred. Here Ratzel followed the lead that had already been conceptually developed in the writings of Herbert Spencer, O. Hertwig, and many others, and identified the political state as the corresponding aggregate-organism (Ratzel, 1897, p. 8). Composed of coalesced homogenous populations of human individuals, he argued, the state not only bore a morphological resemblance to forests or herds, but operated according to the same laws of development (Ratzel, 1897, p. 11; 1899–1912, I, p. 2). The state organism was based on a certain defined territorial expanse – a *Lebensraum* – in which a certain level of sustenance was available. A human society could consolidate and develop on the basis of this level. The correspondence between human population and territory was dynamic, however, and was bound to be upset as the former grew, resulting in an increased demand for sustenance, which meant a greater space-need (Ratzel, 1896, p. 98; 1923, p. 90). The ubiquitous response to this circumstance was a "flowing over" of excess population beyond the formal political boundaries of the state (Ratzel 1899–1912, I, p. 121; 1923, pp. 70, 90). Under optimal conditions, the state would then itself physically expand to adjust to the new level of need, acquire additional *Lebensraum* and once again consolidate on the newly enlarged state territory. If, however, the state were either unable to attempt acquisition of new lands, or if its attempts should prove unsuccessful – if, in short, it did not expand – then it would necessarily exhaust its sustenance base and decline (Ratzel, 1899–1912, I, p. 72).

The problem remained, of course, that while this imperative for territorial growth was shared equally by every state, still the Earth's surface was finite and offered only a limited amount of territory for this purpose (Ratzel, 1901–02, II, p. 590). Moreover, as states grew larger through history, this available territory became ever more limited, and as this happened states were forced to compete ever more directly and aggressively with each other for territorial advantage. Generalizing upon this circumstance, Ratzel suggested that the notion of a *Kampf ums Dasein* or "struggle for existence" popularized by the Social Darwinists could be put more concisely and meaningfully as a *Kampf um Raum*, or "struggle for space." "As organic life first began to develop on earth," he explained, it

> was quickly able to [spread and] take over the territory of the earth's surface as its own, but when it reached the limits of this surface it flowed back, and since this time, over the entire earth, life struggles with life unceasingly for space. The much misused, and even more misunderstood expression "the struggle for existence" really means first of all a struggle for space. For space is the very first condition of life, in terms of which all other conditions are measured, above all sustenance (1901, pp. 153, 165–68).

The ultimate expression of this struggle for space was the contemporary imperialist competition, for Ratzel understood overseas colonial acquisition as the only remaining means by which the European states – by the late nineteenth century already hopelessly overpopulated in their own native *Lebensräume* – could further expand territorially (Ratzel, 1898, pp. 143–4; 1899–1912, II, p. 191; 1923, pp. 106–7, 257, 308). The European continent itself he viewed as effectively occupied and thus unavailable for new settlement (Ratzel, 1906b, p. 376; 1923, p. 270), a perspective which the geopoliticians who were to follow him did not share.

Thus, the direction of Ratzel's argument is transparent. Germany's Second Reich was best understood as a biological entity, which like all other organisms had a specific set of life requirements. Chief among these was territory, which had to increase in equal measure as the state grew and its population swelled. In order to increase its territory and secure adequate *Lebensraum* for future generations, the German state had to look abroad, and join in the on-going struggle with the other European powers for territorial advantage in the non-European colonial realm. The existential choice facing the nation could not have been more grave, and Ratzel characterized it tellingly as the option of being either a hammer or an anvil.

> Whether [we Germans] become one or the other depends on [our] recognizing in good time the demands which the world situation presents to a nation which is struggling to rise. Prussia's task in the 18th century – to win for itself a position as a major power in the middle of the European continental powers – was different from that of Germany in the 19th century: to win a place among the world powers. This task can no longer be solved in Europe alone; it is only as a global power that Germany can hope to secure for its people the land which it needs for its growth. Germany must not remain apart from the transformations and redistributions taking place in all parts of the world if it does not want to run the risk . . . of being pushed into the background for generations (Ratzel, 1906, pp. 377–8).

Determinism, Nationalism, and History

Since the eighteenth century, doctrines of national identity have stressed the factor of environmental influences very strongly. The vital connection between a people and its physical environment was already explicit in the emphasis which nascent doctrines of national identity put on the natural rootedness of the national group in a defined home region or homeland. From here, it was only a small, and entirely intuitive, step to assume that this organic connection with the land was in some way important in shaping national characteristics. Indeed, the argument for environmental influences offered a very special appeal for nationalists, who appreciated how effectively it could help the popular imagination transform a notional construct into the desired vision of the nation as a natural and eternal or "primordial" entity. Environmental evidence could be presented in various ways to specify precisely how physical–geographic conditions were important for the life of the nation, but most effective by far was to weave these conditions into chronicles of national history, in order to explain the genesis of the nation and the main contours of its developmental process. In this manner, geography could be identified as a determining agent at the very moment the nation came into being, and this determining influence was maintained over a protracted course of historical evolution, effectively down to the present. Thus, the studies of "national history" that came increasingly to dominate the agenda of academic historiography throughout the nineteenth century would characteristically begin with a chapter or section setting out the environmental context in which the epic tale would unfold, the implication being that this geographical arena itself was implicated in vital ways in the life-story of the nation. A methodological foundation for this sort of analysis was provided in the influential writings of Carl Ritter or those of his yet-more influential colleague at Berlin University, Hegel, and the works which they inspired – for example, the histories of Michelet for France or Sergei Solov'ev for Russia – figured among the greatest historiographic achievements of the day.

This determinist cast became much more pronounced in the latter decades of the century, as nationalist historiography sought to give its subject a more rigorous analytical foundation. To accomplish this, historians ended up following precisely the same path that we have traced above in the example of political geography – that is to say, by adapting their historical studies to the epistemological framework of the social sciences, which in their turn were based on the premises and principles of the natural sciences. This seemed to promise a methodological rigor that would be thoroughly "scientific" and an objectivity of research results that would be incontestable. Now history as well joined in the quest after those universal laws which governed or determined the development and activities of the social phenomena they studied, and these phenomena themselves were increasingly understood in terms of concepts taken more or less directly from the sciences of the natural world. Human societies, in other words, were now taken to be biological organisms, and as such they could be studied in terms of the universal laws of growth and development valid for all organic life that were identified by the physical sciences. Needless to say, such a perspective fit the needs of nationally-minded historians quite well, for they were anxious to secure a scientific foundation for their personal visions of the nation, and organicist imagery offered precisely this opportunity. Herbert Spencer was particularly influential in developing an organismic perspective in a way that was useful for historians, and his belief in the universality of social laws was expressed in his grand scheme of evolutionary stages of historical development through which all individual society-organisms would eventually pass (Spencer, 1863). This appropriation of natural science by a self-proclaimed "scientific history," however, served to bring out tensions, which – somewhat paradoxically – were to make geographic determinism even more important to the positivist historiography of the late nineteenth century than it had been to its Ritterian or Hegelian antecedents.

An excellent example of these points can be seen in the work of the Wisconsin historian Frederick Jackson Turner, who in his celebrated "frontier hypothesis" of 1893 sketched out what virtually overnight became the single most popular and enduring historical explanation of the genesis and character of American nationhood ever offered. Turner was a nationalist historian *par excellence*. This was apparent, to begin with, in the very nature of his scholarly project, which from the outset was exclusively concerned with studying the American ethos. "We do not know ourselves," he complained on behalf of all his countrymen in 1891 (cited in Bensen, 1980, p. 22), and he accordingly took his fundamental mission to be the explication of the life story or biography of the nation. Until such a historical record was produced, he observed, "we shall have no real national self-consciousness"

(Turner, 1965a, p. 72). Turner was a nationalist moreover in terms of his subjective evaluation of the qualities of the nation in question, for he was animated by the unshakeable conviction that America was the loftiest accomplishment to date of world civilization, in a variety of important respects. It was entirely fitting that he should have selected the occasion of a national festival organized to celebrate the manifold glories of a century of progressive accomplishment – Chicago's Columbian Exposition of 1893 – for the first proclamation of his theory of American history, and the description of the American national character he offered therein fairly gushed with the passion of a fervent nationalist. "That coarseness and strength combined with acuteness and inquisitiveness; that practical, inventive turn of mind, quick to find expedients; that masterful grasp of material things, lacking in the artistic but powerful to effect great ends; that restless, nervous energy; that dominant individualism, working for good and for evil, and withal that buoyancy and exuberance which come with freedom" (Turner, 1965a, p. 57) – all this comprised the unique charisma of the American. Such indeed was the intensity of Turner's homily that he managed to cast an aura of grandeur over even his country's shortcomings.

Turner was an enthusiastic adherent of the precepts of the positivist "scientific" history of his day, and he gave way to none in his eagerness to appropriate ideas and theories from the natural sciences – in particular those from Darwinian teachings on organic evolution – for the purposes of his own historical analysis. This included, most prominently, the notion that a society, or more specifically the nation whose life-story was the preoccupation of the historian, could be analyzed as a biological organism (cf. Coker, 1910; Hertwig, 1899). His belief in society as an evolving organism was fundamental to his entire understanding of the past (Billington, 1971, p. 18), and his historical writings were filled with engagingly vivid anatomic and biological images of American society. The development of America he considered as nothing less than "a history of the origin of a new political species," while American society itself was a "protoplasm," the spread of which could be likened to "the steady growth of a complex nervous system" (Turner, 1897, p. 284; 1931a, p. 206; 1963, p. 37).

Yet however powerful the confidence of Turner and his colleagues may have been in the possibility of engaging scientific history for their biography of the national organism, this option brought with it a contradiction which could not ultimately be overlooked. Simply put, the assumption of the uniformity of social organisms and the developmental laws that they obeyed were entirely at odds with the nationalist frame of mind which fundamentally inspired their work. How was it possible to reconcile the nationalist temper – which rested after all on the assumption not only of fundamental differences between nations, but commonly also of the superiority of some over others – with an analytical perspective which preached precisely the opposite, that is to say the basic uniformity of all social organisms and the similarity of their developmental patterns? A universalizing social science wanted to explain how all societies were essentially similar and how they moved through a single progression of developmental stages, while exceptionalist, nationalist historiography was concerned with demonstrating how and why societies did in fact differ, and indeed naturally. Certainly the latter was true for Turner, whose aim was to demonstrate how a North-American offshoot of European civilization could develop not along the same lines of universal evolution, but rather into something that was radically different and emphatically non-European. Out of this implicit contradiction, the notion of geographical determinism took on an exaggerated appeal, for environmental influences offered a way of sorts out of the dilemma. Rather than focus on the inherent differences in the social organisms themselves, which would have been immediately disruptive for their "scientific" premises, historians could concentrate instead on naturally occurring and entirely obvious differences in the geographic milieux in which these organisms developed. Armed with an adequate theory of environmental causation, it would then be possible to explain in principle how differences between organisms came from the differential influences of the differing geographic constellations, and specifically, find satisfactory explanations for just why a given nation possessed the particular – and unique – qualities that it did.

This particular approach was foreshadowed in the enormously popular work of the English historian Henry Thomas Buckle, whose *History of Civilization in England* (1857–61) was one of the great prototypes of "scientific history" and it was adopted by Turner as well. His entire hypothesis of American national development was founded on the assumption that the external natural milieu exercised a pre-eminent influence upon the evolution and attributes of the human social organism. In developing this perspective for his own purposes, he in fact drew heavily on the anthropogeography developed by Ratzel and his American disciple Ellen Churchill Semple (Turner, 1905, p. 34; 1908, pp. 43–8; Billington, 1971, pp. 96–8, 173–4, 268), and he was entirely receptive to their suggestion of a direct causal relationship between the development of organic phenomena and the conditions of the natural environment. And as it turned out, environmentalism proved to be eminently congenial to his particular explanatory needs. There was hardly a better way to demonstrate how a national organism had developed the unique qualities that were America's than to point to the formative influence exerted by the unique conditions of the New World environment on an evolving and hence entirely malleable social organism. American history was thus the record of "European germs developing in an American environment," in which one could trace "the evolution and adaptation of organs in response to changed environment" (Turner, 1931a, p. 206; 1963, p. 29). "Into this vast shaggy continent of ours," he wrote in 1903, "poured the first feeble tide of European settlement. European men, institutions, and ideas were lodged in the American wilderness, and this great American West took them to her bosom, taught them a new way of looking upon the destiny of the common man, trained them in adaptation of the conditions of the New World, to the creation of new institutions to meet new needs" (Turner, 1931a, p. 267). He left no doubt that American society was ultimately to be understood as a product of its environmental context, for as he wrote in one of his most famous passages: "The existence of an area of free land, its continuous recession, and the [resulting] advance of American settlement westward, explain American development" (Turner, 1963, p. 27). In America's frontier the nationalist Turner located the source of the country's great national virtues: its egalitarianism and rugged individualism, its elemental energy, and, above all, its democratic inclinations.

Turner's environmentalism was clearly apparent in his argument that the empty continental spaces of the American wilderness had themselves given rise to a social specimen superior to its European antecedents. America was a fresh and a new nation, energetic, unrestrained, and above all unencumbered by the "tyranny of Old World custom and traditions" (Turner, 1965b, p. 140). The development of the New World was to be understood in terms of a frontier which, as it advanced ever further to the west, had engendered "a steady movement away from the influence of Europe, a steady growth of independence on American lines" (Turner, 1963, p. 30), and for Turner this geographic and attendant cultural distancing was a veritable condition *sine qua non* for the emergence of a distinctive and great nation. The American settlers "turned their backs on the Atlantic Ocean, and with grim energy and self-reliance began to build up a society free from the dominance of ancient forms." The challenge of the frontier had fostered America's cardinal native virtue, the political institution of populist democracy that set America's social order so loftily above the stifling "reign of aristocracy" that dominated in Europe (Turner, 1931b, p. 253).

The spiritually invigorating influence of the western frontier had acted moreover as a protective agent at those moments when American society threatened to revert to European norms and patterns: "And ever as society on [America's] eastern border grew to resemble the Old World in its social forms and its industry, ever, as it began to lose faith in the ideal of democracy, [the western frontier] opened new provinces, and dowered new democracies in her most distant domains with her material treasures and with the ennobling influence of that fierce love of freedom, the strength that came from hewing out a home, making a school and a church, and creating a higher future for his family, furnished to the pioneer" (Turner, 1931b, p. 267). For Turner, national biography was the explanation of the origins of those proud qualities that served to set the US not only apart from but also unquestionably above other nations, in particular those of the European Old World.

An Environmentalist Theory of Society

The appeal of environmental determinism was one that even orthodox Marxists were to find irresistible. In the final quarter of the nineteenth century, leading Marxist theoreticians began to stress the nature of Marxism as a rigorously "scientific" perspective and to identify significant affinities between it and the natural sciences. Marx himself had voiced his strong appreciation of the *Origin of Species* as soon as it appeared, asserting that in it he recognized nothing less than the "the natural-scientific substantiation (*Unterlage*) for class struggle in history" (Marx, 1964a, pp. 131, 578). After his death, the assimilation of contemporary natural science into Marxist doctrine was energetically pursued by Engels and other disciples, and in this process the identification of Marxism as a "science" and of socialism as "scientific" became ever more direct and pronounced. Heavily influenced by the same positivist naturalism of the late ninteenth century that we have been considering (Lichtheim, 1961, pp. 247–8), *fin-de-siècle* Marxism adopted a perspective that was essentially monist, arguing that human society was an integral part of the natural-organic realm and that its historical development was but a chapter – the most recent and most significant, to be sure, but a chapter nonetheless – in the evolution of nature in general. "Nature is universal (*das Allgemeine*)," observed one of the great authorities of scientific socialism, "human society is only a particular case within it" (Kautsky, 1927–9, I, p. 198). Human history, accordingly, was seen as a product of the same causal relationships and was governed by the same fixed laws that determined the operation of the rest of the natural world (Kolokowski, 1978, I, pp. 181, 337, 400; II, p. 36; Lichtheim, 1961, pp. 235, 237–8, 245, 295–6).

This pronounced naturalism had the general effect of encouraging a greater emphasis in Marxist theory on the significance of human individuals and societies as biological entities. The specific implications could be developed in very different ways, but one powerfully appealing option was precisely the environmental-determinist perspective that we have been tracing in this essay. Re-baptized by the Marxists as "geographical materialism," the theory of environmental influences proved to be extremely appealing, for it represented a materialist, causal, and suitably "scientific" explanation, the veracity of which had appeared to be demonstrated in the natural-organic realm and which was eminently useful for anthropological purposes. Karl Kautsky, the outstanding theoretician of the Second International, who had come to Marxism as a Darwinian convert, stressed its quality as a natural-scientific law, affirming that "the movements and evolutionary processes" of plants and many animals "are nothing else but the reactions of the organism to stimuli which come from the external environment, and which often precede an adaptation between the organism and this environment" (Kautsky, 1927–9, I, pp. 129, 187–8). The most important point, however, was the universality of the principle: in other words its validity for human society. "I believe that the general law, upon which human as well as animal and plant evolution is dependent, is that every change in [human] society, just like that of [plant and animal] species, is the result of a change in the environment.... New forms of organisms and social organizations come into being through adaptation to an altered environment" (Kautsky, 1927–9, II, pp. 630–1; Kolakowski, 1978, II, pp. 38, 51–2). Such a perspective was immediately relevant for the historical development of human civilization, which could be understood and explained in terms of influences exerted by its external natural milieu. A principal task of Marxist historical materialism, accordingly, was to study "the dependence of the history of human culture on the physical characteristics of the earth's surface and on society's physical environment" (Woltmann, 1900, p. 6).

A revealing example of how this task was engaged can be followed in the work of G. P. Plekhanov, a brilliant polymath of late nineteenth-century Marxism and the "father" of the Marxist movement in Imperial Russia. Plekhanov was ebullient in his enthusiasm for natural science, arguing that Darwin's and Marx's teachings belonged together as two symmetrical and symbiotic parts of a larger whole. Darwin, he wrote, "succeeded in solving the problem of how plant and animal species (*vidy*) originate in the struggle for existence." Marx, in parallel fashion, "succeeded in solving the problem of how various forms (*vidy*) of social organization arise in the struggle of people for their existence. Marx's examination begins logically at the very point where Darwin's ends... [and] for this reason it is possible to say that Marxism is Darwinism in application to the study of society" (Plekhanov, 1956a, p. 690n; 1956b, pp. 292–3). And for Plekhanov, as for his colleague Kautsky, this meant above all that the study of society should be founded on a consideration of environmental influences (Fomina, 1955, pp. 189–90; Opler, 1962, p. 533; Vucinich, 1976, pp. 186–8).

The external natural environment, *the geographical environment*, its paucity or its wealth, has exerted an unquestionable influence on the development of [human] industry.... [M]an

> receives from the surrounding natural environment the material for the creation of artificial implements, which he uses to conduct his struggle with nature. The character of the external natural environment determines (*opredeliat'*) the character of man's productive activity, the character of his *means of production. The means of production determine [in turn] the mutual relations of people in the social process of production....*

It is the interrelationship of people in this social process of production, he continued, which determines the entire structure of society. "For this reason, the influence of the natural environment on this structure is undeniable. *The character of the natural environment determines the character of the social environment*" (Plekhanov, 1956c, pp. 154–5). "The task of contemporary materialism applied to social science," he concluded elsewhere, "is precisely to demonstrate how the development of humankind takes place under the influence of conditions which are independent of its will" (Plekhanov, 1923, p. 24).

Plekhanov is interesting for our purposes, however, not so much in terms of this theoretical orientation – which as we have seen he shared with the main currents of European Marxism of the day – but rather in the very particular way in which he made use of it in his social and political analysis. Plekhanov "applied" his environmentalist perspective as part of a broad national debate about the problems of modernization and Russia's relation to the West, and so a word about this context is necessary. Ever since the early eighteenth century, it was clear that enormous gaps separated Russia from much of the rest of Europe. In stark contrast to the technologically advanced capitalist West, Russia remained an overwhelmingly peasant–agrarian society. Even in Plekhanov's day, a modern industrial proletariat – Marx's designated bearers of the new social order – was only in the very initial stages of formation, and the bourgeoisie was extremely weak and virtually ineffectual against the overwhelming power of tsarist autocracy. By the late nineteenth century, this contrast with the rest of Europe had begun to call into question the degree to which patterns of Western historical development, and the social theories based on this historical experience, were really relevant to the Russian example. A significant section of those working toward the revolutionary transformation of Russia's autocratic order concluded that they were not. Precisely because of the social conditions which set it so obviously apart from Europe, they argued, Russia could avoid Europe's protracted and difficult period of capitalist development and pass directly from its present backward state to an advanced socialist order. Among the most vociferous proponents of this latter perspective were the Russian populists, among whom Plekhanov had begun his career as a political revolutionary (Malia, 1971, p. 38; Venturi, 1966, p. 150ff).

It was against precisely such an "exceptionalist" perspective on Russia and the West, however, that Plekhanov as a Marxist was constrained to wage a bitter struggle. The scientific basis of Marxism meant that all social development was seen to be governed by laws which were immutable and universal, and thus when Marx postulated a scheme of generic historical–economic stages – from feudalism through capitalism to socialism and communism – through which all individual societies would pass in regular succession, the possibility of exceptions was not part of his vision (Marx, 1964b, p. 9). Indeed, disciples such as Plekhanov were even more insistent than Marx himself on the unilinearity of social evolution (Baron,

1963, p. 297; Opler, 1962, p. 544; Vucinich, 1976, p. 188). By its most basic principles, therefore, their view of social development could not accommodate what the populists and others argued for: namely the notion of a special course of development which would diverge from the standard Marxist progression (Baron, 1963, p. 96ff). Russia, Plekhanov argued, was developing essentially along the same continuum of social development as the West, and in accordance with the same general historical laws. The differences between the two were simply evidence of what he very aptly termed Russia's *evropeiskaia nedocheta*, or "deficiency in Europeanness," in other words the circumstance that Western Europe had moved "much, much further [than Russia] along the path of civilization" (Plekhanov, 1925, p. 87). His task, therefore, consisted in supplying an explanation for precisely how this state of affairs had originated, and it was toward this end that he pressed his environmentalism into ideological service. The retrograde character of Russian social, economic, and political development, he argued, was ultimately to be ascribed to the qualities of the natural milieu in which this development took place.

He developed this analysis in one of his largest and most important works, the monumental three-volume *History of Russian Social Thought* (1914–16). "The [historical] course of events in our country, as everywhere," he wrote,

> was controlled at all times by the conditions of the natural environment. The relative peculiarity of the Russian historical process in fact is explained by the relative peculiarity of that geographical milieu in which the Russian people were constrained to live and operate. The significance of this milieu was extraordinarily great (Plekhanov, 1925, p. 99).

The critical characteristic in Russia's natural milieu which worked to shape the country's development, Plekhanov argued, was the basic physiographic formation upon which this development took place, namely the Great Russian plain. Ancient Slavic tribes had emerged from the south-west onto this expansive stretch of territory in the early centuries of the Christian era, and it was subsequently to serve as the principal geographic arena for the development of the Russian state and nation. Plekhanov asserted that it was in the monotonous and overwhelming uniformity of this vast and sparsely inhabited landmass that the ultimate source of Russia's retarded development and endemic backwardness must be sought. On the one hand, the lack of any significant topographical variation allowed for virtually unobstructed movement in all directions, and, on the other, the seemingly endless stretch of fields and forests offered resources of game and arable land for the rudimentary economies of these primitive Slavic tribes. These physical qualities insured that *migration* and *colonization of new lands* would be constant features of Russian society from the very beginning, and it was through them that the environment worked its pernicious influence and hindered Russia's normal progressive development.

The availability of open lands, Plekhanov argued, set in motion a process of constant resettlement. As population in any one locale increased, and along with it pressure on existing resources, groups could simply migrate to new and unoccupied regions and carry on exactly as they had before, rather than expanding the resource base already at their disposal by improving cultivation methods and diversifying their economic activities. In this way, Russia's external milieu fostered economic formations among the Russians that remained *extensive* and primitive rather than *intensive* and modern (Plekhanov, 1925, pp. 35–6). The empty spaces of the Russian homeland had also directly interfered with progressive development of social relations in Russia. In Western Europe, he explained, the lack of land reserves meant that the out-migration of an over-populated countryside was directed to the urban centers. This circumstance lead to ever-increasing population densities in these centers, which, in turn, produced irresistible pressures to improve the existing means of production. Social tensions and the resulting struggle between classes would necessarily be more clearly articulated and enhanced as a result of greater population concentrations, and these centers consequently represented the most important source of progress in these societies (Plekhanov, 1925, pp. 104–5). In Russia, by contrast, this pattern was completely disrupted by the particular qualities of its natural environment, which offered the option of agricultural resettlement of new lands and thereby effectively thwarted the natural tendency toward ever-greater population concentration. Thus, rather than gathering in cities, dispossessed rural migrants simply dispersed yet more thinly onto remote open lands, precluding the role of Russian cities as centers for the articulation of class antagonisms and the progressive development of society.

> The history of Russia was the history of a country which was colonized under primitive economic conditions. This colonization meant . . . the non-diversified activities and constant mobility of the population, which obstructed . . . the deepening of those class differences which arise as a result of the social division of labor. And this means that, by virtue of these conditions, the internal history of Russia could not be characterized by the intense struggle of social classes. The source of political strength of the upper class – its economic domination over a significant part of the population – could not be stable, and threatened moreover to dissipate due to the constant movement of the population to "new lands" (Plekhanov, 1925, p. 84).

The epistemological and analytical resonances between Turner and Plekhanov are powerful, but they are overshadowed by the diametrical opposition of their respective conclusions. For while Turner used his environmentalism scientifically to explain and celebrate the United States as a superior civilization, Plekhanov used his – no less scientifically – to explain the primitive and backward nature of Russian society and civilization. This Plekhanov accomplished on the basis of a consistent and plausible Marxist analysis. Environmental influences were essential to this analysis insofar as he used them in order to demonstrate the critical point that Russia's evolution was essentially *gesetzmässig*: that is, it proceeded according to the same fixed laws of development as did capitalist Europe, and (implicitly at least) toward the same end. With this, he was able to place Russia back onto the unilinear developmental continuum from which the populists and others had sought to remove it, and at the same time – in a most admirable union of theory and praxis – indicated a program of political action as well. Capitalist technology and social relations had already made a faint beginning in Russia, he asserted, and they must be allowed to mature, for which a bourgeois political order must be established and fostered. Only at the point when this process had run its full course could there be any thought of a socialist revolution and the construction of a socialist society. It was for this reason that Plekhanov and other Menshevik Social Democrats so staunchly opposed the October revolution of 1917, in which Lenin and the Bolsheviks seized power in the name of the industrial proletariat and the peasantry with the avowed goal of pressing immediately forward toward the establishment of a socialist order.

Conclusion

The goal of this essay was to provide some insight into the history of what might be called the "argument from nature:" the argument, in other words, that the past, present, and future constitution of human society is in some way dependent upon and determined by the objective physical–geographic conditions of the natural environment. Insofar as this perspective is still very much with us, we may in conclusion consider what meaning for the present day we might locate in the experience of the late nineteenth and early twentieth centuries. It was stressed at the outset that this would not be a matter of identifying and then tracing direct and specific ideological or doctrinal antecedents across the century, and the material presented in the essay makes abundantly clear why this is so. The theoreticians of the earlier *fin de siècle* obviously were driven by a variety of preoccupations – among them a rather overbearing scientism, a clearly articulated agenda of territorial-imperial extension, and an abiding commitment to nationalist sentiment – which acted to determine, as it were, their own engagement with geographical determinism but which are not really operative today. The affinity, therefore, must be sought on a different level. As already suggested, the essential common ground is in the *ideological* dimension of determinist theorizing: in the fact, in other words, that the argument from nature appears always to be deployed toward a recognizably programmatic end. The specific quality of the ideological entanglement can vary widely, as the contrast between the arch-conservative Ratzel and the Marxist Plekhanov indicates unmistakably, but in neither case is it any less important for that. This ideological dynamic is something we can readily recognize in geo-determinist thinking from periods prior to that we have considered, and we can certainly see it in the determinism which is enjoying a renaissance of sorts in our own day.

BIBLIOGRAPHY

Baron, S. H. 1963. *Plekhanov. The Father of Russian Marxism*. Stanford, CA: Stanford University Press.

Bassin, M. 1993. Reductionism redux, or the convolutions of contextualism. *Annals of the Association of American Geographers*, 83, 156–66.

Bassin, M. 1996. Nature, geopolitics, and Marxism: ecological contestations in the Weimar Republic. *Transactions of the Institute of British Geographers*, 21, 315–41.

Bensen, L. 1980. Achille Loria's Influence on American Economic Thought: including his contributions to the Frontier Hypothesis. In *Turner and Beard: American Historical Writing Reconsidered*. Westport, CN: Greenwood, 2–40.

Bergevin, J. 1992. *Déterminisme et géographie. Hérodote, Strabon, Albert le Grand, et Sebastian Münster*. Sainte-Foy: Presses de l'Université Laval.

Billington, R. A. 1971. *The Genesis of the Frontier Thesis*. San Marino, CA: Huntington Library.

Brzezinski, Z. 1997. *The Grand Chessboard. American Primacy and its Geostrategic Imperatives*. New York: Basic.
Coker, F. W. 1910. *Organismic Theories of the State. 19th-Century Interpretations of the State as an Organism or as a Person*. New York: Longmans, Green, & Co.
Diamond, J. M. 1998. *Guns, Germs, and Steel. The Fates of Human Societies*. New York: Norton.
Fomina, V. A. 1955. *Filosofskie vzgliady G.V. Plekhanova*. Moscow: Gos. Iz-vo Pol. Lit.
Fukuyama, F. 1992. *The End of History and the Last Man*. Harmondsworth: Penguin.
Glacken, C. J. 1967. *Traces on the Rhodian Shore. Nature and Culture in Western Thought from Ancient Times to the End of the 18th Century*. Berkeley, CA: University of California Press.
Hertwig, O. 1899. *Die Lehre von Organismus und ihre Beziehung zur Sozialwissenschaft*. Jena: G. Fischer.
Kautsky, K. 1927–9. *Die materialistische Geschichtsauffassung*, 2 vols. Berlin: Dietz, vol. I, 198.
Kolakowski, L. 1978. *Main Currents of Marxism. Its Rise, Growth, and Dissolution*, 3 vols. Oxford: Clarendon Press.
Landes, D. 1999. *Wealth and Power of Nations. Why Some are so Rich and Some so Poor*. London: Abacus.
Lewthwaite, G. R. 1966. Environmentalism and determinism: a search for classification. *Annals of the American Association of Geographers*, 56,1–23.
Lichtheim, G. 1961. *Marxism. An Historical and Critical Study*. New York: Praeger.
Malia, M. 1971. Backward history in a backward country. *New York Review of Books*, 17 (7 October), 36–40.
Martin, A. F. 1951. The necessity for determinism. *Transactions of the Institute of British Geographers*, 17, 1–12.
Marx, K. 1964a. *Marx-Engels Werke*, 30 vols. Berlin: Dietz.
Marx, K. 1964b. Zur Kritik der Politischen Ökonomie [orig. 1859]. In *Marx-Engels Werke*. Berlin: Dietz, vol. 13, 3–160.
Montefiore, A. and Williams, W. 1955. Determinism and possibilism. *Geographical Studies*, 2, 1–11.
Opler, M. 1962. Two converging lines of influence in cultural evolution theory. *American Anthropologist*, 44, 524–47.
Overbeck, H. 1965. Ritter-Riehl-Ratzel. Die grossen Anreger zu einer historischen Landschafts-und Länderkunde Deutschlands im 19. Jahrhundert. *Kulturlandschaftsforschung und Landeskunde. Heidelberger geographische Arbeiten*; Bd. 14, 88–103.
Peet, R. 1985. The social origins of environmental determinism. *Annals of the Association of American Geographers*, 75, 309–33.
Plekhanov, G. V. 1923. O knige L.I. Mechnikova [orig. 1890]. In *Sochineniia*. Moscow/Leningrad: Gos. Iz-vo, vol. VII, 15–28.
Plekhanov, G. V. 1925. Istoriia russkoi obshchestvennoi mysli. Tom I [orig. 1914]. In *Sochineniia*. Moscow/Leningrad: Goz. Iz-vo, vol. XX.
Plekhanov, G. V. 1956a. K voprosu o razvitii monisticheskogo vzgliada na istoriiu [orig. 1893]. In *Izbrannye filosofskie proizvedeniia*. Moscow: Gos. Iz-vo Politicheskoi Literatury, vol. I, 507–737.
Plekhanov, G. V. 1956b. Pis'ma bez adresa [orig. 1899–1900]. In *Izbrannye filosofskie proizvedeniia*. Moscow: Gos. Iz-vo Politicheskoi Literatury, vol. V, 282–392.
Plekhanov, G. V. 1956c. Ocherki po istorii materializma [orig. 1896]. In *Izbrannye filosofskie proizvedeniia*. Moscow: Gos. Iz-vo Politicheskoi Literatury, vol. II, 33–194.
Ratzel, F. 1869: *Sein und Werden der organischen Welt*. Leipzig: Gebhardt & Reisland.
Ratzel, F. 1885: Entwurf einer neuen politischen Karte von Afrika. *Petermanns Geographische Mittheilungen*, 31, 245–50.

Ratzel, F. 1896. Die Gesetze des räumlichen Wachtums der Staaten. *Petermanns Geographische Mittheilungen*, 42, 97–107.

Ratzel, F. 1897. *Politische Geographie*. Munich/Leipzig: Oldenbourg.

Ratzel, F. 1898. Politisch-Geographische Rückblicke. *Geographische Zeitschrift*, 4, 143–56, 211–44, 268–74.

Ratzel, F. 1899–1912. *Anthropogeographie*, 2nd edn, 2 vols. Stuttgart: J. Engelhorn.

Ratzel, F. 1901. Der Lebensraum. Eine biogeographische Studie. In K. Bücher *et al.* (eds.) *Festgaben für Albert Schäffle zur siebenzigsten Wiederkehr seines Begurtstags am 24 Februar 1901*. Tübingen: Verlag der Laupp'schen Buchhandlung, 101–89.

Ratzel, F. 1901–02. *Die Erde und das Leben. Eine vergleichende Erdkunde*, 2 vols. Leipzig/ Vienna: Bibliographisches Institut.

Ratzel, F. 1906. Flottenfrage und Weltlage. In *Kleine Schriften*. Munich/Berlin: Oldenbourg, vol. II, 375–81.

Ratzel, F. 1923. *Politische Geographie, oder die Geographie der Staaten, des Verkehrs, und des Kriges*, 3rd edn. Munich/Berlin: Oldenbourg.

Sachs, J. 1997. Nature, nurture, and growth. *Economist*, 343 (14 June).

Sachs, J. 2001. Why are the tropics poor? Assessing the roles of politics, economics, and ecology. *Journal of Economic History*, 61, 521–44.

Sachs, J., Mellinger, A., and Gallup, J. 2001. The geography of poverty and wealth. *Scientific American*, March.

Semple, E. C. 1911. *Influences of Geographic Environment, on the Basis of Ratzel's System of Anthropo-Geography*. New York: Henry Holt, 1–2.

Spencer, H. 1863. The Social Organism. In *Essays: Scientific, Political, And Speculative*. London: Williams & Norgate, vol. II, 143–84.

Tatham, G. 1951. Environmentalism and Possibilism. In G. Taylor (ed.) *Geography in the 20th Century*. London: Methuen, 128–64.

Turner, F. J. 1897. The West as a field for historical study. *Annual Report of the American Historical Association for the Year 1896*. Washington, DC: AHA, vol. I, 281–319.

Turner, F. J. 1905. Geographical interpretations of American History. *Journal of Geography*, 4, 34–7.

Turner, F. J. 1908. Report on the Conference on the Relation of Geography and History. *Annual Report of the American Historical Association for the Year 1907*, 43–48.

Turner, F. J. 1931a. The Problem of the West [orig. 1896]. In *The Frontier in American History*. New York: Henry Holt & Co., 205–21.

Turner, F. J. 1931b. Contributions of the West to American Democracy [orig. 1903]. In *The Frontier in American History*. New York: Henry Holt & Co, 243–68.

Turner, F. J. 1963. *The Significance of the Frontier in American History* [orig. 1893]. New York: F. Ungar.

Turner, F. J. 1965a. Problems in American History [orig. 1892]. In W. R. Jacobs (ed.) *Frederick Jackson Turner's Legacy*. San Marino, CA: Huntington Library, 71–83.

Turner, F. J. 1965b. Why did not the United States become another Europe? In W. R. Jacobs (ed.) *Frederick Jackson Turner's Legacy*. San Marino, CA: Huntington Library, 116–41.

Venturi, F. 1966. *Roots of Revolution*. Transl. F. Haskell. New York: Gorsset & Dunlap.

Vucinich, A. 1976. *Social Thought in Tsarist Russia. The Quest for a General Science of Society 1861–1917*. Chicago: University of Chicago Press.

Woltmann, L. 1900. *Der historische Materialismus. Darstellung und Kritik der Marxistischen Weltanschauung*. Düsseldorf: Hermann Michel.

Chapter 3

Spatial Analysis in Political Geography

John O'Loughlin

Unlike its sister disciplines of economics or political science, political geography has a relatively small amount of published research that contains quantitative analysis, or as I shall term it in this chapter, spatial analysis.[1] Political geography has reflected the rest of the geographic discipline in the flow and ebb in spatial quantitative modeling over the past 40 years. Early examples of correlation and regression analysis appeared in the other social sciences before 1945 but it was not until H. H. McCarty's (1954) analysis of the geographic patterns of the vote for Wisconsin's right-wing senator Joseph McCarthy that a spatial methodology for the examination of electoral results was widely introduced. Following McCarty's lead, the use of aggregate socioeconomic variables for geographic units (counties, wards, census tracts, or countries) as predictors of the political outcomes (votes, international behavior, or legislative votes) in a nonspatial regression framework, widely used in political science, was now complemented by a focus on the geographic pattern of the residuals (error terms, indicating the places that did not closely correspond to the general trend). Only in the late 1970s, thanks to the pioneering work of Cliff and Ord (1973) and extended by Anselin (1988) and Griffith (1987), did it become apparent that the classical statistical methodology was almost always inappropriate for geographic data because of their special nature and a new spatial statistical analysis developed in geography. Unfortunately, the misuse of classical statistical methods continues in geography, including political geography, despite two decades of evidence that these models can produce erroneous results.[2]

The "special nature of spatial data" (Anselin, 1988) requires a more complicated and extended modeling procedure than is usually found in basic statistics texts. Moreover, a significant debate about the nature of "context" (the environment in which political behavior is shaped and expressed) between political geographers and political scientists has propelled the search for new methodologies that will clarify whether place matters or (stated baldly) whether political geography as a discipline is sustainable. If contexts (places) matter little except as convenient units to map or visualize political behavior, political geography fits the role assigned to it by the political scientist, Gary King. "(T)hey (geographers) are skilful at pointing out what we do not understand.

Geographical tools are essential for displaying areal variation in what we know, but this is nowhere near as powerful as the role of geography in revealing features of data and the political world that we would not otherwise have considered" (King, 1996, p. 161). In this chapter, I will make the case that in order to remain a vital part of the wider social science enterprise to understand human behavior, political geography has to merge its central theoretical elements and methodological approaches with appropriate spatial and statistical modeling techniques. Failure to do so will consign the discipline to the kind of cartographic *cul-de-sac* that King envisions for the discipline or worse, further isolation from the other social sciences and continued retreat from the quantitative analysis of important social scientific questions.

The reasons for the relative paucity of quantitative work in political geography can be traced to dual trends that have been evident for the past 20 years and that can be easily recorded from a perusal of the contents of the journal, *Political Geography* (founded in 1982) (Waterman, 1998). First, like the rest of human geography, political geography has seen a rise in interest in poststructuralist and humanistic research methodologies as the 1970s heyday of positivism passed. Longley and Batty (1996, p. 4) believe that this trend is because "words are more persuasive than numbers," although it seems more likely that political geography is returning to the *status quo ante* where quantitative methodology is just one of a plethora of options on the research menu. Second, and connected to the first, quantitative geography (and shortly after, Geographic Information Science – GIS) was promoted as a response to the challenges of the day, especially economic stagnation in Western countries. By pursuing spatial analysis and GIS, and later merging these approaches, geography could certify its "scientific" status and show its uses to the corporatist state (Taylor and Johnston, 1995). Longley and Clarke (1995) stress the amount of "technical deskilling" that has occurred in geography in an era in which transferable skills and flexible specialization hold the keys to adaptability and change in a constantly restructuring labor market. Geography's relative abandonment of its spatial analysis/GIS birthright is allowing other disciplines to fill the labor and market niches.

Unfortunately, a gap developed early between GIS technology and spatial analytical methods and only in the past few years has a sustained effort been made to re-link them so that spatial analysis does not remain an afterthought in a GIS environment. The release of Arc 8© in spring 2001 contains a fully integrated module on geostatistics (useful for analysis of point patterns such as earthquake epicenters) but does not yet include regression-based analyses.[3] Longley and Batty (1996, p. 18) correctly identify the important challenge facing geography: "We are now at a crossroads: either we will make a significant effort to understand the workings and representation of spatial entities, locational processes, and system dynamics, or we will retreat to the margins of academic debate, denying the notion that spatial measures and analyses can ever mean anything, and sniping at the successes of non-geographers when even quite rudimentary spatial analytical techniques are shown to be applicable in planning contexts." Political geography stands at a similar junction.

Big Social Science Questions and Political Geography

Spatial analysis obviously requires some sort of spatially-coded data; these are most commonly areal (also called polygonal) data. But a fundamental problem of geographic data is that we usually collect them for existing political units that, despite their historical and governmental meaning, are less than optimal for spatial analysis. In order to answer the key question posed by Ann Markusen (1999) for policy research but relevant for all geographic research – "How will we know it when we see it?" – ("it" is explicitly conceptualized and empirically operationalized research), we need a dual-track approach in political geography that promotes a model-based methodology to tackle theoretical claims and a set of explanatory variables (i.e. data) to test them.[4] The geographic units that we use suffer from the MAUP (modifiable areal unit problem), visible in results that are scale-dependent. For example, if we correlate data on socioeconomic class and voting for the Republican party with the coefficient varying across the scales as a result of the number of data points and the geographic configuration of the districts, we cannot be sure which coefficient is correct (see the review of MAUP in Openshaw, 1996). If we had a realistic choice, we would gather data on the basis of districts that are arranged in a regular geometric pattern, such as on a grid or for a standardized worldwide unit of analysis, say a square kilometer lattice. Not only is political geography research plagued by a paucity of data in some sort of standardized collection scheme but further, we are hostage to data collection schema that are ill-designed for our purposes.[5]

The key concept related to geographic data is spatial autocorrelation. It is rare for a geographic dataset to lack spatial autocorrelation, defined as like objects clustering together in a nonrandomized manner. Spatial autocorrelation is a mixed blessing since without it, geography as we know it would hardly exist because the world would unquestionably be more idiosyncratic. Spatial modeling research is clearly divided into two camps, commonly referred to as "geostatistics" (analysis of point patterns) and "spatial econometrics" (regression analysis of areal data in a spatial framework). Geostatistics is typically concerned with making a generalized map surface from a sample of points (termed kriging) whereas spatial econometrics blends regression analysis with spatial autoregression methods that use geographic data coordinates to check if location has a significant impact on the compositional relationships (e.g. class on voting choice). As Griffith and Layne (1999, p. 469) note, an integration of the two schools of spatial analysis is long overdue since spatial autocorrelation is the "progenitor of both."

Though there are many issues and choices in spatial analysis that could be the subject of debate and discussion in this chapter, I will focus on the five topics that I think are central to political geography and, at the same time, are topical subjects in spatial analysis. I will begin with the contextual debate between geographers and political scientists about research on the use of aggregate data to infer individual behavior. Then I will look at recent developments in local indicators of spatial association (LISAs) and new methods of visualization and display. Finally, I will end with an exposition of multi-level modeling that offers a powerful methodology to political geographers who assert that relationships between scales are what separates our discipline from others and gives us a special role in the social science collective enterprise.

Context debates in political geography

What distinguishes spatial analysis from sociological, political or economic modeling is a consideration of both compositional and contextual elements of the problem.

Political studies typically lack any consideration of the context or environment in which the political process takes place. It is now common practice to see context carefully evaluated in epidemiological or educational studies, because environmental considerations (neighborhood, school, metropolitan area, region) have been dramatically significant in explaining variations in disease rates and school test scores. The contextual approach is not new in social science and Blakely and Woodward (2000) credit the first multi-scalar study to the sociologist Emile Durkheim, whose work on the environmental and personal factors underlying suicide was published in 1898. Over the past hundred years, social scientific and medical research moved away from reductionist environmental explanations of the style that simply adds a contextual variable to a set of compositional factors. In such a model, a dummy variable measuring the setting of the survey respondent or regional location of the geographic unit is added to the right-hand-side of the regression equation to the usual array of compositional factors (class, age, gender, religion, educational status, etc.). At best, such a model can demonstrate that there are "unexplained" effects emanating from environmental settings, but it cannot readily show the relative importance or interactive influence of these effects. A more formal and sophisticated modeling strategy is warranted that allows for interaction between the multiple scales; the effects of the ecological variables might be mediated by intermediate variables at the individual level (Blakely and Woodward, 2000, p. 368).

While geographers have argued that context counts (see Agnew, 1987 for the most complete statement; see also Agnew, 1996a,b; Cox, 1969; Johnston, 1991, 2001; Johnston et al., 1990; O'Loughlin and Anselin, 1991), political scientists have countered that contextual effects are either insignificant or bogus. (A bogus contextual effect is one that evaporates in a statistical analysis that incorporates many compositional elements or has a different functional form – non-linear, for example.) The most direct challenge to the geographers' position has come from Gary King's (1996, p. 161) conclusion that "if we really understood politics, we would not need much of contextual effects... [T]o understand political opinions and political behaviour, we are usually trying to show that context does not matter." King bases his position on the undoubtedly accurate assessment that while the geographic variation in political outcomes (say, percent Republican vote) is large to begin with, after compositional effects are introduced into the model accounting for geographic variation in the characteristics of the voters, there is little left for contextual effects.

There are three possible retorts from geographers to King's important challenge. The first is that one cannot know how important the contextual effects are until they are formally identified and measured; these checks are not usually carried out in political science or sociology. The impact of the context will vary from study to study and unless the contextual variables are considered, it is highly probable that their direct and indirect (mediated by compositional variables) impacts will go unmeasured. The second retort is that in aggregate data analysis, compositional estimates will probably be inefficient, biased, inconsistent, and insufficient (Anselin, 1988; Griffith and Layne, 1999). We cannot retain much confidence in the compositional coefficients if spatial autocorrelation is present, as is usually the case with aggregate geographic data. Third, King's challenge misses the important point that political geographers have reiterated for the past quarter-century. Agnew (1996a) calls this approach the "geo-sociological" model, a sharp contrast to King's concentration on individuals as separate from their environment. In the geo-sociological approach, geographic research focuses on "how individuals are spread around and divided into aggregates . . . [W]e can never satisfactorily explain what drives individual choices and action unless we situate the individuals in the social-geographical contexts of their lives" (Agnew, 1996b, p. 165). Herein lies the central quandary for geographers – although we argue the case for a geo-sociological approach in which individuals are embedded in their contexts, we do not specifically offer a methodology that allows measurement of the relative contribution of the direct and indirect effects of the environment on individual behavior. Until we have the methods and the trained personnel to use them correctly, we will be making an argument that will not carry much weight in the disciplines that are more quantitatively oriented, especially political science and economics. The importance of multilevel modeling (discussed below) as a way to bridge the gap with political science should therefore not be underestimated.

Inferring individual behavior from aggregate data

Related to the context debate, attempts to bridge the political scientists' emphasis on individuals and political geographers' focus on aggregate units are getting underway. The central problem is one of scale and is also related to the MAUP discussed earlier. Geographers usually resort to aggregate statistics and as a result, we have not been able to infer individual behavior from these large unit data. Since the early twentieth century, it has been noted that conclusions deriving from aggregate data often show significant differences to those based on individual data. In the 1950s, the term "ecological fallacy" came into common use and students were steered away from making any kind of inference to individual behavior from analysis of aggregate data. The result of the widespread recognition of the ecological fallacy was twofold. First, political scientists turned strongly to survey methods over the past 50 years to elicit attitudinal and behavioral characteristics of citizens. Second, geographers with recourse predominantly to aggregate data refrained from extending their conclusions to individuals, making generalizations only about populations or regions. A typical conclusion of quantitative geographic study is "Elderly voters in the southwest of the city are more likely to support the Republican candidate." Missing are any specific measures of the level of support over and above some baseline measure (such as all elderly in the city) as the regression coefficients are incapable of conveying this information.

Until the appearance of Gary King's 1997 book *A Solution to the Ecological Inference Problem*, attempts to bridge the aggregate–individual gap suffered from serious statistical and theoretical shortcomings and assumptions. Though King's solution is not a panacea for all methodological problems surrounding the ecological inference problem such as MAUP or spatial autocorrelation, it nevertheless offers a breakthrough for political geographers because it allows inference to individuals on the basis of fairly sparse aggregate data. Unlike the entropy-maximizing method promoted by Ron Johnston and Charles Pattie (2000), King's method does not require an overall system-wide value in order to get the estimates for the individual geographic units. Thus, in the example below, it is impossible to know what ratio of Protestants voted for the Nazi party in Weimar Germany in 1930 as this was an era before public opinion polls were conducted. In most historical circumstances and in many local elections, system-wide values that drive the entropy-maximizing estimating procedure will be unknown. King's method warrants further attention from geographers and although the estimates for individual units can be affected by the overall distributional statistics and should be used only after examination of the confidence bounds, the global estimates have been shown to be reliable.

The ecological inference problem and solution can be explained by illustration. What ratio of the Protestant population in Weimar Germany voted for the Nazi party in 1930? From previous studies, it is well known that the Protestant ratio in a district was positively correlated with support for the Nazi party (O'Loughlin et al., 1994). The data to be used for the inference is the ratio of Protestants in each of the 743 districts in Germany, the ratio of the vote for the Nazis, and the total number of voters in each district. Nationally, the Nazis received 18.3 percent at the 1930 election and the Protestant ratio in Germany was 62 percent. The global estimate will be the national percentage of Protestants that voted for the Nazis and the local estimates are the respective county (*Kreis* in German) ratios. Using King's notation, the independent variable X is the Protestant population and T is the national Nazi vote. For each county, we have the Protestant and Nazi totals from census and electoral archives but not the cell values that must be estimated (see table 3.1). Using the information in the marginals (the totals of each row and column), ecological modeling estimates the values for the question marks for the country as a whole and for each *Kreis*. Any estimates must meet the conditions of the marginals (must sum to these values). King's solution avoids the homogeneity pitfall that plagued Goodman's double regression method; the assumption of homogenous distribution of parameters across all geographic units is an untenable assumption for political geographers.

King's ecological inference method uses an identity from the modified Goodman formula to generate combinations of values for T_i (the Nazi vote in *Kreis i*) and X_i (the Protestant vote in the *Kreis*) in the form of $T_i = \beta_i^b X_i + \beta_i^w (1 - X_i)$. The purpose of the ecological inference modeling is to estimate β^b (the national ratio of Protestant voters who chose the Nazi party) as well as the estimates for the individual *Kreise*, β_i^b. Combined with information about the bounds of each district, found by projecting the line onto the horizontal axis (β_i^b, the Protestant vote for the Nazi party) and the vertical axis β_i^w (the non-Protestant vote for the Nazis), King's method combines the double regression approach with the information on bounds. Clearly the narrower the bounds, the higher the reliability of the estimates is likely

Table 3.1 The ecological inference problem for a typical *Kreis* in Weimar Germany

	Vote		
	Nazi	Non-Nazi	Totals
Protestant	?	?	13,261
Non-Protestant	?	?	6,735
Totals	8,423	11,573	19,996

to be. (Further information is found in O'Loughlin, 2000.) In the case of the 1930 election, the ecological estimate of 22.4 percent of Protestant voters who picked the Nazi party is 3.6 percent higher than the national average of 18.3 percent.

The individual ecological inferences for the 743 *Kreise* of Germany can be used in a further "second-level" analysis as dependent variables; there is significant variation in these ratios across Germany from 2 to 50 percent, showing that the Protestant support for the Nazis varied according to local conditions. The maps of these ecological inferences shows a concentration of high values in scattered locales in Northern Bavaria, Northwest Germany, and Saxony (O'Loughlin, 2002). These contextual anomalies suggest local circumstances that propelled the Protestant population to support the Nazi party far in excess of their national average. Like all methods, King's ecological inference procedure works best (giving most reliable estimates) if the districts are nearly homogenous on the predictor variable (Protestant ratio in this case), the units are small (precincts or some other small geography unit is most suitable) and there is a large number of districts (more than 100). In the USA, racially-homogenous districts are common and, therefore, the method has had its most publicized successes in this context (King, 1997).

Nonstationarity and directional analysis of spatial autocorrelation

In the example above, the mapping of the ecological inferential values for the Protestant support of the Nazi party indicates that a disaggregated approach to the study of political phenomena is valuable. Of course, there is a fine line between total disaggregation to each of the data points (complete uniqueness) and a study that remains at the global (most aggregated) level. Spatial analysis is clearly interested in the social scientific enterprise of drawing generalizations and making inferences to populations from samples but at the same time, geographers remain acutely aware that national-level statistics hide great regional and local variations. A way out of this impasse was suggested by Siverson and Starr (1991) who believe that "domain-specific laws," incorporating important local and regional circumstances under consideration in a general model, offers the most attractive alternative. Thus, in a study of the correlates of the Nazi party vote in 1930 Germany, O'Loughlin et al. (1994) were able to show that the specific mix of supporters of the party varied between six large cultural–historical regions of the country. In some regions, the middle-class was a significant base for the party but in other regions, the coefficients show that support was weak and nonsignificant. What was most evident in this study is that the national average hid great regional variation. Moreover, local effects in the form of small clusters of districts that stood out from surrounding values (high values in generally low-value regions and vice versa) were also visible in a spatial analysis and could therefore be modeled.

The balance between global and local measures and approaches in statistical geography seems to have been resolved strongly in favor of local measures in recent years. Because most geographic datasets have large amounts of nonstationarity (relationships between variables vary across the dataset and are not consistent in all regions), we often tend to find multiple regimes of spatial association, as in the case of Nazi Germany above. We need more than one parameter estimate in these cases and the fitting of models according to a theory-derived regional division is indicated. Nonstationarity in spatial modeling can have a number of underlying causes: random sampling variations, the fact that relationships vary because of regional circumstances, a mis-specified model in which the measures are poor reflections of reality, or possibly because one or more of the relevant variables are omitted or are represented by the incorrect functional form (linear, rather than nonlinear) (Fotheringham, 1997).

Because of the widespread attention to nonstationarity, there has been a significant return to basics in spatial analysis, paralleling the rise in exploratory data analysis in social science in general. Rather than confirmatory procedures, such as regression of a theory-derived model, geographers tend carefully to tease out local trends in the data. To do this, specific indicators of local significance are derived, and as becomes clear in Anselin's (1995) work on LISAs (local indicators of spatial association), there is a clear linkage between global measures of clustering and local indicators. Local statistics are well-suited to (i) identifying the existence of pockets or "hot spots" that are significantly different than the regional or global trend (such as disease clusters or a congregation of supporters of a particular party), (ii) assessing assumptions of stationarity, and (iii) identifying distances beyond which no discernible spatial association is present (Getis and Ord, 1996). After dissecting global statistics to their local constituents, we can produce local statistics that can be mapped. But the dilemma is not resolved just by deriving local measures. As Openshaw (1996, p. 60) notes, "the confirmatory dilemma is as follows; either you test a single whole-map statistic against a null hypothesis or you test N hypotheses relating to zone or locality-specific statistics. In the former, the test is silly from a geographical point of view because of its "whole-map" nature, its dependency on the definition of the study region, and the nature of the underlying globally defined hypothesis. In the latter case, there is the problem of multiple testing." A reaffirmation of the confirmatory hypothesis-testing approach has been achieved by blending modeling procedures with diagnostic, exploratory, and interactive techniques.[6]

As well as being nonstationary, geographic data are often anisotropic. (Isotropic data means that spatial dependence – autocorrelation in other words – changes only with the distance between the values but not with their directional orientation with respect to each other.) In physical geography, prevailing winds in climatology, the spread of beetles in a pine forest from an external source or earthquake fault lines come to mind as examples of directional influences. In political geography, one might expect directional influences to be significant in a pattern that results from a diffusion process. It is plausible, for example, that war spreading directionally across a continent, the growth of a political party from a local core, or the diffusion of the democratic form of government will violate the isotropic assumption. Given these possibilities, it is necessary to identify and account for any anisotropic developments. While mapping the LISAs might conceivably show a directional trend, say a north-west trend caused by the migration of pine beetles in this direction as a consequence of local environmental (terrain or climatologic) conditions, it is better to use methods developed specifically for the measurement of directional bias. We need a statistic that incorporates the geographic coordinates, their angular relations with respect to a fixed bearing (e.g. east) and the values of the item of interest (in this case, the level of tree infestation by beetles) to determine if there is significant directional bias in the pattern.

Most of the direction-based methods come from genetics, animal ecology, and organismic biology, emanating from Oden and Sokal's (1986) introduction of directional spatial autocorrelation techniques by developing "distance/direction classes" to create a windrose correlogram; sectors represent the same distance but different angles grouped together in rings called annuli (Rosenberg, 1999, p. 270). The selection of spatial weights (measuring the attraction or contiguity of places to each other) has bedeviled spatial analysis because no commonly-agreed method for choosing the weights structure is available. In the bearing spatial correlogram, the weight variable incorporates not only the distance or contiguity between points (they could be areal centroids) but also the degree of alignment between the bearing of the two points and a fixed bearing. For each distance-class (predefined based on some theoretical conception of appropriate distance bands for the study), the weights matrix is determined by multiplying the nondirectional weight value (distance between the points) by the squared cosine of the angle between the points and the eastern bearing, or formally as $w'_{ij} = w_{ij} \cos^2 (\alpha_{ij} - \theta)$, where w'_{ij} is the $i - j^{\text{th}}$ entry of the bearing weights matrix, w_{ij} is the distance weights between the capitals, α_{ij} is the angular direction between points i and j measured counterclockwise from due east, and θ is the angular direction of the fixed bearing. We can calculate the standard spatial autocorrelation statistic, Moran's I, in the normal manner using the w'_{ij} weights in the place of the usual nondirectional weights in the measure.[7] Examples of the methodology using an anisotropic lens to the study of political processes are O'Loughlin (2001a) for the diffusion of civil and political rights, and O'Loughlin (2002) for the study of the diffusion of the Nazi party vote in Germany 1924–33.

Visualization and displaying results

With the renewed emphasis on local measures of spatial association (autocorrelation) in recent years, new methods of visualization as a first step in spatial analysis have been proposed to highlight these circumstances. A useful distinction between private and public visualization has been noted by Cleveland (1993). In the early stages of the research, private visualization in the form of graphs, diagrams, maps, and descriptive indicators can be generated and saved as screen captures or low-quality prints. Most of the statistical software packages offer adequate visualization procedures (Q–Q plots for normality tests, histograms or box-plots for distributional displays, etc.), although Stata© and S-Plus© provide suites of trellising options that allow detailed exploration of the data structures. Trellis displays are tools for visualizing multidimensional datasets and trellis graphics display a large variety of one-, two- or three-dimensional plots in an automatically generated trellis layout of panels, where each panel displays the selected plot type for a slice on one or more additional discrete or continuous conditioning variables. Few trellis displays make it to the second kind of visualization, that of public presentation in the traditional print medium or as web documents where the emphasis is on presentation and the dissemination of knowledge. Good examples of trellis graphics are available in Cleveland (1993) while Tufte (1997) provides clear guidelines and magnificent examples of public visualization. In general, the purpose of visualization is to identify geographic clusters of similar data points, identify local and global outliers, and identify trends in the relationships (Fotheringham, 1999).

Three regression-type models are now available to political geographers who wish to build local spatial relationships into the usual compositional models of the political scientists. First, the mixed spatial–structural model adds a spatial autoregressive term to the usual regressors if there are indications in the data that significant spatial autocorrelation is present that is not simply the result of omitted variables. In analysing the distribution of conflict in Africa between 1966 and 1978, O'Loughlin and Anselin (1991) show how a spatial autoregressive term (measuring the effects of neighboring states at war) is an important addition to a regression with other characteristics of states (colonial history, ethnic fractionalization, nature of government, economic status, etc.). While not every research problem in political geography will benefit from the incorporation of a spatial autoregressive term, extensive experience now indicates that every dataset should be carefully checked for the presence of spatial autocorrelation. If spatial autocorrelation is near zero, the traditional statistical model with only compositional variables will suffice but as noted earlier, a significant danger of biased parameters will result from ignoring the presence of sizeable autocorrelation.

Two alternative forms of local spatial measurement in multivariate relationships are now readily available. The expansion method (Jones and Casetti, 1992) allows parameter drift so that if the parameters of the regression model are functions of geographic location (say, latitude and longitude), the trends in parameter estimates over space can then be measured. A more recent alternative is geographic weighted regression (GWR), where localized parameter estimates can be produced and, also, localized versions of all the regression diagnostics can be developed (Brunsdon et al., 1996). GWR is based on the assumption that data are weighted according to their proximity to point i and the weights are not constant but vary with proximity to point i. These parameters can be mapped to see the geographic pattern and possibly lead to further analysis of the residuals. Like the discussion of nonstationarity and the use of multiple regimes, these methods are motivated by the belief that strong evidence of regional heterogeneity will normally be a feature of geographical data.

In the environment of exploratory spatial data analysis (ESDA), one of the motivations behind the visualization push is to redress one of the troubling aspects of quantitative analysis, the growing gap between those who use spatial models and the rest of the discipline. Unlike the situation at the height of the quantitative revolution in Geography, graduate students in the discipline can now finish a Ph.D. without being obliged to pass a course in statistical methods. The splintering of the discipline has led to the acceptance of alternative methods courses (qualitative, feminist, field) in lieu of the quantitative requirement. The development is enforcing an increasingly fractionalized discipline, with a small or no common core of knowledge and a lack of understanding of the language and methods of each sub-discipline. Because the theory and language of spatial analysis is increasingly arcane, not only to fellow geographers but also to colleagues in other social sciences, it places additional pressure on modelers to write in an accessible style and include more materials that present statistical results in a visual manner. Nonlinear modeling generates coefficients that can be difficult to interpret and logit models benefit from conversion of the coefficients by anti-logs to render them meaningful. Too frequently, spatial analysts simply regurgitate the output from their computer packages. Maps are wonderful tools for making sense of complex data, though clearly the choices of metric, color schemes, analytical methods, scale, and symbols are critical in presenting results that can be understood and evaluated. In political science, a similar separation between the methodologists and the rest of the discipline has propelled a re-thinking of the way in which statistical results are presented. Gary King and his colleagues have written a series of programs in Stata© to convert results from nonlinear models into values that can be graphed using a simulation technique.[8] Thus, for example, O'Loughlin (2001b, p. 29) used box-plots of simulated values to show the ranges of the estimated probability of support for the free market by household finances and by region in Ukraine 1996. While households with "better finances" in western Ukraine had a mean probability of supporting the free market at a rate of 0.62, families with poor finances in the south of the country only had a 0.21 probability of supporting the free market. These huge differences by region and family finances are thus easily understandable to readers without statistical training, though the logit coefficients may not be especially meaningful to them.

Multilevel modeling and scale effects

As will hopefully be clear by this point in the chapter, the problems posed by aggregate data organized on a geographic basis are formidable. Not only do issues connected to spatial autocorrelation require attention but for political geographers, scale problems in the form of identification of individual and contextual variations also must be tackled. If all political outcomes are the result of individual choices and behaviors in an atomized world, then political geography is severely under threat. But an atomized world-view is highly implausible and it can be countered by a "geo-sociological" imagination (Agnew, 1996a). While offering a counter model to the political scientists and public opinion pollsters is a start, it is unlikely to carry political geography very far in the face of a sceptical audience that wants statistical evidence of scale and context effects. Recent developments in multilevel modeling allow the calculation of statistical variance at each scale (individual, local, regional) and thus, enable the researcher to determine if the geo-sociological imagination holds any value. The interaction effect (individual-context) offers an additional element of variance explanation and thus, the hypothesis of a geo-sociological imagination can be tested statistically. As Jones and Duncan (1996, p. 80) note, there has been too much stress in spatial analysis on the stereotypical and the average and not enough on variability because the underlying trend has been sought by ignoring difference. The multilevel approach preserves between-place heterogeneity and does not annihilate space as context in a single equation that is fitted for all places and all times.

Multilevel modeling extends the technique of ordinary least-squares (OLS) to explore the variation among units defined at the various levels of a hierarchical structure. I will illustrate using the example of the political attitudes of residents of 17 neighbourhoods (*rayoni* in Russian) in Moscow in March 2000. (The notation and review is modified from Bullen et al., 1997; Goldstein, 1995; and Kreft and de Leeuw, 1998.) The simple regression relationship is expressed as $y_i = \beta_0 + \beta_1 x_i + e_i$, where subscript i ranges from 1 to n_j, the number of respondents in the jth neighborhood in Moscow. For the ith respondent, y_i is the dependent variable (willingness to protest in this case) and x_i is an independent predictor, say age. In the usual single-level model, e_i

is the residual, that part of the dependent variable not predicted, and with only one level, the variation is simply the variance of these e_i.

In the multilevel case, where the 17 *rayoni* (districts) are regarded as a random sample of all neighborhoods in Moscow, we can express the multiple relationships as: $y_{i1} = \beta_{01} + \beta_1 x_{i1} + e_{i1}$, $y_{i2} = \beta_{02} + \beta_1 x_{i2} + e_{i2}$, etc. These equations can be generalized to $y_{ij} = \beta_{0j} + \beta_1 x_{ij} + e_{ij}$, where in the final general expression the subscript j takes values from 1 to 17, one for each *rayon*, and the first subscript now refers to respondent i in *rayon* j. In a multilevel analysis, the level-2 groups (*rayoni*), are treated as a random sample. We therefore re-express the last equation as $y_{ij} = \beta_0 + \beta_1 x_{ij} + u_j$, where u_j is the departure of the jth *rayon*'s actual intercept from the overall mean value. It is thus a level-2 residual. (β_0 has no level subscript, indicating that it is constant across all *rayoni*; β_{0j} is specific to *rayon* j, but is the same for all respondents in that *rayon*.) The full model for actual scores can be re-expressed as $y_{ij} = \beta_0 + \beta_1 x_{ij} + u_j + e_{ij}$. In this equation, both u_j and e_{ij} are random quantities, whose means are equal to zero. The quantities β_0 and β_1 are fixed and must be estimated. The presence of the two random variables u_j and e_{ij} in the last equation make it a multilevel model and their variances, σ_u^2 and σ_e^2, are referred to as random parameters of the model. The quantities β_0 and β_1 are known as the fixed parameters. A multilevel model of this simple type, where the only random parameters are the intercept variances at each level, is known as a variance components model. For political geographers, the real interest is the relative contribution of the second-level variance σ_u^2 to the overall model.

In a multilevel model, between-place differences can be examined in relation to the social characteristics of individuals in combination with the characteristics of places. For example, a voter of low social class may vote quite differently according to the social class composition of the neighborhood in which he or she lives (Taylor and Johnston, 1979). Using the example of Moscow, a person's attitude towards protest (dependent variable) is modeled as a function of (i) the person's characteristics (age, gender, ideology, education, etc.), (ii) the neighborhood in which the person lives, and (iii) the compositional/contextual interactions. Data on the characteristics, civic behavior and political preferences of 3,476 Muscovites in 17 sample neighborhoods were collected in door-to-door interviews in March 2000 just after the Russian Presidential election that elected Vladimir Putin. Four key characteristics of voters (educational level, age, whether they voted for Vladimir Putin, and whether they support the rapid transition to the free market) are used to explain whether the respondent was willing to take part in protest or not. Only 9.9 percent of the 3,476 respondents were willing to take part in protests against falling living standards.

In multilevel modeling, the first stage is to measure the level-2 (neighborhood) variance; in this study of Moscow, the value was rather large and significant. Then, the characteristics of the level-1 units (respondents in this case) are added to the model and, as in the usual regression format, only significant independent predictors are included in the equation. All of the four variables are in the expected direction and significant; the chance of protest increases with age, educational level, voting for Putin and with distrust of the market economy. Overall, the model indicates that the second-level variance contributes 7 percent of the total variance while the interaction term (across the two levels) accounts for 4 percent, and as usual, the overwhelming proportion, 89 percent, is attributed to the individual-level variance.

This study thus supports the claims of geographers that a contextual effect exists over and above the varied distribution of voters among geographic units and that the geo-sociological model which emphasizes interaction effects across the levels is also useful in helping to explain the political choices of citizens. Similar interaction and contextual effects have been identified by Jones et al. (1998) for the Labour vote in the 1992 British election. The multilevel individual–context interaction model parallels the explanation offered by Pattie and Johnston (2000) that extensive and intensive local contacts help to shape political opinions and choices. Contextual effects account for a significant part of the overall explanation and compositional models that ignore context are likely to offer only partial explanations.

Conclusions

This review of developments in spatial analysis in political geography has stressed key developments and challenges. Sceptical challenges to quantitative political geography emanate from two sources: from within the discipline from those who are antithetical to hypothesis-testing and empirical data analysis; and from outside the discipline where, though sympathetic to quantitative analysis, researchers have not yet been persuaded that significant and measurable contextual and geo-sociological effects exist. To answer these critics, political geographers need to develop further training and expertise in the spatial analysis of aggregate data, the collection of survey data, the conversion of statistical results into visual and accessible formats, and the matching of appropriate methodologies to specific research questions. Each of these desiderata are formidable and time-consuming but without their implementation, I fear that political geography will become marginalized in a small discipline and excluded from the social science enterprise.

Political geographers, unfortunately, have come to rely on aggregate data collected by government agencies on the basis of pre-existing geographic units. Not only does this reliance magnify the modifiable areal-unit problem (MAUP), but it also forces political geographers to turn to complex analytical techniques because the usual statistical models are inappropriate for spatial data. Of course, misapplications of OLS models to geographic data continue to appear in the literature, and not only in political geography. For aggregate data, often available in circumstances for which no other information is available like the example of Nazi Germany in the 1930s, it is high time to follow tried and true procedures. Griffith and Layne (1999) list the steps from descriptive statistics and visual plots to measures of local and global spatial autocorrelation to semi-variogram plots for geostatistics, and spatial econometric modeling for aggregate data, and they conclude (p. 478) that "now is the time for all good spatial scientists to begin implementing appropriate spatial statistical specifications."

Many core political geographic questions, however, cannot readily be answered by the use of aggregate data and must be tackled instead through survey methodologies. Few political geographers receive formal training in the design, selection, sampling, analysis, and pitfalls of survey data. Unlike the many large databases and panel data designed for political scientists and economists, political geographic research tends to tweak these data rather than designing specialized surveys from the start of projects. Recently, Shin (1998) and Secor (2000) conducted surveys in

Central Italy and Istanbul, respectively, to elicit information about contemporary political changes in these sites and to determine the role of local contexts in helping to shape opinions and behaviors. Both survey samples were chosen on the basis of neighborhood typologies so instead of sampling randomly, these researchers developed a systematic design that covered the range of possible context effects. Although the time and effort of such enterprises exceed those of mining pre-existing aggregate data (from census offices or archives), they compensate by allowing the researcher to match the methodology to the nature of the research questions.

With the continued growth in the use (and misuse) of GIS technology and the slow integration of GIS and visual displays, it is likely that greater attention will be given to improving the presentation of research results, the public visualization at the end of a project. Undoubtedly, private visualization will allow more insights into the structure of data and help to route scholars around the potholes of inappropriate statistical tools. More use of color, web animation, dynamic links, and free software and data downloads, as well as continued presentation in the print medium, will make research results both more accessible and comparative. (See O'Loughlin et al., 1998 for an example.) Compared to political science, little replication of the research of others or attention to the accumulation of research results occurs in political geography. Hopefully, the trend of isolation will be reversed as standard procedures become more formalized and accepted.

To paraphrase Longley and Batty (1996), quantitative political geography now stands at a junction. Either it will be integrated more intensively with the rest of political geography (this has to be a two-way street and will only succeed if non-quantitative political geographers accept our approaches and research results) and more generally with other quantitative social science, or it will become further isolated. After four decades of development, we now have accumulated expertise and powerful analytical software and display tools to answer many lingering questions regarding the role of place and space in political behavior. Although political geographic theory has raced ahead of empirical tests and statistical expertise over the past 20 years, the gap can be narrowed and many untested theoretical propositions can be checked. As this chapter has shown, political geography is an important part of the enterprise that is trying to understand human behavior; now is the time to challenge the atomizing model and reassert the contextual/geo-sociological one in a hypothesis-testing spatial analytical mode.

ENDNOTES

1. By spatial analysis, I mean the analysis of data that have spatial coordinates or geographic locations such as data for electoral precincts, countries, regions, cities, or locational attributes of voters (street address, work location, personal networks, etc.).
2. In the interests of full disclosure and self-criticism, I admit that I followed McCarty's methodology in my Masters thesis at Penn State (1971), although I introduced a strong spatial focus by close examination of the residuals in the analysis of the Mayoral elections in Philadelphia.
3. Luc Anselin has developed an interface between his spatial econometrics package, Spacestat™, and ArcView3.2©. See Anselin (1999) and the website www.spacestat.com.

4. I agree with Paul Plummer (2001) who makes a similar case for economic geography and who is also responding to Markusen's call for an end to fuzziness and a clearer conceptual base for empirical research.
5. Gary King and his colleagues have engaged in a massive effort to collect, standardize, and make accessible electoral data for the past 20 years in a GIS format. The political units range from precincts to congressional districts in the US. Called the ROAD project (Record on American Democracy), the data are available from the project website www.data.fas.harvard.edu/ROAD.
6. A good example of the multiple options for spatial analysis is Luc Anselin's Spacestat[TM] program.
7. The standard global measure of spatial autocorrelation, Moran's I, is given by $I = (N/S_o) \sum_i \sum_j w_{ij} x_i x_j / \sum_i x_i^2$, where w_{ij} is an element of a spatial weights matrix W that indicates the new bearing weight matrix for i and j; x_i is an observation at location i (expressed as the deviations from the observation mean); and S_o is a normalizing factor equal to the sum of all weights ($\sum_i \sum_j w_{ij}$). PASSAGE (Pattern Analysis, Spatial Statistics, and Geographic Exegesis) is a directional analysis computer program from Michael Rosenberg, available from www.public.asu.edu/-mrosenb/Passage.
8. The program is called CLARIFY and is available from Gary King's webpage at http://gking.harvard.edu. It is described in King et al. (2000).

BIBLIOGRAPHY

Agnew, J. A. 1987. *Place and Politics: The Geographical Mediation of State and Society.* Boston, MA: Unwin Hyman.

Agnew, J. A. 1996a. Mapping politics: How context counts in political geography. *Political Geography*, 15, 129–46.

Agnew, J. A. 1996b. Maps and models in political studies: A reply to comments. *Political Geography*, 15, 165–8.

Anselin, L. 1988. *Spatial Econometrics: Methods and Models*. Dordrecht: Kluwer.

Anselin, L. 1995. Local indicators of spatial association – LISA. *Geographical Analysis*, 27, 93–115.

Anselin, L. 1999. *Spacestat Manual, Version 1.91*. Ann Arbor, MI: TerraSeer Inc.

Blakeley, T. A. and Woodward, A. J. 2000. Ecological effects in multi-level studies. *Journal of Epidemiology and Public Health*, 54, 367–74.

Brunsdon, C. F., Fotheringham, A. S., and Charlton, M. E. 1996. Geographically weighted regression: A method for exploring spatial non-stationarity. *Geographical Analysis*, 28, 281–98.

Bullen, N., Jones, K., and Duncan, C. 1997. Modelling complexity: Analysing between-individual and between-place variation – a multilevel tutorial. *Environment and Planning A*, 29, 585–609.

Cleveland, W. S. 1993. *Visualizing Data*. Summit, NJ: Hobart Press.

Cliff, A. D. and Ord, J. K. 1973. *Spatial Autocorrelation*. London: Pion.

Cox, K. R. 1969. The voting decision in a spatial context. *Progress in Geography*, 1, 81–118.

Fotheringham, S. 1997. Trends in quantitative analysis: Stressing the local. *Progress in Human Geography*, 21, 88–96.

Fotheringham, S. 1999. Trends in quantitative geography III: Stressing the visual. *Progress in Human Geography*, 23, 597–606.

Getis, A. and Ord, J. K. 1996. Local spatial statistics: An overview. In P. Longley and M. Batty (eds.) *Spatial Analysis: Modelling in a GIS Environment*. New York: Wiley, 261–78.

Goldstein, H. 1995. *Multilevel Statistical Models*. London: Edward Arnold.

Griffith, D. A. 1987. *Spatial Autocorrelation: A Primer.* Washington DC: Association of American Geographers Resource Publications.
Griffith, D. A. and Layne, L. J. 1999. *A Casebook for Spatial Statistical Analysis: A Compilation of Analyses of Different Thematic Data Sets.* New York: Oxford University Press.
Johnston, R. J. 1991. *A Question of Place: Exploring the Practice of Human Geography.* Oxford: Blackwell.
Johnston, R. J. 2001. *Electoral geography in electoral studies: An overview on putting voters in their place.* Workshop on Political Process and Spatial Methods, Florida International University, Miami, FL.
Johnston, R. J. and Pattie, C. 2000. Ecological inference and entropy-maximizing: an alternative procedure for split-ticket voting. *Political Analysis*, 8, 333–45.
Johnston, R. J., Shelley, F. M., and Taylor, P. J. (eds.). 1990. *Developments in Electoral Geography.* New York: Routledge.
Jones, J. P and Casetti, E. (eds.). 1992. *Applications of the Expansion Method.* London: Routledge.
Jones, K. and Duncan, C. 1996. People and Places: the multilevel model as a general framework for the quantitative analysis of geographical data. In P. Longley and M. Batty (eds.) *Spatial Analysis: Modelling in a GIS Environment.* New York: Wiley, 79–104.
Jones, K., Gould, M. I., and Watt, R. 1998. Multiple contexts as cross-classified models: The Labor vote in the British general election of 1992. *Geographical Analysis*, 30, 65–93.
King, G. 1996. Why context should not count. *Political Geography*, 15, 159–64.
King, G. 1997. *A Solution to the Ecological Inference Problem: Reconstructing Individual Behavior from Aggregate Data.* Princeton, NJ: Princeton University Press.
King, G., Tomz, M., and Wittenberg, J. 2000. Making the most of statistical analyses: Improving interpretation and presentation. *American Journal of Political Science*, 44, 341–55.
Kreft, I. and de Leeuw, J. 1998. *Introducing Multilevel Modeling.* Thousand Oaks, CA: Sage Publications.
Longley, P. and Batty, M. 1996. Analysis, modeling, forecasting and GIS technology. In P. Longley and M. Batty (eds.) *Spatial Analysis: Modelling in a GIS Environment.* New York: Wiley, 1–16.
Longley, P. and Clarke, G. 1995. Applied geographical information systems: developments and prospects. In P. Longley and G. Clarke (eds.) *GIS for Business and Service Planning.* Cambridge: Geoinformational International, 3–9.
Markusen, A. 1999. Fuzzy concepts, scanty evidence, policy distance: the case for rigour and policy relevance in critical regional studies. *Regional Studies*, 33, 317–70.
McCarty, H. H. 1954. *McCarty on McCarthy: The Spatial Distribution of the McCarthy Vote 1952.* Iowa City, IA: Department of Geography, University of Iowa.
Oden, N. L. and Sokal, R. R. 1986. Directional autocorrelation: an extension of spatial correlograms in two dimensions. *Systematic Zoology*, 35, 608–17.
O'Loughlin, J. 2000. Can King's ecological inference method answer a social scientific puzzle: who voted for the Nazi party in Weimar Germany. *Annals, Association of American Geographers*, 90, 592–601.
O'Loughlin, J. 2001a. Geography and democracy: the spatial diffusion of political and civil rights. In G. Dijkink and H. Knippenberg (eds.) *The Territorial Factor in Politics.* Amsterdam: Amsterdam University Press, 77–96.
O'Loughlin, J. 2001b. The regional factor in contemporary Ukrainian politics: scale, place, space or bogus effect? *Post-Soviet Geography and Economics*, 42, 1–33.
O'Loughlin, J. 2002. The electoral geography of Weimar Germany: exploratory spatial data analysis (ESDA) of Protestant support for the Nazi party. *Political Analysis*, 10, 217–43.

O'Loughlin, J. and Anselin, L. 1991. Bringing geography back to the study of international relations: spatial dependence and regional context in Africa, 1966–1978. *International Interactions*, 17, 29–61.

O'Loughlin, J., Flint, C., and Anselin, L. 1994. The geography of the Nazi vote: context, confession and class in the Reichstag election of 1930. *Annals, Association of American Geographers*, 84, 351–80.

O'Loughlin, J., Ward, M., Lofdahl, C. et al. 1998. The spatial and temporal diffusion of democracy, 1946–1994. *Annals, Association of American Geographers*, 88, 545–74.

Openshaw, S. 1996. Developing GIS-relevant zone-based spatial analysis methods. In P. Longley and M. Batty (eds.) *Spatial Analysis: Modelling in a GIS Environment*. New York: Wiley, 55–73.

Pattie, C. and Johnston, R. J. 2000. "People who talk together vote together": an exploration of contextual effects in Great Britain. *Annals, Association of American Geographers*, 90, 41–66.

Plummer, P. 2001. Vague theories, sophisticated techniques and poor data. *Environment and Planning A*, 33, 761–64.

Rosenberg, M. S. 2000. The bearing correlogram: a new method of analyzing directional spatial autocorrelation. *Geographical Analysis*, 32, 267–78.

Secor, A. J. 2000. *Islamism in Istanbul: Gender, Migration and Class in Islamist Politics*. Ph.D. dissertation. Department of Geography, University of Colorado at Boulder, CO.

Shin, M. E. 1998. *Rossa, ma non troppo: Contextual Exploration into the Geography of Italian Voting Behavior, 1987–1996*. Ph.D. dissertation. Department of Geography, University of Colorado at Boulder, CO.

Siverson, R. M. and Starr, H. 1991. *Opportunity, Willingness and the Diffusion of War*. Columbia, SC: University of South Carolina Press.

Taylor, P. J. and Johnston, R. J. 1979. *Geography of Elections*. New York: Holmes & Meier.

Taylor, P. J. and Johnston, R. J. 1995. GIS and geography. In J. Pickles (ed.) *Ground Truth: The Social Implications of Geographic Information Systems*. New York: Guilford Press, 51–67.

Tufte, E. R. 1997. *Visual Explanations: Images and Quantities, Evidence and Narrative*. Cheshire, CT: Graphics Press.

Waterman, S. 1998. *Political Geography* as a reflection of political geography. *Political Geography*, 17, 373–88.

Chapter 4

Radical Political Geographies

Peter J. Taylor

Introduction: Beyond Conservative Political Geography

In its origins and development, political geography has been conspicuously conservative in orientation. By this I mean that, by and large, political geographers have not been at the forefront of querying the *status quo*, rather they have provided spatial recipes for the powerful. Typically ignored in radical circles, until the last couple of decades there has not been an identifiable radical tradition in political geography. This radical by-pass operation has, as well as reflecting the nature of political geography, also resulted from the nature of the dominant radical tradition, orthodox Marxism with its antagonism towards study of separate political process. Hence, putting "radical" together with "political geography" has been a difficult enterprise, hindered, as it were, from two sides at once.

The conservatism has been very straightforward and has operated largely within two strands of ideas. First, there has been the treatment of the state as a "spatial entity" wherein social processes, especially social conflict, are conspicuous by their absence. From Ratzel's (1969) initial organic theory of the state in the late nineteenth century where the strong devour the weak, through Hartshorne's (1950) explicit omission of the social in his functional theory of the state in the mid-twentieth century, to Gottmann's (1982) final statement of his political geography where the watchword is stability, leading political geographers have created the most traditional of all geography's subdisciplines. Unexciting fare, it is hardly surprising that political geography had many "Chiefs" but relatively few "Indians" in its intellectual development. Second, there have been the various manifestations of the "heartland model," that icon of political geography, usually the only part of the subdiscipline remembered by students passing through its introductory courses. From its originator, the British imperialist Mackinder (1904), through the German Nazi-era *geopolitik* of Haushofer (Bassin, 1987) to the American "cold warrior" polemics of Gray (1988; Dalby, 1990), this particular contribution to statecraft provided a geopolitical world model that could be adapted to the geostrategic needs of any major power without concern for the rights of minor powers. Although well-known within and without geography, the heartland model stimulated prescription and polemic rather than a sustained research agenda so that it suffered from the same "top-heavy" input as the "nonsocial" state theory.

Between them these two dominant strands of political geography created a most unsuccessful conservative subdiscipline. For most of the twentieth century conservative political geography was under-researched, pedagogically incoherent, perennially in crisis, and, not surprisingly, widely ignored by the rest of human geography (see the particularly scathing review by Johnston, 1981). An easy target for radical critics, there was, however, little or no engagement with the new radical geography that emerged in the 1960s. It is tempting to argue that this was because conservative political geography offered *too* easy a target, but no: the reasons are to be found in the form that the radical school took within human geography. This is the subject matter of my first substantive section below. It considers how, in the development of radical geography, there was no place found for a radical political geography. When the latter did appear it was formed through a specific radical theory to create a world-systems political geography. This is the subject matter of my second substantive section. In a final conclusion, the most recent flowering of alternative radical political geographies is briefly described indicating the subdiscipline's final arrival as an integral part of a dynamic human geography.

Political Geography and the Radical Turn

The most prominent engagement between political geography and politically radical ideas in the first half of the twentieth century was Karl Wittfogel's Marxist critique of geopolitics in 1929 (Ó Tuathail, 1996, pp. 143–51). Although vigorously dismissing his target (German *geopolitik*) as a shallow, "bourgeois" materialism, no long-term debate was initiated partly, no doubt, because of the author's later conversion from Marxism. From a different angle, after World War II, leading geographer Griffith Taylor (1946) tried to transform geopolitics into a new "geopacifics" but with even less impact – it is easily the least-known of his books. And with political geography consolidating its position as a moribund backwater of the academy in the third quarter of the century, there was no peg available on which to hook new debate as geography in general became politicized in the radical turn of the world academy from the late 1960s.

Radical geography had its own specialist journal (*Antipode*, founded 1969) fully 13 years before political geography (*Political Geography Quarterly*, founded 1982). This meant that throughout the 1970s there was a journal publishing very political articles in geography but which were not considered to be political geography as such by either their authors or self-ascribed political geographers. Emerging out of the conflicts in the US in the 1960s, this new radical geography built upon concerns for poverty and racism so that its geographical "subdisciplinary home" became urban geography and economic geography (as well as it having a key role in constructing a new "development geography"). Political geography had no urban research tradition and had avoided the sort of economic issues that so concerned the radicals. The result was a curious situation of parallel "political geographies" in human geography. The readings brought together by Peet (1977) illustrate this well:

every chapter in *Radical Geography* is undoubtedly political, but none of the authors would describe themselves, at that time, as political geographers.[1]

Although geography had had a radical anarchist tradition in the nineteenth century, the new radicalism was strongly Marxist. Led by David Harvey in a series of groundbreaking books spatializing Marxist theory (e.g. Harvey, 1982), with Castells (1977), Cockburn (1977) and others, new theory was brought to urban studies. There was also the beginnings of an urban political geography (Cox, 1973) with radical concerns. However, significant inter-flow between radical and political geography only really began with the introduction of Marxist theories of the state into geography. As we have seen, conservative political geography had always been state-orientated, hence state theory was the obvious point of contact. It was, however, no easy task to marry the state as a spatial entity with conflict theories of the state (Taylor, 1983). Although Harvey (1976, 1985) did provide the Marxist theory, it was the work of Johnston, Short, Dear, and Clark that made the transition across from radical geography to a newly invigorated political geography.

Johnston's (1982) *Geography and the State* marks a significant break with political geography's traditional treatment of the state. Although concerned with spatial aspects of the state, Johnston sets this within an overall social context that owes much to Marxism. The development of state forms is described in terms of a history of phases of capitalism and the geography of state forms is described in terms of capitalist world-economy zones. In this particular treatment, the state's role in legitimation is given precedence over its role in accumulation, but the whole point is that the state is deemed necessary for the reproduction of capital. This is made explicit in a later essay where the relations between political geography and Marxist political economy are spelt out (Johnston, 1984). Outlining the base–superstructure model – material base producing the economic outputs, ideological superstructure making the political inputs – Johnston provides a broadly integrationist interpretation that confers critical functions on the capitalist state. Hence, he argued that political geography can prosper by ridding itself of its empiricist past and beginning a new research agenda on the state "as an integral part of the superstructure of the capitalist mode of production" (p. 484).

One basic result of taking this political economy route to political geography is to eschew any thoughts of developing an independent body of political geography knowledge (Johnston, 1984, p. 489). Political economy provides an all-embracing theory of society, an integration of economic, social, and political themes, which leaves no room for autonomous disciplines or subdisciplines. This position is strictly spelt out in Short's (1982) contemporaneous *An Introduction to Political Geography* that explicitly argued that political processes cannot be studied separately from economic processes (p. 1). Thus, for Short, political geography "is not a specific object of inquiry but an indication of the nature of the endeavour" (p. 1). From this position, Short uses a three-level geographical scale organization of his subject matter in the manner typical of most political geography at the time, in his terminology: world order, nation-state, and local state. For instance, the world order begins with the economic – "Uneven development: the capitalist whirlpool" – and then deals with the geopolitical – "The rise of the superpowers: the east-west fulcrum", with the "processes which underline the two dimensions" deemed to be "inextricably linked" (p. 10). The message is clear: a radical political geography investigates spatial power relations within a holistic political economy.

Within this evolving consensus, the researches of Clark and Dear stand out as sustained investigation into the user of radical state theory in geography. Culminating in their book *State Apparatus* (Clark and Dear, 1984), this political geography did not just assert integration into a wider realm of radical scholarship – the authors worked within ongoing debates in state theory and made their own distinctive contributions. As the title of their book suggests, their research broke down the category "state" to investigate systematically the mechanisms through which the state was able to carry out its necessary role in capitalism. Their starting point was distinguishing between "theories of the state in capitalism" that describe state functions and "theories of the capitalist state" wherein the state is integral to capitalism. From the latter perspective a taxonomy of eleven categories in the "capitalist state apparatus" were defined that opened up a huge new research agenda which they could only begin to tackle. Quite simply, Clark and Dear did not borrow from political economy to reinvent political geography, rather they made a deeper integration in the form of political geographic contributions to state theory thus positioning geography within theory of the state literature. This is a more mature political geography, confident in its role as a creator of knowledge on relations between political power and geographical space, developing a coherent and distinctive body of knowledge that Dear was later to proscribe within human geography (Dear, 1988).

Before this discussion of the "radical turn" in geography is concluded, brief mention must be made of the French geographer Lacoste. Viewing geographical knowledge as essentially strategic in nature, he created a radical geopolitics. Although both Marxist and Foucauldian roots can be traced, his was an activist's geopolitics countering the "common-sense" geographies underlying contemporary realist foreign and domestic policy. Despite the founding of his own journal, *Hérodote*, in 1976, his brand of radical political geography had relatively little influence on the Anglo-American mainstream described above, even after translations of his work into English in 1987 (Girot and Kofman, 1987). The relative neglect of Lacoste's radical contribution is itself an illustration of contemporary political geography as an example of US hegemony in the world academy.

World-systems Political Geography

Political geography's partner in its major engagement with radical social theory has been world-systems analysis (Taylor, 1985). A particular variant of radical political economy developed primarily by Immanuel Wallerstein (1974, 1979), it aspired to transcend political economy's tendency to a state-centrism common to all social sciences. Assuming that any social science must be about understanding social change, Wallerstein's fundamental starting point was to ask what is the basic social entity within which this change should be studied. Instead of the taken-for-granted norm of society-cum-state, Wallerstein offered historical systems as his answer to the question. These were defined as social systems in the sense that they had integrated patterns and processes of change but at the same time they were historical in the sense that they were grounded in particular times and places. Today there is but one such system, the modern world-system whose origins he traces back to the "long sixteenth century" (*c.* 1450–1650). In current parlance, over a quarter of a century ago, Wallerstein was insisting that it was necessary to study the global to understand the social. As such, this was a profoundly *geographical* challenge to the social sciences, both conventional and radical, with their multiple "societies" each one contained in its own bit of a worldwide spatial mosaic.

Eschewing state-centrism, does not, at first glance, seem a viable route to a revitalization of political geography, given the latter's traditional focus on the state. Since the social containers which Wallerstein's theory replaced are defined by sovereign territories, how was a world-systems political geography to proceed? Certainly not by ignoring, or even neglecting, state practices. Quite the opposite, in fact, because by freeing the study of states from their unexamined role as containers of societies, world-systems analysis provided a fresh interpretation of how states feature in social change. The key theoretical advantage was avoiding the old political economy conundrum of the "relative autonomy of states." In neo-Marxist analyses, the political superstructure was seen not as a mere epi-phenomenon of the material base, but instead as having its own distinctive role in affecting social change. In a state-centric analysis, this created the question of how the material base – a national economy – was related to the political superstructure – a politics of the state. If the latter was not to be determined by the former, then it must have relative autonomy. But this formula was fraught with difficulties for a materialist theory that all political economy arguments claimed to be. How "materialist" did political economy remain if the politics were, to some unspecified degree, autonomous? World-systems analysis side-steps this question by not equating the spaces of economy and state. The "material base" is a world-economy comprising multiple states so the one-to-one (economy to state) question of autonomy does not arise. Instead, the problem is one of state maneuverability: how do social agents use the state against other social agents using other states in their social conflicts to mould the world-economy to their own perceived material advantage (Taylor, 1993)? Thus, rather than downgrading the study of states – by replacing relative autonomy by maneuverability – world-systems analysis inserts states as dynamic institutions in the production and reproduction of the modern world-system. Clearly not a hindrance, this world-systems interpretation of states had much to offer to a radical overhaul of political geography.

Wallerstein's overall interpretation of the modern world-system or capitalist world-economy is built upon Marxist political economy insights combined with a geography derived from Latin American radical dependency theory and a history derived from the French *Annales* school. The end-result is the on-going creation of a geohistorical theory of how our modern world works. As with other radical social theory, it is no respecter of disciplinary boundaries: world-systems analysis provides a framework for study that integrates economic, social and political themes in an integrated argument. Thus, this is not a theory to be used to construct political geography as a separate subdiscipline (Taylor, 1982); rather, the political geography features as a particular perspective on the modern world-system. Whether this turns out to be a productive perspective will be judged by the quality and distinctiveness of the results of seeing the world-system through political geography lenses. A corollary of this position is that whether world-systems analysis is good for political geography is a relatively minor, intellectually partisan, concern in the overall scheme of things. However, in the spirit of this volume, I will pursue the question of how world-systems analysis did contribute to creating a more coherent political geography with a radical cutting edge.[2]

There are five key ways that world-systems analysis has intervened in the development of contemporary political geography. These are: (i) by providing a geohistorical framework for political geography topics; (ii) by de-mystifying the three-scale organization through a simple relational model; (iii) by treating the state as an institution alongside other institutions through which power is expressed; (iv) by emphasizing the plurality of states as integral to their meaning; and (v) by providing continuity of ideas in the post-Cold War context to bridge the gap to contemporary globalization. I treat each of these in turn.[3]

(i) World-systems analysis integrates time and space into its general social processes. Thus space is not a mere stage on which events unfold, every historical system has a specific spatiality associated with its temporal trajectory. In the case of the modern world-system, the basic spatiality is a world-economy with a core–periphery pattern. At the material base,[4] economic processes are differentiated into two broad categories, core-producing (high-tech, relative high wage) and periphery-producing (low-tech, relatively low wage) creating a spatial polarization across the system. The polarization is not complete, however, in part because some social actors are able to use elements of the superstructure, notably their states, to prevent peripheralization, and sometimes they may even succeed in putting in place mechanisms to facilitate "core-ization." These political processes define a middle zone, the semi-periphery, between core and periphery zones. Thus the conventional core–periphery spatial model is converted into a three-zone structure which is not static but incorporates mechanisms of change.

This change itself is differentiated. The modern world-system does not simply "unfold" in a progressive manner; rather, it is inherently cyclical as a product of contradictions in the development of its material base. The modern world-system is crisis-ridden: rapid growth is followed by slow growth or stagnation in a series of economic cycles. Each cycle is a product of restructuring to resolve a crisis creating new growth followed by the gradual breakdown of the resolution leading to stagnation again and a new crisis. The major structural cycles of the system are known as "Kondratieffs" and last for approximately half a century: an A-phase of about 25 years overall system growth followed by a B-phase of about 25 years overall system stagnation. Even longer cycles are related to the rise and fall of hegemonic states – economic, political, and cultural leaders – which fully integrate superstructure processes with developments in the material base. The three examples of such world hegemony – the Netherlands followed by Britain followed by the USA – provide the modern world-system with a three-fold temporal structure that interweaves with the three-fold spatial structure described earlier. This creates a nine-fold space–time zonal structure in which to locate political geography topics as both producers and expressions of the world-system's spaces and times.[5]

(ii) World-systems analysis brings back the global to centre-stage in social study but for political geography this in itself is not unique given the subdiscipline's geopolitical tradition. What specifically world-systems analysis does offer that had been missing is a relational approach to geographical scales (Taylor, 1982). Political geography, both traditional and radical, had tended to treat the three scales it focused upon – national, international, and sub-national – as constituting separate, even autonomous, bundles of processes. Of course, this cannot be. You can cross a boundary and leave a country but you cannot "leave a scale" behind by moving. All events and actions simultaneously take place in a kaleidoscope of scales. An action might have more repercussions at one scale than another but all human scales of activity will ultimately be touched. Thus, any credible intellectual activity that deals with geographical scales must include an argument for how the different scales are related one to another.

In the case of political geography's three-scale organization, a world-systems interpretation uses the commonplace radical argument that ideology separates reality from experience. In this case, territorial states through their designation as "nation-states" are deemed to represent a scale of ideology. The scale of reality is the system-scale, here global, where the basic structures that define the system are to be found. The remaining scale of experience is then the local scale, the scale at which people experience their everyday lives as producers and consumers. An example is the way an event such as animal disease (say BSE in cattle) percolates through the scales. On receiving reports of the disease, the national government's first reaction will be to minimize its importance to protect consumer confidence in the product. This myth-making will involve inventing and using such abstractions as the "national herd" as prime supplier to the "national market." Such containment of the problem will only last for so long. Once the ideology is exposed, the "national herd" becomes revealed to actually be lots of local herds whose slaughter produces the experience of catastrophe in pastoral farming communities. At the same time, "national market" becomes revealed to be actually a world market with meat supplies imported from safe countries thus producing a restructuring of this small part of the world-economy. Thus, ideology is separating experience from reality, at least for a short time. Overall, we can see that in political geography most key decisions are made where power is concentrated – at the national scale – but they are experienced at the local scale and ultimately feed into the changing structures of the world-economy.

(iii) World-systems analysis's treatment of states, as outlined above, has been widely misunderstood. If the common misapprehension that this form of analysis simply involved replacing states by a world-system were true, then there would be no world-systems political geography. Instead, states are treated as one of four key institutions through which power is transmitted by social agents in the reproduction of the modern world-system (Taylor, 1991). The others are households – the "atoms" of the system, "peoples" – groups with common cultural identity notably nations, and classes – the global economic strata of the system. Again, the first step is to consider the relations, in this case between these institutions. The relational arrangement consistent with the scale model above is that households culturally reproduce nations that in turn legitimize states while simultaneously dividing classes. Alternatively, these institutions can be considered in terms of different combinations to create different forms of politics. For instance, considering households with states leads to a gender politics in which power in households may be challenged by state power, as, for instance, in programs to combat domestic violence. Fourteen such politics have been identified, each of which will have its own particular spatial dimension to create a political geography (Taylor, 1991).

There are two important corollaries of this approach. First, political geographers are encouraged to look for expressions of political power beyond the state itself. The latter might remain the main constellation of such power but other practices of political power, such as in households, should not be ignored. Second, state and nation are separated as analytical categories despite their commonplace merging as "nation-state." This allows us to consider the different sorts of locale each institution constructs (Taylor, 1999). Nations construct places, national homelands to which people identify as the imaginary "home" – a haven – of the "national family." States, on the other hand, construct spaces, sovereign territories in which they organize, manage, regulate, and administer people and property. Spaces are abstract and therefore impersonal locales whereas places are humanised spaces (Tuan, 1977). Therefore, the creation of a "nation-state" out of existing territorial state is to humanize the state, a very important process over the last two centuries. All of which brings space and place creation to the center of political geography (Taylor, 1999).

(iv) World-systems analysis has the inter-state system as the center-piece of its superstructure. Thus, references are not made to "the state" as found in most radical, political economy but always to "states" in the plural. I have already referred to the elimination of the relative autonomy problem with this formula but the plurality of states has further implications for political geography (Taylor, 1995). First and foremost, it allows political geography to straddle the artificial intellectual divide that separates the disciplines of political science (studying domestic politics) and international relations (studying international politics). Although it will be different parts of the state apparatus that conduct home and foreign policies, it will still be the same government with the same executive in the same state making the decisions. World-systems political geography with its integrated scale approach is able to study power processes within and through state apparatuses in the whole.

The focus on multiple states is also important for distinguishing the "international" from the "global". The former, despite its reference to nations, is about a particular expression of the space of flows in the world-economy that links together the states. In diplomacy there are flows of information, in trade there are flows of commodities, and in immigration there are flows of people, but they all have one particular characteristic in common: both origin and destination are countries so that each flow is defined by crossing an international boundary. These "inter-state" flows are fundamentally different from flows that are not controlled by international boundaries, which are termed "trans-state" flows. Pollution is no respecter of political boundaries but neither is the global financial market, albeit in a very different way. The balance between inter-and trans-state flows varies over space and time as per the previous framework, but the distinction is particularly important at present when globalization has led to much talk of the decline, even the end, of the state. Since states are always adapting to change it is not clear how to tell the difference between such relatively routine change during another restructuring and fundamental change that is genuinely undermining the reproduction of states *per se*. Looking at changes in inter- and trans-state processes provides a particular political geography route to facing such vital questions (Taylor, 1995).

(v) World-systems analysis was created during the Cold War but it has not suffered the way other radical political economy has with the demise of the USSR. This is because the "second world" was never interpreted as an alternative socialist world-system in world-systems analysis. Focusing on "one system not two" (Chase-Dunn, 1982), the "communist" challenge was seen as a distinctive semi-peripheral strategy to restructure the world-economy. Its politics-led processes ultimately failed in this objective. The resulting catastrophe is vividly expressed in the former lands of the USSR which have fragmented between a relatively small semi-periphery hanging on but with large swaths converted into new periphery. Thus, in the space–time framework of world-systems analysis, political geography could interpret the end of the Cold War as a geopolitical transition (Taylor, 1990), not a fundamental shift in the nature of the system.

World-systems analysis treats globalization as another matter altogether. The contemporary organization of the world-economy on the back of new enabling technologies (combining communications and computers) has created what is probably an unprecedented level of trans-state processes in the history of the modern world-system. This is signified by the range of processes that are identified with globalization: as well as the financial and ecological processes briefly referred to above, there are the economic production processes of global corporations; the cultural processes of consumption led by advertising agencies and the global media; the political processes of "global governance" promoting a neo-liberal economic agenda across the world; and, the social processes of imagining "one world" society with consequent global social movements. This impressive list covers the whole gamut of activities in the modern world-system and their co-incidence is no coincidence. The current restructuring of the world-economy certainly includes many features common to past restructurings, that is to say it represents another reproduction of the system. But it also includes new elements that suggest a fundamental change of system. If this is the case then political geography is in a prime intellectual position to monitor these unique changes.[6]

Globalization is sometimes viewed as a change from a world dominated by spaces of places to one dominated by spaces of flows (Castells, 1996). In political geography terms, the former is represented by the mosaic of sovereign territories that is the world political map. Any erosion of the concentration of political power in states will result in a diminution of the importance of this mosaic political map. To where is much of this power leaking? The most common answer would be to large corporations, but where are the alternative loci in which power is beginning to concentrate? One answer to that question identifies world cities as nodes in a new network space of flows, trans-state flows of information and knowledge made possible by the new communications/computers combination (Taylor, 2000). Here is a completely new political geography that transcends the scale and institutional models previously set out by world-systems political geography that can engender new life into world-systems political geography.

Conclusion: Alternative Radical Political Geographies

World-systems political geography provides a rare connection between the heyday of radical geography and the global geographies of today. There have been many changes over these decades, not least in intellectual matters. In many ways the world of the academy has become much more complex over this time. Soon after its emergence, radical geography, whether of Peet, of Harvey, of Blaut, or of Bunge, became explicitly Marxist in nature (Blaut, 1975; Bunge, 1973; Harvey, 1973; Peet, 1975). As we have seen, this orientation passed into radical political geography. World-systems political geography was therefore something of an exception in the family of radical geography projects. Nevertheless, it could still be reasonably designated "neo-Marxist" in its ideas, still part of a radical mainstream in the world academy. Only partly as a result of the demise of the USSR, to be a radical scholar today no longer inevitably means taking a Marxist, "neo" or otherwise, approach to understanding social change. In the contemporary academy there is an array of schools of thought that aspire "to turn the world upside down."[7] Inevitably this variety has found its way into political geography.

There are three particular approaches that have become important in radical political geography in recent years. The first is the marriage of international political geography with international political economy as the study of "geopolitical economy" which shares several ideas with world-systems analysis (e.g. cycles and hegmonies) but without taking on the specific geohistorical theory (e.g. Agnew and Corbridge, 1995). Secondly, there is the postcolonial political geography which is concerned with the imposition of "western ideas", including Marxism, on other parts of the world (e.g. Slater, 1997). Thirdly, and closely related, there has been the rise of a "critical geopolitics" that problematizes the whole relation between geography, knowledge and power (e.g. Dalby and Ó Tuathail, 1996; Ó Tuathail, 1996). With other inputs of cultural theories (e.g. Painter, 1995) and a continuing world-systems political geography (Taylor and Flint, 2000), there is a healthy heterodoxy in radical political geography at the beginning of the twenty-first century, which is itself part of a wider pluralism in contemporary political geography as a whole (Taylor and van der Wusten, 2002).

ENDNOTES

1. In discussions at the time this required a distinction which took the form of identifying those contributing to the subdiscipline with capitals – "Political Geographers" – leaving those pursuing political themes in geography as "political geographers" (Taylor, 1983).
2. The question of coherence was of particular concern at this time given the incoherent legacy of conservative political geography (Claval, 1984; Cox, 1979).
3. Although there are some specific references in the following discussion, the ideas below can be followed up in general by reference to the world-systems political text first published in the mid-1980s (Taylor, 1985) and going through two editions (1989) and (1993) before its current fourth edition (Taylor and Flint, 2000).
4. Mindful of the limitations of the simple base–superstructure architectural metaphor, I will use it in my discussion of world-systems analysis to make comparison to other radical geography easier.
5. In the textbook (Taylor, 1985; Taylor and Flint, 2000), Kondratieffs are used for the time framework leading to more space–time zones.
6. They are unique because the modern world-system is unique in the history of world-systems, the others all being world-empires not world-economies. There has never been a transition from a world-economy before.

7. This phrase has its origins in the radicalism of the English Civil War and I find it a useful descriptor of what was simply referred to as "radical" not so long ago. Today the adjective radical is joined by many others – emancipatory, dissident, critical, alternative, oppositional – to describe such intellectual projects. Each adjective carries with it its own intellectual baggage so that it is becoming increasing difficult to find a single label for the "upside down turners."

BIBLIOGRAPHY

Agnew, J. and Corbridge, S. 1995. *Mastering Space*. London: Routledge.

Bassin, M. 1987. Race contra space: the conflict between German geopolitik and national socialism. *Political Geography Quarterly*, 6, 115–34.

Blaut, J. M. 1975. Imperialism: the Marxist theory and its development. *Antipode*, 7, 1–19.

Bunge, W. 1973. The geography of human survival. *Annals, Association of American Geographers*, 63, 275–95.

Castells, M. 1977. *The Urban Question: a Marxist Approach*. London: Arnold.

Castells, M. 1996. *The Rise of Network Society*. Oxford: Blackwell.

Chase-Dunn, C. 1982. *Socialist States in the World-System*. Beverly Hills, CA: Sage.

Clark, G. L. and Dear, M. 1984. *State Apparatus*. Boston: Allen and Unwin.

Claval, P. 1984. The coherence of political geography. In P. J. Taylor and J. W. House (eds.) *Political Geography: Recent Advances and Future Directions*. London: Croom Helm.

Cockburn, C. 1977. *The Local State*. London: Pluto.

Cox, K. R. 1973. *Conflict, Power and Politics in the City: a Geographic View*. New York: McGraw-Hill.

Cox, K. R. 1979. *Location and Public Problems: a Political Geography of the Contemporary World*. Chicago: Maaroufa.

Dalby, S. 1990. *Creating the Second Cold War*. London: Pinter.

Dalby, S. and Ó Tuathail, G. (eds.). 1996. Special issue on "Critical Geopolitics". *Political Geography*, 6/7, 451–665.

Dear, M. 1988. The postmodern challenge: reconstructing human geography. *Transactions, Institute of British Geographers, N.S.*, 13, 262–74

Girot, P. and Kofman, E. (eds. and transl.). 1987. *International Geopolitical Analysis: a Selection from Herodote*. London: Croom Helm.

Gottmann, J. 1982 The basic problem of political geography: the organization of space and the search for stability. *Tijdschrift voor Economische en Sociale Geografie*, 73, 340–9.

Gray, C. 1988. *The Geopolitics of Superpower*. Lexington, KY: University Press of Kentucky.

Hartshorne, R. 1950. The functional approach in political geography. *Annals, Association of American Geographers*, 40, 95–130.

Harvey, D. 1973. *Social Justice and the City*. London: Arnold.

Harvey, D. 1976. The marxian theory of the state. *Antipode*, 8, 80–98.

Harvey, D. 1982. *The Limits to Capital*. Oxford: Blackwell.

Harvey, D. 1985. The geopolitics of capitalism. In D. Gregory and J. Urry (eds.) *Space and Social Structures*. London: Macmillan, 128–63.

Johnston, R. J. 1981. British political geography since Mackinder: a critical review. In A. D. Burnett and P. J. Taylor (eds.) *Political Studies from Spatial Perspectives*. Chichester: Wiley, 11–31.

Johnston, R. J. 1982. *Geography and the State. An Essay in Political Geography*. London: Macmillan.

Johnston, R. J. 1984. Marxist political economy, the state and political geography. *Progress in Human Geography*, 8, 473–92.

Mackinder, H. J. 1904. The geographical pivot of history. *Geographical Journal*, 23, 421–42.
Ó Tuathail, G. 1996. *Critical Geopolitics*. Minneapolis, MN: University of Minnesota Press.
Painter, J. 1995. *Politics, Geography and "Political Geography": a Critical Perspective*. London: Arnold.
Peet, R. 1975. Inequality and poverty: a Marxist-geographic theory. *Annals, Association of American Geographers*, 65, 564–71.
Peet, R. (ed.). 1977. *Radical Geography*. London: Methuen.
Ratzel, K. 1969. The laws of the spatial growth of states. In R. E. Kasperson and J. V. Minghi (eds.) *The Structure of Political Geography*. Chicago: Aldine.
Short, J. 1982. *An Introduction to Political Geography*. London: Routledge.
Slater, D. 1997. Geopolitical imaginations across the North-South divide. *Political Geography*, 16, 631–53.
Taylor, G. 1946. *Our Evolving Civilization – an Introduction to Geopacifics*. London: Oxford University Press.
Taylor, P. J. 1982. A materialist interpretion of political geography. *Transactions, Institute of British Geographers, N.S.*, 7, 15–34.
Taylor, P. J. 1983. The question of theory in political geography. In N. Kliot and S. Waterman (eds.) *Pluralism and Political Geography*. London: Croom Helm, 9–18.
Taylor, P. J. 1985. *Political Geography: World-Economy, Nation-State and Locality*. London: Longman.
Taylor, P. J. 1990. *Britain and the Cold War: 1945 as Geopolitical Transition*. London: Pinter.
Taylor, P. J. 1991. Political geography within world-systems analysis. *Review (Fernand Braudel Center)*, 14, 387–402.
Taylor, P. J. 1993. States in world-systems analysis: massaging a creative tension. In B. Gills and R. Palan (eds.) *Domestic Structures, Global Structures*. Boulder, CO: Lynne Rienner.
Taylor, P. J. 1995. Beyond containers: internationality, interstateness, interterritoriality. *Progress in Human Geography*, 19, 1–15.
Taylor, P. J. 1999. Places, spaces and Macy's: place-space tensions in political geography. *Progress in Human Geography*, 23, 7–26.
Taylor, P. J. 2000. World cities and territorial states under conditions of contemporary globalization. *Political Geography*, 19, 5–32.
Taylor, P. J. and Flint, C. 2000. *Political Geography: World-Economy, Nation-State and Locality*. London: Prentice Hall.
Taylor, P. J. and van der Wusten, H. 2002. Political geography: spaces between war and peace. In G. Benko and U. Strohmayer (eds.) *Human Geography: a Century Revisited*, in press.
Tuan, Y-F. 1977. *Space and Place*. London: Arnold.
Wallerstein, I. 1974. *The Modern World-System*. New York: Academic Press.
Wallerstein, I. 1979. *The Capitalist World-Economy*. Cambridge: Cambridge University Press.

Chapter 5

Feminist and Postcolonial Engagements

Joanne P. Sharp

Recent feminist and postcolonial critiques of geography have insisted that dominant forms of knowledge are the products of particular discursive and institutional contexts. They argue that the prevalence of patriarchy and Eurocentrism have colonized accepted world-views. Drawing on various philosophers, notably Michel Foucault, they have argued that power and knowledge are intimately and inherently relational. Dominant forms of knowledge are inseparable from dominant relationships of power and so are creative of the world, not simply reflective of it (Foucault, 1980). This has resulted in the marginalization and dispossession of other voices and knowledges, particularly those of women and non-Western people.

Sometimes in concert, sometimes antagonistically, feminist and postcolonial voices have challenged the basis of dominant forms of knowledge, offering powerful reflections on subjectivity and identity, the importance of culture, and the nature of politics and resistance. They have insisted on decentering the apparently universal knowledges of the West to demonstrate their situatedness (e.g. Haraway, 1988) in a gendered or placed or historical location. This has led to a complex and ambivalent model of political geography in which there is a tension between a perception of the fluidity of borders and identities and an acknowledgment of the inescapable materiality of both. As a political project, it asks people both to use and refuse who they are, to draw upon their identities as women or as Third World subjects, but also to constantly challenge the definition of these subjectivities. In acknowledging the materiality of boundaries, it asks us to fight against injustices and to celebrate and protect identity. However, it also asks us to think about the power of culture and language, the power to construct identities that label and contain us.

Both feminism and postcolonialism offer challenges to dominant knowledge and therefore to the traditional forms of political geography. Feminist approaches seek to challenge the operations of patriarchal power in the formal sphere of politics and in the supposedly apolitical sphere of the private. Additionally, feminists insist on challenging the division of the world into political and apolitical, and so challenge boundaries. Postcolonialism challenges the dominance of Eurocentric and Western knowledges that place boundaries around peoples and places. Postcolonial critiques of totalizing Western historiography attempt to destabilize the Western canon. It raises questions of for whom knowledge is created and from where. It rejects the universality of modernity, enlightenment, and rationality, and locates them in the history of the West, particularly in the violence of colonialism. It paves the way for a politics of opposition and, importantly for the study of political geography, problematizes the relationship between core and periphery. Postcolonial theorists wish to focus upon positions that have traditionally been marginalized and excluded, privileging the margins over the center and, like feminists, challenging boundaries, sometimes going further to favor fluidity over stability.

Feminist and postcolonial theorists ask political geography to expand its gaze. Feminists famously stated that "the personal is political," and that patriarchy works to exclude women from the political realm. Postcolonial theorists argue that the domination of Western forms of knowledge has similarly marginalized the voices and experiences of those from outside the West. Both require political geography to examine the power relationships woven through everyday life, and to challenge boundaries wherever encountered.

This chapter will discuss some of the most significant challenges offered by feminist and postcolonial critiques to consider the actual and potential impacts of these on the understandings and practices of political geography. By necessity, given the limits of a book chapter such as this, I will create a sense of greater coherence between various feminist and postcolonial positions than is the case, although I have attempted to indicate points of divergence wherever possible.

Breaking Down Boundaries

Challenging binaries

For both feminist and postcolonial critics, knowledge and identity are interrelated and always connected to relationships of power. Feminists have suggested that patriarchal knowledge is established through the expulsion of all that is Other: "woman" acts as the Other of "man." In Western culture, the values that are associated with women, and therefore are antithetical to the definition of masculinity – irrationality, embodiment, subjectivity, and so on – are devalued. French feminist Luce Irigaray (1985) for example has argued that the figure of "woman" acts as a mirror from which man is reflected back as being opposite, characterized by none of the feminine traits of nature and emotion.

In a structurally similar way, in his hugely influential work, *Orientalism*, Said (1978) sees European representations of the Orient as telling more about the Occident than the lands and peoples being represented. The Occident projected into the Orient all that it reviled in itself so that the Orient became an oppositional space which reflected back an image of the Occident that was positive and enlightened. To Said (1978, p. 12), the Orient "has less to do with the Orient than it does with 'our' world." Said's contribution then was to turn attention from the political and economic bases of colonialism towards the prime significance of the cultural realm (although it is important to note that Fanon [1968] had considered the cultural implications of colonialism around ten years beforehand). For Said, examination of the two indivisible foundations of imperial authority – "knowledge and power" – is central to any analysis of the operation of colonial power. Naming is of central importance so as to order the complexity of the political world into something knowable and governable:

> Naming is part of the human rituals of incorporation, and the unnamed remains less human than the inhuman or subhuman. The threatening Otherness must, therefore, be transformed into figures that belong to a definite image-repertoire (Trinh, 1989, p. 54).

Often the terms of producing the spaces of the colonized intersected with representational structures at home. The bourgeois mindset that dominated the production of culture projected its own fear of the working classes onto the non-European others. Similarly, many representations of the Orient were gendered. To many scholars, travellers and colonial administrators, the Orient was imagined as female and seductive or weak and effeminate, in each case reflecting back a strong and moral masculinity in Europe (Lewis, 1996; Phillips, 1996).

An acknowledgment of the cultural basis of colonial power "does not efface the violence of conquest and control" (Blunt and Wills, 2000, p. 181). The structures of meaning and representation projected onto diverse and heterogeneous peoples and places to render them a coherent and negative oppositional presence to the West, was in itself a form of "geographical violence" (Said, 1978). Colonized people were subjected to knowledge that rendered them second-class citizens or even as being inhuman, stripping them of their dignity, culture, and history (Wolf, 1982). Of course, it is impossible to separate out the material from the discursive so that the geo-graphing of various colonial imagined geographies allowed and excused the military and administrative subjugation of peoples around the globe. It also provided the foundation for resistance, a ubiquitous power in any exercise of domination (Sharp et al., 2000, p. 1).

The colonizing aspirations of dominant knowledge, then, have had profound effects on the construction of both female and colonial subjects. This insidious political geography works through the apparently universal characteristics of dominant knowledge and so until feminist and postcolonial critics began to challenge the source and effects of this knowledge, it was seen as beyond the political realm. Given that colonialism was, seen from this perspective, concerned with the suppression of a heterogeneity of subjects, and patriarchy renders gender as an essentialist category (or hides difference under the sign of "universal"), it has been central to postcolonial and feminist work to understand and problematize issues of identity and subjectivity (Goss, 1996, p. 242). In feminist and postcolonial attempts to challenge the binary logic of identity formation, alternative understandings based around fluidity, movement and hybridity have emerged.

Ambivalent geographies

Feminist geographers have considered the nature of the relationships of identity and subjectivity to space and place. Feminism is not an homogeneous approach to politics. The oppositional politics of earlier expressions of feminism ("first" and "second" waves), has more recently been replaced by a more subversive form of gender politics that stresses the importance of an ambivalent stance (Rose, 1991), that accepts difference between different groups of people without essentializing these into natural or timeless differences. While "first-wave" liberal feminists sought to deny difference (in that men and women should have equal access to the public sphere of work, education, and other opportunities), "second-wave" feminists sought to emphasize the differences between men and women. They argued that women should not (indeed, could not) accept masculinist culture as women would never flourish in a society constituted through masculinist values. Global politics – broadly defined to include issues ranging from the environment to population migration – could be interpreted as being shaped by masculinist culture (see Enloe, 1989; Seager, 1993). A more radical change would be required for women's emancipation. This politics of gender opposition has been challenged by more recent "third-wave" feminists who insist upon the social construction of masculinity as well as femininity, and the differences in what it means to be "woman" in different circumstances. Politics here emerges not from opposing the valorization of masculinist traits with a value system arranged around feminism, but instead from a subversion of the binary structure that assumes an essential link between biological sex and gender identities and practices.

It is possible to see the different expressions of the politics of resistance articulated in each of the "waves" of feminism. In the first wave, women resisted a patriarchal state that refused them the vote and equal opportunities in the job market. Second-wave feminists resisted masculinist society, which was seen to offer particular limited roles and possibilities to women. Third-wave feminists have acknowledged the greater complexity of gender relations, not simply operating around a male–female binary, but cross-cut by issues of race, class, and sexuality. The political geographies that emerge from this form of feminism are articulated as a much more "ambivalent" form of politics (Rose, 1991), refusing to be identified – and therefore captured – by a label. French and Italian feminists in particular have resisted attempts to delimit and name the feminine, arguing that femininity is constructed as "that which disrupts the security of the boundaries separating spaces and must therefore be controlled by a masculine force" (Deutsche, 1996, p. 301).

Hybrid identities

Much recent postcolonial theory has also challenged the politics of opposition, if this means a simple reversal of valence (i.e. now it is white, "Occidental," scientific that is the negative sign, and nonwhite, "Oriental," personal narrative that is positive). Instead, like third-wave feminists, postcolonial theorists have celebrated the subversion of the binary logic of "us and them," inside and outside. Oppositional movements have been challenged as too readily accepting the overarching logic of western dominance instead of offering a different model of politics.

This challenge can be seen to run through postcolonial and feminist re-evaluations of theories of nationalism. Theorists have suggested that liberation movements based on national rhetoric can act to reinforce both colonial and patriarchal power. Nations have been presented as being Europe's great gift to the world, as coherent forms of political organization around which independence movements could be forged (Chatterjee, 1993, p. 4). Nations offered "imagined communities" to which each person born in a territory could belong to an equal fraternity[1] of belonging (Anderson, 1991). Certainly nations can be understood as being a product of the colonial period where the introduction of the language of the colonizer along with the institutions of governmentality established clearly defined borders of belonging. Indeed, some would argue that it was the transformation of societies under colonialism into nations with commonly imaginable boundaries – when people could for the first time communicate with those across territory previously too remote or fragmented for such contact – that enabled anti-colonial resistance (Anderson, 1991). So, for example, the production of maps as part of the colonial project of managing new territories (Godlewska, 1994) also acted, ironically, as a catalyst for postcolonial movements. With maps, people from across the territory could for the first time start to imagine a nation of individuals united as a community being oppressed by the colonial structure. Similarly, the creation of "brown Englishmen" and their equivalent around the world produced the qualities required to organize postcolonial governance. The gift of Europe indeed was the mechanism for escape from European power.

But Chatterjee (1993) challenges this imposition of a Western image of nation. For if it is simply the result of the imposition of Western values that allows self-governance, then what kind of imagined community does this facilitate? As he argues, "even our imaginations must remain forever colonised" (Chatterjee, 1993, p. 5). Many commentators (e.g. Fanon, 1968) and cultural producers have noted the poverty of the copy: the Westernized business leaders in Senegalese writer Sembène Ousmane's novel and film *Xala* (1974) demonstrate very clearly not only figures still ambiguously but firmly linked to "traditional ways," but also very obviously "poor" copies of the Western model. Fanon and Ousmane argue that the process of national liberation was a struggle on the colonizer's terms (Goss, 1996, p. 243) which can only end up replacing the colonial elite with a national one (Fanon, 1968). It is also a process that can end up recovering uncritical views of pre-colonial gender relations, rendering new forms of patriarchy in the new nationalisms (Sharp, 1996b).

The conventional political analyst's attention is perhaps too attuned to the formal sphere of politics where opposition parties and protest movements emerge. Chatterjee (1986, 1993) argues that the native national consciousness became established long before the emergence of formal opposition. For Chatterjee, the colonized national imaginary did not live in the sphere of the material, where economy, statecraft, science, and technology were practiced and where the West had predominance. Instead, the national imaginary was protected and developed in the private and hidden sphere of the spiritual, an inner domain bearing the marks of cultural identity where colonial power cannot penetrate. And so, rather than simply copying a pre-existent European model of national identification, "nationalism launches its most powerful, creative, and historically significant project: to fashion a 'modern' national culture that is nevertheless not Western" (Chatterjee, 1993, p. 6): a hybrid form that freely takes from Europe elements but refuses to be bound by the confines of the European experience. The notion of the hybrid is arguably the most significant metaphor for postcolonial theorizing. Hybridity resists political geographies of border drawing at the scales of the nation-state, region, place, and individual subject. Thus, the challenge is not only to processes of power within particular spaces, but also the definition of these spaces – about where the center and margins are located (the colonial center and colonized margins) – but also who decides on these locations: who has the power to draw these maps of political geography and to define one place as center, one as margin, and to define individuals as subjects and as subjugated to the rules of others. This fluid concept celebrates impurity and actively resists the will to power of drawing boundaries and naming.

Bhabha goes back to Fanon to argue that hybridity is the necessary attribute of "the colonial condition" (Loomba, 1998, p. 176). For Bhabha (1994), the importance of this ambivalent condition is that it illustrates the inability of colonial authority to replicate itself perfectly and so undermines its own position. The Westernized native is neither clearly self nor other but an uncomfortable mixture of the two, suggesting through this hybridity that the difference between colonizer and colonized is not so great, not so complete as the binary logic of colonial thought required:

> The mimic man, insofar as he is not entirely like the colonizer, white but not quite, constitutes only a partial representation of him: far from being reassured, the colonizer sees a grotesquely displaced image of himself. Thus the familiar, transported to distant parts, becomes uncannily transformed, the imitation subverts the identity of that which is being represented, and the relations of power, if not altogether reversed, certainly begins to vacillate (Young, 1990, p. 147).

In contrast to Said, who could be criticized for emphasizing the internal coherence of Orientalism, Bhabha has focused on the internal contradictions, fractions, and inconsistencies of colonial representation. He sees a much more fragile assemblage of discourses and practices constantly offering up the possibility of resistance or subversion. Because the colonial relationship is always ambivalent, it generates the seeds of its own destruction as the examples of colonial maps and "brown Englishmen" suggested. This offers a politics in which change comes not through conscious acts of resistance and opposition, but instead is performed through this ambivalent *condition* of colonialism which emerges from the cracks in colonial discourse itself (Bhabha, 1994).

The space of meeting and subversion, Bhabha's (1994) "third space" (the "zone of transculturation" in Pratt's [1992] terminology) is a place of creative possibility. Just as ambivalent feminism subverts a politics of masculine–feminine binary, the existence of third space displaces the neat binary of Occident and Orient, self and other, inside and outside, and so challenges the very logic of Western thought, and not just the extent of the moral cartographies it has drawn. In addition to theorizing, the emergence of postcolonial hybidity can be seen in a range of cultural productions from the generation of new forms of music (collaborations of Indian musician Ravi Shankar to the more recent fusions of bands such as Apache Indian, Joi, and Nitin Sawnhey), expressions of hybridity in postcolonial novels (Salman Rushdie's *The Satanic Verses* and Hanif Kureshi's *The Buddha of Suburbia* are the most oft quoted examples), and films (*My Beautiful Laundrette, East is East*, and *Bhaji on the Beach*, for instance, have examined the hybrid nature of life in the contemporary UK).

Feminist ambivalence and postcolonial hybridity thus present a challenge to political geographies based around territorial or bounded notions of power, and at scales that range from the globe to the body. However, these processes of hybridization are neither ubiquitous nor necessarily positive. There are important political questions regarding who can choose the different cultural "ingredients" to mix together. Often Western artists are accused of using "the rest" as a resource from which to take inspiration to revive their work (from Picasso's inspiration from African "primitive" art to musicians such as Paul Simon drawing on forms of music from around the globe) without changing the balance of power in the cultural industries in question. Thus, feminist and postcolonial critics are often wary of the unquestioned celebration of hybridity.

Recognizing Boundaries and Challenging Imagined Globalizations

Locating resistance

Various feminist imagined geographies have been produced which attempt to overcome the political divisions that have been created through masculinist state culture. The most prominent is Robin Morgan's (1984) collection *Sisterhood is Global* which sought to explain the condition of exploitation and subjugation faced by women in societies throughout the patriarchal world. Contributors from around the globe offered tales which Morgan explained as indicative of the global condition of patriarchy that united women in a sisterhood of oppression. However, this vision has been critiqued by Third-World feminists for its inherent Eurocentrism. Mohanty (1997, p. 83) insists that the image of a global sisterhood is only possible with the erasure of the effects of race and class. Mohanty (1997) sees Morgan as positing a transcendental and essentialist model of "woman" that implies that other identities – those based on race particularly – have no bearing on the construction of gendered identity. As McDowell (1993) has argued, however, identity is not cumulative with elements simply adding up to more or less marginal or privileged identities. The effects of race, class, ethnicity, sexuality, and physical ability are transformative of what it means to be female so that these forms of identification cannot simply be subsumed into an inherited category of woman. For Third-World feminists like Mohanty, the global sisterhood image silences the histories of colonialism, imperialism, and racism from which Western feminists still benefit. She argues that women in the West benefit from the economic development of their countries that was (and still is) predicated on the underdevelopment of the counties of their sisters in the Third World. Second-World – perhaps now more appropriately termed "post-communist" – feminists have similarly criticized Western feminism for its liberal, middle-class assumptions (see Funk and Mueller, 1993). Global oppression, then, cannot simply lead to a global feminist politics. The existence of patriarchy across the globe and the "oppression of women knows no ethnic nor racial boundaries, true, but that does not mean it is identical within these boundaries" (Mary Daly quoted in Trinh, 1989, p.101).

Clearly the danger of this way of understanding gendered politics is that the power of the identity "woman" is eroded by all of these qualifications. Mohanty (1997) argues that rather than thinking in terms of a sisterhood to combat increasingly globalized forms of subjugation and patriarchy, women have to form temporary alliances to fight particular battles. Coalitions are formed not because they are enjoyed but because they are required for survival. Spivak (1988) has talked about "strategic essentalisms" that allow women to recognize certain borders in certain contests only for them to dissolve in later struggles. "Strategic essentialism" is "not a description of the way things are, but . . . something one must adopt to produce a critique of anything" (Spivak, 1990, p. 51) – in order to produce progressive political critique, one must stand somewhere. She acknowledges the need for strength in numbers and the power of an identity to fight for, but also recognizes the inevitable differences and power relations that will emerge in any group so ensuring that some voices are lost.

Mohanty and Spivak do not romanticize women's plights as Morgan's essentialist experience does. Women's identities and experiences are constructed through historically and geographically specific instances of struggle. In contrast, Mohanty believes that Morgan's sisterhood offers a naïve view of what it is to be a woman:

> Being female is thus seen as *naturally* related to being feminist, where the experience of being female transforms us into feminists through osmosis. Feminism is not defined as a highly contested political terrain; it is the mere effect of being female (Mohanty, 1997, p. 84).

There is then a tension between social construction of boundaries and territorial realities. Whilst there is a recognition of the social construction of boundaries, this does not mean that feminist and postcolonial writers ignore the material impacts of the everyday enunciation of these social constructions: they have real consequences in the lives of people the world over. In her writing about life on the US–Mexican border, Anzaldúa (1987, p. 3) is aware of the constructed nature of the boundary and yet also the physical pain of this actual spatial marker, it is "where the Third World grates against the first and bleeds."

The concept of hybridity, notions of thirdspace, fluidity and movement have also been embraced by a number of postmodern commentators (see Jameson, 1984; Rushdie, 1988, 1991; Soja, 1996). Again there is enthusiasm over the possibilities of global mixing, and a dynamic understanding of cultural identity, or "travelling culture" as Clifford (1992) called it. Commentators such as Salman Rushdie see impurities that arise from mixing as a positive outcome, and attachments to "pure" identities a romantic ideal. He argues that it "is normally supposed that something always gets lost in translation; I cling, obstinately, to the notion that something can also be gained" (Rushdie, 1991, p. 17).

However, some of the same criticisms leveled against Western feminists by Mohanty apply to this approach. The image of a shrinking and fluid globe rather than being a new emerging global sense of place is a geopolitical imaginary very much situated in the West, imagined predominantly by white men (Massey, 2000). For those who have the actual and cultural capital, the world is indeed becoming a smaller place linked by jet travel and the electronic communities of the Internet. For others, however, the globe is as large as ever still posing the barriers of nation-state borders, the cost of travel, and the continuing hold of Orientalist discourses of race. Massey (1991) argues that rather than facing a shrinking world where all sorts of possibilities are available, some people are finding their lives becoming ever more bounded and immobile. Outside images, products, and knowledges are forced upon them with a power that distorts the "hybrid" result.

Although recognizing the extra-local nature of definitions of place, Massey (1993) argues that global space is nevertheless subject to the laws of a set of "power-geometries" based in wealth, patriarchy, and Western-centrism. These structurings of global space ensure that mobility is not available to all, and that certain groups are still subject to the constraints of place, and can thus be exploited by the power of capital, which is mobile across the globe. This does not mean that those subject to the confines of place are simply defined by class. Global poverty is increasingly a gendered condition with women now estimated to comprise the majority of the world's poor. Even in cultural matters, global production and consumption do not imply homogenous political geographies: the fluidity and indeterminacy that Rushdie has celebrated of course were silenced by dominant interpretations of his work which demonstrated all too clearly the power/knowledge formations still dominating various global geographical imaginations (see Keith and Pile, 1993; Sharp, 1996a).

Subjectivity and resistance

The boundaries of the subject may also not be so malleable as theorists of ambivalence and hybridity might suggest. The claim of poststructural critics that the subject is dead, that it has fragmented and become fluid has been received with anger by feminist and postcolonial theorists who point out that these marginalized people have never enjoyed singular subjectivity. Furthermore, although this work has been very influential, some feminists have drawn limits to the collaboration that a feminist politics can have with Foucault. His theorization of subjectivity is considered by some as being too passive. As Linda Alcoff suggests, Foucault ignores the fact that, some of the time, "thinking of ourselves as subjects can have, and has had, positive effects contributing to our ability effectively to resist structures of domination" (Alcoff, 1990, p. 73). Other feminists have reacted more forcefully to postmodern and poststructural pronouncements of the "death of the subject" wondering whether this had occurred just when the male, white, subject might have had to share its status with those formerly excluded from subjectivity (see Fox-Genovese, 1986; Mascia-Lees et al., 1989).

Before the emergence of the recent attention to postcolonial theory, Fanon warned of the effects of hybridity. Rather than being a necessarily positive and liberating condition, Fanon (1968) regarded hybridity to be a psychologically damaging state. This emerged from the existence of two facts under colonialism. First was the "fact of blackness" through which the colonized subject would be immediately identified and marked as inferior (Fanon, 1968, 1995). Second, was the constant promotion of the superiority of the colonizer's values that should be copied and aspired to. However, because of the fact of blackness, this copying could never be successful. Colonial authority was reinforced through this invitation to black subjects to mimic white culture but the constant affirmation that this performance would never be successful. As Loomba (1998, p. 173) has explained, "Indians can mimic but never exactly reproduce English values... their recognition of the perpetual gap between themselves and the 'real thing' will ensure their subjection." Whereas Bhabha (1994) sees the mimic man as an inherently troubling figure for colonial authority, Fanon argues that the anxiety over this subjectivity impacted upon the bodies of the colonized, bodies which he treated for psychological disorders in colonial Algeria.

Bhabha's (1994) concentration has been on the textual. For him and many other postcolonial theorists, the word was the preserve of colonial authority. This explains the importance of textual analysis to many postcolonial critics who are trying to understand the forms of resistance and transgression generated through the colonial condition.[2] From this position, politics emerge from the structural position of natives as hybrids rather than any conscious decision to resist. For Fanon (1968), resistance must be conscious and cathartic, liberating the colonized subjects from their subjugation. Indeed Fanon argues for the necessity of violence to cleanse the psychological scars of the colonial process.

Mitchell (1997) argues that there is nothing inherently politically positive or negative in hybridity and it can be constraining and exploiting as often as it is liberating. Abstract liminal spaces are as easily appropriated by reactionary forces as they are by figures of resistance (Mitchell, 1997, p. 533). Similarly, Kortenaar argues that "*neither* authenticity nor creolisation has ontological validity, but both are valid as metaphors that permit collective self-fashioning" (quoted in Loomba, 1998, pp. 182–3). But Bhabha and others influenced by literary critiques see space in an abstract metaphorical sense rather than as also possessing grounded and material values. For Bhabha, spaces of in-betweenness avoid completeness and closure. Thus, the space of the nation, for instance, is also incomplete and always already in process; it is, in his words (Bhabha 1990), "nation as narration," so that we find "the formation of the nation is in the act of formation itself" (Mitchell, 1997, p. 536). Only after the nation is narrated as a bounded space, can the act of crossing and transgressing erase and rewrite its boundaries. Mitchell (1997, p. 537) argues that a central political question must then be "what are the actual physical spaces in which these boundaries are crossed and erased?" This is not a question that Bhabha raises, for his interest is in the abstract discursive spaces within which people are culturally inscribed by the narration of the nation. This flattens out the possibilities, directions, and the intentions of resistances:

> Without context, it is possible to locate resistance in *all* spaces in-between, in every liminal movement and minority discourse that supplements the nation and thus forces a renegotiation of political and cultural authority (Mitchell, 1997, p. 537).

However, it is important, then, to distinguish between the types of marginality or hybridity being celebrated (Slater, 1992, p. 320). Shohat (1997) argues that there is a necessity to discriminate between "diverse modalities" of hybridity. During colonialism, there was a geography to contact so that some never saw a European face whereas for others it was part of daily routine. Similarly, postcolonial hybridity has a geography that needs critical examination. For instance, Mitchell (1997) demonstrates that global corporations and business interests (capital), are particularly adept at exploiting hybrid conditions, of mutating and evolving to suit the coordinates of each new situation.

Leaving the Armchair: the Challenge to Political Geography

Postcolonial and feminist theories have made a significant impact on geography over the last ten years or so. Although less impact has been felt directly on political geography *per se*, the intellectual challenges that have been posed have clear implications for the ways in which political geographers consider definitions of what comprises political acts, where politics are enacted, and how and in whose interests particular political identities are formed and performed. Postcolonial and feminist theorists have directed attention away from sole concentration on the public spaces of politics to the ubiquity of power relations and opened up the possibilities for new political theories of subjectivity, identity, and boundaries.

However, particularly in the case of postcolonial theory, the impact has been at the abstract level of theorization rather than having significant impacts upon the actual practices within the academy.[3] As Duncan (1993) argued in his Progress Report on cultural geography in 1991–2, the rise of postcolonialism in geography has coincided with a declining interest in regional studies. Similarly, Goss (1996, p. 246) argues that despite the amount of critique of texts and representations, there is no action, as this too is based in discourse, so that "we now have no ability whatsoever to speak of the act – only to explain its presence." He suggests that postcolonial critics, "have guaranteed themselves the position of armchair decolonisers" (1996, p. 248).

Said's (1978) critique of the geographical violence of Orientalism has had long-lasting effects as researchers have backed off from studying other places, trying to speak for those outside the academy, or translating other geographies and experiences into the languages of research (Duncan, 1993). Ironically, then, the impact of postcolonial theory has led to an intensification of interest in the West and an increasing marginalization of other voices and places, even in the recoding of regional study. Postcolonial work although superficially studying difference, is more interested in what representations of others says about the self, about the West. On the one hand, this could be seen as a positive move in that Others are no longer being (mis)represented by Western research, but at the same time, they are being further marginalized from the production of so-called global knowledges and from institutions of power. Perhaps it is necessary to reconsider some of the implications that postcolonial challenges might have for geographical research.

The use of postcolonial theorists in mainstream work in recent years might suggest a decentering of knowledge, and acknowledgment of the displacement of the central authority of the West. However, Sparke (1994, p. 113) warns that the use of lists of postcolonial theorists is "increasingly becoming a high-quality currency in contemporary academia" but such lists are removing the specifics of individual's arguments in the homogenizing figure of the postcolonial critic:

> Although the *content* of the work they represent might always carry the promise of interrupting the smooth reproduction of Western authority, the metonymic contradictions that are the names themselves can become, in the hands of the already powerful, simply tokens of exchange in an economy whose only interest is what Spivak calls the "new Orientalism" (Sparke, 1994, p. 113).

Sparke points to the production of an "anemic geography" wherein an instrumental use of spatial referents is used to stage broader arguments where the "non-West" is never examined as a multiply-inscribed self. The heterogeneity of the non-West is ignored (except to note that it is heterogeneous) because, as in the older versions of Orientalism, its role is to mark a debate about the limits of the West (Sparke, 1994, p. 113). Just as artists look outside the West for inspiration for their music, art and literature, so too do theorists. Sparke argues that many Western theorists (and

I would add, some Western educated "Third-World" academics) who are interested in postcolonialism look for differences and new approaches to enliven their own theories and to advance their own careers rather than having any deep commitment to drawing the marginalized and silenced into the heart of academic debate. Despite initial appearances, then, new debates replicate the geographical violence of Orientalism rather than overturning it.

For hooks, it is time that Western theorists realized the potent force of those on the margins. For her, the margins are a site of "radical possibility" (hooks, 1990, p. 341) which rejects the politics of inside and outside as "to be on the margins is to be part of the whole but outside the main body." But, as a result of anemic geography, this resisting power has been domesticated. hooks has felt silenced by Western academics seeking the experience but not the wisdom of the other. She argues that "I was made 'other' there in that space . . . they did not meet me there in that space. They met me at the center" (hooks, 1990, p. 342). The experiences of the marginalized are used in postcolonial theories but without opening up the process of theorizing to the knowledges and wisdom of the marginalized. When there is a meeting, it is in the center – in the (predominantly) Western institutions of power/knowledge (aid agencies, universities, the pages of journals, and so on) and in the languages of the West. So, by approaching the institutions of knowledge, she has been forced to the center – a location both metaphorical in its control of authority and geographical in its physical presence. There is a reluctance to abandon the mark of authority:

> No need to hear your voice when I can talk about you better than you can speak about yourself. No need to hear your voice. Only tell me about your pain. I want to know your story. And then I will tell it back to you in a new way. Tell it back to you in such a way that it has become mine, my own. Re-writing you I write myself anew. I am still author, authority. I am still colonizer, the speaking subject and you are now at the center of my talk (hooks, 1990, p. 343).

Many have argued that rather than offering a significant challenge to the authority of the West, postcolonial critiques have reinforced the centrality of the West in history and knowledge (McClintock, 1995; Childs and Williams, 1997). The West has allowed itself one further privilege in the negotiation of the end of colonialism, "that of painting its own misdeeds in dark colours and evaluating them on its own terms" (Ferro, 1997, p. vii).

The subaltern cannot speak through academic research Spivak argues, because their speaking would automatically involve the false transparency of the intellectual (1988): her words would be forced into the language of academia, dominated and distorted by the norms of Western logic and reasoning. So how do we as political geographers carry out research? It is not possible to be anti-essentialist because the subject is always centered (Spivak, 1990, p. 108). In a discussion with Spivak, Young suggests that Western researchers are now trapped by postcolonial critiques:

> If you participate [in research] you are, as it were, an Orientalist, but of course if you don't then you're a eurocentrist ignoring the problem (quoted in Sparke, 1994, p. 119).

This perception has led to a split between geographers and others involved in development issues "on the ground" in Third-World countries and postcolonial theorists who critique the colonial inspirations and consequences of this work. Any attempt to represent others will inevitably produce mis-representation as difference is interpreted and translated by the Western academic. Spivak, however, counters that the argument is not so simple, that all have a responsibility to be critically aware of their position in research. Not to do so would assume that only members of a community may talk about it, but this will always homogenize a community and deny historical and contemporary geographies of connection and constitution. Spivak argues that individuals from colonized countries can equally fall into Orientalist traps (see also Jones, 2000). Further, Western researchers have a responsibility to challenge their position. She says to one politically-correct, white, male student in her class who feels he cannot speak:

> Why not develop a certain degree of rage against the history that has written such an abject script for you that you are silenced? [...] make it your task not only to learn what is going on there through language, through specific programs of study, but also at the same time through a *historical* critique of your own position as the investigating person, then you will see that you have earned the right to criticize, and you will be heard. When you take the position of not doing your homework – "I will not criticize because of my accident of birth, the historical accident" – that is a much more pernicious position. In one way you take a risk to criticize, of criticizing something which is *Other* – something which you used to dominate. I say that you have to take a certain risk: to say "I won't criticize" is salving your conscience, and allowing you not to do any homework. On the other hand, if you criticize having earned the right to do so, then you are indeed taking a risk and you will probably be made welcome, and can hope to be judged with respect (Spivak, 1990, pp. 62–3).

Ó Tuathail (1996, p. 175) has similarly argued for the inclusion of a more responsible and embodied account of political geography which "establishes a moral proximity with personalized victims." This reworking of political geography, what he calls "an anti-geopolitical eye," emphasizes the links and causalities of historical and contemporary geographies of power, insisting on the connections between scales of meaning and identity ("the personal is the geopolitical") (1996, p. 176).

It is the "hard work of specific analyses" which "interrupts the reifications of anemic geography" (Sparke, 1994, p. 119). To avoid reproducing anemic geographies, political geographers should heed hooks' request that we celebrate "marginality as a site of resistance" not in the center but in the margins. Rather than abandoning regional work after Said, political geography needs to embrace the specifics of placed empirical research. This way, it can rise to the challenge for more embodied and impassioned discussion of the political issues structuring the geopolitics of everyday life.

Acknowledgments

I would like to thank Gearóid Ó Tuathail and John Briggs for their insightful comments on earlier drafts of this chapter.

ENDNOTES

1. Feminist commentators have noted that active national citizenship is available only to men. Women symbolize the nation to be protected by male agency (McClintock, 1995; Sharp, 1996b).
2. It is, of course, also important to consider the impacts of Eurocentrism on the psychoanalytic theory that Bhabha adopts, and, in particular, the representations of women it is centered around (see Blunt and Wills, 2000, p. 189).
3. Many feminists would still critique the academy for masculinist practices and knowledge structures. Political geography is a subdiscipline that seems to be particularly unattractive to women (see Staeheli, 2001).

BIBLIOGRAPHY

Alcoff, L. 1990. Feminist politics and Foucault: the limits to a collaboration. In A. Dalley and C. Scott (eds.) *Crises in Continental Philosophy.* Albany, NJ: SUNY Press.
Anderson, B. 1991. *Imagined Communities: Reflections on the Origin and Spread of Nationalism*, 2nd edn. London: Verso.
Anzaldúa, G. 1987. *Borderlands/La Frontera: the New Mestiza.* San Francisco: Aunt Lute Press.
Bhabha, H. (ed.). 1990. *Nation and Narration.* London: Routledge.
Bhabha, H. (ed.). 1994. Of Mimicry and Man; the ambivalence of colonial discourse. In *The Location of Culture.* London: Routledge.
Blunt, A. and Wills, J. 2000. *Dissident Geographies.* London: Pearson.
Chatterjee, P. 1986. *Nationalist Thought and the Colonial World: A Derivative Discourse.* Minneapolis, MN: University of Minnesota Press.
Chatterjee, P. 1993. *The Nation and Its Fragments.* Princeton, NJ: Princeton University Press.
Childs, P. and Williams, P. 1997. *An Introduction to Post-colonial Theory.* London: Prentice Hall.
Clifford, J. 1992. Travelling Cultures. In L. Grossberg et al. (eds.) *Cultural Studies.* London: Routledge.
Deutsche, R. 1996. *Evictions.* Cambridge, MA: MIT Press.
Duncan, J. 1993. Landscapes of the self/landscapes of the other(s): cultural geography 1991–2. *Progress in Human Geography*, 17, 367–77.
Enloe, C. 1989. *Bananas, Beaches and Bases: Making Feminist Sense of International Relations*, Berkeley, CA: University of California Press.
Fanon, F. 1968. *The Wretched of the Earth.* New York: Grove.
Fanon, F. 1995. The Fact of Blackness. In B. Ashcroft et al. (eds.) *The Post-colonial Studies Reader.* London: Routledge.
Ferro, M. 1997. *Colonialism: a Global History.* London: Routledge.
Foucault, M. 1980. *Power/Knowledge: Selected Interviews and other Writings, 1972–77.* New York: Pantheon.
Fox-Genovese, E. 1986. The claims of a common culture: gender, race, class and the canon. *Salmagundi*, 72, 119–32.
Funk, N. and Mueller, M. (eds.). 1993. *Gender Politics and Post-Communism: Reflections From Eastern Europe and the Former Soviet Union.* New York: Routledge.
Godlewska, A. 1994. Napoleon's geographers: imperialists and soldiers of modernity. In A. Godlewska and N. Smith (eds.) *Geography and Empire.* Oxford: Blackwell, 31–54.

Goss, J. 1996. Postcolonialism: subverting whose empire? *Third World Quarterly*, 17, 239–50.

Haraway, D. 1988. Situated knowledges: the science question in feminism and the privilege of partial perspective. *Feminist Studies*, 14, 575–99.

hooks, b. 1990. Marginality as a Site of Resistence. In R. Ferguson et al. (eds.) *Out There: Marginalization and Contemporary Cultures*. Cambridge, MA: MIT Press.

Irigaray, L. 1985. *Speculum of the other woman*. Ithaca, NY: Cornell University Press.

Jameson, F. 1984. Postmodernism, or the cultural logic of late capitalism. *New Left Review*, 146, 53–92.

Jones, P. 2000. Why is it alright to do development "over there" but not "here"? Changing vocabularies and common strategies of inclusion across the "First" and "Third" Worlds. *Area*, 32, 237–41.

Keith, M. and Pile, S. 1993. *Place and the Politics of Identity*. London: Routledge.

Lewis, R. 1996. *Gendering Orientalism: Race, Femininity and Representation*. London: Routledge.

Loomba, A. 1998. *Colonialism/Postcolonialism*. London: Routledge.

Mascia-Lees, F., Sharp, P., and Cohen, C. 1989. The postmodern turn in anthropology: cautions from a feminist perspective. *Signs*, 15, 7–33.

Massey, D. 1991. A global sense of place. *Marxism Today*, June, 24–29

Massey, D. 1993. Politics and Space/Time. In M. Keith and S. Pile (eds.) *Place and the Politics of Identity*. London: Routledge.

Massey, D. 2000. *Imagining Globalisation*. Lecture at the University of Strathclyde, 23/2/2000.

McClintock, A. 1993. Family Feuds: Gender, Nationalism and the Family. *Feminist Review*, 44, 61–80.

McDowell, L. 1993. Space, place and gender relations. Part II: identity, difference, feminist geometries and geographies. *Progress in Human Geography*, 17, 305–18.

Mitchell, K. 1997. Different diasporas and the hype of hybridity. *Environment and Planning D: Society and space*, 15, 533–53.

Mohanty, C. T. 1997. Feminist encounters: locating the politics of experience. In L. McDowell and J. Sharp (eds.) *Space, Gender, Knowledge: feminist readings*. London: Arnold.

Morgan, R. 1984. *Sisterhood is Global: The International Women's Movement Anthology*. New York: Anchor Press/Doubleday.

Ó Tuathail, G. 1996. An anti-geopolitical eye: Maggie O'Kane in Bosnia, 1992–93. *Gender, place and culture*, 3, 171–85.

Ousmane, S. 1974. *Xala*. Chicago: Lawrence Hill Books.

Phillips, R. 1996. *Mapping Man and Empire: a geography of adventure*. London: Routledge.

Pratt, M. L. 1992. *Imperial Eyes: Travel Writing and Transculturation*. London: Routledge.

Rose, G. 1991. On being ambivalent: women and feminisms in geography. In C. Philo (comp.) *New Words, New Worlds*. Aberystwyth: Cambrian Printers.

Rushdie, S. 1988. *The Satanic Verses*. New York, Viking.

Rushdie, S. 1991. *Imaginary Homelands*. London: Granta.

Said, E. 1978. *Orientalism*. New York: Vintage.

Seager, J. 1993. *Earth Follies: Feminism, Politics and the Environment*. New York: Routledge.

Sharp, J. 1996a. Locating imaginary homelands: literature, geography and Salman Rushdie. *GeoJournal*, 38, 119–27.

Sharp, J. 1996b. Gendering nationhood: a feminist engagement with national identity. In N. Duncan (ed.) *BodySpace: Destabilizing Geographies of Gender and Sexuality*. London: Routledge.

Sharp, J., Routledge, P., Philo, C., and Paddison, R. (eds.) 2000. *Entanglements of Power: Geographies of Domination/Resistance*. London: Routledge.

Shohat, E. 1997. Post-Third-Worldist culture: gender, nation and the cinema. In Alexander et al. (eds.) *Feminist Genealogies, Colonial Legacies, Democratic Futures*. London: Routledge.
Slater, D. 1992. On the borders of social theory: learning from other regions. *Environment and Planning D: Society and Space*, 10, 307–27.
Soja, E. 1996. *Thirdspace: Journeys to Los Angeles and other Real-and-imagined Places*. Oxford: Blackwell.
Sparke, M. 1994. White mythologies and anemic geographies: a review. *Environment and planning D: Society and space*, 12, 105–23.
Spivak, G. 1988. Can the subaltern speak? In C. Nelson and L. Grossberg (eds.) *Marxism and the Interpretation of Culture*. Basingstoke: Macmillan.
Spivak, G. 1990. *The Postcolonial Critic: Interviews, Strategies, Dialogues*. New York: Routledge.
Staeheli, L. 2001. Of possibilities, probabilities and political geography. *Space and Polity*, 5(3), 177–89.
Trinh, Minh-ha 1989. *Woman/Native/Other*. Bloomington, IN: University Press.
Wolf, E. 1982. *Europe and the People Without History*: Berkeley, CA: University of California Press.
Young, R. 1990. *White mythologies: Writing History and the West*. London: Routledge.

Chapter 6

Geopolitical Themes and Postmodern Thought

David Slater

Context and Purpose

For the contemporary period, it can be argued that the construction of political identities increasingly takes place without any attempt being made to ground their legitimacy and action in a universalist perspective (Laclau, 1994). The geographically widespread surfacing of ethnic and cultural particularisms, the erosion of unitary conceptions of emancipation, most visible perhaps in the eclipse of socialist politics and the emergence of new archipelagoes of resistance, expressed in the form of new social movements, give substance to a sense of "new times" and a break from older forms of social struggle.

In times of difference, plurality, and fragmentation, which are frequently seen as key markers of the postmodern, there is a strong tendency to assume that one important trend can be taken as constitutive of the era as a whole. But our period is also marked by the presence of a neo-liberal regime of truth that is driven by a clearly-defined universalist ambition. Viewed from one society of the periphery, the Argentinian philosopher Reigadas (1988) has argued that in the 1980s Latin America was impacted by two waves of Western truth: first by a neo-liberal discourse that purported to provide the sole prescription for development and progress – the only possible horizon, the "sole thought"[1] – and second by a postmodernism that destabilized the ground for any alternative horizon, whilst celebrating an ever-proliferating pluralism. Although Reigadas, unlike other Latin American writers,[2] fails to acknowledge the potentially enabling and oppositional elements of the postmodern, it can be argued that a postmodern perspective that evades any critical consideration of the prevailing modes of neo-liberal thought remains complicit with the established order (Soper, 1991, pp. 99–101). There are two points here that need to be emphasized.

First, although what has been termed a postmodern frame of interpretation in social science has become increasingly influential (Rattansi, 1994), corresponding to the need to understand differences, plurality, instabilities, fragmentations, and processes of deterritorialization and global–local change, nonetheless this postmodern sensibility is juxtaposed to a universalizing and homogenizing discourse of neoliberalism that centralizes and standardizes truth around a limiting number of objectives – privatization, the prioritization of market forces, individualism, and the commodification of social life. Second, within postmodern approaches there are differences and heterogeneities which are important for our analysis of geopolitics, and whilst, as I shall indicate, there is a significant current of "cynical reason" (Sloterdijk, 1988), which is both melancholic and Euro-Americanist, it is also possible to identify what Santos (1999) calls an "oppositional postmodernism" in relation to the search for new forms of radical politics.[3]

Keeping these observations in mind, my purpose in this chapter is to explore some of the intersections between the ways we might interpret geopolitics and our approach to postmodern thinking. A guiding question here is: how might a postmodern frame help us to develop more engaging and critical perspectives on geopolitics, and equally how might a concern with power and spatiality in global times shed light on the vicissitudes of postmodern thinking? This double question will be tackled through the short treatment of three interconnected themes:

- the politics of time–space in relation to the concept of "chronopolitics;"
- the re-imagining of power and the spatial in global times; and
- a call for a decolonization of geopolitical thinking.

First of all, however, I will discuss what I consider to be some key aspects of postmodern thinking through focusing on certain texts of Lyotard and Baudrillard, both highly influential exponents of a postmodern analytical sensibility.

In the Tracks of Melancholic Reason

The French philosopher Lyotard, one of the most influential of theorists associated with the postmodern turn, has recently suggested that the crisis and hell of modernity – the end of hope – is the state of postmodern thought which suffers from a lack of finality and is deeply affected by melancholia (Lyotard, 1997). In the context of distinguishing the differences between fable and narrative, Lyotard suggests that the fable offers no cure but rather an imaginary with no ethico-political pretension but more an aesthetic or poetic status, which is imbued with melancholic sentiment. In a directly political setting, Lyotard contends that the liberal capitalist system under which we live is not subject to radical upheaval but only to revision. Radicalism is becoming rare and in contrast to modern politics, which were characterized by conflicts for legitimacy in the form of civil and total war, "postmodern politics are managerial strategies, its wars, police actions" (Lyotard, 1997, p. 200). Thus, struggles for human rights, for example, are always played by the rules of the game, in consensus with the system, and "politics will never be anything but the art of the possible" (op. cit., p. 193).

There are echoes of these lines of thought in the work of the French social theorist Baudrillard (1994, 1998), where postmodernity is defined in terms of the recycling of past forms and an eclectic sentimentality. For Baudrillard (1994, p. 51), "the political died with the historical passions aroused by the great ideas and the great empires." The present melancholy of the century is for Baudrillard (op. cit., p. 118) a triumph for Walt Disney, "that inspired precursor of a universe where all past or present forms meet in a playful promiscuity, where all cultures recur in a mosaic." There is no end to anything since all things will continue to unfold slowly, tediously, recurrently, and unlike mourning, where there is a possibility of returning, with melancholia we are not even left with the presentiment of an end or a return, but only with a resentment at the disappearance of ends and values. Today there is no revolt any more, no antagonism, but rather a new perverse consensual social contract in which everyone tries to gain their recognition as a victim, and where the unemployed and the various cases for social assistance are now encouraged to look after themselves, to manage their own "enterprise" more effectively. There are no longer any convictions, and all forms of concrete freedom are being absorbed into the only freedom which remains, the freedom of the market (Baudrillard, 1998, pp. 55–7, 65).

Such a vision finds a parallel in Lipovetsky's (1994) idea that for the first time we are living in a society that devalues the self-denial associated with the pursuit of a higher societal ideal (for example, as connected to a religious ethic), and instead systematically stimulates immediate desires, the passion of the ego and materialist and intimate forms of happiness. Contemporary society is witness to the "twilight of duty" and to the prevalence of a postmoralistic logic where individualism, the seductiveness of consumerism, and the continuing differentiation of commodity production constitute salient features of the era.

Notwithstanding differences of conceptual and thematic orientation, Baudrillard, Lyotard, and Lipovetsky, as symptomatic observers of the postmodern, share a vision of neo-liberal times in which there is a continuing present and an eclipse of radical, insurgent politics. Guattari (2000, pp. 41–2), writing at the end of the 1980s, alerted us to the crucial interconnections among society, the psyche and the environment, and strongly argued against the "fatalistic passivity" and "destructive neutralization of democracy" characteristic of certain kinds of postmodern thinking. Similarly, Rancière (1995) takes issue with those writers, Lyotard being a cited example, who undermine any optimistic reading of postmodernity and minimize the continuing relevance of political struggles and a democratic ethos.[4] There is much at stake here, theoretically and politically.

First, it is important to note that trends associated with the postmodern, such as the proliferation of difference, the decentering of the social subject, the plurality of subjectivities and the end of pre-given unitary views of emancipation, do not have to usher in a mode of thinking that abandons attempts at radical reconstruction and passively accepts the neo-liberal nostrum that there is no alternative to the present disposition of power relations. In fact, as Zizek (1999, pp. 352–4) argues, in his comments on postmodernism and political identitites, what is needed is a "re-politicization of the economy" and the development of a new critique of global capitalism that challenges the common acceptance of capital and market mechanisms as neutral instruments and procedures. I shall return to this theme below, in a later section.

Secondly, the tendency to preclude the possibility of political alternatives – assuming that the triumph of liberal capitalism is definitive – is itself another kind of metanarrative, and closes off the full range of social and geopolitical heterogeneities which are still thinkable.

Thirdly, the oppositional nature of postmodernism can be enabling in that it can help us move from monocultural to multicultural forms of knowledge, from knowledge as regulation to knowledge as emancipation and, as intervention, it can engender a move from conformist to rebellious forms of action (Santos, 1999). None of these moves are essentially rooted in postmodern thought; rather, they are more appropriately seen as one possible reading of the postmodern which can be deployed as part of a critique of contemporary neo-liberalism.

Finally, the postmodern can be associated with a re-assertion of spatiality (Benko and Strohmayer, 1997; Jones et al., 1993; Soja, 1989) and the heterogeneous local contra the standardizing global (see, e.g., Escobar, 2001): in this sense, it can be associated with new attempts to think space and politics in terms of difference, plurality, and resistance. And it is here in the arena of the spatial that we again encounter key questions of vision, of interpretation, of the framing of knowledge, and of the geographies of reference. It is also the case, as we shall now see, that the postmodern has been closely linked to an important rethinking of the relationship between space and time and the implications for politics.

Space, Time, Politics

In the analysis of space–time relations, it has been suggested that in a world increasingly marked by simulation, speed, and surveillance, geopolitics is being replaced by chronopolitics (Der Derian, 1990; Virilio, 1986).[5] Virilio (1997, p. 69) has recently reasserted his distinctive analytical position by suggesting that in the realm of territorial development "time" now counts more than "space." Further, he goes on to write that "from the urbanization of the real space of national geography to the urbanization of the real time of international telecommunications, the 'world space' of geopolitics is gradually yielding its strategic primacy to the 'world time' of a chronostrategic proximity without any delay and without any antipodes" (ibid.). In Virilio's vision, we will come to see a world of electronic information highways that will no longer be divided along a North–South axis but rather into two speeds – one absolute and the other relative. There will be an even more radical divide between those who will live in the virtual community of the world city under the "empire of real time" and those more destitute than ever who will survive in the real space of local towns. For Virilio, "the society of tomorrow will splinter into two opposing camps: those who live to the beat of the real time of the global city, within the virtual community of the 'haves', and the 'have-nots' who survive in the margins of the real space of local cities, even more abandoned than those living today in the suburban wastelands of the Third World" (op. cit., p. 74).

Leaving aside Virilio's dystopian vision of the future, a vision which falls well within the realm of "melancholic reason" as outlined above, his perspective on time–space relations re-asserts a key salience for the temporal whilst reducing the significance and heterogeneity of the spatial to an apparently past era. As Ó Tuathail (1997) has appropriately argued, Virilio's perspective, which illustrates a wider tendency to signal "the end of geopolitics," fails to appreciate the richness and complexity of spatial politics and overemphasizes the impact of *one* contemporary trend. What Virilio's interpretative stance fails to include is an awareness of the complex interplay of overlapping but different tendencies and the reworking of older meanings and practices into novel and contradictory combinations.

Whilst there is abundant evidence for the development of a borderless world in the context of financial transactions, cyberspace, movements of commodities, flows of investment, telecommunications, and electronic surveillance, equally national frontiers are being fortified (as with the US–Mexico border) and flows of people are being interrupted, checked, and curtailed, as similarly, within the territories of nation-states, gated communities or "fortified enclaves" (Caldeira, 1996) are the increasingly visible signs of opposition to what is perceived to be a limitless and increasingly chaotic world of flows. Whereas there are tendencies towards fusion, exemplified in formations such as NAFTA and the EU, the splitting of erstwhile territorial unities, and the dissolution of nation-states testify to new forms of fission. Whilst there is evidence of increased global connectivity, within individual societies there are trends towards introversion, as reflected in the decline in coverage of foreign news stories in the US and other Western societies (Moisy, 1996). Whereas the borderless economy may signify a move towards de-territorialization, at the same time, under the remit of "decentralization" or "devolution" struggles for re-territorialization are occurring with the formation of new local and regional governments. Whilst there are trends towards accentuated globalization in the realms of trade, finance, investment, and the media (Giddens, 1999), at the same time there are calls for "localization" (Hines, 2000), the dismantling of the power of transnational corporations, and a re-validation of the significance of local economies and resources. We are encouraged to believe that we live in postcolonial times, yet when the old forms of colonizing power have faded from sight, Third-World scholars refer to new processes of the recolonization of peripheral societies by metropolitan capital and organizations (González Casanova, 1995).

Finally, there are symptomatic divergences over the way time is framed. On the one hand, there is the dominant media frame of a "continuing present" whereby past events are pillaged, de-contextualized, and repackaged into new amalgams of marketable meaning that break the link between past, present, and future (Bourdieu, 1998; Ramonet, 2000). On the other, there are the movements of indigenous peoples that seek to validate a past cultural time as part of a struggle for recognition and emancipation in the present and future. In the fight against what Subcomandante Marcos (2000) calls the "globalization of pessimism," the deployment of previous symbols of political contestation is seen as crucial in opposing the power of contemporary neo-liberalism. Moreover, in an era of accelerated velocity, the neo-liberal meaning attached to the general significance of speed – the removal of barriers to the fastest possible realization of profit – engenders resistance through an emphasis on "slowness" and the most visible signifiers of "fast-ness" such as McDonalds become the target of new forms of protest.[6] As a generalized reaction against fast food there is now a "Slow Food Movement," which originally started in Italy, and now has an estimated 60,000 members in 25 countries: its key objective is to resist the homogenization and globalization of food production in all its forms (*The Guardian Weekend*, September 9, 2000, p. 21).

Consequently, we can say that the meanings associated with time are increasingly becoming a site of conflict, so that, for example, the radical ecologist Wolfgang Sachs (1999, p. 16) writes that "without a wealth of time, there is bound to be less generosity, less compassion, less dedication and less freedom – a sort of modernized poverty." One recent and powerful example of the politicization of time comes from

Subcomandante Marcos' (2001, p. 73) comment, when asked about the nature of negotiations with the Mexican President Vicente Fox, that there is a struggle between a clock operated by a punch card, which is Fox's time, and an hourglass, which is ours. Marcos adds that neither one nor the other will prevail since both of them need to understand that another clock must be assembled that will time the rhythm of dialogue and finally of common agreement.

Overall, the sorts of conflicting juxtapositions and paradoxes that we have noted above, have tended to generate a feeling of analytical vertigo and related calls for new ways of describing and interpreting the complexities of global politics (Luke, 1996, 1999). Such complexities have also stimulated greater interest in those forms of analytical enquiry which go beyond essentialism. Hence, the postmodern emphasis on difference and plurality, as well as the poststructuralist development of discourse analysis, provide increasingly influential guidelines for the ways we might want to rethink and reimagine the processes of geopolitics in relation to both the terrain of analysis and the subjects of knowledge.

Re-thinking the Place of Power in Global Times

Bauman (1999, p. 74), a leading social theorist of postmodernity, and a writer who assigns considerable significance to the spatial in his work, states that a characteristic feature of our times is the "ongoing *separation of power from politics*: true power, able to determine the extent of practical choices, *flows*," and "thanks to its ever less constrained mobility it is virtually global – or rather, exterritorial" (emphasis in the original). For Bauman, all existing political institutions remain "stoutly local," and, crucially, the heart of the contemporary crisis of the political process is the absence of an agency effective enough to "legitimate, promote, install and service any set of values or any consistent and cohesive agenda of choices" (ibid.). Today, the principal agenda setters are "market pressures" which are replacing political legislation. The agenda is seen as neither rational nor irrational; it just exists and there is "no alternative" (ibid.). In a subsequent and related passage, Bauman (op. cit., p. 120) defines a key element of globalization as the progressive separation of power from politics, whereby politics stays as local and territorial and power flows globally: "we may say that power and politics reside in different spaces ...physical, geographical space remains the home of politics, while capital and information inhabit cyberspace, in which physical space is cancelled or neutralized" (ibid.).[7]

Politics, power, and spatiality are central in Bauman's formulation and I want to pursue his argument a little further as a way of introducing some quite central questions for geopolitical analysis.

In Bauman's account, politics is traced back to the classic distinction between the public and private spheres, where one has the household (the *oikos*) and the site of politics or the Assembly of the people, (the *ecclesia*) in which matters affecting all members of the *polis* are discussed and settled. But between these two spheres the Greeks introduced a third domain, that of interaction and communication, where an attempt was made to ensure a functional and continuing interaction between the two primary spheres of the household and the Assembly of the people. This third sphere was called the *agora* (the private/public sphere) – see Castoriadis (1991) – and its function was to bring together the household and the Assembly of the people, in order to guarantee an autonomous *polis*, resting on the independence of its members.

For Bauman, there are two routes from which the *agora* can be endangered. One emanates from a totalitarian tendency, which seeks to annihilate the private sphere and to make independent thinking redundant and irrelevant to the effectiveness of power, as, for example, was witnessed in the period of fascist rule in mid-twentieth century Europe. A second danger, which is more pressing in its contemporary actuality, originates from what Bauman refers to as the separation of power from politics, so that the powers that truly matter, the forces of global capitalism, have cut their ties with the *agora* and exercise power beyond the more limited realms of the territorially circumscribed *polis*. The only way to effectively challenge this neo-liberal hegemony, and make the *agora* fit for autonomous individuals and autonomous society is to simultaneously block the privatization and depoliticization of the public/private interactive sphere and to re-establish the "interrupted discourse of the common good, which renders individual autonomy both feasible and worth struggling for" (Bauman, op. cit., p. 107).

It is clear from this line of argument, as well as from Bauman's work as a whole, that the Greek political philosopher Castoriadis has had a significant influence on Bauman's overall perspective, symptomatically expressed perhaps in one reference to Castoriadis, where it was suggested that the trouble with our civilization is that it has stopped questioning itself (Bauman, op. cit., p. 125). Turning briefly to Castoriadis, it is worthwhile noting that in his discussion of power and politics a number of distinctions were introduced which are relevant to our treatment of geopolitics.

First, it was noted that power in general may be envisaged as the capacity for an individual or institution to induce someone to do (or not to do) that which left to him/herself s/he would not necessarily have done. In this context, it ought to be clear argues Castoriadis (1991, p. 149) that the "greatest conceivable power lies in the possibility of pre-forming someone in such a way that, of *his/her own accord*, s/he does what one wants him/her to do, without any need for *domination* or of *explicit power* to bring him/her to . . . [do or abstain from doing something]" (emphasis in the original). Compared to this "absolute power," explicit power and domination are for Castoriadis deficient and limited. The key notion here is the act of "pre-forming" which Castoriadis defines in terms of the institution by society of a "*radical ground-power*" (emphasis in the original). This radical or primordial power is defined as the manifestation of the "instituting power of the radical imaginary" and is not locatable. It is however, according to Castoriadis, always historical since the "instituting society . . . always works by starting from something already instituted and on the basis of what is already there" (op. cit., p. 150). In this sense, also, the institution of society "wields a radical power over the individuals making it up" and this is how they are socially "pre-formed" and constituted in heteronomy, where the citizen is subject to laws and juridical norms which are made by others, the opposite of autonomy.

Secondly, this instituting ground-power, as Castoriadis calls it, never succeeds in wielding its capacity in an absolute fashion, or there would be no history. There are limits and these limits are related to four factors:

(a) the plurality of essentially different societies, whereby, for Castoriadis, the institution and signification of the others are always a "deadly threat to our own;"
(b) the fact that although society creates the world, there is also a "presocial world" which is always present as an "inexhaustible provision of alterity" and a source of threat for the meaning of society;
(c) the capacity and plasticity of the individual to thwart the incessant schooling imposed upon it; and,
(d) society can never escape itself, so that, as the principal limit to the wielding of a grounding power, beneath the established social imaginary there is always the flow or magma of radical imaginaries which have the potential to create alternative social orders. These factors then call into question and disrupt society's stability and self-perpetuation, but more acutely crime, violent contentions, natural calamities, and wars pose threats to society which explain the need for what Castoriadis calls *explicit power* (op. cit., p. 154).

Thirdly, this explicit power is rooted in both coercion and crucially in the capacity to "interiorize" or implant within the individual the socially produced significations that bind together the institutional and interpersonal webs of the societal order (Castoriadis, 1998, p. 158). For Castoriadis, society institutes itself through representation, affect – its way of living itself and of living the world and life itself – and intention, society's push and drive which is not just aimed at its own preservation and reproduction, but more at the ends and objectives of its future. Here, too, explicit power is rooted in the necessity to decide what are the strategic purposes of society's push and drive towards the future. Explicit power is made synonymous with the dimension of "the political," but explicit power is not identical to the state because societies without the state are by no means societies without power.

Bearing in mind the fact that two of the potentially enabling features of a critical postmodernism concern the epistemological significance of difference and the incorporation of a spatial imagination, the framework for understanding power and politics proposed by Castoriadis might be questioned in the following way.

Initially, the notion of a radical or instituting "ground-power" which is conceived in an aspatial manner and which is historically limited by the plurality of essentially different societies leaves open and undecided the question of our geographies of reference. Apparently the institution and signification of different societies would seem to be a "deadly threat to our own," but this already presupposes an identification with an "our" which is not explicitly stated but presumably is constituted as an "our" that is Western. What is missing here, as in Bauman, is a connection between penetration and founding. In other words, in the process, for example, of the colonial encounter, acts of instituting a grounding power were part and parcel of external penetration and the transgression of sovereignty. We have here, too, the interweaving of time and space since the penetration and invasiveness of the colonial intervention generates a time–space relation that is externally governed. Homi Bhabha (1995, pp. 328–9), a theorist of the postmodern as well as the postcolonial, has a nice phrase for this kind of time–space nexus, noting that "if . . . the past is a foreign country, then what does it mean to encounter a past that is your own country reterritorialized, even terrorized, by another?"

Subsequent struggles to achieve political independence have been concerned with overturning that original act of instituting a grounding power, a grounding power which also provided the basis for the deployment of what Castoriadis calls explicit power, with a combination of coercion and attempts to "interiorize" within the individual the significations that would bind together the institutional and interpersonal webs of the colonialized societal order. The timing and spacing of these struggles, as well as their outcomes, have of course varied enormously from the early nineteenth-century movements for national independence in Latin America to the decolonization struggles of the 1940s and 1950s in Asia and Africa. Despite these cardinal variations, it is always necessary to remember the impact of the coloniality of power in our treatments of time, space, and politics, something which is not always done as I shall mention at the end of the chapter.

What can be constructively emphasized here, in contrast to Bauman's association of power with flows and the "exterritorial," is the intrinsic connection of power and the territorial, and the link with what Castoriadis calls the flow or "magma" of radical imaginaries that have the potential to create alternative social orders. These alternative flows may challenge the geopolitical location of a given society in the global order – as the Zapatista movement has done in the Mexican case – and at the same moment they may encapsulate another kind of power, the power to resist and to generate alternative imaginaries of social change, based, *inter alia*, on the prioritization of radical democracy, social justice, national dignity, and indigenous rights. The presence of an oppositional geopolitical imagination that fuses a range of spatial arenas – i.e. the global, the national, the regional, and the local – can be seen in the Mexican case, where, for instance, Subcomandante Marcos in an interview in 1995 drew out the following three interrelated points (see *La Jornada*, Mexico City, August 27, 1995, pp. 10–11).

First, current processes of globalization have the potential to break nation-states and to accentuate internal regional differentiations, as reflected in the divergence between the northern, central, and southeastern zones of Mexico. Second, in relation to the question of war, Marcos indicates that political confrontation and the battle for ideas has acquired more significance than direct military power, so that as expressed more recently, "the seizure of power does not justify a revolutionary organization in taking any action that it pleases, . . ." since, "we do not believe that the end justifies the means . . . we believe that the means are the end" (Subcomandante Marcos, 2001, p. 76). And third, pivotal importance is given to the role of the means of communication, for if, it is argued, a movement can be made to appear dead or moribund, irrespective of the reality on the ground, this constitutes a greater threat than superior military strength. It is in this situation that the use of e-mail and the Internet have become significant as an alternative means of disseminating oppositional narrative and analysis, giving the Zapatistas, according to certain readings, a postmodern allure (Burbach, 2001). At the same time, and crucially, it is important to take into account that the Zapatista movement is deeply rooted in a long regional history of social struggle and opposition which provide it with a deep political sustenance (Harvey, 1998).

This sense of rootedness, which is allied to an alternative and indigenous affirmation of rights and dignity, the validation of a different grounded power, has been concisely expressed in the Ecuadorian context by a leader of that country's Indian rights movement: "a connecting central theme has been the struggle for the identity of all the nationalities, and within this framework the struggle for territorial rights, and the struggle for cultural recognition, such as language, bilingual education, etc." (quoted in Brysk, 2000, p. 34). What the Ecuadorian movement as well as other indigenous movements in Latin America and elsewhere have done is to put into question an established association between national state power and the territorial, suggesting the possibility of multiethnic identities and plural territorialities (Warren, 1998; Wilmer, 1993). Moreover, not only are power, politics and the territorial connected in struggles to redefine the meanings and practices of their grounding within given national spaces, but also, the flows of opposition and resistance across frontiers, expressed in the networks of transnational activists in the domains of environmental politics, human rights and womens' movements (Keck and Sikkink, 1998), bear testimony to another kind of association between power, flows, and the geopolitical.

Many of these examples, and in particular, in the post-Seattle conjuncture (see, for example, Cockburn et al., 2000) can be seen as part of the current wave of "anti-globalization" protests, which, as Naomi Klein (2001) usefully points out, are more appropriately described as protests and mobilizations against the deepening and broadening of corporate power, or the privatization of every aspect of life, and the commodification of every activity and value. There is also encapsulated in these new struggles a kind of "globalization of hope," which is like a connecting thread of energy that ignites the various sites of opposition and protest. Many of these protests receive a sense of unity through being targeted at the global operations of transnational corporations. For example, because of the global presence of Monsanto, farmers in India are working with environmentalists and consumers around the world to construct direct-action strategies that cut off genetically modified foods in their fields and in the supermarkets, and due to the global impact of Shell Oil and Chevron, human rights activists in Nigeria, democrats in Europe, and environmentalists in North America have united in a fight against the unsustainability of the oil industry (see Klein, 2001).

All of these examples can be taken as a reflection of what Castoriadis refers to as the capacity and plasticity of individuals and groups to thwart the impact of the hegemonic processes of socialization and ordering. This points to a duality of power – the power over and the power to resist – but equally set against the notion of hegemony suggested by Castoriadis, we can posit the possibility of a counter-hegemony; of the struggle for a different set of values and objectives that offer an alternative to neo-liberalism, rooted for example in the principles of radical democracy and collective struggle.

Decolonizing the Geopolitical Imagination

Although authors such as Castoriadis (1991, p. 200) were critical of Western projects of development, pointing, for example, to the importance of Western violence in global politics, Western culture was always envisaged as being in a privileged historical and geopolitical position. One reads in Castoriadis, for example, that Western culture "is the only one to have taken an interest in the existence of other cultures, to have interrogated itself about them and, finally, to have put itself in question..."

(ibid.).[8] This is not a view unique to Castoriadis, and has in fact been echoed in the work of the American philosopher Rorty (1991, 1999), a writer sometimes associated with postmodern thinking.[9] In other words, in these texts, the idea of a thinking, analytical, self-reflexive subject is rooted in Western history, and much of postmodern thought as it has developed in the West has not broken from this predilection, despite the fact that much that is new about the postmodern has had its epistemological origins outside the heartlands of the capitalist West (Anderson, 1998), just as in earlier times, classical Greece was as much Arab–Muslim as it was Latin–Christian (Amin, 1989).

Connecting to the questioning, subversive element of the postmodern, that element which also prioritizes difference and undermines totalizing narratives of modernization and social change, we might want to interrogate the apparently natural attachment of thought and the universal to a pre-given Western subject. As the Latin American philosopher Enrique Dussel (1998) reminds us, a great part of the achievements of modernity were not exclusively European but emerged from a continuing impact and counter-impact between Europe and its peripheries. Moreover, the *ego conquiro* (I conquer), as a practical self, predates the *ego cogito* and the *Discours de la Méthode* of Descartes by more than a century. But the intersections of modernity and Empire are not infrequently by-passed just as the existence of other traditions of thought and reflection are customarily subsumed under a posited Euro-American universality. A significant task for an insurgent postmodernism is precisely to question the origins and continuity of Euro-American notions of the universal subject as rooted in Western history.

For a contemporary geopolitical analysis that is sensitive to the critical potential of postmodern thought, and that seeks to develop a global sense of the changing spatialities and temporalities of power, it is not sufficient to reconfigure the thematic terrain, giving a crucial place to the geographical histories of West/non-West encounters. It is also necessary, as I have argued elsewhere (Slater, 1999), to foreground the question of the agents of knowledge: to what extent, for example, do our own particular locations of thought, and specific inventories of theoretical enquiry in Western academia, give rise to a continuing prioritization of issues and agendas which may be more Occidental than global, and which may well subsume or marginalize issues that are intrinsic to the changing dynamic of North–South relations? In this context, to begin to explore the limits of our enquiry requires honesty, cooperation and a frank recognition of their importance, as reflected, for example, in David Harvey's (2000, p. 94) recent remark that his own work, despite all his geographical interests, has tended to remain Eurocentric.

Critical Themes for New Times

In a world frequently portrayed in terms of flows, speed, instant communication, and the politics of de-territorialization, it may not be out of place to keep in mind the recurring stories of poverty, inequality, and exclusion – the "shock of the old." For example, global inequalities in income in the twentieth century have increased by more than anything previously experienced. The distance between the incomes of the richest and poorest country was about 3 to 1 in 1820, 35 to 1 in 1950, 44 to 1 in 1973, and 72 to 1 in 1992. In addition, a recent study of world income distribution among households shows a marked rise in inequality with the Gini coefficient increasing from 0.63 in 1988 to 0.66 in 1993, whilst the average annual growth of income per capita in 1990–8 was negative in 50 countries, only one of them being an OECD country. Worldwide, 1.2 billion people are income poor, living on less than US$1 a day, and in Third-World countries more than a billion people lack access to safe water, and more than 2.4 billion people lack adequate sanitation. Also worldwide, about 1.2 million women and girls under 18 are trafficked for prostitution each year, about 100 million children are estimated to be living or working on the street. In 1998 there were an estimated 10 million refugees and 5 million internally displaced persons.[10]

Inequalities are also markedly present in the world of cyberspace, where access to the Internet displays a familiar geographical distribution. Whilst low-income economies had 25.7 telephone main lines per 1,000 people in 1995, compared to a figure of 546.1 for the high-income economies, the gap increased when moving to personal computers per 1,000 people – 1.6 for the low-income economies compared to 199.3 for the high-income economies – and finally for internet users per 1,000 people in 1996 the contrast was between figures of 0.01 and 111.0 (World Bank, 1999, p. 63). As the World Bank (op. cit., p. 9) comments, throughout much of the South basic communications technology is available only to the fortunate few, so that whilst South Asia and Sub-Saharan Africa have only about 1.5 telephone lines for every 100 people, the US has, in comparison, 64 lines per 100. However, in other parts of the Third World (for instance in Latin America and Asia and the Pacific) there has been important growth in the products of the world information technology market (e.g. personal computers, data communications equipment, packaged software) and, furthermore, in other areas there have been improvements too, so that whereas in 1900 no country had universal adult suffrage by the end of the 1990s the majority of the world's countries do, or, for instance in the last three decades, life expectancy in the developing countries increased from 55 years in 1970 to 65 in 1998 (UNDP, 2000, p. 4).

Overall, these data on contemporary North–South differentiations can be used to underscore the importance of a certain continuity in an era of rapid global change. Equally, they can be employed to confirm the phenomenon of unevenness, perhaps echoing that notion from an earlier epoch, of the "law of combined and uneven development," intrinsic to international capitalism. A critical postmodernism that is rooted in discourse analysis can be open to the usage of such supposedly obsolescent terms since what is crucial is how given ideas and concepts are incorporated into a framework of interpretation. A postmodern reading, where agency, subjectivity and the determination of human thought are prioritized, would treat Marxist thought as a system with differences within it, and with concepts and ideas that can still be deployed in ways that address the contemporary scene. For example, rather than taking "class" as an inevitable and pre-given point of departure for our diagnosis of collective struggles, we might want to analyse the complex processes of collective action and the formation of collective wills, where "class struggle" can in certain circumstances be one point of arrival – one form of conceptualization of a much more heterogeneous array of struggles. This would lead us into the significant difference between Marx and Gramsci on hegemony and collective wills, and help us move away from the deployment of marxist constructs as if they were fixed and final abstractions. In other words, contra Baudrillard and Lyotard, we would not define Marxist ideas as being immanently obsolete, as part of a machine for totalizing truth, but rather we would introduce the key elements of openness and difference within a system of thought so that there are always possibilities for enabling forms of radical reconstruction within new problematics (Laclau, 2000).[11]

A similar observation can be made in relation to the previous discussion of power and geopolitics, since it seems more useful to point to the multiple ways in which space, power, time, and politics intertwine, rather than make linear separations in our contextualizations. Rather than posit that power has now become "exterritorial," we may suggest that the "power over" and the "power to" have different spatial modalities, and that some new forms of transnational power may well be more de-territorialized than previously was the case, but that other forms of territorial power remain firmly grounded within nation-states and require us to take cognizance of what Castoriadis referred to as an instituting ground-power that gives societies their primary foundation. Rather than argue that what is required is a "transnational politics" to challenge contemporary neo-liberalism, we may want to suggest that struggles for new forms of democratic politics can reach out in all kinds of unpredictable ways, including the transnational, with calls, for example, for the globalization of democracy in institutions such as the World Bank, the WTO, and the IMF. But those local and regional struggles for more democratic forms of governance within nation-states are just as significant, because it is the blending together of micro-politics and macro-politics, of the interlinking of "scales of protest and contestation" that provides the potential for new forms of emancipation. Rather than uncritically championing "localization" as an opposite to an apparently irredemiably oppressive globalization, we might want to critically examine the meanings of the local in political practice, avoiding any romanticization of the local or the regional. What really counts is how the local or the regional are given political meaning – by which social forces and for what purposes.

A contemporary Third-World example here would be the struggles over water provision in Bolivia, where local and regional protests over the privatization of water supply in the Cochabamba region and the connection with a transnational consortium (*Aguas de Tunari*) have led to the formation of a new popular movement that is organized in the form of a Coordinating Committee for the Defense of Water and Life, and which challenges the policies of the local, regional, and national levels of government and their dependency on foreign capital. What has emerged from this struggle has been the articulation of an opposition to the subordination of regional and national needs and resources to transnational capital (a key slogan for example has been "The Water is Ours – damn it"), and at the same time the forging of new connections for making a more democratic society (see Farthing and Kohl, 2001, for a brief discussion). The "water wars" in Bolivia reflect the existence in other parts of the world of new forms of opposition to the politics of privatization, and the perspective developed by the leaders of the Coordinating Committee in Cochabamba, and in particular by Oscar Olivera, with his emphasis on the need to develop a movement without *caudillos* (political bosses), parallels key orientations of the Zapatista movement in Mexico.

In the overall treatment of my three interconnected geopolitical themes – the position of chronopolitics, the re-imagining of power and the spatial, and the call for a decolonization of geopolitical thinking – I have traced out some of the implications for our analysis of postmodern interpretations. The chapter has explored a number of theoretical intersections that are relevant to the way we examine space and politics in global times. I have pointed to the differences within the postmodern and the need to be aware of the subtleties of perspective contained under its sign. Finally, I have signalled the importance of taking into account the role of agents of knowledge in the critique of Western universality and along this pathway the connection to postcolonial imaginations becomes paramount. But that is a discussion for another time.

ENDNOTES

1. The term "sole thought" or "*pensamiento único*" originates in the work of the Spanish social and political theorist Ramonet (1997); for a connected analysis see Estefanía's (1997) critique of neo-liberal thought. The idea here is of a one-track vision – the "one and only perspective." Ramonet defined "sole thought," which is closely related to the Thatcherite notion of "there is no alternative", as the translation into ideological terms, with universalist pretensions, of the interests of transnational capital.
2. Elsewhere I have discussed in some detail the various Latin American approaches to the postmodern (see Slater, 1994).
3. Returning to the theme of differences within the postmodern, Foster (1996) reminds us that postmodernism has always been a disputed notion, but that it is possible to support a type of postmodern thinking which contests reactionary cultural politics and advocates "artistic practices not only critical of institutional modernism but suggestive of alternative forms – of new ways to practice culture and politics" (op. cit., p. 206).
4. For a similar critique of Lyotard's politics, see Drolet (1994) and Slater (1994).
5. In the world of critical anthropological enquiry, Fabian (1983) made a similar argument in his work on time and alterity, although in his case the focus fell on the Eurocentric tendency to deny coevalness to non-Western societies which were always situated in another, separate and less advanced time which for Fabian seemed to be more crucial than differences across space – "chronopolitics" coming to assume more significance than "geopolitics."
6. For example, ten French trade union members of the Agrarian Confederation destroyed a McDonalds restaurant in Millau as part of a protest against fast-food penetration and Americanization in France (see *El País*, July 1, 2000, p. 2, Madrid). This protest, led by José Bové, was also targeted against the import of hormone-treated beef from the US, so that the aim was to mark an opposition to fast as well as unhealthy food.
7. Elsewhere, Bauman (1997, p. 65) makes a link between globalization and territorialization, arguing that territorially weak states are beneficial to the power of globalizing capital in the sense that economic globalization and what he calls political "tribalization" are close allies, conspiring against the chances of justice being done and being seen to be done. Weak states operate as the local agents of globalized capital that flows beyond and across territories (see also Bauman, 1998).
8. Elsewhere, this standpoint is expressed more forcefully with the assertion that it is only Occidental civilization that has developed the capacity to be critical of its own history: all the other civilizations of the world are, according to Castoriadis, deficient in this key element of comparative assessment (Castoriadis, 1998, pp. 94–5).
9. In his recent text, Rorty (1999, p. 273ff) argues against any notion of the superior rationality of the West but goes on to note that it is in the West that there has been a

higher degree of comity and happiness than elsewhere, a kind of "experimental success" story, juxtaposed to a view that there are "lots of cultures" we would be much better off without (p. 276). In his earlier texts he was of the view that the West may have invented the Gatling gun and imposed colonial rule but it was also only in the West that the capacity for self-critique and intellectual reflection evolved so that the world could be alerted to the reality of West/non-West encounters. For a critique, see Slater (1994.)

10. For all these figures see UNDP (2000, pp. 4–6).
11. For an interesting treatment of Marx on democracy where a postmodern sensibility is identified and analysed, see Carver (1998).

BIBLIOGRAPHY

Amin, S. 1989. *Eurocentrism*. London: Zed.

Anderson, P. 1998. *The Origins of Postmodernity*. London: Verso.

Baudrillard, J. 1994. *The Illusion of the End*. Cambridge: Polity.

Baudrillard, J. 1998. *Paroxysm*. London: Verso.

Bauman, Z. 1997. *Postmodernity and its Discontents*. Cambridge: Polity.

Bauman, Z. 1998. *Globalization*. Cambridge: Polity.

Bauman, Z. 1999. *In Search of Politics*. Cambridge: Polity.

Benko, G. and Strohmayer, U. (eds.). 1997. *Space and Social Theory*. Oxford: Blackwell.

Bhabha, H. 1995. In a Spirit of Calm Violence. In G. Prakash (ed.) *After Colonialism*. Princeton, NJ: Princeton University Press, 326–43.

Bourdieu, P. 1998. *On Television and Journalism*. London: Pluto.

Brysk, A. 2000. *From Tribal Village to Global Village*. Stanford, CA: Stanford University Press.

Burbach, R. 2001. *Globalization and Postmodern Politics*. London: Pluto.

Caldeira, T. P. R. 1996. Fortified Enclaves: the new urban segregation. *Public Culture*, 8, 303–28.

Carver T. 1998. *The Postmodern Marx*. Manchester: Manchester University Press.

Castoriadis, C. 1991. *Philosophy, Politics, Autonomy*. New York: Oxford University Press.

Castoriadis, C. 1998. *El Ascenso de la Insignificancia*. Madrid: Ediciones Cátedra, S.A.

Cockburn, A., St.Clair, J., and Sekula, A. 2000. *5 Days That Shook the World*. London: Verso.

Der Derian, J. 1990. The (S)pace of International Relations: simulation, surveillance and speed. *International Studies Quarterly*, 34, 295–310.

Drolet, M. 1994. The wild and the sublime: Lyotard's post-modern politics. *Political Studies*, XLII, 259–73.

Dussel, E. 1998. *The Underside of Modernity*. New York: Humanity.

Escobar, A. 2001. Culture sits in places: reflections on globalism and subaltern strategies of localization. *Political Geography*, 20(2, February), 139–74.

Estefanía, J. 1997. *Contra el Pensamiento Único*. Madrid: Santillana S.A. and Taurus.

Fabian J. 1983. *Time and the Other*. New York: Columbia University Press.

Farthing, L. and Kohl, B. 2001. Bolivia's New Wave of Protest. *NACLA Report on the Americas*, XXXIV (5, March/April), 8–11.

Foster, H. 1996. *The Return of the Real*. Cambridge, MA: MIT Press.

Giddens, A. 1999. *Runaway World*. London: Profile.

González Casanova, P. 1995. *O Colonialismo Global e a Democracia*. Rio de Janeiro: Civilização Brasileira.

Guattari, F. 2000. *The Three Ecologies*. London: Athlone (first published in France 1989).

Harvey, D. 2000. Reinventing geography. *New Left Review*, 4 (July/August), 75–97.

Harvey, N. 1998. *The Chiapas Rebellion*. Durham, NC: Duke University Press.
Hines, C. 2000. *Localization – A Global Manifesto*. London: Earthscan.
Jones III, J. P., Natter, W., and Schatzki, T. R. (eds.). 1993. *Postmodern Contentions*. New York: The Guilford Press.
Keck, M. E. and Sikkink, K. 1998. *Activists Beyond Borders*. Ithaca, NY: Cornell University Press.
Klein, N. 2001. Reclaiming the Commons. *New Left Review*, 9 (May/June), 81–9.
Laclau, E. 1994. Introduction. In E. Laclau (ed.) *The Making of Political Identities*. London: Verso, 1–8.
Laclau, E. 2000. Structure, History and the Political and Constructing Universality. In J. Butler et al. (eds.) *Contingency, Hegemony, Universality*. London: Verso, 182–212 and 281–307.
Lipovetsky, G. 1994. *El Crepúsculo del Deber*. Barcelona: Editorial Anagrama.
Luke, T. W. 1996. Governmentality and contragovernmentality: rethinking sovereignty and territory after the Cold War. *Political Geography*, 15 (6/7, July/September), 491–507.
Luke, T. W. 1999. Environmentality as Green Governmentality. In E. Darier (ed.) *Discourses of the Environment*. Oxford: Blackwell, 121–51.
Lyotard, J-F. 1997. *Postmodern Fables*. Minneapolis, MN: University of Minnesota Press.
Moisy, C. 1996. *The Foreign News Flow in the Information Age*. The Joan Shorenstein Center, Harvard University, Discussion Paper D-23, November.
Ó Tuathail, G. 1997. At the end of geopolitics? Reflections on a plural problematic at the century's end. *Alternatives*, 22 (1, January–March), 35–55.
Ramonet, I. 1997. *Un Mundo sin Rumbo*. Madrid: Temas de Debate.
Ramonet, I. 2000. *La Golosina Visual*. Madrid: Temas de Debate.
Rancière, J. 1995. *On the Shores of Politics*. London: Verso.
Rattansi, A. 1994. "Western" Racisms, Ethnicities and Identities in a "Postmodern" Frame. In A. Rattansi and S. Westwood (eds.) *Racism, Modernity and Identity*. Cambridge: Polity, 15–86.
Reigadas, M. 1988. Neomodernidad y Postmodernidad: preguntado desde América Latina. In E. Marí (ed.) *Posmodernidad?* Buenos Aires: Editorial Biblos, 113–45.
Rorty, R. 1991. *Objectivity, Relativism and Truth*. Philosophical Papers, Vol. 1. Cambridge: Cambridge University Press.
Rorty, R. 1999. *Philosophy and Social Hope*. Harmondsworth: Penguin.
Sachs, W. 1999. Rich in things, poor in time. *Third World Resurgence*, 196 (September/October), 14–16.
Santos de Sousa, B. 1999. On Oppositional Postmodernism. In R. Munck and D. O'Hearn (eds.) *Critical Development Theory*. London: Zed, 29–43.
Slater, D. 1994. Exploring Other Zones of the Post-Modern: problems of ethnocentrism and difference across the North-South divide. In A. Rattansi and S. Westwood (eds.) *Racism, Modernity and Identity*. Cambridge: Polity, 87–125.
Slater, D. 1999. Situating Geopolitical Representations: inside/outside and the power of imperial interventions. In D. Massey et al. (eds.) *Human Geography Today*. Cambridge: Polity, 62–84.
Sloterdijk, P. 1988. *Critique of Cynical Reason*. London: Verso.
Soja, E. W. 1989. *Postmodern Geographies*. London: Verso.
Soper, K. 1991. Postmodernism and its discontents. *Feminist Review*, 39, 97–108.
Subcomandante Marcos. 2000. El Fascismo Liberal. *Le Monde Diplomatique*, Edición Española, Año V No 58–59, Setiembre, 25–8.
Subcomandante Marcos. 2001. Punch card and hourglass. *New Left Review*, 9 (May/June), 69–79.
UNDP. 2000. *Human Development Report 2000*. Oxford: Oxford University Press.

Virilio, P. 1986. *Speed and Politics*. New York: Semiotext.
Virilio, P. 1997. *Open Sky*. London: Verso.
Warren, K. B. 1998. Indigenous Movements as a Challenge to the Unified Social Movement Paradigm for Guatemala. In S. Alvarez et al. (eds.) *Cultures of Politics, Politics of Cultures*. Boulder, CO: Westview Press, 165–95.
Wilmer, F. 1993. *The Indigenous Voice in World Politics*. Newbury Park: Sage.
World Bank. 1999. *World Development Report 1998/99*. New York: Oxford University Press.
Zizek, S. 1999. *The Ticklish Subject*. London: Verso.

Part II Essentially Contested Concepts

Chapter 7

Power

John Allen

Introduction

Geography, we are often told, is about power and political geography is about the use of power to administer, control, and fix territorial space. In many respects, this could be the end of the story and for some it probably remains so. For many nowadays, though, the recognition that fixed territories and bounded states no longer possess the last word on power and authority has altered their understanding of the political landscape. Even though it is hard to let go of the fact that power is something which is distributed intact to authoritative locations from an identifiable center, this view now happily coexists with the idea that power may be diffused, decentered, and networked. Equally, whilst it is hard to get beyond the notion that power is always exercised at someone else's expense, this understanding of power runs alongside the view that power is merely a means for getting things done, a general facility for realizing outcomes.

In this chapter, I want to survey these different meanings of power and to move between them in ways that reflect how they have been understood and taken up within political geography and its associated areas in recent times. I should say at the outset that what follows is not an exhaustive attempt to map the field of geography and power, but rather one that foregrounds, first, the *slippage* between instrumental and facilitative understandings of power, and, second, the awkward *tension* between centered and diffuse notions of power that characterize a growing number of political geography tracts.

As such, the essay falls into two parts. In the first I draw upon the thinking of those such as Max Weber, Hannah Arendt, Talcott Parsons, and Michael Mann to illustrate the contrast between accounts which stress power instrumentally, as something which is held *over* others, and those which emphasize its collaborative side, the power *to* secure outcomes. Following that, I explore some of the more recent influences within political geography, from Michel Foucault to Gilles Deleuze, and the manner in which their thought has been used to loosen a hitherto somewhat rigid account of territoriality, borders, and political power. Throughout, I try to show the implications for political geography of the different readings of power and the twists and turns that they have been subject to as different authors adapt them to their own concerns and interests.

In the conclusion, I suggest that an instrumental conception of power, one focused upon the ability to bend the will of others, remains a political geography benchmark, as does the entrenched view that politics has more to say about how power works over and across space than geography ever could.

Power as Constraint/Power as Enabling

Exercising "power over" others rather than exercising the "power to" act is perhaps another way of highlighting the difference between instrumental and facilitative notions of power. The difference between the two senses of power is critical to an understanding of what power is deemed to be for and how it works. The idea that power is something that is held over others and used to obtain "leverage" is a rather different conception of power from one that thinks of it as a medium for getting things done. Whereas the former considers power to be an instrument of domination and constraint, the latter stresses its potential for empowerment. Where one starts from the position that power is all about shaping the will of others, the other thinks of it as a means of enablement (Allen, 1999; see also Agnew, 1999). In practice, the two senses of power are often blurred, as for example when the process of collective mobilization leads to the furtherance of one group's set of interests at the expense of another. Yet both remain influential accounts within political geography at large.

Exercising power over others

In suggesting that an instrumental conception of power remains influential within political geography, I do not wish to imply that this is always the result of a conscious, deliberative choice. More often than not, the adoption of an instrumental framework of power is one that, forgive the pun, goes with the territory. There is a familiar, everyday sense in which politics and power are wrapped up with, on the one hand, conflict, opposition and disorder and, on the other, authority, control, and compliance. If the former reveals much about the context of power, the latter is suggestive about where power lies and how it is exercised.

It is, after all, commonplace to ask the question, "who holds power?" in political circles or at least a version of it, and then seek to locate it, almost as a reflex, in people and institutions. Few of us, for example, would have any misgivings if we were told that various local political elites or a certain charismatic national political figure "had" power and used it to shape and influence events. We may quibble about the extent of their power or ask questions about who, behind the scenes, really has the power to set the agenda, but the scenario itself remains a plausible one. That we can accept this is, in part, to do with the fact that, all things being equal, power *must* lie somewhere. It would, after all, almost beggar disbelief to think that power has no reference point at all.

Once we fall into this position, a whole string of assumptions come into play that have their roots in the line of thought that runs from the political philosopher, Thomas Hobbes, to the mainstream sociologist, Max Weber. They are, first and foremost, that power is a *possession*; it can be held, delegated or distributed. Moreover, it is held as a *capacity*, insofar as it may be latent or potential in its effects. From the long-standing coercive powers of the legal profession, to the economic powers of the big finance houses, or to the disciplinary powers of the many and varied state agencies, the capacity for domination is generally conceived as present and capable of being exercised should the need arise or circumstances dictate. Such bodies are said to "hold" power, regardless of whether or not they actually use it. In that sense, it is the possession of power that reveals its location, as something separate from the exercise of that capability.

From this, it is but a short step to talk about powers "held in reserve" or as a resource capability "located" in the apparatus of the state or an economic body. In this view, power is something that is *delegated* or *distributed* from a *centralized* point to authoritative locations across any given territory. At its simplest, power is transmitted intact in a relatively straightforward manner down through an organizational hierarchy under clear lines of authority. In which case, either the rules, regulations, and constraints imposed by the center are successful in meeting their goals or their organizational impact is minimized and deflected by the degree of resistance met *en route*. Either way, the force of a unitary centralized power remains intact and its "store" of capabilities present and awaiting distribution.

Although somewhat sparse as a diagram of power, it is nonetheless one that underpins the "state-centred" versions of power criticized by John Agnew (1994, 1999; see also Agnew and Corbridge, 1995) in his attempt to map political power beyond state boundaries. His target was the conventional understandings of the geography of political power held by mainstream international relations theorists and relatedly those of a political realist persuasion who considered the state to be a unitary and singular actor. At the nub of his concerns was the simple equation of state territoriality with a stable, bounded set of power relations that "contains" all that really matters politically. Distinguishing (following Walker, 1993) between an internal, domestic space in which governments exercise power in an orderly fashion through the distribution of their powers, and an external, international domain defined by the absence of order, the "territorial state" in mainstream international relations is represented as a uniform political community maintained and controlled from an identifiable center. As such, the spatial organization of rule-making authority is portrayed as an almost effortless process whereby power radiates out from the center to select elites and bureaucratic agencies.

In this view, power is almost akin to a solid "bloc," whose certainties are then spread from the center outwards, and neither distance nor dispersal problematizes its reach. Barry Barnes (1988), in particular, has questioned whether such traditional forms of delegation really do work in that manner. In the first place, he argues that an element of discretion is built into the very exercise of power: once authority is devolved, delegates are empowered and able to make independent use of their new-found capabilities. As such, the dispersal of (positions of) authority in this manner opens up the possibility for ambiguity and displacement to take hold in the so-called "chain of command" and renders any rule or judgement provisional.

Moreover, he goes on to argue that with the dispersal of power there is also more opportunity at the many points of interaction with other bodies for agents to mobilize other resources, other sets of interests, and to shift the line of discretionary judgement in unanticipated ways, or even to break with it. While there is no necessary reason why the line of authority should be broken, the simple fact that there is a larger number of interests and views to negotiate should alert us to the need to be cautious about claims which portray power at-a-distance as an unproblematic affair.

Of late, there has been greater recognition among political analysts that the exercise of power in a globally re-ordered world is less than straightforward. Increasingly, there is talk of "multi-tier" or "multi-level" governance, where power is no longer seen to operate in either a top-down or center-out fashion, but rather upwards and downwards through the different scales of political activity, both transnational and subnational (see especially, Rosenau, 1997; Newman; 1999). In such accounts, there is a greater recognition of the larger number of interests involved in any instrumental power formation, with multiple sites of authority dotting the political landscape, from numerous quangos and private agencies to local administrative units and other subnational actors. In place of the conventional assumption that the state is the only actor of any real significance, the playing field is now shared with nongovernmental organizations, multinational enterprises, and other supranational, as well as interstate, organizations. In this more complex political geography, power is largely about the re-organization of scale, insofar as it is redistributed to take account of overlapping sites of authority and multidimensional boundaries.

And yet, for all the recognition that the workings of power involve more than a simple vertical or horizontal reallocation over space, the vocabulary of power employed is still largely one of capacities "held" and the "movement" of power between the different levels, albeit in a complex and contradictory manner. Whilst power is not so much mediated as relativized, it is hard to avoid the impression that even in this decentered, territorially re-ordered world, power is more or less distributed intact within each "contained" level. The idea that power may be something other than a capacity, or that power may not be as much distributed over space as constituted by the many networked relationships which compose it, has yet to be fully absorbed by this literature.

For that, we need to draw on a different understanding of power, one that starts from the position that power is a means of enablement, not a tool to achieve order or constraint.

Exercising power with others

In contrast to an instrumental view of power, the idea that power is simply a means to an end has less of a foothold in our everyday thinking. It is, for instance, not a straightforward reflex action to think of power as an effect *produced* through the actions of people or institutions pooling their resources to secure certain outcomes. In this view, it makes little sense to talk of power as "contained" within territories or "stored" as a capability ready for use. If power is an effect which is generated through the actions of groups or organizations, then it is not something that may be "held in reserve": it can only be mobilized on what often appears to be a rather tenuous basis (see Allen, 1997).

Far from appearing solid in form, therefore, as part of the organizational apparatus, so to speak, power on this account is understood as a rather fluid *medium* which can expand in line with the resources available to collective ventures, or it can diminish once collective, short-term goals have been achieved. Alternatively, if we extend this insight, it may disappear overnight should alliances or tenuous collaborations fall apart. Such an understanding of power is present, in different guises, in the writings of various political philosophers and social theorists, and has proven to be immensely influential.

In one particular guise, the roots of such a conception are evident in Niccoló Machiavelli's stress upon the contingent strategic ability of an able Florentine statesperson to exercise power effectively, as it is in Hanna Arendt's (and, following her, Seyla Benhabib's, 1992) concern that power be treated as something which is rooted in mutual action, designed to further common purposes, and as empowering in and of its own right. Benhabib's stress upon the collaborative, enabling dimension to power is viewed as a positive gesture in which all those taking part benefit in some way (that is, a positive rather than a zero-sum scenario), but only for so long as the effective mobilization lasts. When power is not sustained by mutual action, quite simply, it passes away. It evaporates. This associational view of power has much to offer in understanding the actions of nongovernmental organizations and the constitution of social movements, as we shall see shortly.

In another guise, an understanding of power as a fluid medium, as something intrinsic to all forms of social interaction, is present in the social theorizing of those such as Talcott Parsons, Anthony Giddens, and Michael Mann. In line with Parsons, Giddens (1977, 1984) wishes to preserve a sense of power as a general facility for enabling things to get done, where power itself is not conceived as a resource but as something generated by the employment and application of resources over tracts of time and space. In contrast to Parson's rather benign view of power, where the satisfaction of all parties concerned is met, Giddens pointedly argues that power should refer to any range of social interventions, including those that lead to domination and constraint. As such, the "power to" do things may be directed as much towards the collective-minded bending the will of their less collective brethren, as it is directed towards mutually beneficial ends.

In this assessment, power may still be described as an exercise in facilitation, but one that is about the constraint of social action as much as it is the enablement of it. For Giddens (1985, 1990), in a world of disembedded relations and institutions, the mobilization and retrieval of resources over space, especially those of information, represents a modern, facilitative means of securing and controlling distant outcomes. Action at-a-distance, as for example in the case of state surveillance, is said to enable centralized governments to administer and control the detail of people's daily lives through the routine storage and monitoring of information on anything from health, education and housing to political, criminal and other illicit activities (see Painter, 1995). The effect, it is argued, is to "stretch" government across its sovereign territory as a form of distanciated power.

As with the somewhat sparse "state-centered" diagram of power, however, there appears to be little else happening in this action at-a-distance scenario besides that of an extension of centralized authority and resistance to it from the communities at large. For a more considered account of how power is organized and transmitted across space, Mann's (1986, 1993) work is more illuminating.

The distinctive twist to his account comes through the recognition that the mobilization and control of resources actually takes place through various *networks* of extensive and intensive social interaction. In brief, the expansion of power and its consolidation are said to take their shape from a series of networks organized over space which cut and overlap one another, the most important of which stabilize around four types of (re)sources – broadly defined as economic, ideological, political, and military. At their simplest, such networks are formed through patterns of association and interaction that bind people together in the pursuit of certain ends. An array of institutions and practices, from the broad, sweeping alliances of geopolitical institutions and their equally international economic counterparts, transnational firms, at one end of the spectrum, to the more regional associations of culture, religion or political practice, at the other, connect people and places across the networks. Differences in the make-up and dynamism between the networks ensure that they reach out across space in different ways and to varying extents, in some cases transcending established social boundaries, in other cases heightening or consolidating them.

Among the most powerful networks, those centered on the most effective (re)-sources, a more stable shape and form of organization is assumed to take hold. In laying one network over another, however, as in some kind of "entwinement," the dynamism of each is said to fuse and modify the other's pattern of interaction so as to bring forth all manner of unintended consequences. The result, for Mann, is a view of history and power as a complex "mess" and a geography that eschews any simple notion of societies as territorially bounded. In the absence of any monolithic power associated with a particular resource, Mann notes that the most effective institutions are those which blend the different forms of organizational reach. Indeed, it is his recognition that institutions may enhance their spatial reach through extensive and intensive networks, combining both authoritative and diffused techniques of organization to achieve far flung goals, that adds to an understanding of power as a medium.

Within political geography at large it is probably fair to say that his work has been picked up less for its geographical insights and more for its political distinctions (see, for example, Muir, 1997; Painter, 1995; and also Ó Tuathail, 1996.) The different sources of power and the distinction he drew in earlier work between despotic and infrastructural power (Mann, 1984, which foreshadows his later efforts on organizational reach) have led others to address the extent to which state bureaucracies can "penetrate" into the furthest recesses of their territories. Indeed, the idea of networked power itself seems to have attracted less attention and by virtue of that the precise role that spatiality plays in constituting power. For whereas Arendt foregrounds mutuality as the constitutive force of associational networks, it is easier to read Mann's account of networked power as little more than a series of conduits through which organizational resources flow. Clearly, he recognizes that mobilized powers are not always transmitted outwards from an identifiable center, but nonetheless there is every impression that power generated in one part of the network, at different sites and locations, is transmitted intact across it.

If there are fewer implications for political geography of Mann's grasp of power than anticipated, however, the same is not true of Arendt's (1970, 1975) associational understanding. Among the more obvious examples of late is the proliferation of nongovernmental organizations, in particular, those focused upon development, health, environment, and human rights issues. What is illuminating about such

NGOs is that the collective act of mobilization is itself regarded as a resource. The activation of moral and political energies, to draw upon Arendt's lexicon, and the ways in which they are channeled to influence and appeal to wider audiences, illustrates nicely the powers of association. Moreover, given that such networks of alliance are organized across national borders, often orchestrated through the Internet, such energies may be directly supportive and enabling for those seeking to bring about political change in their own countries. The relative success of the Jubilee 2000 debt relief campaign, for example, and the extent of the empowerment (rather, that is, than any simple act of resistance) provides a useful illustration of a loose coalition of interests acting "in concert" to achieve a specific – of the moment – outcome.

Post-power

Thus far, I have tried to show that even though power may be understood in both its instrumental and facilitative modes, there is frequent slippage between the two different meanings involved as collaboration spills over into opposition and enablement becomes the language of gaining influence at the expense of others. In a changing world of nation-states, local elites, NGOs, multinational firms, and transnational political organizations, it takes little to blur the sense in which power may be thought of as a range of inscribed capacities at the center of a diverse and wide ranging series of networks (see Johnston, 2000). If little thought is given to how power works across space (as opposed to its unalloyed flow or simple extension over space), then the room for such slippage is evident.

More recent thinking on power, largely from a poststructuralist perspective, has tended to remain agnostic to such theoretical crossovers, exhibiting a certain ease over the fluidity of meaning that is reflected in their willingness to blur geographical boundaries, scales and territories. Michel Foucault is arguably the predominant influence on contemporary accounts of power and space within political geography and its related fields, although the writings of Edward Said, Jacques Derrida, and Gilles Deleuze have all played a part. How such figures have been read and absorbed, however, has made for rather different interpretations and inflections of power. In the remainder of the essay, I look at three different adaptations of poststructuralist thought, looking first at the geo-graphing of power undertaken by exponents of critical geopolitics and then at the more spatial governmental approach adopted by the likes of Nikolas Rose and others. Finally, I turn briefly to the recent work of Michel Hardt and Antonio Negri to show what a deterritorializing apparatus of rule might look like as an approach to power and political space.

Geo-power, or writing political space

Geo-power is a term coined by Gearóid Ó Tuathail (1994, 1996) to convey the kinds of representational practices used by statespersons, elites, and policy writers to proclaim certain "truths" about how and why political space is ordered, occupied, and administered in the way that it is. Drawing extensively upon Foucault's work on the relationships between power and knowledge, attention is focused upon the geopolitical discourses that make it difficult to think about, say imperialist expansionism, the boundary between Islam and Christianity, or global climatic change, in ways other than those laid down by the "voices" of authority. In relation to the administration and disciplining of territories, for instance, the right of a political body "to speak" sovereignty over particular spaces is made sense of by intellectuals, institutions, and practising statespersons who mobilize geographical understanding in such a way that its "obviousness" is there for all to see. The power here, then, derives from the politics of *geo-graphing space*; that is, writing or representing it in ways that justify a particular group's authority over a subject population.

Monopolizing the right to speak authoritatively about particular places and regions, especially when done by invoking the "national interest" or, following Said (1978), designating familiar spaces as "ours" and unfamiliar territory as "theirs" (that is, beyond "ours"), is thus a means of enframing spaces within particular regimes of "truth" (see Dalby, 1991). In this vein, the practice of foreign policy making, for instance, appears as primarily a collection of scripts which combine various coded geographical assumptions and descriptions about "far-away" places which are then used to narrate geopolitical events and legitimize a particular course of action (Dodds, 1993, 1994). In writing such scenarios, geographical metaphors and tropes come into play, such as the identification of "rogue states" recently deployed by the US and its allies in their "war" against "terrorism," as well as rhetorical proclamations such as the "clash of civilizations" and the "end of history" (see Dodds, 2000; Dodds and Sidaway, 1994; Ó Tuathail et al., 1998). Such devices are thus the means through which power is exercised in the production of both formal and popular knowledges.

Indeed, much of the critical geopolitics literature is taken up with making explicit what is implicit in the writing and representation of political spaces. In doing so, their work draws attention to the relations of power embedded within geopolitical discourses and their contested, as well as potentially influential, nature (see also Dalby et al., 1998). The possibility of alternative political imaginaries drawn up in opposition to the dominant political discourses is recognized, although broadly understood in terms of resistance to domination rather than empowerment through association, to follow Arendt's lead. An "anti-geopolitics" is stressed in contrast to an emphasis upon collective, integrative action as an end in itself, yet one which speaks to a wider political audience and sets of interests (see Routledge, 1998).

On balance, then, the exercise of geo-power takes its cue from Foucault's sense of discursive power as a normalizing rather than a repressive force, although in doing so the exponents of critical geopolitics have little to say about the enabling side to Foucault's grasp of power. In terms of spatial imagery, what they do take from Foucault is a topological sense of power; that is, one which is less concerned with so-called scales of power – from the local and regional through to the national and supernational (as practised, say, in world-systems theory; see Taylor and Flint, 2000) – and more concerned with the varied points or sites of power and the relations between them. But that still leaves the Foucaldian-type question as to how people govern themselves at-a-distance.

Governing the self

The idea that power acts as a guiding force which does not show itself in an obvious manner is central to Foucault's idea that government works indirectly to limit the possible range of people's actions. In his brief outline of the art of governmentality, it is possible to see how power functions through people "working" on their own conduct to bring themselves into line according to what they imagine to be the most appropriate or acceptable forms of behavior (see Foucault, 1991). This *enabling* side to power lies at the heart of Nikolas Rose's attempt to resist, much like those writing on critical geopolitics, a "state-centered" conception of authority that views power as something imposed by the center from the top-down. In *Powers of Freedom* (1999), he sets out the kinds of political apparatus required to effectively deliver government of conduct at-a-distance. Although limited in geographical terms, its topological approach does mean that it shares with critical geopolitics a concern to establish the sites and connections through which power is exercised (as opposed to a fixation with scalar politics). Rose's preoccupation with the practice of government through freedom, however, takes his analysis in a quite different direction.

Stressing the art of government in the liberal sense as an activity which intervenes to regulate the behavior of widely dispersed populations, people, on Rose's understanding, bring themselves to order through obligation and self-restraint. The willingness of "free" subjects to transform themselves in a certain direction is said to hold the key to how control is exercised at points remote from their day to day existence. What appears to hold the apparatus of government together and give recognition and due credence to the various political inscriptions, pronouncements, edicts, judgements, aspirations, and protocols in circulation are the many and varied "centralized" authorities in play. The extension of authority over particular zones of social activity, for example through the centralized standards and assessment practised by social welfare agencies, is what enables the governable spaces of the family, the school, the clinic and so forth to take shape. A multitude of "experts," dispersed through a variety of state and quasi-autonomous agencies in the private, voluntary, and informal sectors, as well as parents, teachers, doctors, and social workers, are networked both to one another and to an "authorizing" center through which all translations must pass (Rose, 1994).

All this distancing by independent authorities and experts, however, is not something that once set in train works itself out in accordance with a kind of immutable logic drawn up at a "distant" center. Rose, drawing upon the work of Bruno Latour and Michel Callon (and actor network theory generally: see especially Callon, 1986; Callon and Latour, 1981; Latour, 1987), is aware of the role of translation in forging loose and flexible networks between experts of different hues and colors, so that any particular welfare objective, say over the yardstick used to calculate efficiency in social welfare programs or how value-for-money is defined in healthcare services, would necessarily involve attempts to broker an understanding which experts and skilled professionals alike could subscribe to without loss of face or judgement. In this way, through a process of mobilization, the truth claims of a range of accredited authority figures – under the guise of neutrality and efficiency – set out the norms of conduct that enable distant events and people to be governed at arm's length.

How far individuals in a variety of settings come to see themselves as responsible welfare citizens is a moot point, however. This particular style of governing based upon self regulation may on occasions possess sufficient reach, but the scope of the appeal and its intensity are likely to remain limited. The assumption by Rose that authority integrates individual choice and the techniques of a responsible self into its own order may work for those predisposed towards such options, but its basic message is likely to pass unabsorbed by those who are not, especially if they are distant in both space and time.

In that respect, Rose's attempt to outline the basis of government's decentered rule suffers ironically from not thinking enough about space. In his account, power may not be something that is distributed intact from place to place in a loose (realist) fashion, but the spread of certainties from the dispersed "centers" tends to be assumed not evidenced, judged by their intended effects rather than by their actual effects. When power is judged to be an immanent force like this, that is, as something inseparable from its effects (as Foucault conceives of it), the effects themselves, somewhat surprisingly, appear pre-scripted, as if they have been read-off from an already given set of political spaces.

Power as a deterritorializing apparatus of rule

In Michael Hardt and Antonio Negri's recent work, *Empire* (2000), the idea that power as an immanent force works itself out across a scripted set of spaces is taken to its ultimate conclusion, but not without a provocative adaptation of Foucault and Deleuze's thinking to the question of globalization and sovereignty today.

Contemporary power in a global age, for Hardt and Negri, takes the kind of amorphous form that anti-globalization protestors from the *tutte bianche* (all white) movement to Naomi Klein (the author of *No Logo*) characterize as being so everywhere, it seems nowhere. It is, they argue, a decentered, deterritorialized apparatus of rule that has no center and no edges. In this kind of rhizomatic, topography of power "a series of national and supranational organisms united under a single logic of rule" – orchestrated by (but not centered in) the US in coalition with any number of "willing" states, transnational corporations, supranational institutions, and NGOs – exercise "imperial" control through open, flexible, modulating networks of "command."

The style of "command" that they have in mind, however, is not the kind of delegated rule-making machinery that, in "state-centered" versions of power, rely on fixed boundaries and defined territorial edges to know the limits of their sovereignty. Rather, imperial sovereignty as seen through the lens of globalization, we are told, has "no outside." In Deleuze and Guatteri's (1988) fashion, it operates as an *immanent, deterritorialized force* where the "core" of the imperial apparatus is itself caught up in the intense whirl of political and economic activity, much of which passes by it, some of which transforms it, but without which the whole apparatus could not sustain itself. As a form of "command," then, it is less about disciplinary constraint and more about the diffusion of moral, normative, and institutional imperatives across the globe.

As a diagram of power, imperial sovereignty is likened to the operation of the world market which is organized on the basis of free subjects, yet who find themselves constrained by the direction and influence of the economic and monetary frameworks in play. Without the slightest hint of obligation and perhaps despite a reluctance to comply, we bring ourselves into line with the interests of the markets simply because we have no choice but to do so (domination "by virtue of a constellation of interests," as Max Weber [1978] liked to call it). In equivalent political terms, the institutional processes of normalization, including the tenets of neoliberalism, are assumed to reach so far into the lives and subjectivities of individuals that it is no longer possible to discern their points of application, nor even begin to question their "taken-for-grantedness." Thus in the open networks of "command," power is not seen as something that is applied by the likes of a "superpower" such as the US, rather the networks themselves are constitutive of the very power that enables the US and its allied governments and organizations to act. Everything is, as it were, "bundled" together (in the sense of Microsoft "bundling" software inside its operating systems) so that global trade, open markets, human rights, democracy, freedom, and much more are inseparable elements of rule: if you "buy" (or are immersed in) the logic, you have no choice but to take its disparate elements. They are part of a seamless logic – a "smooth space" of rule, where the surfaces of everyday life appear uniform precisely because they are criss-crossed by so many sets of relations.

However, in casting globalization and the new political order in the mould of a new form of imperial sovereignty that leaves nothing outside of its reach, it could be argued that such an impression leaves no room, or rather no space, for political alternatives to take hold. Yet Hardt and Negri's novel twist is that they are able to argue for an alternative form of political organization, a counter-Empire, so to speak, which has its roots within the confines of the new sovereign order. In challenging globalization, the emergence of an alternative, more inclusive order comes about through people pitting one form of global sovereignty against another, not from any number of localized struggles at the margins. Instead of mobilizing to "escape" the influence of the new imperial order, they stress the need to build a movement that takes politics directly through it and out the other side. In this respect, today's anti-globalization protests from Seattle, Chiapas, Genoa and beyond can be seen as a new form of global sovereignty "in the making," an associational politics in the republican tradition, rather than a series of discrete local actions.

Needless to say, the outlines of such a political project remain sketchy, but perhaps that matters less than the manner in which "global" activism is conceived. The claim that anti-globalization protests are themselves global is a bold one, prone to exaggeration at the best of times. Yet the argument is not that a series of localized struggles add up to something greater or indeed that, individually, they resonate beyond their local context. Rather, each struggle for an alternative, whether it emanates from the developed or less developed world, addresses the "center" of imperial power directly, as a kind of *virtual* target. In the absence of a locatable center of authority, anti-globalization protest mirrors the "non-place" of power by choosing targets that symbolize global imperial sovereignty. MacDonalds is an obvious choice, as is Nike or Microsoft, but so too is the World Trade Organization, Washington, and Kyoto. Thus, for Hardt and Negri, the immanent nature of global power today has brought forth not only a counter political movement in its own diffuse image, but it has also, somewhat paradoxically, led to the appearance of virtual "locations" of power with an all too familiar capacity to dominate space.

Conclusion

It would be misleading to suggest that Hardt and Negri's account of power has made any real headway in political geography overall, in part because of its recent origin, but also because more conventional accounts of global power, such as world-systems theory (and its assessment of US hegemony), still hold sway in many quarters (see Taylor, 1993; Taylor and Flint, 2000). While it remains to be seen in what ways the waters of political geography will become muddied, it is clear that such a process is already underway. In terms of a willingness to blur geographical boundaries, scales, and territories, for example, the topological sketches of critical geopolitics or the organizational reach of Mann's networks or indeed the reorganization of scale brought about by "multi-level" governance, all speak to agendas beyond that of "state-centered" versions of power.

Yet despite this problematization of "centered" accounts of political power in each of the above adaptations, geography, arguably, is still assumed to have less to do with the way that power *works* over space than does politics. Ultimately, it seems that notions of sovereignty, hegemony, domination, coercion, discipline, authority, surveillance, as well as political rule, organization, and administration, comprise the core vocabulary of power for many within political *geography.* Whereas distance and proximity, diffusion, and distanciation, or even territory and scale, make up the supporting glossary – if not the backcloth to power, then part of its rich texture – as if space makes little difference to the way that power works (see Allen, 2002).

This is obviously a contentious claim and one likely to be disputed, whereas the claim that an instrumental conception of power predominates in much of the political geography literature is likely to prove less controversial. For the sense in which power acts as a constraint, an instrument of domination rather than an enabling force, runs through a great number of international relations, world systems, critical geopolitics, and political geography tracts. This is neither altogether surprising, nor wrong in essence, but it is partial. From the work of Hannah Arendt, but also from Michel Foucault, it is possible to see how an alternative understanding of power, one based upon enablement and association, can help us understand the political landscape in different ways. Such an understanding may lend itself to an explanation of the mobilizing actions of NGOs or protest movements in general, but it is by no means restricted to such accounts.

BIBLIOGRAPHY

Agnew, J. 1994. The territorial trap: the geographical assumptions of international relations theory. *Review of International Political Economy*, 1(1), 53–80.

Agnew, J. 1999. Mapping political power beyond state boundaries: territory, identity, and movement in world politics. *Millennium*, 28(3), 499–521.

Agnew, J. and Corbridge, S. 1995. *Mastering Space: Hegemony, Territory and International Political Economy.* London: Routledge.

Allen, J. 1997. Economies of power and space. In R. Lee and J. Wills (eds.) *Geographies of Economies*. London: Arnold, 59–70.

Allen, J. 1999. Spatial assemblages of power. In D. Massey et al. (eds.) *Human Geography Today*. Cambridge: Polity, 194–218.
Allen, J. 2002. *Lost Geographies of Power*. Oxford: Blackwell.
Arendt, H. 1970. *On Violence*. San Diego, CA: Harvest.
Arendt, H. 1975. *The Human Condition*. Chicago: University of Chicago Press.
Barnes, B. 1988. *The Nature of Power*. Oxford: Blackwell.
Benhabib, S. 1992. *Situating the Self: Gender, Community and Postmodernism in Contemporary Ethics*. Cambridge: Polity.
Callon, M. 1986. Some elements of a sociology of translation: domestication of the scallops and the fishermen of St Brieuc Bay. In J. Law (ed.) *Power, Action, and Belief*, London: Routledge and Kegan Paul, 196–233.
Callon, M. and Latour, B. 1981. Unscrewing the Big Leviathan: how actors macro-structure reality and how sociologists help them to do so. In K. Knorr-Cetina and A. Cicourel (eds.) *Advances in Social Theory and Methodology: Towards an Integration of Micro- and Macro-Sociologies*. Boston, MA: Routledge and Kegan Paul, 277–303.
Dalby, S. 1991. Critical geopolitics: discourse, difference, and dissent. *Environment and Planning D: Society and Space*, 9, 261–83.
Dalby, S. and Ó Tuathail, G. (eds.). 1998. *Rethinking Geopolitics*. London: Routledge.
Deleuze, G. and Guattari, F. 1988. *A Thousand Plateaus: Capitalism and Schizophrenia*. London: Athlone.
Dodds, K. 1993. Geopolitics, experts and the making of foreign policy. *Area*, 25, 70–4.
Dodds, K. 1994. Geopolitics in the Foreign Office: British representations of Argentina 1945–1961. *Transactions of the Institute of British Geographers*, 19, 273–90.
Dodds, K. 2000. *Geopolitics in a Changing World*. Harlow: Prentice Hall.
Dodds, K. and Sidaway, J. D. 1994. Locating critical geopolitics. *Environment and Planning D: Society and Space*, 12, 514–24.
Foucault, M. 1991. Governmentality. In G. Burchell et al. (eds.) *The Foucault Effect: Studies in Governmentality*. Hemel Hempstead: Harvester Wheatsheaf, 87–104.
Giddens, A. 1977. *Studies in Social and Political Theory*. London: Hutchinson
Giddens, A. 1984. *The Constitution of Society: Outline of a Theory of Structuration*. Cambridge: Polity.
Giddens, A. 1985. *The Nation State and Violence*. Cambridge: Polity.
Giddens, A. 1990. *The Consequences of Modernity*. Cambridge: Polity.
Hardt, M. and Negri, A. 2000. *Empire*. Cambridge, MA: Harvard University Press.
Johnston, R. J. 2000. Power. In R. J. Johnston et al. (eds.) *The Dictionary of Human Geography*, 4th edn. Oxford: Blackwell.
Latour, B. 1987. *Science in Action*. Cambridge, MA: Harvard University Press.
Mann, M. 1984. The autonomous power of the state: its origins, mechanisms and results. *Archives Européennes de Sociologie*, 25, 185–213.
Mann, M. 1986. *The Sources of Social Power, Vol. I: A History of Power from the Beginning to AD1760*. Cambridge: Cambridge University Press.
Mann, M. 1993. *The Sources of Social Power, Vol. II: The Rise of Classes and Nation States, 1760–1914*. Cambridge: Cambridge University Press.
Muir, R. 1997. *Political Geography: A New Introduction*. Basingstoke: Macmillan.
Newman, D. (ed.). 1999. *Boundaries, Territory and Postmodernity*. London: Frank Cass.
Ó Tuathail, G. 1994. (Dis)placing geopolitics: writing on the maps of global politics. *Environment and Planning D: Society and Space*, 12, 525–46.
Ó Tuathail, G. 1996. *Critical Geopolitics: The Politics of Writing Global Space*. London: Routledge.
Ó Tuathail, G., Dalby, S., and Routledge, P. (eds.). 1998. *The Geopolitics Reader*. London: Routledge.

Painter, J. 1995. *Politics, Geography and "Political Geography"*. London: Arnold.
Rose, N. 1994. Expertise and the government of conduct. *Studies in Law, Politics and Society*, 14, 359–97.
Rose, N. 1999. *Powers of Freedom: Reframing Political Thought*. Cambridge: Cambridge University Press.
Rosenau, J. N. 1997. *Along the Domestic – Foreign Frontier: Exploring Governance in a Turbulent World*. Cambridge: Cambridge University Press.
Routledge, P. 1998. Anti-geopolitics: Introduction. In G. Ó Tuathail et al. (eds.) *The Geopolitics Reader*. London: Routledge, 245–55.
Said, E. 1978. *Orientalism*. New York: Pantheon.
Taylor, P. J. (ed.). 1993. *Political Geography of the Twentieth Century: A Global Analysis*. London: Bellhaven.
Taylor, P. J. and Flint, C. 2000. *Political Geography: World Economy, Nation State and Locality*, 4th Edn. Harlow: Prentice Hall.
Walker, R. B. J. 1993. *Inside/Outside: International Relations as Political Theory*. Cambridge: Cambridge University Press.
Weber, M. 1978. *Economy and Society: An Outline of Interpretive Sociology*. G. Roth and C. Wittich (eds.). Berkeley, CA: University of California Press.

Chapter 8

Territory

Anssi Paasi

. . . territory is a compromise between a mythical aspect and a rational or pragmatic one. It is three things: a piece of *land*, seen as a sacred heritage; a *seat* of power; and a functional *space*. It encompasses the dimensions of *identity* (. . .) . . . of *authority* (the state as an instrument of political, legal, police and military control over a population defined by its residence); and of administrative bureaucratic or economic *efficiency* in the management of social mechanisms, particularly of interdependence. . . . The strength of the national territorial state depends upon the combination of these three dimensions. (Hassner, 1997, p. 57)

Introduction

Territory is an ambiguous term that usually refers to sections of space occupied by individuals, social groups or institutions, most typically by the modern state (Agnew, 2000). As the previous citation from Pierre Hassner shows, several important dimensions of social life and social power come together in territory: material elements such as land, functional elements like the control of space, and symbolic dimensions like social identity. At times the term is used more vaguely to refer at various spatial scales to portions of space that geographers normally label as region, place or locality. Because contemporary territorial structures are changing rapidly, all of these categories imply many politically significant questions, above all, whether we should understand territories, places, and regions as fixed and exclusively bounded units or not (Massey, 1995). This forces us to reflect the responsibility of researchers in defining and fixing the meanings of words that may contain political dynamite. This has been an important question in the history of political geography and geopolitics, where the interpretations of concepts such as territory and boundary have been always simultaneously expressions of the links between space, power and knowledge (Agnew, 1998; Ó Tuathail, 1996; Paasi, 1996).

The term territory may also be used in a metaphoric sense. Becher (1989), for instance, speaks about "academic territories," referring to the way disciplines have their own internal power structures and "boundaries," and links to external "territories." The tradition of geopolitics illustrates that these academic territories, in the sense of different academic vocabularies, may be crucial contexts in the production of the language that can be used in the interpretation of the spatiality of the world.

Only a few major studies have been written on territory by Anglo-American political geographers (Gottmann, 1973; Sack, 1986; Soja, 1971), in spite of its significance to social life and even though it has been among the primary sources of conflicts. As Gottmann (1973, p. ix) aptly reminds us, "Much speech, ink, and blood have been spilled over territorial disputes." Geographers have traced the meanings of territories and territoriality for the state and societies, and have expanded the reductionist views of ethologists and sociobiologists. The latter have often understood territoriality as an expression of the "basic nature" of human beings in organizing their social life, while geographers have in common stressed the social and cultural construction of territories and the power relations that are part of this construction.

One background for this conceptual vagueness is the fact that people simply mean different things when discussing the idea of territory: these ideas are contextual. One more problem is that territory is implied in many other keywords of political geography, such as nation, state, nationalism, and boundary, and it is practically impossible to write on these keywords without reflecting concomitantly the meanings of territory. Furthermore, the etymology of the term is also unsettled and different views exist about what "territory" originally meant. The *Oxford English Dictionary* (1989) states that it is usually taken as a derivate of *terra* – the Earth – but the original form has suggested derivation from *terrere* – to frighten – which implies that territory and power are inextricably linked. Further specifications in the OED express, or at least imply, social control, administration, governance, politics, and economy at various spatial scales. The modern meaning of territory is closely related to the legal concept of sovereignty which implies that there is one final authority in a political community (Taylor and Flint, 2000, p. 156). This also means that territory and the strategies that are used in the control of territories – different forms of territoriality – are two sides of the same coin.

This chapter considers territories as social processes in which social space and social action are inseparable. Territories are not frozen frameworks where social life occurs. Rather, they are made, given meanings, and destroyed in social and individual action. Hence, they are typically contested and actively negotiated. As Knight (1982, p. 517) has pointed out "territory is not; it becomes, for territory itself is passive, and it is human beliefs and actions that give territory meaning." Spatial organizations, meanings of space, and the territorial uses of space are historically contingent and their histories are closely interrelated. Sack (1986) has studied the history of human territoriality and concludes that two historical transformations have seen the greatest changes in territoriality: first, the rise of civilizations, and, secondly, the rise of capitalism and modernity. In the former, territoriality was taken into use to define and control people within a society and between societies; in the latter, territoriality was used to create images of emptiable space, impersonal relations, and to obscure the sources of power (p. 217).

This chapter goes as follows. After mapping the ideas of territoriality, it traces how the ideas of territory became significant along with the rise of the modern nation-state and nationalism and how they have become an almost self-evident part of current understanding of the spatialities of power. Therefore, the meanings of territory will be reflected in relation to such categories as state, nation, and boundary. Territory became a popular term in the social and political sciences during the 1990s but it is understood differently in different contexts. It is still crucial among the categories introduced in political geography and political science textbooks, but is hotly contested in the fields of critical geopolitics, international relations, and economic and cultural geography. While the most extreme voices proclaim how territories and nation-states are vanishing from the globalizing world, most geographers have been more sensitive to the changing spatialities. For them the functions of territories and the meanings of sovereignty may change but states still remain major actors in the global constellation of power – while being an increasingly integral part of the global political economy (Agnew, 1998; Amin and Thrift, 1995). Cultural geographers have questioned the often-supposed homology between the state, nation, and society, and the belief in the existence of exclusive national cultures. On the one hand, the argument is that future social spaces and identities will be increasingly transnational and that new political networks will emerge. On the other, the rise of nationalism and ethno-regional activism suggests that people are also looking inwards in their states. These challenges for territory and territoriality, and their implications for political geography, will be discussed in the final section.

Human Territoriality

Most contemporary authors in social sciences make a clear distinction between human territoriality and other forms of territoriality, emphasizing that most portions of space occupied by persons, social groups or states are made into territories in a multitude of social practices and discourses by using abstract, culturally laden symbolism. This occurs in all social contexts, from local neighborhoods and gangs to nation-states and supra-state territories. Territories are always manifestations of power relations. The link between territory and power suggests that it is important to distinguish between a place as territory and other types of places (Sack, 1986). Whereas most places do not, territories – especially states – require perpetual public effort to establish and to maintain.

Sack (1986) outlined how different societies use different forms of power, geographical organization, and conceptions of space and place. Hence, territories are historically contingent while territoriality as a social practice seems to be based on some common principles. Sack (1986) defined territoriality as a strategy that human beings employ to control people and things by controlling area. Similarly territoriality is, he argues, a primary geographical expression of social power. Territoriality is an effective instrument to reify and depersonalize power. This is particularly obvious in the case of states, which exploit territoriality in the control of their citizens and external relations. This control occurs by using both physical and symbolic power (ideologies). While territoriality is in operation at a variety of spatial scales, at the societal level territoriality is instrumental in the regulation of social integration (Smith, 1986). Territoriality is crucial in defining social relations, and location within a territory partly shapes membership in a group (Sack, 1986).

Sack (1986) argues that the formal definition of territoriality not only tells us what territoriality is, but also suggests what territoriality can do. This effect is based on three interrelated relationships, which are contained in the definition. First, territoriality must involve a form of classification by area, i.e. categorization of people and things by location in space. Secondly, territoriality is based on communication and particularly significant is the communication of boundaries. Thirdly, territoriality must involve an attempt at enforcing control over access to the area and to things within it or to things outside of it. Territoriality, as a component of power, is not only a medium of creating and reproducing social order, but is also a medium to create and maintain much of the geographic context through which we experience the world and give it meaning (Sack, 1986).

The territoriality of states in particular is deeply seated in the (spatial) division of labor: some actors concentrate on the production of the symbolic and material dimensions of territoriality (e.g. administration, economy, army) using their power as part of the social division of labor, whereas most people are rather reproducers. Key actors in the production of territoriality are politicians, military leaders, police, journalists, teachers, and cultural activists, for instance. The roles of these groups of actors may differ according to the spatial scale at which they act, but in the case of state territoriality their power is obvious. These actors may also mediate between activities occurring at different spatial scales. The organization of police and military forces as well as education and media usually effectively combine local-scale activities with national values (Herbert, 1997; Paasi, 1999; Schleicher, 1993; Schlesinger, 1991).

Territoriality is not, however, a stamp that is mechanistically put on social groups "from above," since processes occurring at different spatial scales come together in territories. Herbert's (1997) study of the Los Angeles Police Department shows how social processes occurring at various spatial scales and motives originating from different sources may come together in a territory. Police forces are – together with the army – one part of the "repressive sub-apparatus" (Clark and Dear, 1984) that modern states exploit in the control of spatial behavior by controlling space, spatial representations, and narratives. This occurs typically at the local scale. Herbert's study shows that the control of space is a fundamental source of social power and that origins of control may emerge from different sources and spatial scales.

Territories as Social Constructs

Instead of defining with a sentence or two what territories are and how they operate, it is more useful to understand them as social processes, which have certain common characteristics. The process during which territorial units emerge as part of the socio-spatial system and become established and identified in social action and social consciousness, may be labeled as the "institutionalization of territories" (Paasi, 1991, 1996). This process may be understood through four abstractions that illustrate different aspects of territory formation. These aspects can be distinguished analytically from each other, but in practice they are entirely or partly simultaneous. The first is a territorial shape – the construction of boundaries that may be physical or symbolic ones. Boundaries, along with their communication, comprise the basic element in the construction of territories and the practice of territoriality. Encompassing things in space or on a map may identify and classify places or regions, but these become territories only when their boundaries are used to control people (Sack, 1986).

Traditional political geography has taken the link between territory and boundaries very much for granted and boundaries have been understood as neutral lines that are located between power structures, i.e. state territories. It is, however, crucial to realize that the power of territoriality is based on the fact that boundaries – as lines of inclusion and exclusion between social groups, between "us" and "them" – do not locate only on border areas but also are "spread" – often unevenly – all over the state territory. Boundaries penetrate the society in numerous practices and discourses through which the territory exists and achieves institutionalized meanings. Hence, it is political, economic, cultural, governmental and other practices, and the associated meanings, that make a territory and concomitantly territorialize everyday life. These elements become part of daily life through spatial socialization, the process by which people are socialized as members of territorial groups. The emergence of the Finnish state and nation since the nineteenth century, for example, shows how spatial socialization requires effective mechanisms, such as symbolism and institutions, that will bind people together (Paasi, 1996).

Hence the second crucial element in territory formation is the symbolic shape which includes (a) dynamic, discursively constructed elements (like the process of naming), (b) fixed symbols such as flags, coats of arms and statues, and (c) social practices in which these elements come together, such as military parades, flag days, and education. These practices and discourses point to the third crucial element, the institutional shape. This refers to institutionalized practices such as administration, politics, economy, culture, communication, and the school system through which boundaries, symbolism and their meanings are produced and reproduced. Institutional shaping is typically very complex and the operation of one institution often supports several others. Fourthly, territories may gain an established position in the larger territorial system, i.e. have an "identity," narratives that individuals and organizations operating in the area and outside use to distinguish this territory from others. The institutionalization of territories at different scales is often an overlapping process. The institutionalization of the Finnish state, for instance, was based on the simultaneous creation of state, regional, and local institutions and symbols, and social practices, such as education and media, that ultimately fuse previous scales and draw people as part of the nation (Paasi, 1996). When territories are identified as historical processes, they may also come to the end, i.e. de-institutionalize. This holds also in the case of the most naturalized territory of the modern world, the state. The most dramatic recent examples have been the dissolution of the former Soviet Union into separate states and the merging of East and West Germany.

State Territoriality

Most theories of the state identify territory as one basic element of the state and sovereignty is typically related to a bounded territory. States have constructed international law and a state can usually acquire a territory only under this law (Biersteker and Weber, 1996). The territorial framework of state sovereignty has for a long time included a model of citizenship and territorially-based narratives of identity that typically draw on the past. The state uses its territorial power in control of its citizens and, increasingly, those who have not achieved citizenship, such as refugees, immigrants, and displaced people.

For nationalists, sovereignty is the keyword and the state is seen as the primary political expression of community (Anderson, 1991). Loyalty to the state (patriotism) can either reinforce or conflict with nationalism and it is only within "real" nation-states that patriotism and nationalism support each other. The share of such states is only ten percent (Connor, 1992) which means that the institutionalization of state territories is typically a contested process. This is most obvious in the struggles of minority groups, such as Basques or Kurds, that are not satisfied with their cultural and economic position inside a state or several states.

The rise of the first "states" in Mesopotamia can be traced back 5000 years (Soja, 1971). However, while the ideas of dividing the land are very old, most ancient cultures and civilizations have left very little mark on the territorial organization of the current world. Different opinions exist on the relationships between the bounded territories of the past and those of the present-day world. Malcolm Anderson (1996, p. 13) argues that while the cosmologies in which old ideas of territory have been rooted are totally different from modern secular thought, some modern ideas of territoriality – e.g. on international frontiers and sovereignty – are based in part on Roman ideas of territoriality that were transmitted through the Catholic Church, rediscovered by political theorists during the Renaissance period, and regarded useful by jurists in the early modern period of European history. Isaac (1990, p. 417), for his part, has pointed out that the rulers of ancient empires (such as Rome) were not interested in defining the frontiers of their empires in terms of fixed boundaries and that territory was not so important as the control of people and cities.

The modern state system that emerged in Europe after the Treaty of Westphalia (1648), helped to establish the dominance of a horizontal, geostrategic view of the space of states and brought together territory and sovereignty. Rigid spatial boundaries became crucial only when the sovereignty of the state and citizenship came together. The emergence of states was related closely to the rise of capitalism and industrialism but the nation-state system cannot be reductively explained in terms of their existence. Instead, the modern world has been shaped through the complex interaction of the nation-state, capitalism, and industrialism (Giddens, 1987). The development of the abstract, metrical space went in hand with capitalism's need to increase production and consumption (Sack, 1986, p. 218). James Anderson (1996, p. 144) argues that the medieval era was characterized by spatial fluidity and mobility but temporal fixity in that change through time was typically seen as cyclical. The modern era that followed, for its part, was characterized by a more "mobile temporality" and time was associated with development and progress while space became more fixed, particularly with respect to the politics of states.

Territoriality became an institutionalized principle after the turn of the seventeenth century (Holsti, 2000). Practices defining the exact contents of bounded administrative units became one part of state building processes. Particularly within the Western nation-states, effective mechanisms of coordination, social integration, and administration were created. They stabilized the formerly dynamic functional spatial organization into a system of rigid, clearly delineated territorial units. State boundaries began to define the boundaries of society and polity. Sahlins (cited in

Soja, 1971, p. 15) wrote that "the critical development was not the establishment of territoriality in society, but the establishment of society *as* a territory." Along with this process, the membership in the state system – citizenship – was increasingly defined by birth or residence in states. While both the principles of territoriality and sovereignty are social constructs with a long history, it was only at the turn of the twentieth century that state territoriality and sovereignty began to manifest them in fixed boundary lines that were generally established instead of the former, more or less loose frontiers (Taylor and Flint, 2000).

The number of states has been continually increasing. Whereas about 50 states existed at the turn of the twentieth century, and some 80 in the 1950–60s, their current number is more than 190. Almost 120 new states have emerged since World War II as a result of decolonization (95 states), federal disintegration (20 states), and secessionism (2 states) (Christopher, 1999). Since the mid-1990s only a few conflicts between states have occurred, whereas the number of internal conflicts has been in the order of 26–28 per year. Some 500–600 groups of people identify themselves as nations, which means that territorial disputes will be with us also in the future. Christopher (1999) suggests that the current potential for new states is perhaps 10–20.

During this long process the state has become the most significant body in the control of territoriality that also effectively mediates between processes occurring at diverging spatial scales. The ability to exercise sovereign power over a defined area is the hallmark of a state, so laws as its instruments to exercise power are territorial too (Johnston, 1989, 1991). The state has one overwhelming advantage over other territorial entities – the monopoly of force and power. Several scholars have identified the territorial organization of the state as one precondition of state power. Giddens (1987) has famously defined the state as a bordered "power container" that is organized territorially. Similarly, Mann (1984, p. 198) reminds us how only the state is inherently centralized over a delimited territory over which its authority and power extends; territoriality is necessary for the definition and operation of the nation-state, and also for its autonomy in capitalist society.

State power is exploited both in the internal control of the society ("nation") and the state's external relations. While foreign and domestic affairs are, in practice, inseparable, there is a qualitative difference in how territoriality is exploited in these fields. More than any other institution the modern state exploits territoriality in its foreign policy through the principles of sovereignty and self-determination. The importance of boundaries for these principles becomes clear in the fact that the history of states is characterized by boundary disputes, which involves one further dimension of nation-states – military power. As to the internal control of the state, territoriality is present in the operation of the institutions and channels that Mann (1984) calls infrastructural power. This refers to the ability of the state to penetrate daily life within civil society, implement political decisions, and provide public goods and services among the citizens. While the state functions and the instruments to create images and narratives of nation – national education and media in particular – have emerged mainly since the nineteenth century and the widening of modern political and social citizenship took place during the early twentieth century, the power of the state and its capacity for intervention in social life have increased remarkably since 1945 (Smith, 1992).

Mann (1984, p. 208) argues that the greater the infrastructural power of the state, the greater will be the territorializing of social life. Agnew (1998) states that power is present in all relationships among people and the power of state relies on several sources it can tap into. Hence, infrastructural power is present at other spatial scales, too. Local and supra-state governments effectively use these mechanisms and accentuate the fact that state territoriality is not the only framework of power.

Territory and Identity

Like the ideas of sovereignty, ideas of national territory have also been in perpetual transformation. During the nineteenth century in particular, the ideas of the symbolic roles of the national territory changed fundamentally. A major medium for this change was nationalism, an ideology that slowly emerged in Western Europe during the eighteenth century and spread elsewhere with European colonialism. The basic factor in nationalism was to transfer group loyalty from kinship to local and other territorial scales (Anderson, 1988; Knight, 1982). Nationalism and romanticism influenced the new interpretations so that ideas of the link between land and nation became increasingly important. Nationalist discourses introduced expressions like "homeland," fatherland, and motherland that included a distinct territorial division between "us" and "the Other." Several scholars have shown how the songs, music, poetry, literature, and national figures – at times real people, at times allegories – are impregnated with territorial meanings (Murphy, 1996; Paasi, 1996). Territory became one of the key markers of national identity in this process and simultaneously changed from a pure bounded commodity – that can be sold and bought on the market – to a constituent of the national history, culture, identity, and political order (Holsti, 2000). More than ever, the state also entered into the everyday life of individuals in the form of mechanisms that again helped to create an image of what Anderson (1991) has labeled an "imagined community," a group of people who identify themselves with a collective while not knowing each other. National education, in particular, became a key institution in the socialization of citizens into national–territorial thinking. In spite of this fact, nationalism's relationships with territory have been ignored in research. Anderson (1988) states that nationalism is territorial in the sense of claiming specific territory but it is also partial, since "national interests" may be more in the interests of some part of a nation than others. Hegemonic groups may use space, boundaries, and various definitions of memberships (or citizenship) effectively to maintain their position and to control others inside the territory. This may occur by generating and maintaining social fragmentation as has been the case in some areas in Israel (Yiftachel, 1997). However, every nation is only a small part of humanity and visions inside one nation may differ radically from those of others.

"National identity" brings together the complex dimensions of nationalism and the national state. It is typical to see territory as one of the constitutive "ideas" of national identity (Knight, 1982; Smith, 1991; Williams and Smith, 1983). This is based on the implicit idea of the link between nation and state (and hence sovereignty). While noting the significance of territory (or "homeland") among such constituents of identity as common myths and historical memories, a common mass public culture, common legal rights and duties, a common economy, and territorial mobility (see Smith, 1991, p. 14), scholars have not been interested to the same extent in the social and discursive construction of territory and territoriality, or in how these become a part of the historical narratives and myths of a nation and of local daily life or what Billig (1995) calls "banal nationalism."

Territorial identification is not usually based merely on territory itself but this requires elements that integrate people living in different parts of the territory. Various abstract symbols are needed to express physical and social integration. Territoriality may be hidden in this symbolism, i.e. in many cultures symbols are associated with more or less abstract expressions of power, group solidarity, and authority. Interestingly enough, territorial symbols often depict ideas and symbols of power (such as wild animals), not people (Duchacek, 1975).

Governance and administrative practices, media, and education (national socialization) provide a common horizon for "identity" and for understanding the spatial "reality" that surrounds social groupings. The development of "nations" is indicative of this. Most scholars who have analysed the formation of national identity remind us that nations usually require a territory, which they share with their larger social groups (Smith, 1991). Hence, the state has been very effective in the production of not only the physical infrastructure for its reproduction but also social practices and institutions (education, research, media, statistics, mapping, military, etc.) to create an image of itself as the most significant territorial entity that most people also effectively identify with.

Deterritorialization of the Contemporary World

The link between state, territoriality, and sovereignty – all symbolized by an idea of the existence of exclusive boundaries – has been so dominating in the spatial imagination of international relations scholars that it is possible to talk about a "territorial trap," a state-centered account of spatiality, which has tended to link state power and territorial sovereignty intimately together (Agnew, 1994). Taylor (1996) speaks about "embedded statism" where states have come to dominate over the ideas of nation and ultimately both categories have become naturalized as a major framework to human life.

According to Agnew (1994), the territorial trap is based on three analytically distinct but invariably related assumptions. First, it suggests that modern state sovereignty requires clearly bounded territorial spaces. Secondly, it assumes a strict distinction between inside and outside, and this suggests that there exists a fundamental opposition between domestic and foreign affairs. Thirdly, it assumes that the territorial state acts as the geographic container of modern society. These three assumptions take for granted an idea of the world as consisting of bounded, exclusive territories, without noticing that these elements are socially constructed and contested.

Academic scholars have been in a key position in the production of the territory-centered outlook on the world and in shaping the practices and discourses through which the current system of territories is perpetually reproduced and transformed. Most of the literature simply assumes statehood, without identifying the basic elements of state, not to talk about challenging them (Biersteker and Weber, 1996; Knight, 1982).

While the state is the main example of how territoriality and territory are exploited in organizing social relations, in practice state territoriality has always been unbundled by agreements and alliances between various territorial and nonterritorial bodies. The state has become more powerful both as an international actor and in its relations to society within its boundaries. Also the number of nongovernmental international organizations has increased perpetually. These processes mean that the territorial "pattern" of various spatial practices and representations has become more complex and these elements are increasingly overlapping. In Europe, some scholars have been ready to talk about a "New Medievalism," a situation characterized by overlapping authorities and administrative structures (J. Anderson, 1996). The re-articulation of international political space would thus lead to the "unbundling of territoriality" (Ruggie, 1993)

State territoriality is challenged by numerous actors that cross and question the boundaries of formal state territories. Movements aimed at promoting the emancipation of women, human rights or environmental questions cross the boundaries of existing territories forming new transnational social spaces. Economic flows cross boundaries at an increasing speed. New forms of communication (cyberspace and Internet) affect the roles of the state and its functions at a variety of scales. Current economic spaces of flows centered on some major world cities, ideas of cosmopolitan dimensions of place, etc. all challenge visions of the world as a grid of bounded territories. Jessop has characterized the current situation as follows:

> ...we now see a proliferation of discursively constituted and institutionally materialized and embedded spatial scales (whether terrestrial, territorial or telematic), that are related in increasingly complex tangled hierarchies rather than being simply nested one within the other, with different temporalities as well as spatialities.... There is no pre-given set of places, spaces or scales that are simply being reordered. For in addition to the changing significance of old places, spaces, scales and horizons, new places are emerging, new spaces are being created, new scales of organization are being developed and new horizons of action are being imagined (Jessop, 2000, p. 343).

Political geographers and political scientists have increasingly called for openness in interpreting what territory, boundaries or place mean, arguing that there is no need to comprehend these categories as closed, strictly bounded entities as politicians, academics, and other actors have been used to doing (Agnew, 1994; Ó Tuathail, 1996; Shapiro and Alker, 1996). Taylor (1994) has reflected on the meanings of state territoriality and concludes that the state has different orientations. As a power container it tends to preserve existing boundaries; as a wealth container it strives towards larger territories; and as a cultural container it tends towards smaller territories, especially when the "nation" consists of diverging cultural groups that become increasingly consciousness of themselves.

Visions of territorially bounded national cultures have also been challenged. Identity and power are inextricably linked: identities are not neutral or naturally given but constructed for specific purposes (Jackson and Penrose, 1993). Cultural researchers in particular have challenged the myths of "national cultures" as "closed cells," entities that would be culturally homogeneous and exclusive. Instead, identities are "hybrids" that draw together influences that function across borders

(Yuval-Davis, 1997). This forces reflection on the links between political communities, identity, and the cosmopolitan elements of territory (Entrikin, 1999), as well as to identify transnational social spaces.

Discussion

Scholars operating in various fields are unsure of the current meanings of territory and boundaries. This uncertainty is based on tendencies that seem to challenge the dimensions of territory mentioned in the citation by Hassner (1997) at the beginning of this chapter. These tendencies are based first on the changing meanings of state territory as a seat of power and authority and how these elements are re-scaled between different territorial scales. Secondly, the dimensions of identity are becoming more complicated and are often linked with such questions as land ownership. Some scholars, most visibly Ohmae (1995), have been ready to argue that we are living in a borderless world in which the nation-state is taking its last breaths while new forms of economic regionalization will become significant.

These doubts on the future of territory are not a new phenomenon. As Keating (1998, p. ix) reminds us, "the end of territory as a factor in social and political life has been predicted regularly over the last hundred years, yet somehow it keeps on coming back." Geographers have had a more versatile perspective on the "future of state" and territory. James Anderson (1996) has observed that in the currently "fluid" situation scholars tend to overgeneralize the effects and tendencies of globalization. He opines that much of cultural, political, and economic life retains a relative fixity in space. Financial speculation and diplomacy, for instance, ultimately rest on the spatial fixity of factories, states, and "national interests."

The meanings of territory and identity are hence diversifying so that at the one extreme territorial identity is highly significant, while at the other, it is less relevant (Rosenau, 1997). Most people are "in-between" and are increasingly able to shift their identities. Rosenau contends that all along the continuum – including the two extremes – territory is not necessarily equated with nation-state boundaries. But while new transnational (and sub-national) communities, identities, and forms of citizenship are emerging, traditional ones (like the nation, state, and territory) are not disappearing; rather, they are changing their forms (Hassner, 1997). The major political problem still remains: how can we best give political recognition to various, often suppressed, identities (Knight, 1982)? Many of these identities are deeply territorially rooted, even if the identities of places are never "pure" (Sibley, 1995). While identity always seems to be based on differentiation from Others, this differentiation does not have to be based on hard boundaries between "us" and "them" (Massey, 1995).

While many of the challenges of the existing territorial order are based on the globalization of economy and increasing flows of information, these elements can also partly motivate new forms of territoriality that are linked with the past. The "first nation" movements in Canada and elsewhere are fitting examples of new challenges for territorial thinking. Often supporting traditional community and cultural identities, environmental values, and people's rights to land and old territories, these movements struggle to affect legislation and the forms of territorial governance that have been established by the hegemonic groups in the society. In many cases these activities and interests cross existing state borders – often by using modern information technology. This border-crossing also characterizes social movements that bring together, for example, workers, poor people, women, and environmentalists, and resist the uncritical acceptance of neo-liberal attitudes and practices behind the current trends in globalization.

Future democratic societies will inevitably require increasing openness and "crossings" of cultural, symbolic, legal, and physical boundaries between territories at a variety of spatial scales, from the local to the global. Researchers, for their part, should be ready to deconstruct the constitutive, at times mystified, elements of territory, territoriality, boundaries, and identity narratives. It is obvious that territoriality is to an increasing degree turning into a continuum of practices and discourses of territorialities which may be, to some extent, overlapping and conflicting. They may be linked or networked partly with the past, partly with the present, and partly with a utopian imaginary of the future forms of territoriality. The examples discussed in this chapter clearly suggest that new territories and territorialities may supercede the established political categories and identities at various spatial scales, and yet partly be linked with them. All this will provide an interesting challenge for the geographic imagination of political geographers and others dealing with the spatialities of power.

BIBLIOGRAPHY

Agnew, J. 1994. The territorial trap: the geographical assumptions of international relations theory. *Review of International Political Economy*, 1(1), 53–80.

Agnew, J. 1998. *Geopolitics: Revisioning World Politics*. London: Routledge.

Agnew, J. 2000. Territory. In R. J. Johnston et al. (eds.) *The Dictionary of Human Geography*. Oxford: Blackwell.

Amin, A. and Thrift, N. 1995. Territoriality in the global political economy. *Nordisk Samhällsgeografisk Tidskrift*, 20, 3–16.

Anderson, B. 1991. *Imagined Community*. London: Verso.

Anderson, J. 1988. Nationalist ideology and territory. In R. J. Johnston et al. (eds.) *Nationalism, Self-Determination and Political Geography*. London: Croom Helm.

Anderson, J. 1996. The shifting stage of politics: new medieval and postmodern territorialities? *Environment and Planning D: Society and Space*, 14, 133–53.

Anderson, M. 1996. *Frontiers: Territory and State Formation in the Modern World*. London: Polity.

Becher, T. 1989. *Academic Tribes and Territories: Intellectual Enquiry and the Cultures of Disciplines*. Stratford: Society for Research into Higher Education & Open University Press.

Bierstaker, T. J. and Weber, C. (eds.). 1996. *State Sovereignty as Social Construct*. Cambridge: Cambridge University Press.

Billig, M. 1995. *Banal Nationalism*. London: Sage.

Christopher, A. J. 1999. New states in a new millennium. *Area*, 31, 327–34.

Clark, G. L. and Dear, M. 1984. *State Apparatus: Structures and Language of Legitimacy*. Boston, MA: Allen & Unwin.

Connor, W. 1992. The nation and its myth. *International Journal of Contemporary Sociology*, 33, 48–57.

Duchacek, I. 1975. *Nations and Men*. Hinsdale, IL: Dryden.

Entrikin, J. N. 1999. Political community, identity and cosmopolitan place. *International Sociology*, 14, 269–82.
Giddens, A. 1987. *Nation-State and Violence*. Cambridge: Polity.
Gottmann, J. 1973. *The Significance of Territory*. Charlottesville, VA: University Press of Virginia.
Hassner, P. 1997. Obstinate and obsolete: non-territorial transnational forces versus the European territorial state. In O. Tunander et al. (eds.) *Geopolitics in the Post-Wall Europe: Security, Territory and Identity*. London: Sage.
Herbert, S. 1997. *Policing Space: Territoriality and the Los Angeles Police Department*. Minneapolis, MN: University of Minnesota Press.
Holsti, K. J. 2000. Territoriaalisuus (territoriality). *Politiikka*, 42, 15–29.
Isaac, B. 1990. *The Limits of Empire: the Roman Army in the East*. Oxford: Oxford University Press.
Jackson, P. and Penrose, J. (eds.). 1993. *Constructions of Race, Place and Nation*. London: UCL Press.
Jessop, B. 2000. The crisis of the national spatio-temporal fix and the tendential ecological dominance of globalizing capitalism. *International Journal of Urban and Regional Research*, 24, 323–60.
Johnston, R. J. 1989. The state, political geography, and geography. In R. Peet and N. Thrift (eds.) *New Models in Geography*. London: Unwin Hyman.
Johnston, R. J. 1991. *A Question of Place*. Oxford: Blackwell.
Keating, M. 1998. *The New Regionalism in Western Europe: Territorial Restructuring and Political Change*. Cheltenham: Elgar Press.
Knight, D. 1982. Identity and territory: geographical perspectives on nationalism and regionalism. *Annals of the Association of American Geographers*, 72, 514–31.
Mann, M. 1984. The autonomous power of the state: its origins, mechanisms and results. *European Journal of Sociology*, 25, 185–213.
Massey, D. 1995. The conceptualization of place. In D. Massey and P. Jess (eds.) *A Place in the World*. Oxford: Oxford University Press.
Murphy, A. 1996. The sovereign state system as political-territorial ideal: historical and contemporary considerations. In T. J. Biersteker and C. Weber (eds.) *State Sovereignty as Social Construct*. Cambridge: Cambridge University Press.
Ó Tuathail, G. 1996. *Critical Geopolitics*. London: Routledge.
Ohmae, K. 1995. *The End of the Nation-State*. New York: Free Press.
Paasi, A. 1991. Deconstructing regions: notes on the scales of spatial life. *Environment and Planning A*, 23, 239–56.
Paasi, A. 1996. *Territories, Boundaries and Consciousness. The Changing Geographies of the Finnish-Russia Border*. Chichester: Wiley.
Paasi, A. 1999. Nationalizing everyday life: individual and collective identities as practices and discourse. *Geography Research Forum*, 19, 4–21.
Rosenau, J. N. 1997. *Along the Domestic-Foreign Frontier*. Cambridge: Cambridge University Press.
Ruggie, J. G. 1993. Territoriality and beyond: problematizing modernity in international relations. *International Organization*, 47, 139–74.
Sack, R. D. 1986. *Human Territoriality. Its Theory and History*. Cambridge: Cambridge University Press.
Schleicher, K. (ed.). 1993. *Nationalism in Education*. Frankfurt: Peter Lang.
Schlesinger, P. 1991. *Media, State and Nation*. London: Sage.
Shapiro, M. and Alker, H. R. (eds.). 1996. *Challenging Boundaries*. Minneapolis, MN: University of Minnesota Press.
Sibley, D. 1995. *Geographies of Exclusion*. London: Routledge.

Smith, A. D. 1991. *National Identity.* Reno, NV: University of Nevada Press.

Smith, A. D. 1992. National identity and the idea of European unity. *International Affairs*, 68, 55–76.

Smith, G. 1986. Territoriality. In R. J. Johnston et al. (eds.) *The Dictionary of Human Geography.* Oxford: Blackwell.

Soja, E. 1971. *The Political Organization of Space.* Washington, DC: AAG, Commission on College Geography.

Taylor, P. J. 1994. The state as container: territoriality in the modern world-system. *Progress in Human Geography*, 18, 151–62.

Taylor, P. J. 1996. Embedded statism and the social sciences: opening up to new spaces. *Environment and Planning A*, 28, 1917–28.

Taylor, P. J. and Flint, C. 2000. *Political Geography.* London: Prentice Hall.

Williams, C. H. and Smith, A. 1983. The national construction of social space. *Progress in Human Geography*, 7, 502–18.

Yiftachel, O. 1997. Nation-building or ethnic fragmentation? Frontier settlement and collective identities in Israel. *Space and Polity*, 1, 149–69.

Yuval-Davis, N. 1997. *National Spaces and Collective Identities: Borders, Boundaries, Citizenship and Gender Relations.* Inaugural Lecture Series, The University of Greenwich.

Chapter 9

Boundaries

David Newman

Definitions
(the *Merriam-Webster OnLine Dictionary*)

border:	an outer part or edge
borderland:	a territory at or near a border
boundary:	something that indicates or fixes a limit or extent
edge:	the line where an object or area begins or ends
frontier:	a border between two countries
	a line of division between different or opposed things

Boundaries, the lines that enclose state territories, have constituted a major theme in the study of political geography. If there is anything that belies notions of a deterritorialized and borderless world more, it is the fact that boundaries, in a variety of formats and intensities, continue to demarcate the territories within which we are compartmentalized, determine with whom we interact and affiliate, and the extent to which we are free to move from one space to another. Some boundaries may be disappearing, or at the very least are becoming more permeable and easy to traverse, but at the same time many new boundaries – ranging from the state and territorial to the social and virtual – are being established at one and the same time. Boundaries are not only static, unchanging, features of the political landscape, they also have their own internal dynamics, creating new realities and affecting the lives of people and groups who reside within close proximity to the boundary or are obliged to transverse the boundary at one stage or another in their lives. Neither are boundaries simply territorial and geographic phenomena. Social, economic, political, and virtual boundaries all create compartments within which some are included and many are excluded. Boundaries are hierarchical: a person's location within the society–space frame is determined by the many boundaries within which he/she is enclosed, some of them being crossed with ease, others retaining features which make it more difficult, in some cases impossible, for the lines to be crossed.

This chapter surveys the place of boundary studies in political geography, past and present, focusing on the importance of lines in creating the spaces and territories within which we reside and which also provide us with identities and affiliations at a variety of spatial scales. The majority of boundary studies have focused almost exclusively on the territorial and the state, and have been descriptive in nature, giving rise to an accumulated knowledge of boundary case studies and territorial change as the world political map has itself experienced a number of major transformations and territorial reconfigurations during the past hundred years. Recently, the focus has began to shift to the notion of "boundary" as a line that separates, encloses, and excludes, at a number of spatial and social scales, thus moving away from the exclusive focus on hard international borders. Notwithstanding this, there is, as yet, no solid conceptual or theoretical framework for the holistic study of the boundary/border phenomenon, linking both spatial scales and alternative disciplinary approaches. In a world impacted by globalization, political rapprochement, and cyberspaces, a deeper understanding of boundaries, beyond the traditional political geographic analyses, can only be attained by recourse to a cross-disciplinary analysis, beyond the exclusive confines of geography, which takes into account the different meanings that boundaries and borders have for different people. Thus, while this chapter focuses on the changing dynamics of territorial boundaries, it broadens the discussion to include the wider phenomenon of "bounding" through which lines are drawn not only around the sovereign territories of states and municipal jurisdiction areas, but also around nations, groups, religions, and individuals, creating a series of bounded compartments within which most of us are contained and from which few of us are able to cross to neighboring compartments with ease. This is the case particularly where boundaries are tied up with the politics of identity and where crossing from one compartment to another requires a change, or at the very least a dilution, of the group identities with which we affiliate.

Interest in boundaries, as reflected by the literature of the past decade, has been on the increase (Blake, 2000b; Kolossov and O'Loughlin, 1998; Newman, 1999; Newman and Paasi, 1998; Paasi, 1996; Thomas, 1999; Waterman, 1994). There is a growing interest in the boundary as a geographic and/or social construct. This chapter begins with a survey of the traditional political geography literature on boundaries. It then develops a number of contemporary themes that have become part of the broader boundary discourse during the past decade. The traditional concepts are not discarded altogether: many of the terminologies originally developed by traditionalist boundary scholars – such as demarcation, permeability, frontiers, and border landscapes – are shown to have relevance for other notions of boundary. This, it will be argued, lays the basis for the development of a conceptual base for the study of boundaries at a multidimensional and hierarchical perspective.

The Study of International Boundaries: Traditional Themes

The study of boundaries has been central to political geography if only because they enclose the territory which defines the spatial extent of the state (Minghi, 1963; Prescott, 1987). In the pre-globalized era of the Westphalian State, the boundaries defined the area within which sovereignty was exercised by the state, a sovereignty which has become increasingly challenged as boundaries have become more permeable and impacted by trans-boundary movement of goods, people and ideas (Hudson, 1998; Paasi, 1998). While new themes have emerged in recent years, the traditional themes remain central to much of the contemporary boundary literature, if only because this is the sort of information that interests government officials engaged in boundary demarcation and/or they are the types of lines which are most recognizable and easily quantifiable to the impartial observer (Blake, 1999, 2000a).

Boundary typologies

Early studies of boundaries were replete with typologies and classifications of boundary types. Of these, Hartshorne's use of terms borrowed from fluvial geomorphology back in the 1930s remains strongly entrenched in the minds of many first-year students of political geography (Hartshorne, 1936). His use of such terms as "primary," "antecedent," and "subsequent" boundaries, were designed to distinguish between boundaries which were delimited in virgin unsettled (*sic*) territories and which determined the spatial distribution of settlement which came at a later stage, differentiated in its cultural and ethnic characteristics by the already existing line of separation, as contrasted with boundaries that were demarcated in accordance with the already existing patterns of human settlement and which consolidated and compartmentalized the ethnic and cultural spatial differentiation into separate political territories, normally states. Using the terminology of today, we would say that boundaries are both determined by, and in turn determine, the formation of separate group identities, identities which are expressed through the processes of spatial compartmentalization, on the one hand, and strong levels of residential segregation, on the other.

Of particular importance to the contemporary discussion of decolonization and postcolonialism was the classification of the superimposed boundary, the lines that had been drawn up by the colonial powers in Europe in the late nineteenth and early twentieth centuries. They had been superimposed upon the colonial landscapes in Asia and Africa as part of the process which brought European notions of fixed territories to regions whose populations were tribal and semi-nomadic, and whose seasonal patterns of movement were at odds with political notions of territorial fixation. These lines, often drawn up in the European capitals using inaccurate maps and with little knowledge and/or care for the spatial patterns of ethnic and tribal distribution, were, more often than not, characterized by long straight "geometric" lines, bearing little relation to the natural features of the human or physical landscapes. Tribes and ethnic communities were, in many cases, split between separate political territories within each of which they now constituted no more than an ethnic minority, while in other cases a number of tribal groups found themselves within a single political territory, now known as a state, competing for dominance and hegemony, subject to the artificial socialization processes of a constructed national identity and, in some cases, bringing about strife and civil war. The legacy of the superimposed boundaries in colonial Asia and Africa remains with us today as these regions, particularly Africa, have still not managed to find a means of bringing these alternative patterns and conceptions of territorial behavior into harmony with each other (Lemon, 2001; Ramutsindela, 1999).

Nowhere was the deterministic nature of early boundary studies reflected more than in the distinction which was made between "natural" and "artificial" boundaries (Boggs, 1940). By "natural" boundaries, it was suggested that the physical features of the landscape (such as rivers, mountain ridges, valleys and so on) determined the ultimate course and demarcation of the boundary line. This form of environmental determinism was later dismissed in favor of the approach which stated that all boundaries are artificial in that their demarcation is determined by people – politicians, planners, and decision-makers (Kristoff, 1959). The fact that natural features of the landscape may be used as a convenient means of demarcation is appropriate for as long as this is convenient for both sides and does not raise any problems in terms of the distribution of ethnic groups, control of physical resources, and other political objectives. Where such objectives demand a deviation from the natural course, be it through agreement or warfare, the line of the boundary will be modified accordingly. But just as we dismiss the very notion of the "natural" boundary, we cannot ignore the fact that environmental features played a more prominent role in the determination of human spatial patterns in the pre-technological eras, when society was less able to manipulate and change the landscape in accordance with its social and political objectives. The spatial distribution of different ethnic and national groups, mountain, and valley cultures, were determined, in part, by the boundaries that separated them from neighboring cultures and groups in the past, in periods when mobility was severely limited and when natural topographical features constituted major obstacles in the way of any movement, interaction or diffusion that did take place. As such, pre-modern boundaries did, in some areas, play a major role in the formation of separate national and ethnic identities within states and other forms of territorial compartments whose spatial definition may have been determined, at least in part, by the existence of environmental features acting as barriers and obstacles to movement.

The functional impact of boundaries

The study of the impact of boundaries on landscape formation marked a step away from the simple description and categorization of boundaries to a more functional approach. Two related concepts in the early boundary literature discuss the functional impact of boundaries upon landscape evolution. First, there is the distinction that has traditionally been made between the notion of "boundary" or "border" and that of "frontier." The former is the line, demarcated and implemented by a government, while the latter is the area or region in close proximity to the line and within which development patterns are clearly influenced by its proximity to the boundary. The second is the notion of boundary permeability and the extent to which interaction takes place in borderland regions on both sides of the boundary. This, in turn, reflects the nature of the political relations between neighboring states and the extent to which trans-boundary interaction can facilitate peaceful political relations.

Political frontiers are expected to be more apparent along "closed" or even "sealed" boundaries, with differential patterns of development taking place on both sides subject to different economic and planning systems, as well as government policies aimed at either developing and "bolstering up" the border landscape, or equally deciding to neglect those areas which may, at a future date, be the scene of military confrontation and resulting devastation. In extreme cases, such as that which occurred along the East–West Germany boundary during the forty-year period of the iron curtain, or on both sides of the "green line" boundary between Israel and the West Bank between 1949 and 1967, or on both sides of the North–South Korea line of division, mono-ethnic areas find themselves dissected into two distinct political territories, completely cut off from each other, and subject to vastly different patterns of socioeconomic development, thus giving rise to differentiated landscapes in a relatively short period of time. These three recent examples from contemporary history indicate just what a powerful impact a sealed boundary can have, as territories are affected by different political and socioeconomic regimes in a relatively short period of time. The opening of these sealed boundaries during the past decade (in the case of Germany, the collapse of the boundary, in the case of Israel/Palestine the opening of a trans-boundary dialog, and in the case of Korea the first signs of some rapprochement) have been the result, in no small part, of globalization processes – the wider understanding that ethno-national conflicts do not have a place in a postindustrial high-tech world. Equally, sealed boundaries of the nonspatial type, such as those separating religious groups and affiliations, are also characterized by completely differentiated border landscapes on both sides, with cultural norms, habits, and rituals contrasting strongly with each other.

Attempts to construct a theory of political frontiers have been few and far between. Of particular note was the work of John House, particularly his notion of "double peripherality," the idea that political frontiers suffer from both geographic and political peripherality at one and the same time (House, 1980). But this is contingent on a perception of the boundary as a line of separation and division, rather than one which can promote contact and cooperation between cultures and societies on both sides of the line. The idea of "frontier" has gradually been replaced with the notion of "borderland," a less evocative term, within which diverse patterns of trans-boundary interaction may take place, ranging from confrontation and exclusion to cooperation, integration and inclusion (Blake, 2000b; Pratt and Brown, 2000; Rumley and Minghi, 1991). In his study of the US–Mexican boundary, Oskar Martinez develops the notion of a borderland milieu, in which trans-boundary interaction may be even greater than interaction between the border region and the central government (Martinez, 1994a, 1994b). In his discussion of the "borderlanders," Martinez develops a continuum of borderland types, ranging from "integrated" to "alienated" borderlands, with intermediate positions being taken up by "co-existent" and "interdependent" borderlands.

Studies of frontier and borderland are, by necessity, linked to notions of boundary permeability. Much of the recent literature has focused on the increasing permeability of boundaries in an era of globalization, characterized by growing levels of trans-boundary movement and cooperation (Gallusser, 1995). This is particularly characteristic of the Western European experience as this region moves ever closer towards a federated political union, with the borders between states gradually being transformed into administrative boundaries. The active promotion of trans-frontier regions, both within the European Union itself, and along the geographic margins of the Union with countries that aspire to be part of the Union in the future, have given rise to a renewed interest in boundary research among European scholars (Ganster et al., 1997; Keating, 1998).

Henrikson (2000) goes further, arguing that trans-boundary diplomacy in the borderland/frontier regions can filter down into the center of the state, from periphery to center, positively affecting the nature of interstate political relations. This type of interaction does not have to be determined by state policies, but can be bottom-up, initiated by localized economic and social relations between the people themselves. Where ethnic minorities straddle the boundaries of neighboring states but do not feel marginalized or desire territorial secession, the common cultural identity is a factor that facilitates trans-boundary coexistence and, in turn, political stability.

Two important themes of contemporary relevance emerge from this short survey of the geographic narratives on territorial boundaries. The first is that the boundary phenomenon is dynamic, rather than static and passive, and that the demarcation of lines – be they spatial or social – affect people's lives and the way in which communities identify themselves and interact with those that are located beyond their own specific compartment. The second is that many of the traditional themes which have been used in the strict territorial context – such as "demarcation," "frontier," "borderland," and "superimposition" – can all be transposed to take account of the other types and scales of boundaries which have become part of the contemporary discourse through which a deeper understanding of the bounding phenomenon is being sought.

The Contemporary Study of Boundaries: Emerging Themes

The study of boundaries has re-emerged as a strong theme during the 1990s (Newman, 1999; Newman and Paasi, 1998; Prescott, 1999). On the one hand, there has been a renewed interest in the hard territorial lines which are constantly being redrawn and redemarcated between states, while on the other, there is a growing interest in the nature of bounding and the way in which people and groups are enclosed within a variety of social and spatial compartments. These have been two parallel, but rarely touching, discourses. This section of the chapter looks at the main themes which have emerged out of the recent boundary narratives, in an attempt to draw the parallels between the different discourses and laying the foundations for a single conceptual framework for future study.

The national and the local: boundary hierarchies

The study of boundaries in political geography has, by default, been concerned with the study of international boundaries, the lines that separate state territories. The study of administrative and municipal boundaries has generally been seen as a separate topic, if only because these lines do not determine the spatial extent of sovereignty. These lower level boundaries do not restrict movement of people or goods, there are few physical signs of the existence of the boundary, and most people are completely unaware of the fact that they may be crossing from one jurisdiction area into another as they go about their daily lives. Yet administrative and municipal boundaries affect the daily lives of most citizens much more than international boundaries, especially those citizens who do not travel beyond the confines of the country within which they reside. Municipal rates and taxes, registering one's children for local schools, police, health, and welfare authorities, are all organized along spatial lines, overlapping with each other in a complex system of geographic hierarchies.

Municipal boundaries are hotly contested, particularly the right to expand the limits of the jurisdiction area, as this affords local authorities the rights to develop the area under their control for new residential and/or commercial purposes, thus raising the tax base and the perception of the area as an attractive place within which to reside or set up one's business. The complexity and proximity of municipal and administrative boundaries gives rise to trans-boundary externalities, spillovers, and free rides. This is a form of boundary permeability that has always existed, with events and/or movement originating on one side of the boundary affecting what happens on the other side. Environmental and pollution spillover can adversely affect life on one side of a boundary, particularly when noxious facilities – be they heavy industrial plants, sewage treatment, highways or chemical plants – are located by one municipality (or country) in close proximity to the boundary. Positive externalities can also occur where attractive features, such as parks, high-quality educational establishments or medical facilities, are located close to the municipal boundaries. Residents of neighboring areas can enjoy the benefits of such facilities, much as they suffer negatively from the close proximity to noxious facilities.

It is at the local level where boundaries may often be more perceived than real. The lines separating one urban neighborhood or group's turf from another may not necessarily be compatible with the formal municipal and jurisdiction lines drawn up by the city planners and engineers. But in terms of the daily movements of peoples within their own microenvironments, these perceived boundaries often take on a much more important role. These invisible lines may determine where members of one group are prepared to move, where they are prepared to shop and where they are prepared to interact with neighboring populations. A road, a piece of wasteland, or a factory or cinema frequented by "other" groups may constitute the perceived boundary marker or frontier zone, regardless of whether this has any formal or administrative function. Perceived boundaries express the geographies of fear and safety which people feel when moving beyond their own territorial areas. Trans-boundary movement is reflected as much, perhaps even more, in shopping in a food store of a different ethnic group than it does in showing one's passport at the customs point of transit into another country.

With the breaking down of rigid notions of territorial sovereignty, there is less reason to maintain the traditional distinctions between the study of "international" boundaries and all other types of boundary. Administrative boundaries may, in time, become transformed into state boundaries, whether or not they reflect ethnic residential patterns. Many of the administrative boundaries in Eastern Europe have, in the wake of the collapse of the Soviet Union, become the boundaries of the new states (Kolossov, 1992; Ratner, 1996), although many of them were state boundaries prior to the Soviet occupation. Their continued existence as administrative boundaries has only served to strengthen the territorial images of homeland held by local populations. The "green line" boundary between Israel and the West Bank is a good case in point (Newman, 1994). This boundary was artificially superimposed in 1949, was formally removed as a result of Israeli occupation in 1967, but remained in force as a powerful administrative boundary of separation and national division between 1967 and 2000. As a result, it is this line which has become the default boundary for all political negotiations between Israel and the Palestinian Authority in trying to put an end to the conflict and the establishment of an independent Palestinian State (Newman, 1998).

The functional impact of the boundary on the behavioral patterns of the people who are enclosed by these lines is common to all types of boundary, regardless of the spatial scale at which the bounding process takes place. Even where there is interaction, or sovereignty is not an issue, the lines, however artificially demarcated in the first place, by definition of their very existence separate two entities and create distinctions between the people on each side of the line. As such, they create their own geographical realities, whether they be state or municipal boundaries. The longer they remain *in situ*, the harder they are to remove or to change.

Inclusion, exclusion, and the politics of identity

By virtue of their existence, the lines that are boundaries enclose spaces and groups. They demarcate the extent of inclusion and exclusion of members of different groups, ranging from the national to the neighborhood (Sibley, 1996). As such, the notion of lines that separate play a prominent, somewhat deterministic, role in the contemporary discourse on the politics of identity. The sequential process through which boundaries were originally conceived, delimited, and eventually implemented on the ground was a topic of study by boundary scholars during the first half of the twentieth century (Boggs, 1940; Holdich, 1916; Jones, 1945). However, they never went beyond a technical description of the delimitation process, ignoring the political and ethnic contexts and the fact that people actually lived in and around these boundary areas and were impacted by the decision to create a line of separation and exclusion in close proximity to their communities. The linkage between territorial demarcation and the formation of ethnic and/or national identity is a "chicken and egg," mutually enforcing, relationship. The existence of lines and territorial compartments in the form of states, creates a territorial frame within which the social construction of national identity has an important territorial dimension. Such boundaries define the contours within which places are imbued with historical and mythical meaning in terms of the nation and collective memory (Paasi, 1996). The social construction of homeland and national territorial identities may take in spaces and territories beyond the state confines, but rarely will it consciously focus on a smaller area than that within the state boundaries. But as identities are becoming increasingly multicultural (within the state) and global (beyond the state), the relationship between national identity and territorial absolutism is weakening (Taylor, 1996). The question of just who is included and, by definition, who is excluded from the social and spatial compartments is much more complex. At the same time, the politics of identity cannot be totally deterritorialized, if only because identity, as is the case for power, cannot be divorced from its territorial base (Wilson and Donnan, 1998). As the state becomes weaker, the focus on identity switches to local and global, religious and cultural, virtual and aspatial, units of affiliation, the majority of which are determined by some form of territorial compartmentalization (Eskelinen et al., 1999). But what is common to them all is that they produce hierarchical and overlapping group identities, coupled with the fact that each retains its own specific membership, including those who are accepted by, or are able to be part of, the "club" (virtual and cybercommunities are, in this sense, some of the most exclusive in that only those with access to computer technology and the ability to use it, can be part of that particular community), while the majority continue to be excluded. Gaining membership, the crossing of the boundary, requires acceptance by those who are already enclosed by the line, a process that, in many cases, is far more difficult than crossing the territorial line separating one state from another.

The Israel–Palestine conflict again exemplifies the way in which notions of boundaries have changed over time and provides a good example of the need to understand the multidimensionality of borders/boundaries, taking into account both the territorial and the identity dimensions (Newman, 1998, 2001, 2002). Geographers have traditionally viewed this conflict from a territorial perspective, focusing on such issues as the demarcation of physical lines, the impact of boundaries on changing settlement patterns, and the position of boundaries in relation to strategic sites and/or scarce water resources. Conflict and peace discourses have both traditionally focused around the notion of territorial boundaries and the need to demarcate lines of territorial separation which meet the various security, resource and settlement needs of the respective sides. But increasingly, the search for "good" boundaries (Falah and Newman, 1995) has demonstrated the need to equally take into account the needs for borders which satisfy the identity requirements of both Israelis and Palestinians, over and beyond the simple lines of territorial demarcation. These become complex as national territories and identities do not overlap in such a way as to enable each side to create lines of maximal separation, resulting in the residual of national minorities residing in the territory dominated by the "other." The formation of national identity for both Israelis and Palestinians is strongly tied up with the nature of territorial separation and sovereignty, each demonstrating, time-after-time, their preference for respective nation-states rather than a single bi-national entity in a single, small, territory. This is particularly evident at local levels, whereby Jews and Arabs reside in their strongly segregated villages, townships, and urban neighborhoods, rarely interacting with each other beyond the economic marketplace and creating a host of invisible and perceived boundaries which are hardly ever crossed by members of the two national groups.

The management of boundaries

Much of the early boundary literature focused on the way in which boundaries come into being, but stopped short of progressing to the next stage of just how boundaries are managed once they exist. The establishment of new territorial boundaries is today tied up with the way that these boundaries will be controlled and managed as part of government policies aimed at encouraging or limiting the extent to which trans-boundary movement takes place. Efficient boundary management can contribute to stable and peaceful trans-boundary relations, but at the same time may result in the prevention of trans-boundary movement or entry to other peoples and groups. Understanding the processes by which boundary management takes place is as important for boundaries of conflict as it is for borders which have become permeable and around which there is a significant increase in trans-boundary cooperation.

This relates not only to the movement of people and migrants, but also to other features, notably environmental hazards which equally affect the quality of life of people on both sides of the line.

The most notable form of boundary management concerns entry procedures for migrants (Sigurdson, 2000). Such management does not only take place at the physical port of entry to the country; it is also rooted in the agencies of control through which migrants are granted basic rights, such as social welfare, education for their children, and the opportunity to involve immediate family members as support. As such, the process of management is closely linked to the nature of exclusion/inclusion, as stricter management procedures strengthen the extent to which groups are excluded from the host society, regardless of their precise location in geographic space which may, as is often the case, be in the very heart of the capital city.

Territorial boundaries can be jointly managed and, as such, are often the catalysts for the creation of a regional identity and awareness that straddles the lines separating states. This is often the case with environmental and physical features, particularly water basins in regions of scarce water resources, such as the Middle East. But it has increasingly come to include human activities, such as a single employment market, or peace parks (Kliot, 2001), all of which create a trans-boundary infrastructure of interdependence that promotes peaceful relations and normalization rather than conflict and warfare. The boundaries separating groups and religions are usually managed by only one side: the ability to pass from one space into another – such as from one religion to another – is fraught with entry procedures which are almost impossible to overcome. The necessary visas often consist of ritual behavior or a particular form of lineage, requiring difficult – almost impossible – processes of conversion, requiring the exchange of one identity or affiliation with another, rather than an adoption of both. Contextually, trans-boundary interaction such as intermarriage or socioeconomic integration while retaining cultural norms, a sort of borderland region in itself, is often greeted with rejection by both core areas, a form of double exclusion, rather than constituting a bridge between the two spaces. Rigid boundary management procedures are particularly carried out by neo-nationalist and orthodox religious groups, sealing their boundaries from infiltration from the outside.

The case of the US–Mexico boundary has figured prominently in the boundary literature relating to both issues of identity and management (Ackleson, 1999). The geographic proximity of populations, the free movement for some and the stringent restrictions for others, and the impact of this boundary on differentiated social and economic systems within a few hundred meters of each other are sobering examples of the impact of a territorial line, even where it borders a postindustrial Western nation. Contextually, the "opening" of boundaries is selective – relevant to some, but not to others. Crossing the boundary results in a change in both cultural and economic status for Mexicans, creating a distinct borderland identity for these trans-boundary migrants (Martinez, 1994b). For Americans, the boundary remains a clear line of separation, between a world of "order" and taxation and one of economic adventurism, and/or (particularly for travelers and tourists) between the "First World" homeland and the "quaint" exoticism of a "Third World" neighbor.

The "borderless" world

The impact of globalization is associated with notions relating to the "end of the nation-state" thesis and, by association, notions of a deterritorialized and borderless world. The fact that boundaries have become increasingly permeable and are not able to prevent the unrestricted movement of goods, people, and ideas from one territory to another has, so it is argued, marked an end of the Westphalian State model in which the complete and absolute territorial integrity and sovereignty of the state was determined by the lines demarcating the territorial extent of political power and control (Albert, 1998; Dittgen, 2000; Kohen, 2000). Such political power has shifted away from the state towards global and virtual associations (Shapiro and Alker, 1996), notably corporate markets and multinational firms, as well as global political associations such as the European Union, NATO, and/or the United Nations, each of which adopts policies that infringe upon the sovereignty of the individual state. Power has, so the proponents of this argument continue, become deterritorialized at least inasmuch as the state is no longer the central locus of that political power.

From a geographer's perspective, this argument is untenable (Newman, 2000). While the world is undergoing significant territorial reconfiguration and re-territorialization, it is not becoming deterritorialized, simply because human activity continuous to take place within well defined territories. Even within the sphere of global capital and corporate markets, boundaries continue to have an impact (Ó Tuathail, 1999; Yeung, 1998) while the diffusion of information and ideas through cyberspace creates new forms of electronic landscapes in place of the old, fixed, territorial ones (Brunn, 1998). Whereas, in the past, the main locus of political power rested with the state, this power is now shared with non-state organizations and territories, be they local or global, giving rise to a world whose territorial compartments are hierarchical and multidimensional. Human society lives in a world defined by many different boundaries, some of them territorial and easy to demarcate and draw, others determining the outer extent of the social contours of the group – be they religious, national, cultural, economic or any other – and much more difficult to actually define in concrete terms.

Notions of the "borderless" world tend to be both culture- and discipline-specific. It is a discourse which has emerged from the Western experience, one which actively promotes boundary permeability and trans-boundary movement as a means by which political rapprochement is achieved. The lines demarcating the territory of the state may still retain political significance, but this, at least as far as Western Europeans are concerned, is diminishing in the face of new political, economic and information trends. This is by no means the case in many parts of the "Third World," notably the African continent, where, after fifty to one hundred years of boundary superimposition by the European and colonial authorities, the state system is only just beginning to come to terms with the notion of fixed boundaries and territorial sovereignty. And just when this is happening, along comes the Western world and tells Africa to forget about fixed boundaries and rigid forms of territorial sovereignty, because globalization is making the world into a borderless space.

The fact that ethno-territorial conflicts continue to take place throughout the globe is ample evidence for the importance of hard, territorial, dimensions of state power. Many of these conflicts, such as those in Cyprus, Israel/Palestine, Bosnia and Yugoslavia, are resolved only by recourse to territorial partition and separation, accompanied by long drawn-out processes of boundary demarcation (Waterman, 1987). Few of these take place in the Western world, the spatial core of the "borderless world" thesis, but territorial dispute is by no means absent from these regions altogether. As globalization impacts the traditional barrier role of state boundaries, so the focus of political power has also shifted to the regions and intra-state areas, such as Catalonia, Scotland, and Sicily, creating new life for territorial demarcators which were, until recently, considered largely redundant.

Conclusion: Towards a Theory of Boundaries and Bounding

The main argument in this chapter is that the process of bounding – drawing lines around spaces and groups – is a dynamic phenomenon, of which the boundary line is, more often than not, simply the tangible and visible feature that represents the course and intensity of the bounding process at any particular point in time and space. A deeper understanding of the bounding process requires an integration of the different types and scales of boundaries into a hierarchical system in which the relative impact of these lines on people, groups, and nations can be conceptualized as a single process. But the study of boundaries continues to take place within separate and distinct realms, be it the geographic and the state, be it the social and the group affiliations, or be it the political and the construction of ethnic and national identities. What is sorely lacking is a solid theoretical base that will allow us to understand the boundary phenomenon as it takes place within different social and spatial dimensions. A theory which will enable us to understand the process of "bounding" and "bordering" rather than simply the compartmentalized outcome of the various social and political processes.

A conceptual framework for the study of the boundary phenomenon would have to take three dimensions into account, all of which have been addressed in this chapter. First, the hierarchical nature of boundaries needs to be recognized allowing for the different types of territorial boundary – national and local, state and municipal – to be studied as part of a single body of theory that may vary in the specificities and intensities of the demarcation process and/or the socio-spatial outcomes, but which recognizes the inherent commonality of the bounding or bordering process. Secondly, it is particularly important for geographers to recognize that while their perception of boundaries is rooted in the organization and partition of territory, social and other aspatial boundaries are equally part of this process. For some groups, territory plays a significant role in the way in which their identity is expressed, for other groups it plays no role whatsoever, while for others, still, it is but one component of a complex socio-spatial dynamic. Thus, a solid conceptual framework for the understanding of boundaries must link the territorial with the nonterritorial ways through which the process of compartmentalization takes place. Thirdly, and linked to the previous point, boundary concepts must be understood as multidisciplinary phenomena. Geographers do not have a monopoly over the definition of just what is a boundary, just as non-geographers cannot understand the boundary phenomenon without recourse to the spatial and territorial dimensions of this dynamic. In order for boundaries to be understood more fully, it requires, in the first place, for boundary scholars to undertake their own form of trans-boundary movement into the disciplines and concepts of the other, those others who have traditionally been excluded from the exclusive boundary discourse practiced by the separate and different academic disciplines.

BIBLIOGRAPHY

Ackleson, J. 1999. Metaphors and community on the US-Mexican border: Identity, exclusion, inclusion and "Operation Hold the Line". *Geopolitics*, 4(2), 155–79.

Albert, M. 1998. On boundaries, territory and postmodernity. *Geopolitics*, 3(1), 53–68.

Blake, G. H. 1999. Geographers and international boundaries. *Boundary and Security Bulletin*, 7(4), 55–62.

Blake, G. H. 2000. State limits in the early 21st century: observations on form and function. *Geopolitics*, 5(1), 1–18.

Blake, G. H. 2000. Borderlands under stress: some global perspectives. In M. Pratt and J. Brown (eds.) *Borderlands Under Stress*. London: Kluwer Law International, 1–16.

Boggs, S. 1940. *International Boundaries: A Study of Boundary Functions*. New York: Columbia University Press.

Brunn, S. 1998. A Treaty of Silicon for the Treaty of Westphalia? New territorial dimensions of modern statehood. *Geopolitics*, 3(1), 106–31.

Dittgen, H. 2000. The end of the nation state? Borders in an age of globalization. In M. Pratt and J. Brown (eds.) *Borderlands Under Stress*. London: Kluwer Law International, 49–68.

Eskelinen, H., Liikanene, I., and Oksa, J. (eds.). 1999. *Curtains of Iron and Gold: Reconstructing Borders and Scales of Interaction*. Aldershot: Ashgate Press.

Falah, G. and Newman, D. 1995. The spatial manifestation of threat: Israelis and Palestinians seek a "good border". *Political Geography Quarterly*, 14, 189–206.

Gallusser, W. (ed.). 1995. *Political Boundaries and Coexistence*. Bern: Peter Lang.

Ganster, P., Sweedler, A., Scott, J., and Dieter-Eberwein, W. (eds.). 1997. *Borders and Border Regions in Europe and North America*. San Diego, CA: San Diego University Press.

Hartshorne, R. 1936. Suggestions on the terminology of boundaries. *Annals of the Association of American Geographers*, 26(1), 56–7.

Henrikson, A. 2000. Facing across borders: the diplomacy of bon voisinage. *International Political Science Review*, 21(2), 121–47.

Holdich, T. 1916. *Political Frontiers and Boundary Making*. London: Macmillan.

House, J. 1980. The frontier zone: a conceptual problem for policy makers. *International Political Science Review*, 1(4), 456–77.

Hudson, A. 1998. Beyond the borders: Globalization, sovereignty and extra-territoriality. *Geopolitics*, 3(1), 89–105.

Jones, S. 1945. *Boundary Making: A Handbook for Statesmen, Treaty Editors and Boundary Commissioners*. Washington, DC: Carnegie Endowment for International Peace.

Keating, M. 1998. *The New Regionalism in Western Europe: Territorial Restructuring and Political Change*. Cheltenham: Edward Elgar Press.

Kliot, N. 2001. Transborder Peace Parks: the political geography of cooperation (and conflict) in borderlands. In C. Schofield et al. (eds.) *The Razors Edge: International Boundaries and Political Geography*. London: Kluwer Law Academic, 407–34.

Kohen, M. 2000. Is the notion of territorial sovereignty obsolete? In M. Pratt and J. Brown (eds.) *Borderlands Under Stress*. London: Kluwer Law Academic, 35–48.

Kolossov, V. 1992. *Ethno-Territorial Conflicts and Boundaries in the Former Soviet Union*. Boundary and Territory Briefing No. 2. Durham: International Boundaries Research Unit.

Kolossov, V. and O'Loughlin, J. 1998. New borders for new world orders: territorialities at the fin-de-siècle. *Geojournal*, 44(3), 259–73.

Kristoff, L. 1959. The nature of frontiers and boundaries. *Annals of the Association of American Geographers*, 49, 269–82.

Lemon, A. 2001. South Africa's internal boundaries: the spatial engineering of land and power in the twentieth century. In C. Schofield et al. (eds.) *The Razor's Edge: International Boundaries and Political Geography*. London: Kluwer Law Academic, 303–22.

Martinez, O. 1994a. The dynamics of border interaction: new approaches to border analysis, In C. H. Schofield (ed.) *World Boundaries Vol I: Global Boundaries*. London: Routledge, 1–15.

Martinez, O. 1994b. *Border People: Life and Society in U.S.-Mexico Borderlands*. Tucson, AZ: University of Arizona Press.

Minghi, J. 1963. Boundary studies in political geography. *Annals of the Association of American Geographers*, 53, 407–28.

Newman, D. 1994. The functional presence of an "erased" boundary: the re-emergence of the "green line". In C. H. Schofield and R. N. Schofield (eds.) *World Boundaries Vol II: the Middle East and North Africa*. London: Routledge.

Newman, D. 1998. Creating the fences of territorial separation: The discourses of Israeli-Palestinian conflict resolution. *Geopolitics and International Boundaries*, 2(2), 1–35.

Newman, D. 1999. Into the millennium: the study of international boundaries in an era of global and technological change. *Boundary and Security Bulletin*, 7(4), 63–71.

Newman, D. 2000. Boundaries, territory and postmodernity: towards shared or separate spaces? In M. Pratt and J. Brown (eds) *Borderlands Under Stress*. London: Kluwer Law International, 17–34.

Newman, D. 2001. Boundaries, borders and barriers: on the territorial demarcation of lines. In M. Albert et al. (eds.) *Identity, Borders, Orders: New Directions in International Relations Theory*. Minneapolis, MN: University of Minnessota Press.

Newman, D. 2002. The geopolitics of peacemaking in Israel-Palestine. *Political Geography*, 21(5), 629–46.

Newman, D. and Paasi, A. 1998. Fences and neighbours in the post-modern world: boundary narratives in political geography. *Progress in Human Geography*, 22(2), 186–207.

Ó Tuathail, G. 1999. Borderless worlds: Problematising discourses of deterritorialization in global finance and digital culture. *Geopolitics*, 4(2), 139–54.

Paasi, A. 1996. *Territories, Boundaries and Consciousness: The Changing Geographies of the Finnish-Russian Border*. New York: Wiley.

Paasi, A. 1998. Boundaries as social processes: territoriality in the world of flows. *Geopolitics*, 3(1), 69–88.

Pratt, M. and Brown, J. (eds.). 2000. *Borderlands Under Stress*. London: Kluwer Law Academic.

Prescott, J 1987. *Political Frontiers and Boundaries*. Chicago: Aldine.

Prescott, J. 1999. Borders in a borderless world: Review essay. *Geopolitics*, 4(2), 262–73.

Ramutsindela, M. 1999. African boundaries and their interpreters. *Geopolitics*, 4(2), 180–98.

Ratner, S. 1996. Drawing a better line: Uti Possidetis and the borders of new states. *American Journal of International Law*, 90(4), 590–624.

Rumley, D. and Minghi, J. (eds.). 1991. *The Geography of Border Landscapes*. London: Routledge.

Shapiro, M. and Alker, H. (eds.). 1996. *Challenging Boundaries: Global Flows, Territorial Identities*. Minneapolis, MN: University of Minneapolis Press.

Sibley, D. 1996. *Geographies of Exclusion: Society and Difference in the West*. London: Routledge.

Sigurdson, R. 2000. Crossing borders: Immigration, citizenship and the challenge to nationality. In M. Pratt and J. Brown (eds.) *Borderlands Under Stress*. London: Kluwer Law International, 141–62.
Taylor, P. J. 1996. Territorial absolutism and its evasions. *Geography Research Forum*, 16, 1–12.
Thomas, B. 1999. International boundaries: lines in the sand (and the sea). In G. Demko and W. Woods (eds.) *Reordering the World: Geopolitical Perspectives on the Twenty First Century*. Boulder, CO: Westview Press, 69–93.
Waterman, S. 1987. Partitioned states. *Political Geography Quarterly*, 6, 151–70.
Waterman, S. 1994. Boundaries and the changing world political order. In C. Schofield (ed.) *World Boundaries Vol 1: Global Boundaries*. London: Routledge, 23–35.
Wilson, T. and Donnan, H. (eds.). 1998. *Border Identities: Nation and State at International Frontiers*. Cambridge: Cambridge University Press.
Yeung, H. 1998. Capital, state and space: contesting the borderless world. *Transactions of the Institute of British Geographers*, 23(3), 291–310.

Chapter 10

Scale

Richard Howitt

Contested ideas about space and scale have been influential in important recent debates in social science. Divergent concepts of space and ideas about its implications for social processes have been widely canvassed and hotly debated. Indeed, the emergence of new ideas about space is widely credited with challenging the previous dominance of historicism in the social sciences. While ideas of space remain important debating points, ideas of scale emerged in the 1990s to challenge dominant understandings of social and political processes. There has been vigorous debate about scale and its implications within political geography in particular. It is clear that scale certainly matters for critical geopolitics. This is particularly clear when one considers the words written about globalization, the nation-state, regionalism, and localism. Yet, for all this, scale remains a troubling and even chaotic concept.

There is a wide consensus amongst human geographers that the social construction of scale affects cultural and political landscapes. This is particularly obvious in the debates about both globalization and localism. Within economic geography, the dominance of a production-centered discourse has often reduced "politics" to consideration of the ways in which states and corporations have constructed scales for their economic or strategic benefit – at the expense of workers or others. In this discourse, issues of social reproduction, cultural dimensions and noneconomic issues of identity politics have been relatively unexplored. Yet in a wider notion of politics and political geography, it is these same issues that have gained prominence in the 1990s. The assertion of a "cultural turn," for example, was accompanied in many studies by a return to consideration of localism, specificity, and diversity. It is tempting – indeed, many have been tempted – to deal with this tension between economic and cultural discourses as a binary, and to conflate it with the simplified global–local scale binary. Discussing the politics of scale in this framework becomes a relatively simple matter, identifying the ways in which relatively local groups constitute their identity within a relatively local politics, and how they seek to counteract disempowerment by jumping scales to assert their specific concerns at a wider, more general scale. This seems attractive. For activist politics, it provides a way of engaging with the challenge of thinking globally and acting locally. Yet, like all binaries, this one has its limits. Conflating the global/economic/general and contrasting it with the local/cultural/specific obscures important dimensions that an alternative approach to scale might bring to critical geopolitical analysis, and responses built from it.

Part of the problem facing any contemporary discussion of scale issues in political geography, however, commences with an effort to explain just what this powerful concept actually means. While there is clarity about the nature of social construction, there is much less clarity about just what sort of a thing scale might be. This chapter reviews the ideas of scale that have emerged in political and economic geography, and their implications for critical geopolitics. It argues that one implication of the discipline's increased awareness of the "politics of scale" is that those trying to understand, participate in or influence spatial politics, need to conceptualize and analyse interconnections between scales and the simultaneity of those connections. This chapter considers in turn the implications of contested notions of scale for the critical geopolitics of environment, difference, place and power. Using the example of indigenous people's efforts to secure recognition of their rights and to influence contested cultural landscapes, it argues that a critical geopolitics that engages with the scale politics of power, identity and sustainability offers dispossessed, marginalized, and disadvantaged peoples a better framework for political action across and between multiple scales. This, in turn, requires geopolitical analysis to articulate and apply more sophisticated approaches to questions of scale.[1]

The Idea of Scale

Within human geography, there has been a robust discussion of the concept of scale in recent years. Two figures dominated discussion of scale in the 1980s: Peter J. Taylor (e.g. 1982, 1987, 1988, 1993, 1994, 1999, 2000a, 2000b) and Neil Smith (e.g. 1984, 1988, 1989, 1992, 1993; Smith and Ward, 1987). Both argued that scale was a fundamental concept in political geography, and their ideas have strongly influenced the terms of more recent debate. Drawing on Wallerstein's world systems theory, Taylor advocated a three-level model of scale in geopolitics. He identifies "world-economy," "nation-state," and "locality" as the three critical scales at which the processes of the world economy are manifest (e.g. 1993, pp. 43–8). Smith, who maintains Taylor's notion of a hierarchy of scales, highlights urban, regional, national, and global as the critical scale categories in his analyses. In their contributions, Taylor and Smith both advocated a politics of engagement that was oriented to a practical geopolitics consistent with Harvey's earlier advocacy of an "applied people's geography" (Harvey, 1984). For both, however, scale categories remained rather more fixed than more recent debate has suggested. Agnew (1993) argued against reification of specific scales as distinct levels of analysis, but acknowledged that because different disciplines had come to specialize in analysis at different scales, integration of analysis across scales had become increasingly difficult. It is precisely this issue – undertaking meaningful analysis across scales or at multiple scales – that has been so troubling in operationalizing scale as a fundamental concept with practical rather than merely rhetorical value.

In contrast to the rather rigid concepts advocated previously, more discursive and relational notions of scale have emerged since the early 1990s. Howitt (1993a) rejected the idea that scale categories are ontological givens, and questioned the previously unquestioned assumption that scale was necessarily a matter of nested hierarchies. An editorial in *Society and Space* (Jonas, 1994) marked a new point of departure for discussing social relations as an element of scale. Jonas emphasized that the political dimensions of spatiality constitute a core issue in conceptualizing scale. Taking up Massey's challenge (1992) to develop a dynamic concept of the spatial in the domains of politics, he sought to untangle the links between "scale as abstraction" and "scale as metaphor," pointing out that the tension between globalization and locality research was often a product of research frameworks that were having trouble dealing with the simultaneity and complexity of power relations, identity and difference that Massey saw as challenging naïve notions of space. Jonas' piece clearly reflected a rapidly growing momentum to move beyond rigid scale labels and naïve conceptualizations of scale itself. His call for a move towards a more sophisticated discussion of the "scale politics of spatiality" was quickly added to by both theoretical and empirical contributions.

In 1997, the journal *Political Geography* ran a special issue under the title "Political Geography of Scale" (Delaney and Leitner, 1997a). Guest editors Delaney and Leitner suggested "scale is a familiar and taken-for-granted concept for political geographers and political analysts" (Delaney and Leitner, 1997b, p. 93). They opened with a confident definition of scale as "the nested hierarchy of bounded spaces of differing size, such as the local, regional, national and global" (p. 93) and asserted that scales are periodically transformed and constructed. The four papers in this special issue advocated a "constructivist" approach to scale and taken together they provide a powerful opening in what the editors saw as "a theoretical project that necessarily involves attention to the relationships between space and power" (p. 96). But despite their best efforts, they found that scale remained elusive:

> The problematic of scale in this context arises from the difficulties of answering the question: once scale is constructed or produced, where in the world is it? Scale is not as easily objectified as two-dimensional territorial space, such as state borders. We cannot touch it or take a picture of it (Delaney and Leitner, 1997b, pp. 96–7).

Since that special issue, scale has been an almost constant presence in the pages of *Political Geography.* Some eighteen papers have considered scale as their theoretical focus. Clearly, scale has been accepted as a central and contested idea in both the journal and the discipline. In 1998 scale was a major concern of Cox's contribution and a series of commentaries (Cox, 1998a, 1998b; Jones, 1998; Judd, 1998, M. P. Smith, 1998). In 1999, a paper by Morrill raised considerable comment concerning the role of jurisdictional issues in mediating conflicts across scales (Fainstein, 1999; Martin, 1999; Morrill, 1999a, 1999b; Swanstrom, 1999). In 2000, Taylor's paper on "world cities and territorial states" in conditions of globalization raised important issues of the role of nation-states and trade blocs as a "nexus of power which straddles geographical scales" (Taylor, 2000a, p. 28; see also Douglass, 2000; Shapiro, 2000; Taylor, 2000b; Varsanyi, 2000).

Cox (1998a) pointed out that scale is a central concept in political discourse. In seeking to clarify the "spaces of engagement" that constitute local politics, he also sought to unsettle previously dominant concepts of scale (also 1993, 1997, 1998b; Cox and Mair, 1989, 1991). His paper argued that there is a scale division of politics in which it is relationships between scales rather than just jumping between them that offers a new view of local politics. Commentary on Cox's paper highlighted the importance of context in dealing with ideas of scale. K. Jones (1998) considered the way that jumping scales really involves a politics of representation, with local groups "actively reshaping the discourses within which their struggles are constituted (and) discursively re-present(ing) their political struggles across scales" (1998, p. 26). She also notes the epistemological concerns about scale categories, and the way that certain concepts of scale render some questions simply unaskable. Judd (1998) responds by reminding us that the power relations that are constructed by the state's construction of scales in material forms through jurisdictional, administrative and regulatory structures, restricts the flexibility of resistance considerably more than Cox allows. M. P. Smith (1998) takes Cox to task for being too vague in terms such as "more global." He criticizes Cox for relegating the "global" to a conflated presence with "scales like the regional and the national" (1998, p. 35). He draws on his own work on cross-border, transnational migrant identities (e.g. M. P. Smith, 1994) to remind us that it is the social construction of networks, identities and relationships that constitute the scaled spaces of engagement that Cox highlights.

In the same journal, Morrill considered how different jurisdictional scales are harnessed by powerful vested interests to their own purposes. In particular, Morrill was concerned to address the question of "whether there is an optimum or appropriate level of decision-making or balance of power across geographical scales" (1999a, p. 1). Using a case study of decision-making about future uses of the Hanford nuclear reservation site in Washington state, Morrill argues that in the US higher levels of government are increasingly harnessed (usually by capital) to pre-empt local decision-making and impose "metropolitan values and preferences" (1999a, p. 2). He points out that federal regulation of the nuclear industry circumscribes local autonomy at Hanford from the start, but that planning processes generally favoring metropolitan priorities over rural concerns reinforces this. Swanstrom (1999) contradicts Morrill's conclusion by suggesting that the absence of local planning and land-use regulation from central authorities characterizes the decision-making process in the US, and suggests that Morrill's policy suggestions to support local autonomy are flawed. Martin (1999) unpacks the assumed congruence of local interest groups and local government, advocating a view of cross-scale relationships that is based on a more careful consideration of multiple interests and social identities at each scale implicated in a decision-making chain. Fainstein (1999) suggests that Morrill has misread some aspects of the Hanford case as demonstrating the power of higher levels of government, because the outcomes at Hanford represent a reduction of federal control of the site. Perhaps the most interesting issue emerging from the discussion of Morrill's paper is the assertion that one scale (the local) does or does not warrant privileging as more politically or environmentally "correct" (Morrill, 1999b, p. 48).

Ideas into Practice: Empirical Studies of Scale

This debate about the practical implications of theoretical work is perhaps one of the most important issues of debate in the recent scale literature. Even before there is consensus on the "what" of geographical scale, there is plenty of heated discussion of the "so what" questions. Many commentators have struggled with the apparent paradox of scale – that it matters, but is almost meaningless as a stand-alone concept: it only matters in context – as a co-constituent of complex and dynamic geographical totalities. This paradox leads us back to the issue of "social construction," and a number of studies that seek to clarify the ways in which scale jumping strategies allow us to better conceptualize the construction of scale. Drawing on N. Smith's work on the social production of scales, Swyngedouw argues that "theoretical and political priority... never resides in a particular geographical scale, but rather in the process through which particular scales become (re)constituted" (1997a, p. 169). Unlike Smith, however, Swyngedouw incorporates Massey's innovative ideas of the "geometry of power" (e.g. Massey, 1993a, see also 1992, 1993b, 1994, 1995) and takes seriously the considerable tension between the economic, political and cultural domains in relation to the social construction of scale (see also Swyngedouw, 1992, 1997b). His awkward neologism – "glocalization," the simultaneous and contested shift up-scale towards the global and downscale to the local as a response to changing economic, political and cultural pressures – is one of many he coins to meet the needs of a new scale vocabulary. Swyngedouw's great contribution has been his insistence that the nature of scale politics is to be found not in a theoretical discourse, but in the real-world practices of social conflict and struggle. Although he maintains an unexplained commitment to nesting of scales,[2] Swyngedouw's effort to provide a new vocabulary of scale has been extremely helpful.

In moving from the abstract discourses of a "theory of scale," there have been many efforts to clarify the sort of impacts scale has in practice. Adams' investigation of the way telecommunications create new linkages across space (1996) emphasized the importance of networks of relations rather than areally bounded and hierarchically nested places as a constituent of scales. He considers the scale at which protest, resistance, autonomy, and consent might be constructed. He considers the networks and flows of information, recognition, and support constructed through telecommunications technology in Tiananmen Square in 1989, in the popular movement for democracy in the Philippines in the mid-1980s, and the US civil rights movement in the mid-1950s. Each of these examples demonstrates the ways in which scaled and territorially bounded jurisdictions are unable to contain or control protest movements' abilities to jump between scales. The paradoxical and simultaneous harnessing of and harnessing by mass media constructs new audiences for (and supporters of or participants in) protest. There is no nested hierarchical vision of scale relations in Adams' account. Kelly also rejects the idea of hierarchy in his investigation of the place-based politics of a power station in Manila, Philippines, to advocate the case for a view of scales as constructed rather than absolute categories (1997). His paper offered the sort of detailed reading of the "translation of the globalization discourse into development policies" (1997, p. 151) needed to get beyond a rhetorical consideration of scale in the emergent discourse of globalization. In contrast, Leitner adopts a "constructionist perspective" on scale, understood as a "nested hierarchy of political spaces" (1997, p. 125) to consider the institutional context of migration in Europe. Herod and Agnew have also provided widely cited empirical studies. Herod's work on the scale politics of labor restructuring in the US (e.g. 1991, 1997a, 1997b) and Agnew's work on post-1992 political restructuring in Italy (e.g. 1997) have both cast considerable light on the processes referred to as "social construction." Herod considers the way in which organized labor's approach to contract bargaining in the eastern US in the 1970s constructed new geographical scales. In the first instance, inter-union rivalry and technological change in the late-1950s produced a national-scale bargaining strategy which pushed the International Longshoreman's Association's focus upscale from regional agreements to a master national agreement. By the mid-1980s, employer reorganization and changes in working conditions around the US produced a scale politics in which the use of non-union labor in southern ports undermined the power of the master contract to meet the needs of many of its intended beneficiaries. Herod's analysis provides a powerful demonstration of how it is particular relationships, developed in specific institutional, technological, political and economic contexts that constitute the scales which themselves become institutionalized as self-evident and embedded in real-world economic geographies. Rather than organized labor, Agnew focuses on political parties and how they are implicated in "writing the scripts of geographical scale" (1997, p. 101), emphasizing the role of political parties in linking individuals to collective action by articulating goals around which people can be mobilized. The institutional organization of electoral processes link parties, policies, and populations to particular places in particular ways, and bring them together in organized political relationships. Their mediation and utilization of the politics of difference, identity and territoriality contribute to the constitution of the state – whether this is in terms of local, regional, national or supra-state governance. The collapse of old-style parties and the emergence of new-style parties in 1994 defined new scale relationships, even if they fitted within the old spatial boundaries of the nation.

Although less cited than work from North America and Europe, Fagan (e.g. 1995, 1997; Sadler and Fagan, 2000), Howitt (e.g. 1993a, 1998a), McGuirk (1997) and others have forged an Australian perspective on scale issues which advocates a radically relational approach. Howitt's (1993a) critique of the dominant thinking about scale suggested that the idea of scale as a set of nested hierarchies was totally inadequate for understanding scale politics, and that the widespread conflation of scaled ideas had produced conceptual confusion in many presentations. He advocated an empirically grounded dialectical approach to investigation of scale issues. Fagan (1995) offered a powerful critique of the difficulty geographers were having in handling the idea of globalization and its implications for action, resistance, and responses at other scales – and geographers' analysis of and contributions to them. He pointed out that the very processes that were being rhetorically constructed as fundamental to an irresistible globalization "can be constructed as *local*" (1995, p. 7, his emphasis). His careful examination of "the region as political discourse" provided a scaled analysis of political economic changes in Australia and the Asia–Pacific region that considered the nature of power relations within and across scales as critical to political process and real-world geographies. Howitt et al. (1996) argued that indigenous and other resistance to the New World Order advocated by the US in the Gulf War was, in large part, a contestation of resources, identity, and territory and was producing new geopolitical relationships across scales. Such shifts in scale produce new analytical interest in scales, places, and relationships that were previously of only marginal interest to political geography (Goldfrank, 1993; Routledge, 1996). McGuirk (1997) was also concerned to move beyond rhetorical discussion and applied a relational view of scale to her careful analysis of urban planning issues in the western suburbs of Adelaide. Contra the widely advocated view that globalization was driving development processes on the ground and subordinating local relationships, McGuirk's paper demonstrates just how wrong it is "to regard localities and regions as being at the mercy of external uncontrollable and mythologised global forces, because they are themselves a formative part of global processes" (1997, p. 493). Fagan (1997) reflected on the way in which the local–global debate in academic circles was paralleled in political circles. His examination of restructuring the Australian food processing industry returned to his theme of the need to integrate global and local analyses in a nondeterministic and politically informed way.

Social Construction: the Consensus View of Scale

Domestic scale and social constructionism

A recent review by Marston focused on the issue of social construction. She argues, correctly, that much of the recent literature reinforces the separation of the economic domain, and specifically a productionist view of the economic, from wider issues of social reproduction and consumption. Indeed, despite the so-called "cultural turn" in geography, attention to cultural (and cross-cultural) issues in the discussion of scale remains limited. In advocating a view of scale as having (at least) three dimensions – size, level, and relation – Howitt (1998a) re-emphasized the importance of social relationships in space as fundamental in constructing geographical scales. Following Howitt, and Swyngedouw (1997a), Marston (2000) offers an expanded concept of scale that encompasses the domains of reproduction and consumption as well as production, as a synthesis of the recent debates.[3] Her presentation of gender dynamics points out that changes in women's roles in social reproduction and consumption in the late nineteenth and early twentieth centuries in the US not only created new spaces – new domestic spaces, new retailing spaces, new social spaces – but also new scales by organizing social relationships in new ways. Marston convincingly explores the ways that the dynamics of these social processes and networks are embedded in the changing relationships between public and private domains, between retailing, production, media, politics, and the institutions of governance. She suggests that we can see these processes producing new scales such as "home" and "neighborhood" in ways that echo loudly not just in the political geographies of the US in the early twentieth century, but throughout the contemporary world.

Social constructionism and the scales of justice

Since the Marxist and behaviorist challenges in the 1970s to positivist efforts to constitute geography as a spatial science, social justice has been a key concern of politically engaged geographers (Swyngedouw, 2000). The traditional image of "blind justice" finding solutions to conflict without fear or favor using mechanical scales offers a fortuitous starting point for our discussion of geographic scale. The image of scale as a relatively straightforward mechanism that juxtaposes, compares or relates phenomena in space and time is consistent with the image of geographic scale as a set of distinct platforms upon which geopolitics (and other social phenomena) are performed. Building on this image, it has been easy to privilege one scale or another as the pre-eminent platform for political action. International relations, for example, posits the nation-state and its interaction with other nation-states as pre-eminent, as did conventional geopolitics. World-systems theory posits the global sphere as the most significant scale (Taylor, 1988, 1993, 2000). Locality studies have privileged the local as the scale at which meaning or lived-experience is constructed. The paradoxical positions taken on local, national, and global scales was a starting point for much of the critical discourse on scale referred to above.

In contradiction of the neat schemas of scale as a nested hierarchy, neither geopolitics nor social justice are reducible to a single dimension – in space, in time, or in cultural relations. Peoples' struggles for justice, their efforts to construct new geographies of justice, are always multifaceted. They always reflect (at least) economic, cultural, and environmental politics. In her seminal paper on social justice, Fraser (1995, also 1997) used a bipolar tension between the old-style socialist (economic) politics of distribution and the new post-socialist (cultural) politics of identity to make the point that a new, "post-socialist" dynamic had to be addressed in the social justice movement. In proposing a contradiction between the strategies of redistribution and recognition, Fraser's analytical framework lost sight of geography, territory and scale as key constituents of political relationships in the real world. Like the "level playing field" of the free market imaginary, Fraser's placeless analytical framework has powerful pedagogic and rhetorical value, but it misses the point that concrete social relationships are always placed and scaled. Critical geopolitics has sought to meet the challenge of dealing simultaneously with issues of justice, equity, and sustainability at multiple scales.

No simple schema that privileges a singular scale as the essential scale at which justice can be achieved is reasonable. And no schema that excludes the scale politics of place, territory, and power will adequately address the nature of geopolitics or the struggle for social and environmental justice. Again, these concerns return us to the issue of the social construction of scale. Harvey tackles this issue, and follows N. Smith in conceptualizing social processes operating in a way that produces "a nested hierarchy of scales (from global to local) leaving us always with the political–ecological question of how to 'arbitrate and translate between them'" (1996, pp. 203–4, quoting N. Smith, 1992, p. 72). Harvey usefully goes on to discuss the ways in which social conceptions of space and time are constructed in social processes and simultaneously become objectified as pervasive "facts of nature" (1996, p. 211) that regulate social practices. Yet neither Smith nor Harvey is clear why the social construction of scale produces a nested hierarchy of scales. Howitt (1993a) argued against the notion of both nesting and hierarchy as adequate metaphors for geographic scale, suggesting that it was in cross-scale linkages, awkward juxtapositions and jumps, and non-hierarchical dialectics that the nature and significance of scale is to be found. Swyngedouw (1997a, b) follows a similar approach, but retains a notion of "nesting" while rejecting some aspects of "hierarchy."

Reconsidering social construction: indigenous and environmental issues

So, what are the mechanisms of social construction of scale? Using struggles for social and environmental justice, let us take a step deeper into this issue. Cox (1998a) suggests that it is not the social construction of scale that matters, but the social construction of the politics of scale. Using a focus on the institutions of local governance, Cox identifies a "scale division of politics" (1998a, p. 1). He advocates a shift away from an "areal concept of scale" (1998a, p. 19) to a view of scale as the spatial form of social networks. Marston's weaving of social reproduction and consumption into her ideas about the construction of scale, alongside issues of economic and political processes, accuses those who have focused on the political and economic dimensions of scale of telling "only part of a much more complex story" (2000, p. 233). She emphasizes also the way in which the social construction of this less-than-local scale in turn influenced the practices of social reproduction and consumption in ways that were quite profound, and which "reached out beyond the home to the city, the country and the globe" (2000, p. 238).

Silvern takes another US example – the efforts by Wisconsin Ojibwe to utilize treaty rights to influence natural resource conflicts – to consider how the scale at which sovereignty is constituted reflects an ongoing struggle "over the control of territory and the political construction of geographical scale" (1999, p. 664). In Silvern's study, as in Marston's, scale is simultaneously constructing and reflecting the spatial form of social relations. In Marston's study it is gender politics that takes priority in understanding the construction of the domestic scale in emergent US capitalism, while in Silvern's study it is the ethno-politics of conquest and dispossession that underscores the creation of Federal and state sovereignties in US legal proceedings, while denying the legitimacy of tribal sovereignty. Despite the long-standing doctrine of a tribal sovereignty, constrained by European legal principles of "discovery," derived from the decisions of US Chief Justice Marshall in the 1820s and 1830s (see e.g. Canby 1988), Silvern reports that the state of Wisconsin sought to restrict the exercise of tribal rights to co-manage natural resources by severely circumscribing the geography of Ojibwe treaty rights. Like Marshall's court 150 years earlier, the state's courts found that it was nontribal principles that defined their jurisdiction and the scope of their capacity to recognize a sovereign entity constructed external to that jurisdiction. Despite some success in securing co-management standing through the courts, the Ojibwe were unable to establish what Silvern refers to as "scale equivalency to the state when it came to management of... resources" (1999, p. 661). Notzke (1995), from a Canadian perspective, similarly sees questions of co-management rights as representing a challenge to jurisdictional and constitutional sovereignty. McHugh (1996) suggests that indigenous peoples' efforts to establish recognition of tribal sovereignty in New Zealand, Canada, and Australia has established significant constraints on the institution of the Crown in those jurisdictions. Following Silvern's reasoning, this affects the ability of state, provincial, and national governments in those countries to construct hierarchical scale systems that exclude or deny the existence of "tribal" as a geographic scale. It is colonial (and postcolonial) states that have assembled instruments of power and institutions of state administration into nested, hierarchical geographical scales that "facilitate the power of the dominant society to control, exclude and marginalize native populations" (Silvern, 1999, p. 665).

Jhappan (1992) offers another example, this time at the level of international relations, of the ways in which the indigenous peoples movement has succeeded in upsetting such taken-for-granted nested hierarchies of control, exclusion and marginalization, and in the process, have challenged the dominant view of scale as an areal concept (scale as size) or a hierarchical concept (scale as level). Drawing on alliances with organized labor, international organizations within the United Nations and European Union, environmental organizations and consumer groups in other jurisdictions, and diverse political alliances within Quebec, Canada, and the international indigenous peoples movement, the James Bay Cree lobbied to stop the massive Great Whale River hydroelectric project. Weaving together a potent combination of local tribal governance and political action, jurisdictional standing as regulators based on modern treaty rights, and effective provincial, national, and international campaigning, Jhappan sees the Cree as modifying Quebec and Canadian government policy options, and, in the process, challenging the "nation state's uncontested sovereignty over domestic policy" (1992, p. 61). Cohen (1994) offers an account of the cross-border alliances between the Cree and environmental and energy consumer groups in the northeast US and the ways in which institutions developed as part of the 1975 treaty settlement provided the vehicle for a new tribal scale to influence the fate of the Great Whale project twenty years later, while Puddicombe (1991) suggests that Inuit institutions developed in the same way adopted a very different scale politics in response to the Great Whale project.

Williams (1999) uses scale as a tool to explore the politics of environmental racism in the USA, and suggests that scale is not only socially produced, but also produces social outcomes (socially generative). He identifies a scale politics that "centers on an antagonistic relationship between a societal problem and its political resolution" (1999, p. 56). The acceptance of common ground between environmental justice advocates and the objects of their criticism often focuses on ideological notions of procedural fairness and equitable distribution of costs and benefits. Williams notes, in relation to distributional issues affecting environmental risk, that the ability of powerful institutions to convince regulators and a wider public that they have followed fair procedures allows an impersonal (and highly valued) "market" to justify distributional outcomes – reducing critics to rather self-interested and locally myopic players. In the process, Williams suggests, industry "wins" the scale politics of environmental justice (1999, p. 66). It is worth observing that powerful institutions (governments, corporations, even some social justice groups) are also able to re-write the local scale politics as constituted by much wider scale forces – recall, for example, Fagan's suggestion (1997) of the way that manipulation of brands by global food-producing companies reconstitutes a powerful global corporation as a local heritage value.[4]

Silvern (1999) also opens a window on an environmental politics of scale in his consideration of tribal efforts to argue that they have regulatory standing in state decision-making about mining and other environmentally degrading activities. This parallels Morrill's concern (1999a) with the issue of jurisdictional scale in planning and land use decisions. M. Jones (1998) takes up similar issues in relation to the changing nature of local governance in the UK, calling for "relational theory of the state" to adequately address the shift from local government to local governance. MacLeod and Goodwin suggest that many of the institutional responses to globalization, regional restructuring and localism in Europe, have failed to problematize scale and consequently "appear to treat as ontologically 'pre-given' the scalar context" of their work (1999, p. 711).[5]

Social Construction as Social Action: Lessons About Scale from the Indigenous Rights Movement

The need for a scaled analysis in critical geopolitics is particularly obvious in the case of indigenous rights, where the construction of postcolonial nation-states was predicated on the dispossession and marginalization of indigenous peoples. The construction of territorial authority over indigenous domains has involved the construction of specific scales of social control (the mission, the community, the reservation), political representation and participation (the "tribe," defined by government), service delivery, governance and recognition (the department, the Bureau, the Commission). At national and sub-national levels, governments established legislation and systems of social control that sought to define indigenous peoples as people without geography (Dodson, 1994; Howitt, 1993c). At the scale of the body, indigenous people were disciplined to conform or be punished. Disciplined through banishment or integration, indigenous identities were subject to the most invasive levels of control – removals of children from families, outlawing of languages and cultural practices, replacement of names, imposition of mission- or government-arranged marriages, and special controls on wages, movement and activities. At the scale of family and clan, indigenous peoples were disciplined by processes of territorial domination, displacement and relocations, threats and exercise of force, and the spatial discipline of new settlements and "communities." At the scale of the nation, indigenous societies were disciplined by dishonored treaties,[6] legal frameworks which denied the existence of their ancient jurisdictions and took their ancestral domains from them, political systems which simply excluded them from democratic process, and economic practices that ignored or bypassed their property rights, skills, and aspirations. At the scale of the international system, the club of nation-states that had dispossessed them established new institutions that restricted their access to international arenas for legal and political redress.

Despite this, indigenous politics provides many examples of the harnessing of scale analysis to the purposes of social transformation – to simultaneously pursue the economic politics of redistribution, the cultural politics of recognition, and the environmental politics of sustainability. In my own experience, the issue of just what scale is has been greatly clarified in my work alongside indigenous colleagues in actually rebuilding the scales of family, clan, language group, tribe, and peoples in the wake of Australia's unacknowledged genocide (Tatz, 1999; see also Human Rights and Equal Opportunity Commission, 1997). In Australia, Aboriginal groups have long struggled to overcome the legacies of the colonial and postcolonial fragmentation of their traditional domains. This has never been just a matter of jurisdictional recognition of property rights. Official indifference to more radical aspects of a reconciliation agenda – including a naïve and self-interested assertion that negotiating treaties in Australia would divide national sovereignty – have left little room for political maneuver. Yet it is in the scale politics of identity, difference, territory, and governance that opportunities can be found.

In rebuilding indigenous governance, the process of social construction of geographical scales is laid bare. To construct the means of new forms of social, economic or political participation, the networks and relationships that bring people together must undergo transformation through their confrontation with, marginalization from and interpenetration by the institutions, relations and processes of existing complexes of territory–governance–identity. In Quebec, for example, the 1975 negotiation of the James Bay and Northern Quebec Agreement provided for new institutional arrangements for local governance and participation among the Cree and Inuit peoples. In 1971, these communities were brought together for the first time as a people – in response to Hydro-Quebec's proposal to regulate every single river in the north:

> For the first time in history, the Cree sat down together to discuss their common problem – the James Bay Hydroelectric Project. But we found out much more than that – we found out that we all survive on the land and we all have respect for the land. Our Cree Chiefs also found out that our rights to land, our rights to hunt, fish, and trap and our right to remain Crees were considered as privileges (not rights) by the governments of Canada and Quebec. (Billy Diamond, then Cree chief at Rupert House and later lead negotiator in the James Bay and Northern Quebec negotiations, quoted in Feit, 1985, p. 40)

Twenty-four years later, the Cree institutions established through negotiation of Canada's first modern treaty were advocating secession from Quebec if Quebec seceded from Canada (Grand Council of the Crees, 1995). By bringing together cultural, territorial, environmental, economic, and jurisdictional concerns – and by doing this in the context of on-going transformational relations with provincial, federal and international authorities (see e.g. Cohen, 1994; Jhappan, 1992; also Howitt, 2001) – the Cree succeeded in constructing a new scale.

This scale of tribal governance is clearly *not* an ontological given. It never appears in the standard scale lists of "local," "national," and "global."[8] It does not even appear in those extended lists that include the scale of "the body," "home," and "infinity" (see Howitt, 2002). In recent work in South Australia, the reality of constructing such a new scale has been driven home to me yet again. In preparing for negotiations with the state government of South Australia, native title claimants[9] have been brought together as a congress to discuss how they might construct a way of negotiating with a united voice that does not subsume their local autonomy as distinct groups, with distinct traditions, values, histories, and experiences. As I have noted elsewhere (Howitt, 2002), in pursuing "regional agreements" to recognize native title, the "region" cannot be assumed. The spatial, social, and political scale (which are best seen as co-constituents of "geographical scale," perhaps along with ecological, economic, and other dimensions of scale) must be dealt with as concrete relationships of mutual recognition, accountability, and acceptance if the idea of scale is to become a meaningful vehicle for indigenous peoples engaging with the transformational politics of negotiation with state, corporate or other interests about native title, reconciliation or sovereignty. Current discussion of a single national treaty in Australia is doomed to fail until there is some success in realizing the national scale as a meaningful scale of indigenous identity. In South Australia, the nascent state "Congress" (Agius et al., 2001) will rely upon a dialectical engagement with its own Aboriginal constituents, and its state and industry negotiation partners. A group that claims representativeness without the concrete network of relationships that constitute a geographical scale of "state-wide indigenous congress" will soon find itself criticized as discredited in the communities. Similarly, a well-developed state-wide network that is not recognized by the state and other powerful groups, will soon fall prey to fragmentation and division.

In other words, the social and political construction of scale is precisely social action – the concrete processes of organizing a political response, a vehicle for participation, recognition, and change. This is always, as so much of the work cited above demonstrates, a matter of links within and across scales to provide opportunities for transformation of existing power relations. What is crystal-clear in indigenous politics is the need to link social, cultural, territorial, and institutional relations in constructing geographic scales at which social action may occur. For other groups, access to existing institutions has perhaps masked some aspects of the political construction of scale. Or, as Marston (2000) points out in relation to women's actions in the construction of new geographic scales, the blindness of the dominant productionist paradigm has rendered their action virtually invisible. But of course, Herod's trade unions, Agnew's political parties, Fagan's food corporations, McGuirk's urban planners, Kelly's Philippine activists, and the other scale-builders whose actions are to be glimpsed through the literature trying to make more sense of their activity, are all engaged in the same sorts of processes. They seek to mobilize social networks, political institutions, economic resources, and territorial rights to the task of creating new geographies – new landscapes of power and recognition and opportunity.

Conclusion: Scale and Critical Geopolitics

If critical geopolitics is about some form of "critical engagement" (Routledge, 1996) or "situated engagement" (Suchet, 1999) and supporting dissent – understanding it, fomenting it, participating in it, responding to it – it is apparent that scale is an important issue because both analysis and dissent are necessarily engaged in addressing and crossing scales. Whether it is a question of organized labor seeking an appropriate forum to contest employers' privileges in setting wages, working conditions or other issues; or marginalized indigenous peoples seeking to rewrite the rules of engagement with postcolonial societies and states; or environmentalists seeking to curtail the impacts of globalization on ecological sustainability – relationships and issues at one scale are actually reconstituted and need new tools of engagement, analysis, and response.

The challenge of scale in contemporary political geography is that it presents a paradox. On the one hand, it seems self-evident. Scale is a term that easily slips into our discussion because the scaled processes of "globalization," "national sovereignty" and "local action" that are the taken-for-granted focus of so much political geography are so obvious. Similarly, it is equally obvious that scales are socially and politically constructed. Yet, when one tries to offer a definition of just what is being constructed, most attempts are unsatisfactory. In the 4th edition of the *Dictionary of Human Geography*, N. Smith (2000, p. 727) takes 2½ pages to arrive at the statement that the "question of scale will become one of mounting theoretical and practical relevance," but does not provide a definition. The nature of scale, then, is paradoxical. But the recent literature on scale has rendered the reason for this much clearer. For a long time, it was assumed that scale was a question of either size or level (e.g. of complexity). What emerges from the recent literature is that scale is preeminently a matter of relation, and that approaches which seek to summarize this dimension with the gloss of labels such as "global" or "local" without engaging with what is actually encompassed in context by the term, will actually miss the substance of the term and the phenomenon it represents. Like another quintessentially geographical term "place," "scale" is rendered most meaningful in its development as an empirical generalization – a concept made real by building up an understanding of complex and dynamic relationships and processes in context. As a theoretical abstraction the risk is that "scale" is reduced to a set of meaningless labels that say something about size and complexity, but which hide precisely the terrain with which critical geopolitics is most interested – the terrain of real landscapes in which spaces of engagement offer a myriad of transformational opportunities at a myriad of scales.

What is paradoxical, perhaps, is not the nature of scale, but geographers' efforts to theorize scale in some way that divorces it from its geographical context. If the role of our theory is to better equip us for our situated engagement in struggles for justice, sustainability, and transformation, then theory divorced from the scaled landscapes of change is probably of limited value.

ENDNOTES

1. Despite the broad literature drawn upon in geographers' discussions of scale, it remains a poorly understood concept within the discipline, and virtually unacknowledged beyond it. For example, a recent literature review in *Ecological Economics* (Clark et al., 2000) limits its consideration to quantitative concerns, citing only the rather naïve Meyer et al. (1992), Harvey (1969), Jammer (1954) and Forer (1978) to represent discussion of scale issues in geography!
2. For example, in the conclusion to his 1997 paper in Cox's collection (1997), Swyngedouw refers to "a nested set of related and interpenetrating spatial scales that define the arena of struggle, where conflict is mediated and regulated and compromises settled" (1997a, p. 160). The inclusion of the term "nested" in this passage is not supported by much of his previous discussion and seems more a legacy of earlier assumptions than a product of the reflection presented in this paper.
3. It is not only in ecological economics that much of the literature Marston reviews is missed. In their 1999 discussion of geopolitics, identity and scale, Herb (1999) and Kaplan (1999) refer to none of this conceptual debate, other than Taylor's work. Their interesting discussion of the interdependence of territoriality, identity and geopolitics, and their reliance on the idea of scale, ultimately reproduces a complex nesting metaphor, in contrast to the relational notions of contested sovereignties discussed below.

4. I have made a parallel point in discussion of the scale politics of social and environmental impact assessment, where benefits of a proposed development are aggregated to present persuasive "state" or "national" benefits, while social and environmental costs are often represented as "merely local" and parochial (Howitt, 1993b). See also a similar point about the political tension between "vested" and "representative" interests in the Australian mining industry (Howitt, 1991).
5. A related question is raised by Wilson et al. (1999) in their consideration of "scale misperceptions" in the management of social–ecological systems. The imposition of conservation area and other jurisdictional boundaries on the development of ecological relations such as nonhuman populations, clearly affects management options. This common mismatch has increased the pressure for bioregional planning as a way of matching ecological and administrative boundaries (see, e.g., Brunkhurst, 2000).
6. The irony, of course, is that the revolutionary pariah state of another century was the US, whose existence was first acknowledged in international law by treaties with First Nations that were later to be subsumed as "domestic dependent nations" (see, e.g., Williams, 1990).
7. Australia is a federal state, with national sovereignty already divided between six colonial states, each of which retains a direct link to the Queen of the Australian Commonwealth, who is, of course, also the Head of State of the UK. This division of sovereignty never troubles conservative and racist criticism of Aboriginal and Torres Strait Islander efforts to reassert their own sovereignty. Indeed, a recent referendum on a shift from monarchy to republic status for the Commonwealth was rejected. These issues are taken up in more detail in Howitt (1998b).
8. Indeed, in one of the key early texts that raises issues of scale, N. Smith makes First Nations completely invisible in his rendering of the American landscape as "poetic nature" (1984, p. 7). In a later paper Smith reinforces his marginalization of First Nations in a throwaway reference where he suggests that "the whole Lower East Side, not just the park, had become 'Indian Country'" (1993, p. 93). This did not mean that there had been a recognition of tribal ownership of the neighborhood. Indeed, Smith's reference is a careless reinforcement of indigenous invisibility. It parallels Soja's cacophonous blindness to the First Nations of California in his influential account of the history of Los Angeles which, like so much ostensibly "radical" geography, places indigenous peoples quite literally outside geography! (see also Howitt, 1993c).
9. Native title rights were formally recognized in Australia by a High Court decision in 1992. Indigenous claimants must lodge native title claims for adjudication under new legislation, enacted in 1993 and amended in ways that many, including the United Nations Committee on the Elimination of Racial Discrimination (W. Jonas, 2000, chapter 2), found racially discriminating.

BIBLIOGRAPHY

Adams, P. C. 1996. Protest and the scale politics of telecommunications. *Political Geography*, 155, 419–41.

Agius, P., Davies, J. Howitt, R. and Johnson, L. 2001. *Negotiating Comprehensive Settlement of Native Title Issues: building a new scale of justice in South Australia*. Paper presented to the Native Title Representative Bodies Legal Conference, Townsville, Qld. [available on request from the author].

Agnew, J. 1993. Representing Space: space, scale and culture in social science. In J. Duncan and D. Ley (eds.) *Place/Culture/Representation*. London: Routledge, 251–71.

Agnew, J. 1997. The dramaturgy of horizons: geographical scale in the "Reconstruction of Italy" by the new Italian political parties, 1992–95. *Political Geography*, 162, 99–122.

Brunkhurst, D. J. 2000. *Bioregional Planning: Resource Management Beyond the New Millennium*. Amsterdam: Harwood Academic.

Canby, W. C. Jr. 1988. *American Indian Law in a Nutshell*. St Paul, MN: West Publishing.

Clark, C. C., Ostrom, E., and Ahn, T. K. 2000. The concept of scale and the human dimensions of global change: a survey. *Ecological Economics*, 32, 217–39.

Cohen, B. 1994. Technological colonialism and the politics of water. *Cultural Studies*, 81, 32–55.

Cox, K. R. 1993. The local and the global in the new urban politics: a critical view. *Environment and Planning D: Society and Space*, 11, 433–48.

Cox, K. R. (ed.). 1997. *Spaces of Globalization: Reasserting the Power of the Local*. New York: Guilford.

Cox, K. R. 1998a. Spaces of dependence, spaces of engagement and the politics of scale, or: looking for local politics. *Political Geography*, 17(1), 1–23.

Cox, K. R. 1998b. Representation and power in the politics of scale. *Political Geography*, 171, 41–4.

Cox, K. and Mair, A. 1989. Levels of abstraction in locality studies. *Antipode*, 21(2), 121–32.

Cox, K. and Mair, A. 1991. From localised social structures to localities as agents. *Environment and Planning A*, 23 (New perspectives on the locality debate), 197–213.

Delaney, D. and Leitner, H. (eds.). 1997a. Special Issue: Political Geography of Scale. *Political Geography*, 162, February.

Delaney, D. and Leitner, H. 1997b. The political construction of scale. *Political Geography*, 162, 93–7.

Dodson, M. 1994. The end of the beginning: re(de)fining Aboriginality (The Wentworth Lecture). *Australian Aboriginal Studies*, 1994/1, 2–13.

Douglass, M. 2000. The rise and fall of world cities in the changing space-economy of globalization: comment on Peter J Taylor's "World cities and territorial states under conditions of contemporary globalization". *Political Geography*, 191, 43–9.

Fagan, R. H. 1995. *The Region as Political Discourse*. Plenary Paper, Institute of Australian Geographers Conference, Newcastle Town Hall, Newcastle, NSW, September 1995.

Fagan, R. H. 1997. Local Food/Global Food: globalization and local restructuring. In R. Lee and J. Wills (eds.) *Geographies of Economies*. London: Arnold, 197–208.

Fanstein, S. S. 1999. Power and geographic scale: comments on Morrill. *Political Geography*, 18, 39–43.

Feit, H. 1985. Legitimation and autonomy in James Bay Cree responses to hydro-electric development. In N. Dyck (ed.) *Indigenous Peoples and the Nation-state: Fourth World Politics in Canada, Australia and Norway*. St Johns: Institute of Social and Economic Research, Memorial University of Newfoundland, 27–66.

Forer, P. 1978. A place for plastic space? *Progress in Human Geography*, 22, 230–67.

Fraser, N. 1995. From redistribution to recognition? dilemmas of justice in a "post-socialist" age. *New Left Review*, 212, 68–93.

Fraser, N. 1997. *Justice Interruptus: Critical Reflections on the "Postsocialist" Condition*. New York: Routledge.

Gibson, C. C., Ostrom, E., and Ahn, T. K. 2000. The concept of scale and the human dimensions of global change: a survey. *Ecological Economics*, 322, 217–39.

Goldfrank, W. L. 1993. Peripheries in a changing world order. In J. Kakonen (ed.) *Politics and Sustainable Growth in the Arctic*. Aldershot: Dartmouth Press, 2–14.

Grand Council of the Crees. 1995. *Sovereign Injustice: Forcible Inclusion of the James Bay Crees and Cree Territory into a Sovereign Quebec*. Nemaska, Quebec: Grand Council of the Crees.

Harvey, D. 1969. *Explanation in Geography*. London: Arnold.
Harvey, D. 1984. On the history and present condition of geography: an historical materialist manifesto. *Professional Geographer*, 36, 1–11.
Harvey, D. 1996. *Justice, Nature and the Geography of Difference*. Oxford: Blackwell.
Herb, G. H. 1999. National Identity and Territory. In G. H. Herb and D. H. Kaplan (eds.) *Nested Identities: Nationalism, Territory and Scale*. Lanham, MD: Rowman and Littlefield, 9–30.
Herod, A. 1991. The production of scale in United States labour relations. *Area*, 231, 8–88.
Herod, A. 1997a. Labor's spatial praxis and the geography of contract bargaining in the US east coast longshore industry, 1953–89. *Political Geography*, 162, 145–70.
Herod, A. 1997b. Notes on a spatialized labour politics: scale and the political geography of dual unionism in the US longshore industry. In R. Lee and J. Wills (eds.) *Geographies of Economies*. London: Arnold, 186–96.
Howitt, R. 1991. Aborigines and restructuring in the mining sector: vested and representative interests. *Australian Geographer*, 222, 117–19.
Howitt, R. 1993a. "A world in a grain of sand": towards a reconceptualisation of geographical scale. *Australian Geographer*, 241, 33–44.
Howitt, R. 1993b. Social Impact Assessment as "applied peoples' geography". *Australia Journal of Geographical Studies*, 312, 127–40.
Howitt, R. 1993c. People Without Geography? Marginalisation and indigenous peoples. In: R. Howitt (ed.) *Marginalisation in Theory and Practice*. ERRRU Working Papers, 12. Sydney: Economic and Regional Restructuring Research Unit, Departments of Economics and Geography, University of Sydney, 37–52.
Howitt, R. 1997. Getting the scale right: the geopolitics of regional agreements. *Northern Analyst*, 2, 15–17.
Howitt, R. 1998a. Scale as relation: musical metaphors of geographical scale. *Area*, 30(1), 49–58.
Howitt, R. 1998b. Recognition, reconciliation and respect: steps towards decolonisation? *Australian Aboriginal Studies*, 1998/1, 28–34.
Howitt, R. 2001. *Rethinking Resource Management: Sustainability, Justice and Indigenous Peoples*. London: Routledge.
Howitt, R. 2002. Scale and the other: Levinas and geography. *Geoforum*, 33, in press
Howitt, R., Connell, J., and Hirsch, P. (1996) Resources, Nations and Indigenous Peoples. In R. Howitt et al. (eds.) *Resources, Nations and Indigenous Peoples: Case Studies from Australasia, Melanesia and Southeast Asia*. Melbourne: Oxford University Press, 1–30.
Human Rights and Equal Opportunity Commission. 1997. *Bringing Them Home: National Inquiry into the Separation of Aboriginal and Torres Strait Islander Children from their Families*. Sydney: HREOC.
Jammer, M. 1954. *Concepts of Space*. Cambridge MA: Harvard University Press.
Jhappan, C. R. 1992. Global community? Supranational strategies of Canada's aboriginal peoples. *Journal of Indigenous Studies*, 3(1), 59–97.
Jonas, A. 1994. The scale politics of spatiality. *Environment and Planning D: Society and Space*, 12(3), 257–64.
Jonas, W. 2000. *Aboriginal and Torres Strait Islander Social Justice Commissioner: Native Title Report 1999*. Sydney: Human Rights and Equal Opportunity Commission.
Jones, K. T. 1998. Scale as epistemology. *Political Geography*, 17(1), 25–8.
Jones, M. 1998. Restructuring the local state: economic governance or social regulation? *Political Geography*, 17(8), 959–88.
Judd, D. R. 1998. The case of the missing scales: a commentary on Cox. *Political Geography*, 17(1), 29–34.

Kaplan, D. H. 1999. Territorial identities and geographical scale. In G. H. Herb and D. H. Kaplan (eds.) *Nested Identities: Nationalism, Territory and Scale*. Lanham, MD: Rowman & Littlefield, 31–49.

Kelly, P. F. 1997. Globalization, Power and the Politics of Scale in the Philippines. *Geoforum*, 28(2), 151–71.

Leitner, H. 1997. Reconfiguring the spatiality of power: the construction of a supranational migration framework for the European Union. *Political Geography*, 16(2), 123–44.

MacLeod, G. and Goodwin, M. 1999. Reconstructing an urban and regional political economy: on the state, politics, scale, and explanation. *Political Geography*, 18, 697–730.

Marston, S. A. 2000. The social construction of scale. *Progress in Human Geography*, 24(2), 219–42.

Martin, D. G. 1999. Transcending the fixity of jurisdictional scale. *Political Geography*, 18, 33–8.

Massey, D. 1992. Politics and space/time. *New Left Review*, 196, 65–84.

Massey, D. 1993a. Power-geometry and a progressive sense of place. In J. Bird et al. (eds.) *Mapping the Futures: Local Cultures, Global Change*. London: Routledge, 59–69.

Massey, D. 1993b. Politics and space-time. In M. Keith and S. Pile (eds.) *Place and the Politics of Identity*. London: Routledge, 141–61.

Massey, D. 1994. Double Articulation: a "place in the world". In A. Bammer (ed.) *Displacements – Cultural Identities in Question*. Bloomington and Indianapolis: Indiana University Press, 110–119.

Massey, D. 1995. Thinking radical democracy spatially. *Environment and Planning D: Society and Space*, 13(3), 283–88.

McGuirk, P. M. 1997. Multiscaled interpretations of urban change: the federal, the state, and the local in the Western Area Strategy of Adelaide. *Environment and Planning D: Society and Space*, 15, 481–98.

McHugh, P. G. 1996. The legal and constitutional position of the Crown in resource management. In R. Howitt et al. (eds.) *Resources, Nations and Indigenous Peoples: Case Studies from Australasia, Melanesia and Southeast Asia*. Melbourne: Oxford University Press, 300–16.

Meyer, W. B., Gregory, D., Turner, B. L. III, and McDowell, P. F. 1992. The local-global continuum. In R. F. Abler et al. (eds.) *Geography's Inner Worlds: Pervasive Themes in Contemporary American Geography*. New Brunswick, NJ: Rutgers University Press, 255–79.

Morrill, R. 1999a. Inequalities of power, costs and benefits across geographic scales: the future uses of the Hanford reservation. *Political Geography*, 18, 1–23.

Morrill, R. 1999b. The tyranny of conventional wisdom? A response. *Political Geography*, 18, 45–8.

Notzke, C. 1995. A new perspective in aboriginal natural resource management: comanagement'. *Geoforum*, 26(2), 187–209.

Puddicombe, S. 1991. Realpolitik in Arctic Quebec: why Makivik Corporation won't fight this time. *Arctic Circle*, Sept.–Oct., 14–21.

Routledge, P. 1996. Critical geopolitics and terrains of resistance. *Political Geography*, 15 (6–7), 509–31.

Sadler, D. and Fagan, R. H. 2000. *Australian Trade Unions and the Politics of Scale: Reconstructing the Spatiality of Industrial Relations*. Paper presented to the Global Conference on Economic Geography, National University of Singapore, December 2000.

Shapiro, M. J. 2000. Commentary on Peter Taylor's essay. *Political Geography*, 19(1), 39–41.

Silvern, S. E. 1999. Scales of justice: law, American Indian treaty rights and political construction of scale. *Political Geography*, 18, 639–68.

Smith, M. P. 1994. Can you imagine? Transnational migration and the globalisation of grassroots politics. *Social Text*, 39, 15–34.
Smith, M. P. 1998. Looking for the global spaces in local politics. *Political Geography*, 17(1), 35–40.
Smith, N. 1984. *Uneven Development: Nature, Capital and the Production of Space*. Oxford: Blackwell.
Smith, N. 1988. Regional adjustment or regional restructuring. *Urban Geography*, 9(3), 318–24.
Smith, N. 1989. The region is dead! Long live the region! *Political Geography*, 7(2), 141–52.
Smith, N. 1992. Geography, difference and the politics of scale. In J. Doherty et al. (eds.) *Postmodernism and the Social Sciences*. London: Macmillan, chapter 3, 57–79.
Smith, N. (1993. Homeless/global: scaling places. In J. Bird et al. (eds.) *Mapping the Futures: Local Cultures, Global Change*. London: Routledge, 87–119.
Smith, N. 2000. Scale. In R. J. Johnston et al. (eds.) *The Dictionary of Human Geography*. London: Blackwell, 724–7.
Smith, N. and Ward, D. 1987. The restructuring of geographical scale: coalescence and fragmentation of the northern core region. *Economic Geography*, 63(2), 160–82.
Suchet, S. 1999. *Situated Engagement: a critique of wildlife management and postcolonial discourse*. Ph.D. Dissertation, Department of Human Geography, Macquarie University.
Swanstrom, T. 1999. The stubborn persistence of local land use powers: a comment on Morrill. *Political Geography*, 18, 25–32.
Swyngedouw, E. 1992. The Mammon quest. "Glocalisation", interspatial competition and the monetary order: the construction of new scales. In M. Dunford and Kafkalasm G. (eds.) *Cities and Regions in the New Europe*. London: Belhaven, 39–67.
Swyngedouw, E. 1997a. Neither Global nor Local: "glocalization" and the politics of scale. In K. R. Cox (ed.) *Spaces of Globalization: Reasserting the Power of the Local*. New York: Guildford Press, 137–66.
Swyngedouw, E. 1997b. Excluding the Other: the production of scale and scaled politics. In R. Lee and J. Wills (eds.) *Geographies of Economies*. London: Arnold, 167–76.
Swyngedouw, E. 2000. The Marxian Alternative: historical-geographical materialism and the political economy of capitalism. In E. Sheppard and T. J. Barnes (eds.) A *Companion to Economic Geography*. Oxford: Blackwell, 41–59.
Tatz, C. 1999. *Genocide in Australia*. Canberra: Australian Institute of Aboriginal and Torres Strait Islander Studies.
Taylor, P. J. 1982. A materialist framework for political geography. *Transactions of the Institute* of *British Geographers*, 7, 15–34.
Taylor, P. J. 1987. The paradox of geographical scale in Marx's politics. *Antipode* 19(3), 287–306.
Taylor, P. J. 1988. World-Systems analysis and regional geography. *Professional Geographer*, 40(3), 259–65.
Taylor, P. J. 1993. *Political Geography: World-Economy, Nation-State and Locality*, 3rd edn. Harlow: Longmans.
Taylor, P. J. 1994. The state as a container: territoriality in the modern world-system. *Progress in Human Geography*, 18(2), 151–62.
Taylor, P. J. 1999. *Modernities: a Geographical Interpretation*. Cambridge: Polity.
Taylor, P. J. 2000a. World cities and territorial states under conditions of contemporary globalization (1999 Annual Political Geography Lecture). *Political Geography*, 19(1), 5–32.
Taylor, P. J. 2000b. Theory and practice. *Political Geography*, 19(1), 51–3.
Varsanyi, M. W. 2000. Global cities from the ground up: a response to Peter Taylor. *Political Geography*, 19(1), 33–8.

Williams, R. A. Jr. 1990. *The American Indian in Western Legal Thought: the Discourses of Conquest*. New York: Oxford University Press.
Williams, R. W. 1999. Environmental injustice in America and its politics of scale. *Political Geography*, 18, 49–73.
Wilson, J., Low, B., Costanza, R., and Ostrom, E. 1999. Scale misperceptions an the spatial dynamics of a social-ecological system. *Ecological Economics*, 32(2), 243–57.

Chapter 11

Place

Lynn A. Staeheli

Place. Such a simple term. One might think that such a concept would be relatively easy to define and describe. Yet "place" could easily be one of the most contested terms in human geography. Perhaps this is because of the feelings and emotions evoked by the term – home, rootedness, order, setting, context. Perhaps it is because the term has been used in different ways by proponents of various epistemologies and theories, so that this simple term becomes a code word for a host of other arguments. Or perhaps it is because the meaning or significance of a place seems to depend on one's social role in that place. For example, a home may be a place of refuge and security or a place of labor and danger depending on the responsibilities one faces there and one's relations with other people in the home.

In this chapter, I examine the different ways in which place is understood and mobilized in political studies. In the first section of the chapter, I describe some of the definitions of place, and discuss them in terms of their implications for understanding the spatiality of politics. In the final section, I examine the ways in which place is mobilized in political action. I argue that the definition and understanding of place that is deployed in a particular study has its own politics: a politics that reflects political and social goals.

Defining Place

If place is such an important concept for geographers, then one might expect that it would be defined in basic geography texts. But one would be wrong! A review of eight new textbooks published in 1999 and 2000 produced only one definition of place in their glossaries. Terry Jordon-Bychkov and Mona Domosh define place as:

> A term used to connote the subjective, idiographic, humanistic, culturally oriented type of geography that seeks to understand the unique character of individual regions and places, rejecting the principles of science as flawed and unknowingly biased (1999, p. 535).

This is an interesting definition in many respects, not the least of which is that it seems to define place as a type of geographic analysis, rather than as a concept in its own right.

Another source for a definition might be basic geographic references. Here, the *Dictionary of Human Geography* (Johnston et al., 2000) seems a likely source. The entry begins "A portion of geographic space" (p. 582). This definition is simple and concise, but also so sterile as to eliminate any sense that this might be an important concept. The rest of the entry provides some sense of the debate, but mainly by referring to other terms and concepts. Place itself is not defined in much more detail.

Ironically, the *Merriam-Webster Dictionary* provides a better entrée to definitions of place than do geographic sources. Here, place is defined in several ways.

place / *n* **1** : SPACE, ROOM **2** : an indefinite region : AREA **3** : a building or locality used for a special purpose **4** : a center of population **5** : a particular part of a surface : SPOT **6** : relative position in a scale or sequence; *also*: high and esp. second position in a competition **7** : ACCOMMODATION; *esp*: SEAT **8** : JOB; *esp* ; public office **9** : a public square

place / *vb* **1** : to distribute in an orderly manner : ARRANGE **2** : to put in a particular place : SET **3** : IDENTIFY **4** : to give an order for <~ a bet > **5** : to rank high and esp. second in a competition

Place is defined as a context or setting, in relational terms, as an outcome or product of processes, and as something active and dynamic. This multifaceted definition suggests the complexity of the term as used by geographers, as well as the reasons people might use the term in vastly different ways. In this section of the chapter, I will expand this definition of place as it is used by geographers. In particular, I will examine five conceptualizations of place:

- place as physical location or site;
- place as a cultural and/or social location;
- place as context;
- place as constructed over time;
- place as process.

Place as physical location or site

This is the most obvious definition of place, as it identifies a material "thing" – something one can point to on a map or take a walk through. Place is often conceptualized as material, grounded, and bounded, and the focus of geographic research is on the particular, and sometimes unique characteristics of places. The definition provided by Jordon-Bychkov and Domosh draws from this way of thinking about place. To the extent this conceptualization is incorporated in political studies, it has often been in a way that reduces place to a backdrop, as for example in comparative studies of revolution (Tilly, 1978).

Place as a physical location is often contrasted with more abstract notions of *space*. If place is grounded and particular, space is understood as abstracted from the particular. As such, the study of space and the relationships that connect discrete places has been the focus of spatial science. Some studies accomplished this in a way that seemed to erase all the particularities of place and assumed an "isotropic plane," a featureless surface on which political activity occurred; the significance of any physical location was reduced to its position relative to other locations (e.g. Guest et al., 1988; O'Loughlin et al., 1998). These studies focus on the spatial relationships that underlie phenomenon, and the authors' intent is to identify general patterns without considering the particularities of place. Sometimes, as in O'Loughlin's diffusion study, the examination of general patterns is merely the first step of an analysis, and a different understanding of place is involved in case studies in subsequent stages of the analysis.

Place as a cultural or social location

This second sense of place is often framed in metaphorical terms, and so it stands in contrast to the material framing of place as physical location. The perspective is often associated with researchers who draw from feminist and cultural studies and who have a concern with the social locations of people and social groups. From this perspective, people are located within webs of cultural, social, economic, and political relationships that shape their identities, or posititionalities.

Geographers argue that these social locations are not simply metaphorical, but that they are associated with real, physical places within cities and regions. Tim Cresswell (1996), for example, writes of moral landscapes, in which certain kinds of people or activities may be thought of as "belonging" to certain kinds of places – as being "in place." In Western cities, for example, women were historically located within private places of the home, church, and neighborhood. This was identified as an appropriate location (materially and metaphorically) for them, given the social construction of gender roles and relationships at the time. By contrast, women who ventured into spaces of commerce, industry, or politics were "out of place," transgressing both social and physical boundaries and locations (see Bondi and Domosh, 1998). As more and more women transgressed the boundaries of public and private places, the physical spaces of the city were gradually transformed in order to accommodate women. At the same time, the transgression reshaped social relations and identities, as the fact of being in "public" gradually came to change the ways that women in public were viewed – both by women and by others viewing them. Meghan Cope (1996) discusses this transformation and the impacts on women's identity in terms of the re-creation of "identities-in-place."

As noted, this conceptualization is particularly common in research that draws from cultural studies, feminism, and identity politics. Geographers have looked to the arguments about "place" in these studies with some bemusement. On the one hand, geographers have been pleased to see the incorporation of spatial ideas such as "borders," "location," "mapping," and "place" taken into social theory. These metaphors draw on familiar and commonsensical ideas to explain complex social relationships in a way that highlights the centrality of space and place to those relationships. On the other hand, geographers have argued that spatial metaphors are not just metaphors – they should be grounded (Smith and Katz, 1993) in ways that draw attention to the role of space and place in shaping and giving meaning to action.

Place as context

If place is a contested term in geography, "context" is contested in political studies. At first blush, then, defining place in terms of context seems to be asking for trouble!

But that is precisely how Nicholas Entrikin defines place: "The geographical concept of place refers to the areal context of events, objects and actions" (1991, p. 6). In many respects, this definition shares the concern for social, political, economic, and cultural relations that characterizes the previous definition. But when geographers talk about context, they are talking about how those relations attach first to space and place, and only secondarily to people; place, in this sense, describes the social positionality of an areal unit and the milieu provided through that areal unit that shapes political action (Agnew, 1987). One could almost think about context as *identity-of-place*, in contrast to the ideas of identity-in-place described previously.

To a geographer, the definition of place as context is relatively uncontroversial. We argue that context provides the setting or milieu for social action, and it is one of the reasons we claim *Geography Matters!* (Massey and Allen, 1984). Yet it so happens that this idea is hotly contested in political studies. Political scientists, for example, have devoted a great deal of attention to the study of "compositional effects" on political behavior. These compositional effects stem from demographic characteristics – such as education and income – that are said to influence political attitudes and behavior. At an individual level, these characteristics have been shown to be strong predictors of political behavior, such as voting. This is consistent with the methodological individualism that has dominated political science in the past several decades. Some political scientists go so far as to argue that any influence that might be attributed to context or to neighborhood effects reflect incorrect or incomplete measurement – that what we call context is simply a residual or error (King, 1996)

To this argument, geographers respond first that places are located in the same webs of power relations as are individuals, but that the characteristics of places are not simply an aggregation of compositional characteristics of the people living there. Second, they argue that compositional effects only take on meaning in local contexts – in places (Johnston, 1991). For example, it may mean something different to be a low income mother in subsidized, rental housing in a middle-class neighborhood than to be a middle-class homeowner in the same neighborhood. One's social positionality with respect to an area may influence political attitudes, outlets for political action, and the effectiveness of action. Context, then, is implicated in political behavior in two ways: first, it shapes meanings, or interpretive frames, of events for different actors, and second, it provides resources for action. In short, geographers argue that both compositional and contextual effects are important, and that place mediates between individuals, social groups, and broader political structures. It is for this reason Entrikin (1991) argues for the "betweenness of place" and that Kirby (1985) argues that we need to understand "action-in-place."

Place as socially constructed through time

The preceding definitions suggest the complexity of the concept "place." When these definitions are combined, it becomes possible to think of the interactions between the geographical and physical characteristics of places, the ways the place is located within webs of broader relations, and the ways individuals are located with respect to the broad relations and to the place itself. The result is a conceptualization of place as socially constructed through time. In this perspective, place is dynamic and changing.

The idea of place as a social construction is common to structurationist approaches, but it has been difficult to demonstrate empirically the structuration of place. One of the more initutively accessible approaches is provided by Doreen Massey (1979), who has suggested a geological metaphor in which years of human activity construct the built and social forms that constitute place. These forms provide both resources and barriers to those who seek to respond to changes created by shifts in broader economic and/or political structures. This metaphor implies a recognition that the elements of context described previously are not just a surface on which politics are played. Rather, a place is the result of the layering of activities that constantly make and remake it. It is with this understanding of place, for example, that Patricia Martin (1999) examines the efforts of economic development leaders on the US–Mexican border. She demonstrates the ways that place promotional efforts drew on ideas about the stability of local power structures and the leaders' ability to ensure a labor force that was capable of responding to economic change. In this situation, stability and flexibility were held in tension in the economic development strategies of local elites.

This way of understanding place builds in a sense of history, a sense that the social, economic, and political processes involved in place-making of times past are significant to, yet not determining of, place-making in contemporary periods. As such, a tension between contintuity and change is implied in this conceptualization (see also Harvey, 1985; Smith, 1984). Yet the invocation of a geological metaphor can imply that the construction of place is somehow natural and inevitable. To the extent that naturalness is accepted, it also shapes the ways in which place is – or is not – mobilized in political action. Further, the layering metaphor implies a stratigraphy within place in which one can dig down a level or two and analyse a layer. This, of course, is not what real stratigraphers do, and it is not what is intended with the metaphor. But to the extent that the metaphor suggests analysis of distinct layers that can be read from the physical record, it is misleading. The idea of place as socially constructed is centrally concerned with interactions between "layers" and with the uses and meanings to attached to place as they influence political behavior. So, for example, Caroline Nagel (2000) examines the politics of the reconstruction of Beirut through an explication of the meaning of Beirut and the social–political–religious relations that shaped the city through time. She argues that the rebuilding of the city after Lebanon's civil war was an attempt both to present the history of the city in a particular way and to direct its future.

Place as a social process

The preceding definitions all imply that place is an outcome, or a product. The final conceptualization of place that I want to consider eschews the idea of place as outcome and emphasizes place as process, as always "becoming" (Pred, 1984).

John Agnew (1987, p. 28) presents one of the clearest ways of thinking about place as a process. He conceptualizes place as involving three elements:

> *Locale*, the settings in which social relations are constituted (these can be informal or institutional); *location*, the geographical area encompassing the settings for social interaction as defined by social and economic processes operating at a wider scale; and *sense of place*, the local "structure of feeling".... A key tenet is that the local social worlds of place (locale) *cannot* be understood apart from the *objective* macro-order of location and the *subjective* territorial identity of sense of place.

While this conceptualization draws from the previous conceptualizations, it highlights several interconnected issues that others may have downplayed or not incorporated. Most importantly, this conceptualization highlights the interaction between processes operating at different scales – from the macro-economy to the individual – into the processes of place. This has the advantage of situating places within the global economy and nation, as well as with regard to other places. Thus, place is not "discrete," "merely particular," or "merely local." Place is seen as intricately binding locales with broad processes and with other locales – bindings, processes, and places that are themselves constantly in flux (Massey, 1994). This, in turn, provides a way to analyse the webs of power – the power geometries, in Massey's terms – that connect locales, as well as the ways that locales and institutions operating at different scales provide opportunities for political action (Cox, 1998).

This definition incorporates a rich, complex understanding of how place is constructed and mobilized in politics. It understands place as incorporating physical and social locations, as incorporating the sense of identity-in-place with a sense of identity-of-place, and as both a context for action and the object of action that is continually in the process of being made. But while this definition can be thought of as incorporating all of the previous definitions, it is surprisingly difficult to use in an analytical, or empirical, sense. Perhaps because of this, political studies that incorporate this definition of place tend to be theoretical and, in some ways, to seem strangely ungrounded. For example, this definition of place is often invoked in structurationist approaches and in theories of radical democracy. Both of these approaches are themselves, often criticized for being overly abstract (Jones and Moss, 1995; Wilson and Huff, 1994). And the approach is one that defies generalizability, a commonly accepted marker of social sciences.

Yet some political analysts have found this to be a useful way to approach place. For example, feminists who are concerned to discern the shifting boundaries between public and private – a concern in understanding how political agents, political spaces, and political spheres are constructed – rely on such a definition (Staeheli, 1996). Michael Brown has also relied on this conceptualization of place in his analyses of the politics surrounding AIDS in Vancouver, British Columbia (1997). And John Agnew deploys it in his studies of politics in Scotland and Italy (1987). But while there is an appreciation of the need to understand place as process and incorporate this understanding into empirical studies, it is probably fair to say that this approach is most often discussed in theoretical terms and with less in the way of systematic, empirical analysis. As such, this definition implies a certain theoretical perspective, only parts of which may actually be examined in any particular study. These studies provide a broad overview and are important for what they do, but the nitty-gritty of daily life and politics that are part of the process of place are almost necessarily too difficult to incorporate in any detail.

Place and Political Studies

With all this attention to the conceptualization of place, one might be tempted to think that ideas of place have been central to political studies of all forms. Once again, one would be wrong! Through the 1980s and early 1990s, geographers bemoaned the general lack of attention to place in political studies or complained that place was understood to be static, fixed, almost a container (e.g. Agnew, 1987; Harvey, 1996; Kirby, 1985; Massey, 1994; Smith, 1984; Soja, 1989). At some level, the debate over the conceptualization of place almost lends itself to the conclusion that contested nature of place is limited to the academic realm – something that intellectuals in the ivory tower worry about, but that is not the stuff of real politics. A quick glance at the newspaper, however, suggests that place itself is often the object of political struggle and that the various definitions of place become part of those struggles. Territorial conflicts (whether over Kosovo, Jerusalem, or Kashmir), electoral college strategies in US election campaigns, and the siting of demonstrations and protests all involve contests over place. These conflicts all occur *in* place, but they also may involve conflicts over the control of place and about what *should be* in place. In other words, they may be about the "moral landscape" of place (Cresswell, 1996).

In this section of the chapter, I want to demonstrate the significance of the various definitions of place to political behavior and conflicts. To do this, I focus on four ways in which place is implicated in political struggle: politics *about* place, politics *in* place, politics in the *construction* of place, and politics that *deploy* or *transgress* place. Several chapters in this book provide extended treatments of these roles of place in political struggle, so the examples I present here are fairly cursory. But whereas many of the chapters will draw on "big politics" – involving macro-scale processes, nation-states, elites, and so forth – most of the examples provided here will be of smaller politics, or what Mann (1994) terms "micropolitics." In all cases, I want to argue against a tendency to fear the role of place in politics that stems from the critique of place as particularistic and "merely local" (e.g. Harvey, 1996). At the same time, I want to avoid creating an impression that place-based politics necessarily mobilizes a progressive politics. Instead, I want to examine the ways that place is mobilized in politics, recognizing that the goal of political action should not be read directly from the strategy employed.

Politics about place

Nationalist and territorial conflicts are conflicts in which control over place is implicated in struggles over the allocation of other political resources. But there are other struggles that have to do with issues of control over place that occur at smaller scales of everyday life. Some of the struggles are identified as "turf conflicts," often involving a not-in-my-backyard (NIMBY) attitude.

Turf politics are widespread, but we often "locate" them in cities and urban areas. Realistically, most individuals have relatively little control over their fate; their sense of losing control is exacerbated by processes of globalization in which decision-making seems increasingly removed from locales and the "average person" (Greenberg, 1995). In the face of such changes and pressures, individuals often strive to keep some control over what they can, and this may mean control over the places in which they live. This struggle takes many forms – from marking territory through graffiti (Ley and Cybriwsky, 1974) to gated communities and neighborhood designations (Davis, 1990) to NIMBYism (Lake, 1993). As Cox and McCarthy (1981) have argued, the politics of turf – or in my terms, politics about place – is often about maintaining the illusion of control in the face of forces that reduce the ability of most people to shelter their lives from the buffeting of larger-scale forces.

To the extent that the above is true, politics about place often involve efforts to defend place as a physical location. That physical location, however, signifies cultural and social locations that are reproduced and restructured over time. Seen from this perspective, we can think of NIMBYism as an attempt to forestall changes that are generally believed to work to the disadvantage of current residents in an area. These efforts may reinforce the *status quo* and block efforts to address social inequities, and so take on negative, reactionary connotations (see Harvey, 1996). But politics about place may also be part of strategies to block changes that may be destructive of place and social relations (Porteous, 1989). So while not progressive, they may also not be purely reactionary, particularly when coupled with other agendas that *do* include a concern for social justice. Some elements of the environmental justice movement, for example, include an element of NIMBYism when activists fight the location of noxious facilities in poor neighborhoods or in neighborhoods with a high proportion of residents of color. Yet organizations such as Mothers of East Los Angeles (Pardo, 1990) and the Labor/Community Strategy Center (Pulido, 1994) link their politics about place with strategies for greater racial and economic inclusion. In these cases, politics about place can be seen as progressive, not reactionary.

Politics in place

This form of politics is closely aligned with the definition of place as context. Politics in place, or as Kirby (1985) terms it "activity-in-place," recognizes the resources that places provide as people make decisions and go about their daily (but still political) lives. Huckfeldt and Sprague (1996), for example, analyse the networks of information that are constructed as individuals go about their daily activities and use these networks to trace the sources of political information – information that shapes both political attitudes and participation. Geographers have dubbed this the "friends and neighbors" effect, an important aspect of context.

Huckfeldt and Sprague examine political networks in instrumental terms, but others use the idea of politics in place to advance a more communitarian, expressive politics. Daniel Kemmis (1990, p. 122), for example, argues that politics in place can be a way of highlighting the common interests that residents share. In this sense, place is a physical location and also a context for action. He argues that a recognition of "a common inhabiting of a common place" breeds habits for politics that can be cooperative and progressive in a way that politics separated from place (e.g. cyberspace or the federal government) rarely are.

Kemmis' arguments are attractive to those who want to reconnect politics with daily life, but who are concerned about the fragmented and often exclusionary aspects of community-based politics. Yet many worry about the equation of locality or place with community that is implicit in communitarian arguments such as Kemmis' (e.g. Agnew, 1987). While it may not be the intent of Kemmis or other communitarians to exclude those who feel either no affinity or a different affinity with place, the assertion of commonality may have the effect of marginalizing some political agents or attitudes. In short, when place identity is assumed to be equivalent to social identity or social positioning, the communitarian version of politics in place may have the effect of masking social differences, rather than understanding or working through them.

Politics as the construction of place

Nationalist struggles are often examples of attempts to create place by equating place and social identities – people become a social unity through the places that they create. As such, some aspects of nationalist struggles can be thought of as political efforts to construct place. Similarly economic development strategies can be thought of as efforts to make or remake places to attract investment and to compete in the global economy (see, e.g., contributions in Jonas and Wilson, 1999; Martin, 1999; Nagel, 2000; Sorkin, 1996). Nationalism and economic development both imply a social constructivist definition of place, and both reveal a politics behind that construction.

But there are other political struggles over the construction of place that involve struggles over meaning and the ways in which place is constructed over time; it is in this context that we can think about public and private places. Ideas about publicity and privacy are central to many debates over democracy and citizenship. Geographers have argued, though, that much of this debate has ignored the spatiality of public and private. Put simply, one cannot operationalize abstract notions of a "public sphere" unless there are places or settings in which a public can come together. It is often the characteristics of places – the physical and social characteristics and meanings – that deny or limit access to certain types of people and certain types of behavior at certain times and that thereby limit or constrain the "public."

Feminists have long noted the ways that certain places putatively constructed as "public" often deny access to women. For example, Habermas' (1969) ideas of the public sphere were first described in terms of eighteenth century European society. Yet the spaces of eighteenth century society were gendered, and there were few places in which women could join the public sphere. Streets, places of commerce, and places of governance were all places to which women's access was limited. These limitations were constructed by practice more often than by law, but the limitations were real. Cooper et al. (2000), for example, argue that the absence of restrooms (toilets) for women in New Zealand cities at the turn of the nineteenth century was a key means by which women were excluded from public spaces. They trace concurrent changes in women's standing as citizens with public provision of restrooms for women, as the physical construction of restrooms entailed a reconstruction of ideas about public places. Similar political struggles over public places occur with respect to sexuality (Bickell, 1999), the construction and reconstruction of public parks (Mitchell, 1995), and provision of facilities and services for homeless people (Smith, 1996; Wolch and Dear, 1993). In all of these examples, the physical and social constructions of place are central to political processes and to achieving political goals.

Politics deploying place

The final role of place in politics is closely aligned with the idea of constructions of place, but I have separated it out for a reason. Politics that deploy place involve a conscious effort at transgression – at what Tim Cresswell (1996) refers to as a conscious effort at disrupting the moral landscape of place as a way of making a political argument. For example, while public and private places have been constructed over time and through political struggle, politics deploying place involve efforts to expose and to challenge the gendered construction of public places.

Protests are some of the most visible ways of transgressing places. For example, Cresswell (1996) argues that the reaction to the women's peace camp at Greenham Common was not simply to the women's opposition to nuclear weapons; in and of itself, this was unremarkable. Rather, the horror expressed by the British public and press stemmed from the tactics of transgression employed by the women protesters. Women camped for months, living in public, often away from their children and families. They were in a muddy, outdoor place, rather than in the neat, clean, indoor places where they ostensibly "belonged." They trespassed onto the military base, rather than staying outside the perimeter fencing. They hung tampons on the fencing, rather than keeping them hidden. These small acts of spatial transgression – seemingly inconsequential to debates over nuclear weapons – drew attention to the protests, though not necessarily in ways that protest organizers intended. In a similar fashion, the women who protested in the Plaza de Mayo in Buenos Aires transgressed place as they marched in the plaza to protest the disappearance of their family members. They marched *as mothers* to protest a brutal regime (Radcliffe, 1993). As such, these two groups of women challenged a moral landscape based on power and brutality by deploying an alternative moral landscape based on place and nurturing.

Conclusion

Place. It really is a messy concept, and it is no wonder that introductory texts have generally eschewed easy definitions of it. One reason the term is contested is that it is bound up with the methodological, epistemological, and ontological perspectives of those who use it. This is seen in the tendency to conceptualize place as a physical location by those who use the methods of spatial science (but see O'Loughlin, chapter 3, this volume). Similarly, approaches that highlight connectivity and the social construction of place proceed from a belief that the social world is constructed through the interactions of structures and human agency; place is only one aspect of the broader processes of structuration. So the search for a definition of place leads to a consideration of more fundamental issues of philosophy, as the definition provided by Jordon-Bychkov and Domosh (1999) suggests.

There is also a *politics* behind the ways in which place is deployed in political struggles. The idea of place as socially constructed is inconsistent with nationalists who claim primordial rights to place or to homeowners who defend property rights irrespective of the broader social implications of their actions. In these struggles, it makes political sense to claim that place is something real, material, and fixed. If, on the other hand, one wants to claim that the way place is organized reflects the power relations of society, then a definition of place that highlights its social construction and connections with the processes that shape identity may make more sense. The difference in definitions of place, then, reflects political strategies and goals.

Finally, the difference between the strategies of political agents and the philosophies of researchers also contributes to the contested (or perhaps confused!) conceptualization of place. Researchers do not always use the same definitions as the people they study. Some studies of nationalism, for example, employ a social constructionist definition, even as the people involved in the actual struggles do not. In other cases, researchers who are critical of social relationships may want to *change* those relationships. Identification of a moral landscape of place is not their end goal; transgression and the construction of new, perhaps less exploitive landscapes may be. So once again, the definitions of place and the purposes to which place is put have their own politics.

Thus, place may be mobilized in many ways in any given context. This is what makes it such a messy concept. Place includes both subjective meanings and structural locations, and it is a process as much as an outcome. Ideas about place, its meanings, and its importance are deeply ingrained in many people. It is this deeply held, and often conflicted, attachment to place in combination with the resources place offers that makes place such a powerful motivation for and shaper of political action, and an effective tool or strategy in political struggle.

BIBLIOGRAPHY

Agnew, J. 1987. *Place and Politics: The Geographical Mediation of State and Society.* Boston: Allen & Unwin.

Bickell, C. 1999. Heroes and invaders: gay and lesbian pride parades and the public/private distinction in New Zealand media accounts. *Gender, Place and Culture*, 7, 163–78.

Bondi, L. and Domosh, M. 1998. On the Contours of Public Space: tales of three women. *Antipode*, 30, 270–89.

Brown, M. 1997. *RePlacing Citizenship: AIDS and Radical Democracy.* New York: Guilford.

Cooper, A., Law, R., Malthus, J., and Wood, P. 2000. Rooms of Their Own: public toilets and gendered citizens in a New Zealand city, 1860–1940. *Gender, Place and Culture*, 8, 417–33.

Cope, M. 1996. Weaving the everyday: identity, space and power in Lawrence, Massachusetts, 1920–1939. *Urban Geography*, 17, 179–204.

Cox, K. 1998. Spaces of dependence, spaces of engagement and the politics of scale, or: looking for local politics. *Political Geography*, 17, 1–23.

Cox, K. and McCarthy, J. 1981. Neighbourhood Activism as a Politics of Turf: a critical analysis. In K. Cox and R. Johnston (eds.) *Conflict, Politics and the Urban Scene.* New York: St. Martin's Press, 196–219.

Cresswell, T. 1996. *In Place/Out of Place* Minneapolis, MN: University of Minnesota Press.

Davis, M. 1990. *City of Quartz.* New York: Verso

Entrikin, N. 1991. *The Betweenness of Place.* Baltimore, MD: Johns Hopkins University Press.

Greenberg, S. 1995. *Middle Class Dreams.* New York: Times Books.

Guest, A., Hodge, D., and Staeheli, L. 1988. Industrial affiliation and community culture: voting in Seattle. *Political Geography Quarterly*, 7, 49–73.

Habermas, J. 1989. *The Structural Transformation of the Public Sphere.* Cambridge, MA: MIT Press.

Harvey, D. 1985. The Geopolitics of Capitalism. In D. Gregory and J. Urry (eds.) *Social Relations and Spatial Structures*. New York: St. Martin's Press, 128–63
Harvey, D. 1996. *Justice, Nature, and the Geography of Difference*. Oxford: Blackwell.
Huckfeldt, R. and Sprague, J. 1987. Networks in context: the social flow of political information. *American Political Science Review*, 81, 1197–216.
Johnston, R. J. 1991. *A Question of Place*. Oxford: Blackwell.
Johnston, R. J. et al. 2000. *Dictionary of Human Geography*. Oxford: Blackwell.
Jonas, A. and Wilson, D. (eds.). 1999. *The Urban Growth Machine*. Albany, NY: State University of New York Press.
Jones, J. P. and Moss, P. 1995. Democracy, identity, space. *Environment and Planning D: Society and Space*, 13, 253–7.
Jordon-Bychkov, T. and Domosh, M. 1999. *The Human Mosaic: A Thematic Introduction to Cultural Geography*, 8th edn. New York: Longman.
Kemmis, D. 1990. *Community and the Politics of Place*. Norman, OK: University of Oklahoma Press.
King, G. 1996. Why context should not count. *Political Geography*, 15, 159–64.
Kirby, A. 1985. Psuedo-random thoughts on space, scale and ideology in political geography. *Political Geography Quarterly*, 4, 5–18.
Lake, R. 1993. Rethinking NIMBY. *APA Journal*, 59, 87–93.
Ley, D. and Cybriwsky, R. 1974. Urban graffiti as territorial markers. *Annals of the Association of American Geographers*, 64, 491–505.
Mann, P. 1994. *Micro-politics: Agency in a Post-feminist Era*. Minneapolis, MN: University of Minnesota Press.
Martin, P. 1999. On the frontier of globalization: development and discourse along the Rio Grande. *Geoforum*, 2, 217–35.
Massey, D. 1979. In what sense a regional problem? *Regional Studies*, 13, 233–43.
Massey, D. 1994. *Space, Place, and Gender*. Minneapolis, MN: University of Minnesota Press.
Massey, D. and Allen, J. 1984. *Geography Matters!* Cambridge: Cambridge University Press.
Mitchell, D. 1995. The end of public space? People's park, definitions of the public, and democracy. *Annals of the Association of American Geographers*, 85, 108–33.
Nagel, C. 2000. Ethnic conflict and urban redevelopment in downtown Beirut. *Growth and Change*, 31, 211–34.
O'Loughlin, J. et al. 1998. The diffusion of democracy, 1946–1994. *Annals of the Association of American Geographers*, 88, 545–74.
Pardo, M. 1990. Mexican American Women Grassroots Community Activists "Mothers of East Los Angeles". *Frontiers*, 9, 1–7.
Porteous, J. D. 1989. *Planned to Death*. Manchester: University of Manchester Press.
Pred, A. 1984. Place as historically contingent process: structuration and the time-geography of becoming places. *Annals of the Association of American Geographers*, 72, 279–97.
Pulido, L., 1994. Restructuring and the contraction and expansion of environmental rights in the United States. *Environment and Planning A*, 26, 915–36.
Radcliffe, S. 1993. Women's Place/El Lugar de Mujeres: Latin America and the politics of gender identity. In M. Keith and S. Pile (eds.) *Place and the Politics of Identity*. London: Routledge, 102–16.
Smith, N. 1984. *Uneven Development*. Oxford: Blackwell.
Smith, N. 1996. *The New Urban Frontier*. London: Routledge.
Smith, N. and Katz, C. 1993. Grounding Metaphor: towards a spatialized politics. In M. Keith and S. Pile (eds.) *Place and the Politics of Identity* London: Routledge, 67–83.
Soja, E. 1989. *Postmodern Geographies*. Oxford: Blackwell.
Sorkin, M. (ed.). 1992. *Variations on a Theme Park*. New York: Hill and Wang.

Staeheli, L. 1996. Publicity, privacy and women's political action. *Environment and Planning D: Society and Space*, 14, 601–19.
Tilly, C. 1978. *From Mobilization to Revolution*. Reading, MA: Addison-Wesley.
Wilson, D. and Huff, J. 1994. Introduction: Contemporary Human Geography – the emergence of structuration in inequality research. In D. Wilson and J. Huff (eds.) *Marginalized Places and Populations*. Westport, CT: Praeger, xiii–xxv.
Wolch, J. and Dear, M. 1993. *Malign Neglect: Homelessness in an American City*. San Francisco, CA: Jossey-Bass.

Part III Critical Geopolitics

Chapter 12

Imperial Geopolitics
Geopolitical Visions at the Dawn of the American Century

Gerry Kearns

The First World War (1914–18) was promised by many as the "war to end all wars" and by others as "the war for civilization." Even before the conflict was ended, with its death toll of at least ten million, experts were busy trying to design a stable peace. Some of these investigations were sponsored by belligerent governments. In the USA, the House Inquiry was established to gather the geographic intelligence needed to design "a new rational, political geography for post-war Europe" (Heffernan, 1998, p. 88). In France was established a *Comité d'Études* to collect the social, geographic, and linguistic justifications for using history to repatriate to France territory claimed by Germany after the earlier war of 1870. For the French, the Americans were idealists for ignoring historical realities. For the Americans, the French were idealists for ignoring political science. Against both, Robert William Seton-Watson (1879–1951) wanted "[t]he creation of a new Europe upon a mainly racial basis" (quoted in Blouet, 1987, p. 161). Each claimed to be offering realism or objectivity. Each criticized the views of others as idealism or self-interest. Clearly, the postwar political map of Europe could be drawn in many different ways. Different people saw different realities and drew different conclusions. In this paper, I explore three contrasting visions of global political realities. I believe that the discourses of races, classes, and ethnic-nations provided alternative perspectives that continue to animate imperial geopolitics.

Normative Geopolitics

Geopolitics is a discourse that describes, explains, and promotes particular ways of seeing how territorial powers are formed and experienced. In geopolitical terms, the twentieth century might fairly be called the American century in recognition of the military and economic hegemony achieved then by the USA. In 1900, it was not clear to all that, over the coming century, the world order would be dominated by the USA. World War I provided the century's first great test of expectations about the global distribution of military and economic power. This period also saw the century's first great setback for British imperial power with the nationalist revolution in Ireland (1916). It was the occasion of the world's first socialist revolution (Russia, 1917). The postwar settlement saw the USA dominate both the redrawing of political boundaries in Europe and the framing of a new institution of international cooperation and recognition, the League of Nations. With hindsight, World War I saw the dawning of the American century and revealed many of its main themes with global capitalism under the direction of the USA facing anti-colonial struggles and socialist challenges.

As commentators and activists tried to understand and shape the new world order ushered in as the sun began setting on the British Empire, they made use of a variety of geopolitical ways of viewing the world. We might call such a world picture, a geopolitical vision. The geopolitical vision is never innocent. It is always a wish posing as analysis. We see the world in a certain fashion because we want to highlight salient dimensions of a new world order we hope is emerging. Explanation is always normative (Kearns, 1998). In this paper, I will explore three contrasting geopolitical visions. Each hoped for a different version of the new world order. The differences were grounded in contrasting conceptions of what I term here the geopolitical subject.

The geopolitical subject is the basic agent shaping global political and economic relations. Among other entities, races, peoples, and classes have been taken to be the fundamental building blocks of geopolitical structures. These correspond quite neatly to the three sets of society-shaping forces identified by Sack (1997) as nature, meaning, and social relations, respectively. By the early twentieth century, most commentators were agreed that on the surface the world order was made up from the actions and reactions of countries. However, in explaining international relations, they made reference to other, more fundamental, agents. This does not mean that someone emphasizing class, for example, did not think races or nations existed. However, if in thinking about, say, the "national question" Marxists would generally seek to show its primary reliance upon more fundamental class relations (see the discussion in Forman, 1998). Similarly, racists did not ignore class but they often saw it as a confounding set of relations that obscure more fundamental racial realities.

Geopolitical commentators recognized that the relations between countries could be stabilized and the common interests among groups of states could be advanced through international institutions. Again, while these international groupings were made up from sovereign states, they were taken to be promoting the essential interests of the underlying geographical subjects, be they classes, peoples or races. Similarly, although international relations were thought to display certain distinct trajectories, depending upon the particular geopolitical vision through which they were viewed, these destinies were actually expressive of the global missions of particular peoples, classes or races. In fact, while these alliances were presented as egalitarian, these institutions were shaped by the military and economic inequalities among the countries that were their members. Furthermore, it was all too tempting for the dominant country to identify itself so strongly with the collectivity that it conflated national and international priorities seeing itself as expressive of the essential interests of the geopolitical subject.

The geopolitical vision, then, is often organized around a distinctive geopolitical subject. To advance the interests of this global subject, various institutions are created. These institutions often come to be dominated by a hegemonic nation-state. I want, now, to consider how World War I posed a challenge to these geopolitical visions and shaped their institutions. It had to be understood, and the postwar settlement had to be addressed, within the particular emphases of the contrasting normative geopolitical discourses. By taking this moment at the dawn of the American century, I hope to illustrate how some geopolitical discourses discharged their normative obligations. I might add that these world views have certain modern resonances, which I will briefly indicate in the conclusion.

Racial Conflict and the Imperial Order

Race has been an important element in many geopolitical visions. In the late nineteenth century, a certain version of neo-Darwinian biology sustained a view of international relations as essentially a conflict between rival races. Halford Mackinder (1861–1947) was a British geographer with a biological world-view. His geopolitical writings were intended as an "aid to statecraft" (Parker, 1982). Geopolitical realities, for Mackinder, were an amalgam of the biological inheritance of races, the effect of environmental influences and the role of imperial strategy (the making and breaking of alliances). Over the very long term, according to Mackinder, races adjusted, through natural selection, to their environment. Certain environments selected for distinct sets of characteristics. The finely grained and temperate environments of Britain, for example, produced "John Bull," a "genus" of the human species noted for its commitment to freedom and civilized values (see Kearns, 1985). In contrast, the open steppes of Russia produced a Slav human-type, easily accepting of despotic rule. Over the shorter term, races move into new environments, taking with them characteristics formed in their hearths. The Anglo-Saxon could, thus, bring civilization to Africa, whereas African people would find it hard to adjust to the strains of British civilization. Over the very long run, African peoples might adapt to a European climate and live up to the demands of the local civilization. Mackinder thought this might take several centuries. In the short term, then, the world should accept the tutelage of the already civilized peoples, the Anglo-Saxons. The danger was, of course, that the less-civilized races might fail to recognize this. The Anglo-Saxons, then, had to impose this most desirable outcome by force. In Darwinian terms, they had to show that they were the fittest by being the strongest.

Although the superiority of the Anglo-Saxons was expressed in cultural terms, it had to be exercised in more martial terms. Nations would expand into the territory of inferior peoples until they reached natural borders or the line of advance of a more-or-less equal. This is where alliances became important. Mackinder identified three types of association. In the first case, there were the vertical relations between a superior race and their colonized peoples. By offering subject peoples benevolent administration, the British could attach to themselves such peoples as the Indian race. These peoples were thought to be incapable of independent existence in a world of competing empires and it was clear, at least to Mackinder, that they could do no better than place themselves under British rule. The British would protect Indian people from both the anarchy of internal rebellion and the disruption of conquest by alternative imperial powers. The British created a space in which they could administer justice and foster economic growth while retaining the option to deploy local resources and people in defence of the broader Empire as the need arose.

The second type of association was also imperial but was between Britain and the imperial dominions. The British Empire included several white settler colonies and here a less hierarchical relationship was required. With its vast resources, a country like Canada might eventually be a more suitable metropolitan core for the British Empire. In the meantime, such white settler colonies had to be accepted as equals within the Empire. A high-handed approach had already lost Britain the valuable colonies that became the USA. This mistake should not be repeated. The Empire needed a sort of imperial parliament through which these horizontal relations within the Anglo-Saxon race could be institutionalized. In order to secure these dominions to Britain, they should be given preferential access to British markets through the raising of tariff barriers against other countries.

The third sort of alliance was with Anglo-Saxons outside the British Empire and here Mackinder was concerned mainly with the USA. He wanted a special relationship that would see that federation identify its global interests most closely with those of the British Empire in the cause of Anglo-Saxon world dominion.

The geopolitical subject in this world-view is clearly the Anglo-Saxon race. Its most coherent institutional expression is taken to be the British Empire. The temptation to conflate the interests of the Anglo-Saxon race with the British nation's is irresistible. Mackinder was convinced of the racial purity of the English and because the people of William Shakespeare (1564–1616), Francis Bacon (1561–1626), Isaac Newton (1643–1727) and John Locke (1632–1704) expressed the best of human civilization, their sway in world affairs could not be allowed to decline along with their relative economic status. The moral integrity of such a pure example of the Anglo-Saxon race might, it was hoped, persuade others to accept its leadership of the free world. World War I was a challenge to this leadership. The German nation was seeking to establish its own empire within central Europe and was also pursuing further colonies in Africa. In Germany, at this time, the political geographer Friedrich Ratzel (1844–1904) was, as Mark Bassin shows in chapter 2 of this volume, justifying German expansion using a very similar argument about the German people as a collective organism needing living space. For Mackinder, the danger lay in Germany establishing a dictatorship over the Slav peoples of central and eastern Europe. Then, a land power would be established from which well-resourced colonial adventures could be launched in competition with the British Empire. At the end of the war, Mackinder wanted a European settling of scores that would irreversibly weaken Germany. Furthermore, Bolshevik Russia needed to be contained and thus he advocated a set of viable buffer states for eastern Europe. In this way the British leadership within the Anglo-Saxon world could be perpetuated and the Anglo-Saxon supremacy over the Slav ensured.

Springtime of the Peoples: Self-Determination and Citizenship

Mackinder was not involved in the postwar settlement of the political boundaries in Europe. Other geographers were. The place of Paul Vidal de la Blache (1845–1918) in the French delegation (Heffernan, 1998) and of Isaiah Bowman (1878–1950) in that of the USA (Smith, 1994) was not extended to Mackinder by the British government. His blood-and-soil racism was dominant neither in the USA nor in Great Britain. The official ideology of the war effort was more liberal. It was identified in particular with the views of the American president Woodrow Wilson (1856–1924). Each people having coherent territorial expression and large enough to be able to defend itself had a right to be a self-governing state. A people were defined in cultural rather than biological terms. A people were a group that had been characterized by intense internal interaction such that it shared a distinct history and set of traditions. This distinctiveness was thought to be most clearly expressed by the possession of a language distinct from that of neighboring peoples. In this world-view, the age of Empires, which Mackinder saw as eventuating in the sway of a single World Empire, was over. The old empires, be they Austro-Hungarian, Ottoman or, but whisper it, British, could not expect to survive the noon-day heat of the springtime of the peoples. National self-determination was the order of the day. In ruling themselves, peoples would adopt some model of citizenship similar to that expressed in the constitution of the USA. The geopolitical subject was the people, expressing itself as a nation-state.

This new global regime, likewise, required alliances and institutions. The primary vehicle for recognizing nation-states was the League of Nations, created in 1919 at the Paris peace conference. This was to be an international body to which peoples could apply for recognition of their right to exist as nation-states. The dictat of an occupying power was no longer to suffice as the principal voice in world affairs. A people that had shown determination, distinctiveness and defensibility should be offered the hand of international fellowship. Colonialism should be curtailed. Multilateral international alliances should deal with rogue states in their own world region. In time, the League of Nations might evolve as a sort of world government of nation-states.

World War I showed the necessity for these new international institutions because the failure to nip in the bud the imperial aggression of Germany had pulled the whole of the civilized world into a global conflict. Effective alliances could have given Belgium such security that Germany would have been scared off. The postwar settlement was intended to replace empires with nation-states. The redistribution of population over space was resorted to in order to create a better fit between cultural and territorial units so that monolingual nation-states could be established in central and eastern Europe. This engineering of a fit between ethnicity and space became the model for many postcolonial settlements in the years to come.

The USA presented itself as the paradigm of the self-governing nation-state with its constitution and territorial integrity. It offered itself as the guarantor of national self-determination in its own hemisphere. It was very easy to believe that what was good for the USA was what was best suited to the interests of the entire "free world." Free democracies came to be seen as those open to American businesses. With Germany and Great Britain reeling from the demographic and economic consequences of the Great War, the early 1920s were the first time when the world looked like the USA's oyster. The prosperity of the USA was presented as further proof of the desirability of the new forms of ethno-territorial citizenship. Empires appeared decadent and now in terminal decline. Colonialism would soon follow. Race did not appear as meaningful grounds of distinction in this geopolitical vision; although, as I discuss below, it continued to be a pertinent dimension of American domestic politics.

Proletarian Solidarity and Capitalist Slavery

The third geopolitical subject that I want to discuss is class. Karl Marx (1818–83) and Friedrich Engels (1820–95) said that all hitherto existing human history was the history of class struggle (Marx and Engels, 1992 [1848]). Slave and master, lord and serf, and now capitalist and worker confronted each other in a zero-sum game that could only end with the abolition of the exploitation of the second by the first and with the creation of a new social order. In capitalist society, there was a large group without property – the proletariat – who could only make a living by submitting themselves to be employed by the class of capitalists who owned the tools of work – the means of production. As Engels, memorably explained in his account of early Victorian Manchester (Engels, 1987 [1844]), under free market conditions, capitalists competed with each other for market share and workers were induced to undercut each other in pursuit of employment. Nation-states were managed by the capitalist class to promote their collective interests against their workers, against capitalists of other countries and, sometimes, in self-restraint to sustain the longer-term viability of the system in the face of its self-destructive, competitive anarchy. The internal contradictions of the capitalist system were expressed as booms and slumps, and a long-term tendency towards a declining rate of profit. These could be alleviated, temporarily, by what Harvey (1982) has called a spatial fix. Imperialism offered national capitals cheaper raw materials, new markets and super-exploitable workers. Imperialist competition, in turn, created the conditions for world wars. These would issue in an orgy of destruction, readjusting the hierarchy of imperialist powers and clearing the way for new cycles of accumulation.

This world promised workers bread and circuses but with unending exploitation and no real freedom. As it cooperated in the making and remaking of this world, the proletariat forged its own chains. But, as capital consolidated into larger units, as the technical demands of modern industry promoted education, and as the detailed management of modern economies fostered state intervention, the capitalist system was digging its own grave. Educated workers thrown together would come to recognize a common interest and would learn to use the institutions of the modern state to abolish private property and create democratic economies under collective ownership. The common interest of workers would be seen first in their defensive, trade-union activity but increasingly in offensive actions of class solidarity on behalf of workers in other industries and in other places. The workers of the world should unite through institutions such as the International of communist parties. When Marx lost control of the First International to what he saw as unscientific socialists he ensured its abolition. It was a Second International of communist and socialist parties that confronted World War I.

For Marxists, the geopolitical subject is class. The global trajectory, then, was of the organized working-class confronting a world of economic cartels and imperialist conflict. Scientific Marxists, such as Rosa Luxemburg (1871–1919), saw World War I as a purely internal affair of the capitalist system. The workers should not involve themselves. This was also the line taken by Lenin (1870–1924) and by James Connolly (1868–1916). The war also, however, created an opportunity for revolutionaries to seize power because the repressive institutions of the state were strained in all cases by the demands of fighting the war and in some cases by the shame of losing a war. With the exceptions of those from Ireland, upon Connolly's advice, and from Russia, under Lenin's direction, the socialist and communist parties in the Second International placed national survival before socialist revolution and suspended the class struggle to support the patriotic cause. Worker fired upon worker as the Second International failed its greatest test. When a new, Third International was formed, there had been a socialist revolution in Russia and with the socialist revolutions in postwar Europe defeated, the defense of "socialism in one country" became its goal. Thus, the cause of the international proletariat was conflated with the national survival of the Soviet Union. National communist and socialist parties of the Third International were to subordinate their local interests to the geopolitical needs of the advanced guard of world revolution, the imperial USSR.

Deconstruction, Contradiction, and Context

The principal differences between these three geopolitical visions are summarized in table 10.1. They each have a different conception of the basic agents of geopolitical change and of the underlying trajectory, or teleology, of geopolitical change. Each proposes a suitable international institutional support for achieving desirable geopolitical change and each is drawn to accept the interest of one specific nation-state as best embodying the true interests of the geopolitical subject.

Each of these geopolitical visions is based on an essentialism that tries to naturalize its world view and thereby devalue competing presentations of the nature and purpose of geopolitical change. This is how geopolitics pursues its normative goals: see the world like this and you can only imagine its future like that. Are the major world issues related to race survival or national self-determination or proletarian liberation? Each vision keeps other utopias off the agenda. However, these essentialisms are unstable. They were always ripe for deconstruction. In claiming a monopoly for their view of the world, they over-reach themselves. They define away

Table 10.1 Geopolitical visions

	Mackinder	Wilson	Lenin
Geopolitical subject	Nature: races	Meaning: ethnic nations	Social relations: classes
Teleology	Consolidation of empires, emergence of single world empire	Break up of empires, emergence of mosaic of ethnic nation-states	Collapse of capitalism, diffusion of socialist revolution
Institutional support	British Empire	League of Nations	Third International
Conflation of state with geopolitical subject	Great Britain	USA	Soviet Union

heterogeneity within the state. They deny the important structuring effects of other dimensions of difference. They also often appeal in a rhetorical fashion to categories that the strict letter of their essentialisms should exclude. Thus, for example, Mackinder's vision of the Anglo-Saxon race implied a set of qualities that would have resonated with class-based meanings (noblesse oblige, fair-play, and a stiff upper-lip were claimed as the self-image of the aristocracy). These traits idealized the English as an aristocratic race when the vast majority of the population were proletarian. Positing the Soviet Union as the vanguard of the proletarian revolution evoked associations with the old "civilizing" mission that the Russians had claimed to bring to the lesser peoples distributed around the edges of their imperium. Within the Soviet Union, the Russians did indeed see themselves in this way, making, as we will see below, the defense of the socialist revolution in Russia the primary goal of the Soviet federation thereby denying other nations an effective right of secession.

The conflation of state with geopolitical subject rests upon unexamined inconsistencies and contradictions. These problems produce empirical embarrassment for the account of the world given within each geopolitical vision. There are factors and phenomena that are not easily reduced to the guiding essentialisms. In some cases, this disparity between theory and context is addressed by changing, violently, the context.

The Multiracial and Multinational British Empire

All states are heterogeneous. Mackinder's Britain was not purely Anglo-Saxon. The demographic weight of the Celtic fringe may have been weaker in 1918 than a century earlier but it was still significant. The Irish, Scots, and Welsh were a large part of the workforce in England's industrial and commercial cities. To pretend that even metropolitan England had remained ethnically pure since Norman times was absurd. To treat economic growth as a tribute to Anglo-Saxon inventiveness was to ignore the central importance of Hugenot weavers, Scots scientists, and Dutch drainage engineers, among many others. For Mackinder to praise the justice and fairness of the English was to deny the role of coercion in securing Ireland and Scotland as part of the Union. In the early twentieth century, Mackinder was one of the Conservatives most opposed to the breaking of the Union between Britain and Ireland. It was all very well for Mackinder to believe that the Empire should be held together by bonds of respect and mutual affection, but in principle he believed the British had a right to hold it together in the absence of any such bonds. Thus, the Irish would have to test by might their right to self-determination. This meant that the Irish were to be disappointed in Woodrow Wilson and the postwar settlement of political boundaries in Europe. Surely, Irish nationalists argued, the Irish were a people with a distinct identity, a language and culture of their own, and a defensible territory. But, Wilson wanted the support of the British for his plans for mainland Europe. Thus the peoples within the British Empire were placed to one side when considering the rights of small nations.

Imperialists such as Mackinder were able to continue seeing the British Empire as a family with the Irish and the Indians as children needing paternal direction. This, of course, ignored the structured inequality of economic relations between Britain and the colonies and, patronisingly, dismissed colonial demands for independence as misguided. If the goal of the British Empire was to secure the dominance of the

Anglo-Saxon race, then, it was not clear why the Celts or the Indians should cooperate. Furthermore, if colonies felt themselves to be exploited, it was not clear why they should not be allowed to be the best judge of their own development priorities. The British coerced their colonies into imperial line and, by treating this as a matter of internal British politics, the other world powers stood by and watched. Yet the contradictions are glaringly obvious. The Empire, although multiracial, is presented as the embodiment of the best interests of the Anglo-Saxon race. The relations between metropole and colony are explained as matters internal to the metropolitan nation-state. Colonialism is thus off the agenda for geopolitical discourse, which is about international relations. This complex rescaling of issues was accepted in 1918 because Britain could insist upon it. After World War II, the line was not so easily held.

The Multicultural and Capitalist United States

The USA was itself the most glaring exception to the rule that each nation should consist of a homogenous people. "Out of many, one" was always a pious hope. A federation that rested upon the violent taking of land from native peoples could hardly embrace the vision of a primordial mosaic of ethnic difference being given political expression as a series of ethnic-nation states. In terms of citizenship, race and gender injustices still disfigured political rights in the USA. The anti-Germanism visited upon second- and third-generation German–Americans during World War I revealed how ethnic difference was easily reconfigured as national difference and thus as treachery. The pious hope that ethnic difference might be suspended in national citizenship was not an unworthy one, yet it sat uneasily with the promotion of ethnic national self-determination for the rest of the world. The liberties of the American constitution were often enforced at the point of a bayonet, be it against native people driven to reservations, or Utah Mormons dragooned into monogamy, or Southern states coerced into outlawing slavery. The federal system was also claimed as legitimating local choices, around Jim Crow laws for example, that flagrantly disavowed any suspension of racial or ethnic discrimination. At this very moment, of course, the United States army was fighting in the war to end all wars with race-segregated army units. The development of the USA was, indeed, an ironic rejection of the teleology of Wilson's vision.

If Mackinder's geopolitical vision naturalized colonialism, then, Wilson's naturalized capitalism. The question of capitalism simply never arose. Private property was accepted as the basis of citizenship. The place of the proletariat in a property-owning democracy was simply not addressed. Capital–labor relations in early twentieth century USA were often violent, with both state and federal troops involved against strikers. But, again, like the violence used in the British colonies, this was an internal matter for the nation-state. Furthermore, the diffusion of liberal citizenship throughout the world was assumed to involve the replication of economies based upon private property and minimal tariff protection. Of course, as the depression of the 1930s took hold, national economies raced to protect local enterprise behind customs barriers thereby producing the catastrophic collapse of world trade that followed. After the World War II, with capitalism more secure in the USA, the question did indeed become part of the agenda of international relations and dominated foreign affairs during the so-called Cold War. However, in 1918 the consolidation of capitalism in the USA was seen as a purely domestic matter. By making ethnic difference the essential marker of national identity, class issues never appeared on the geopolitical agenda.

Awkward Classes in the Multiethnic Soviet Union

Marxists believed that capitalism was polarizing society into two classes, capitalists and proletarians. Socialism would involve expropriating the property of the former and leaving it to the democratic control of the latter. In Russia, things were not as simple as that. For some, this meant, simply, that Russian capitalism was as yet immature in 1917. The question arose, then, of whether Russia could leapfrog the later stages of capitalism to reshape itself as premature socialism. Lenin was convinced that it must. Three somewhat contradictory imperatives imposed themselves. The first was the ideological need to introduce at once a socialist economy elaborated from the sketchy utopian remarks of Marx and Engels (Kornai, 1992). The second was the military need to secure national political unity in the face of external threats (in which Mackinder was himself involved; Blouet, 1976). The third was the economic need to provision the cities. The second and third were clearly mutually dependent. The collectivization of agriculture proceeded unevenly across the territories of the Soviet Union. Under Lenin's New Economic Policy (1921–8), it was recognized that the immediate need for food would have to be met in large part from peasant farms under private ownership. Thus, pricing policies were introduced to stimulate rural production (Smith, 1989). This ensured that the rural sector received a decent share of national resources and under Stalin (1879–1953) this policy was abandoned in favor of diverting all available resources to urban industries in order to militarize rapidly. In 1928, four years after Lenin's death, barely 1.7% of peasant households had been collectivized. Within nine years Stalin had enforced collectivization to the extent that 93% of peasant households were now collectivized. Stalin identified a rich peasant class, the kulaks, as the scapegoat for the continued failure of Soviet agriculture to provide sufficient and cheap food for the industrial towns. They were expropriated. Millions starved and millions more were imprisoned as collectivization marched across the Soviet countryside. The class structure was being simplified by force. By identifying the Soviet state with the proletariat, the kulaks became traitors on their own land.

Class was not, however, the only dimension of difference that mattered in the Soviet Union. Ethnicity and nationalism remained important. In ideological terms, they should have faded as a proletarian identity trumped all others in the socialist utopia. Nominally, the Soviet Union was a voluntary federation. However, the conflation of nation with class operated here too. Russia was identified as the essential heartland of the proletarian cause and thus nations could secede only if by doing so they did not endanger the stability of the proletarian revolution in Russia. Since the constituent non-Russian nations ringed Russia as a defensive buffer, none were ever able to exercise this right (Smith, 1999). Minority rights, then, were constrained by the identification of Russia with the proletariat as a world geopolitical subject and, yet further, by the identification of the Russian proletariat with the Politburo of the Russian Communist Party.

The nature of socialism was asserted to be a domestic matter for the Russian Communist Party. This meant that the unexamined issue at the heart of the Third International was the nature of communism and even of late capitalism. Class conflict and its evolution were naturalized. Capitalism would inevitably fall. Socialism or barbarism would ensue. Socialism was Stalinist collectivization and forced industrialization. A war economy defined the prospects for socialism. For those communist parties in other countries remaining faithful to the Third International, this Soviet domination crippled debate about local priorities and eventually disabled the heroic part played by communists in anti-fascist resistance in 1930s Europe. Geopolitical discourse was monopolized by Stalin. In other countries, talented communist political scientists devoted their energies to justifying each shift in Soviet foreign policy as serving the essential interests of global proletarian revolution.

The Legacies of Race, Ethnicity, and Class

These three geopolitical visions suffered the censorship of context (Therborn, 1980). In other words, the claims they made about the world were subject to empirical embarrassment. The world did not appear as simple as it was thought to be. Global geopolitical realities did not always develop in ways anticipated by the commentators. Furthermore, these geopolitical visions themselves evolved and I have commented upon some of these developments above. However, I think that some of the ways they have directly or indirectly influenced current debates deserve a brief concluding comment. These influences are to be found both in discourses and in institutions.

There are few geopolitical discourses as openly racist as Mackinder's; although there are some. More significant, however, has been a translation of biological racism into cultural racism. The salient continuity lies in the denial of national multicultural realities in favor of the identification of the nation with one ethnic identity, now conceived in cultural rather than biological terms. In a work such as Huntington's (1996) we see the same amalgamation of nations into broad homogenous units, now called civilizations rather than races. We also see a similar assertion of the absolute incompatibility of difference; now viewed as cultural dissonance rather than biological miscegenation. We also see the same teleology of the consolidation and apocalyptic clash of empires, or civilizations. There is the same conflation of one nation-state with the global best interests of humanity, although in this case the USA has replaced Great Britain. Furthermore, it is also clear that subordinated peoples who have been oppressed as "races" come to identify their own common interests in racial terms and construct world-views organized around similar racial concepts. A "planetary humanism" would need to step away from race both as a category of superiority and as a goal of resistance (see Gilroy, 2000).

The avowedly liberal discourse of Wilson has many heirs. One of the most interesting is the way Sen (1999) has taken the discourse of citizenship and applied it to development issues. Here, we see an important modification of the earlier position because Sen argues that substantive and not purely formal rights are necessary for freedom. In other words, development enables dignity, enables choice, and thus allows civil society to function effectively. Of course, this vision of development is about income distribution and not just national product. As such, it has implications for rich countries too. These implications are spelled out by Rawls (1973) and by Baker (1987). Civil society is disabled where rich individuals are able to buy indignity for other much poorer individuals. Again, we see here the same difficulties about the scaling of politics. Many admit that the rights of the unenfranchisable, children for example, are a legitimate concern of international institutions, few seem willing to embrace the rights of the poor.

Marxist geopolitical discourse still thrives. However, it has taken up issues raised most effectively outside its class-analysis through the political action of the new social movements. Some, such as Harvey (1996), have tried to argue that the interrelations between various dimensions of difference (such as race, class, and gender) mean that it is still possible to see social justice as primarily disfigured by the class relations of capitalism. However, this over-determination is so deferred and contingent that effectively a multi-causal analysis has replaced the earlier essentialisms. Harvey (2000) is surely right to argue that class inequality still matters but the challenge now is to think about socialist policies that do not return us to the agenda of comprehensive nationalization along anything like Soviet lines. Capitalist societies are varied and cannot be arranged along any teleological continuum. The global economy does not have a discernable trajectory.

The institutions that sustained these three geopolitical visions have undergone several ironic developments. The British Empire of one nation-state and many dominions and colonies became a Commonwealth of many nations and a few small dependencies. With majority voting at Commonwealth congresses, the stage has been set for a new sort of postcolonial fellowship. In the recent past, some African countries used the Commonwealth to put pressure on apartheid South Africa through sports boycotts. The association of Mozambique, never a British colony, with the Commonwealth is an intriguing development. Perhaps Mackinder's Empire can become a postcolonial Commonwealth capable of addressing the continuing inequalities installed by earlier colonialisms.

The League of Nations dissolved as it failed to halt fascist aggression in Europe. Now a whole series of international institutions from the World Bank to the United Nations and a cluster of international agreements, most notably the Declaration of Human Rights, create the possibility of defining a form of global citizenship (Archibugi and Held, 1995). It is against this background that Sen's insistence on the importance of substantive rights is so important. Basic human needs must be addressed if the asymmetrical world economy is not to render hollow all forms of cosmopolitan citizenship.

The Third International has suffered the harsh judgement of historical hindsight. However, the Second International continues to function and makes a significant contribution to socialist debate within the European Union. As the European Union prepares to include several former communist countries within its economic and political union, it becomes ever more important for welfare issues to claim a place on the European political agenda. The triumphalism of free market capitalism in American ideology after the destruction of the Berlin Wall has received a forceful check from the environmentalists of Europe. Perhaps Europe's socialists may also have something to say about the responsibilities of global commerce.

If the geopolitical visions of the dawn of the American century remain important at its close, then, it is in large part because the normative issues they articulated continue to animate modern political debate. Colonialism and imperialism still persist within the unequal exchanges of international trade. Citizenship and ethnicity continue to set the agendas for national constitutions. Class and exploitation remain significant dimensions of the global political economy. Geopolitical discourses see these issues rescaled between national and international levels as citizenship itself gets reformulated in global as well as national terms. These geopolitical visions were also nation-state visions. They served to declare a division of labor. Certain issues were domestic matters for nation-states whereas others were of international concern. As I have shown above, the importance of geopolitical visions lay as much in what they left behind for nation-states to deal with as in what they put on the table for international arbitration. In the new world order of a world that some declare "postcommunist," these continue to be contentious matters with institutions such as the World Trade Organisation serving to discipline new forms of division between national and international rights, and new geopolitical visions.

Acknowledgments

Thanks to Millie Glennon, Mike Heffernan, Simon Reid-Henry, and Gerard Toal for their advice.

BIBLIOGRAPHY

Archibugi, D. and Held, D. (eds.). 1995. *Cosmopolitan Democracy: an Agenda for a New World Order.* Cambridge: Polity.
Baker, J. 1987. *Arguing for Equality.* London: Verso.
Blouet, B. W. 1976. Sir Halford Mackinder as British High Commissioner to South Russia. *Geographical Journal*, 142, 228–36.
Blouet, B. W. 1987. *Halford Mackinder: a Biography.* College Station, TX: Texas A&M University Press.
Engels, F. 1987. *The Condition of the Working Class in England.* Harmondsworth: Penguin. (Original German edition 1844; English translation 1886.)
Forman, M. 1998. *Nationalism and the International Labor Movement: the Idea of the Nation in Socialist and Anarchist Theory.* University Park, PA: Pennsylvania State University Press.
Gilroy, P. 2000. *Between Camps: Nations, Cultures and the Allure of Race.* London: Penguin.
Harvey, D. 1982. *The Limits to Capital.* London: Blackwell.
Harvey, D. 1996. *Justice, Nature and the Geography of Difference.* Oxford: Blackwell.
Harvey, D. 2000. *Spaces of Hope.* Edinburgh: Edinburgh University Press.
Heffernan, M. 1998. *The Meaning of Europe: Geography and Geopolitics.* London: Arnold.
Huntington, S. 1996. *The Clash of Civilizations and the Remaking of the World Order.* New York: Simon and Schuster.
Kearns, G. 1985. Halford Mackinder. *Geographers: Biobibliographical Studies*, 9, 71–86.
Kearns, G. 1998. The virtuous circle of facts and values in the New Western History. *Annals of the Association of American Geographers*, 88(3), 377–409.

Kornai, J. 1992. *The Socialist System: the Political Economy of Communism*. Oxford: Clarendon Press.

Marx, K. and Engels, F. 1992. *Communist Manifesto*. Oxford: Oxford University Press. (Original Edition, 1848.)

Parker, W. H. 1982. *Mackinder: Geography as an Aid to Statecraft*. Oxford: Clarendon Press.

Rawls, J. 1973. *A Theory of Justice* Oxford: Oxford University Press.

Sack, R. D. 1997. *Homo Geographicus: a Framework for Action, Awareness and Moral Concern*. Baltimore, MD: Johns Hopkins University Press.

Sen, A. 1999. *Development as Freedom*. Oxford: Oxford University Press.

Smith, G. E. 1989. *Planned Development in the Socialist World*. Cambridge: Cambridge University Press.

Smith, G. E. 1999. *The Post-Soviet States: Mapping the Politics of Transition*. London: Arnold.

Smith, N. 1994. Shaking loose the colonies: Isaiah Bowman and the "de-colonisation" of the British Empire. In A. Godlewska and N. Smith (eds.) *Geography and Empire*. Oxford: Blackwell, 270–99.

Therborn, G. 1980. *The Ideology of Power and the Power of Ideology*. London: Verso.

Chapter 13

Geopolitics in Germany, 1919–45 Karl Haushofer, and the *Zeitschrift für Geopolitik*

Wolfgang Natter

It was Haushofer, rather than Hess, who wrote *Mein Kampf* and who furnished the backbone for the Nazi bible and for what we call the common criminal plan. Geo-politics was not merely academic theory. It was a driving, dynamic plan for the conquest of the heartland of Eurasia and for domination of the world by the conquest of that heartland. . . . Really, Hitler was largely only a symbol and a rabble-rousing mouthpiece. The intellectual content of which he was the symbol was the doctrine of Haushofer (Jacobson, 1987, pp. 568–9).

So wrote Sidney Alderman in September, 1945, reporting to Justice Jackson's war crimes tribunal about a study he had read and now wished to recommend to the Chief Council's office as offering positive reasons why Karl Haushofer should be put on the list of major war criminals.

The Karl Haushofer (1869–1946) under investigation for the Nuremberg Trials after World War II was a former general who, after WWI, became an honorary professor who helped launch and served as an editor of the *Zeitschrift für Geopolitik (Journal of Geopolitics)* throughout its entire run (1924–44). As World War II ended, Haushofer was interrogated, placed under house arrest, but ultimately not added to the list of war criminals put on trial at the Nuremberg Military Tribunals. A conservative German nationalist, he had devoted two careers – the first as a military officer (with a retired rank in 1919 of Major General), the other as a prolific writer and academician (some 40 books on the general topic of geopolitics, honorary professor at the University of Munich) to his vision for Germany. With the German Reich's unconditional surrender in May, 1945, suffering from illness and advanced age, and the murder of his son Albert by a roaming SS unit in the waning days of the war, Haushofer ended his own life by suicide, leaving behind numerous questions for a postwar public about his significance for the development of Nazi Geopolitics specifically, and the conceptual development of geopolitics more generally.

In the aftermath of World War II and the full discovery of the extent of Nazi atrocities, it is unsurprising that geographers both in Germany and the USA sought to isolate the "bad" contamination of German *Geopolitik* from the academic traditions of "good" geopolitics and political geography. This was evident in Germany in the critiques of geopolitics offered by the political geographers Carl Troll and later Peter Schöller. This strategy was also evident in the USA, beginning with the explication offered by Edmund Walsh, the founder of Georgetown University's School of Foreign Service, and the Allied representative who most extensively interrogated Haushofer (Schöller, 1957; Troll, 1946; Walsh, 1948). From the present vantage point, this postwar effort in Germany and the USA to distinguish between "the good" and "the bad" could succeed only partially because it now seems inarguable that the emergence of *Geopolitik* is inseparable from the conditions out of which academic geography emerged more generally in the half century preceding 1933 (Kost, 1988; Rössler, 1990; Schultz, 2001; Wardenga, 2001). Furthermore, neither academic geography nor geopolitics of the period in Germany demonstrated much interest in the concerns of the oppressed, except to the extent that Germany as an entire state formation was viewed to number or potentially number among the ranks of these. For Walsh, who would shortly deploy his version of geopolitical science in the crusade against international communism, there was much of scientific merit in Haushofer's work but it had been directed to illegitimate ends.

Scholarship reflective of various disciplinary–historical and poststructuralist impulses has, therefore, come to doubt the limits of a formal distinction between German geopolitics and a presumed "pure" scientific Allied version. Examination of *Geopolitik's* narration in the USA offers a particularly telling illustration of the stakes involved in avoiding the opposite conclusion. Research has pointed to the sometimes curious maneuvers employed, but also avoided, by high and middle texts in containing Nazi Geopolitics (Ó Tuathail, 1996). Recent criticism has also suggested more broadly that all fields of geography ought be approached as forms of geopolitics, which is to say, geo-power. Geopolitics, like other discursive formations that articulate geo-power, would be seen to function as an ensemble of technologies of power concerned with the production and management of territorial space. The work of disciplinary historians of geography has demonstrated the extent to which the demarcation of geography seems inseparable from the history of war, imperialism, and quests for national identity (e.g. Edney, 1999; Godlewska and Smith, 1994; Hooson, 1994; Livingstone, 1990; Ó Tuathail, 1996). Geopolitics, thus, would mark a particular, but in no way separable (and hence containable) geopolitical deployment of geo-power (Natter, 2000).

Inevitably, numerous new questions arise, first of all as they imbricate an historical understanding of the very *concept* of Nazi Geopolitics. It becomes necessary, for reasons elucidated further below, to distinguish between at least two primary meanings signified by this concept. Study of Nazi Geopolitics in the first sense entails analysis of the conceptual frameworks that emerged in the work of leading German academic geopoliticians and statecraft intellectuals during and proceeding the years of National Socialist hegemony in Germany, their media output and institutions, and the substantive outcomes they developed in their various regional geographies. The second primary meaning of Nazi Geopolitics refers to the statecraft exercised in the name of the German Reich during the years in which National Socialism ruled Germany (and occupied various neighboring territories). For it, the primary actors and institutions that merit attention are those who enacted a foreign policy which produced or attempted to produce territorial space in its name, with ancillary attention directed to the intellectual arbitrators of geopolitical ideas. Importantly, by necessarily referring to other figures, disciplines, and institutions normally located outside either geopolitics and geography, this second meaning attached to the concept of Nazi Geopolitics serves to remind us that geopolitics in the first sense is an important, but by no means unitary content of Nazi Geopolitics (Boehm, 2001). Academically, the work of numerous historians, political scientists, archeologists, legal scholars, anthropologists and practitioners from other disciplines merit attention along with that of geographers in assessing the development of Nazi Geopolitics. Reflection on Nazi Geopolitics, as a formalized discourse, and geopolitics, as the exercise of state-centered geopolitics, would then properly also be conjoined to an analysis of how both are experienced and lived by victims and perpetrators outside the domain of expert knowledge. Such an appropriately all-encompassing analysis of the production of space by Nazi Geopolitics, reflective of mediations between representational space, representations of space, and space as lived (Lefebvre, 1990) remains a research desideratum.

Karl Haushofer: "Father of Geopolitics"

Research over the past decades has permitted further clarification of various important points related more specifically to German *Geopolitik*, the significance of the *Zeitschrift für Geopolitik* and the role of Karl Haushofer in fostering geopolitics in Germany. These two issues are sufficently important for the general topic of Nazi Geopolitics to merit fuller attention here. General understanding has moved well past the immediate certainty of the war years and immediate postwar period, when for the English-speaking world in particular the term *Geopolitik* referred to something seemingly clear. German geopolitics was fundamentally a form of pseudo-scientific intellectual activity distinct from geography proper which was used to justify Nazi expansion, imperialism and, ultimately, genocide. Furthermore, *Geopolitik* was to be understood not only as an intellectual exercise, but also as a form of geo-power directly engaged in statecraft. Lastly, mediating both of these levels of geopolitics was one Karl Haushofer, editor of the *Zeitschrift für Geopolitik*, who was taken to be the master strategist of this science and its policies and thus one of the Third Reich's most important figures. These three understandings found erroneous expression in May, 1945, on the Allied side that they would find Haushofer surrounded by a staff of hundreds if not thousands, of fellow geopoliticians planning the war's strategy from his Geopolitical Institute in Munich. A number of war-time Allied reports, including popular ones published in *Life* magazine and visualized by Hollywood, had conveyed just such an impression (Ó Tuathail, 1996). This fantasy ended when advancing American troops reported finding an elderly man in a normal university office. Haushofer was arrested, interrogated and then released, although, as Alderman's post-interrogation report attests, doubts remained whether Haushofer's early release from custody had been wise. Alderman's assessment seems notable in another regard: it points to the overwhelming desire to locate a progenitive source – an individual – whose punishment would begin to bring justice to its victims, in the absence of the primary individual on whose intentions Nazi genocide and other war atrocities could most obviously be attached (i.e. Hitler). Karl Haushofer, *Geopolitik*, and, at another more diffuse level, instrumentalized spatial thinking in Germany more generally, fulfilled a certain postwar need to identify a personification of evil. However, attributing the whole social structure and knowledge systems brought into being in Nazi Germany to one professor is an example not only of an "intentionality fallacy" (the idea that Haushofer intended or directed what Nazism and World War II became) but of a logic that avoids analysing Nazism as a social order of multiple institutions, agencies and practices.

The question of Haushofer's influence on Nazi policy remains to some extent a matter of debate, though most research, following Jacobsen's important two volume biography and study, has tended to strongly relativize its importance and scope (also Herb, 1996). This research has sharply circumscribed the "intentionality fallacy" demonstrating in detail Haushofer's loss of any even modest direct policy input following the flight of Haushofer's former student and friend Rudolf Hess to England. It has also reflected on the prior institutional history of Haushofer's losing battles within various state organizations in which he was active in the 1930s, and has further pointed to Hitler's own personal disregard for Haushofer's views, decisively after their last face-to-face meeting (November, 1938) following the Munich Accord. During it, Haushofer had argued that Germany should now be satisfied with its foreign policy achievements. In all, the two met perhaps ten times between 1922 and 1938. Hitler did not regard Haushofer as a National Socialist (indeed, he was not a party member), even if he thought some of the latter's theses could be made of use. Hitler was also well aware that Haushofer's wife was "half Jewish," a fact of considerable importance in his eyes. Furthermore, the weight of evidence suggests that Hitler's world view was well formed before he ever met Haushofer. He was not someone fundamentally open to new developments of thought (Jäckel, 1969, but for a contrary view, Hipler, 1994). As with so many others who thought they could become the master teacher of Hitler – Heidegger, Schmitt – Haushofer learned instead that National Socialist reality was one in which Hitler would use the thought of others as it suited him, period. When Hess flew to England, Hitler railed amongst his advisors against Haushofer, that "relative of Jews," whom he inferred had tainted Hess (Jacobsen, 1987, p. 451, quoting Engel, 1974). In sum, characterizing Haushofer as the "spiritual father" of National Socialism's war goals misses the mark.

And what of Haushofer's attitude toward Hitler? It seems to have been ambivalent throughout, opportunistic, marked by his loyalty to Hess, and fueled early on by the hope that his own version of geopolitics would indeed become the master political philosophy of German foreign policy. Indeed, Haushofer could find much confluence in the general positions of the Nazi Party as articulated already beginning in the early1920s and his own national-conservative stance: the demand for restitution of German territories "robbed" by the Versailles treaty, the demand that Germany be returned to full sovereignty as a nation alongside all others, and the demand that Germany be accorded sufficient *lebensraum* (living-space) to support its population. A major point of disagreement between Haushofer and the Party from the beginning, however, was National Socialism's emphasis on race as the lever of human history above and beyond any conception of space and geography independent of race (see, e.g., Bassin, 1987; Jacobson, 1986, and my discussion below). Haushofer's geographical materialism before 1933 principally foregrounded the role of the environment, not race, yet here too, he like other writers associated with the journal (but see Hennig, discussion below), opportunistically began writing about race and space, blood and earth (*Blut und Boden*), even if not to the full satisfaction of party officials.

This is not to say that Haushofer's and various of his colleague's efforts on behalf of *Geopolitik* did not overlap with numerous fundamental positions embraced by the emergent Nazi Party as these were formulated starting in the 1920s. Nor is it to diminish the importance of *Geopolitik* and the journal in popularizing a sensibility in the 1920s and 1930s that geopolitics mattered decisively in understanding Germany's political situation. Even though the weight of recent research has pointed to the need to differentiate between Haushofer, the writer and editor, and Haushofer, the policy maker, and has made more apparent the points of disagreement between Nazi Geopolitik in the concept's two primary meanings, it has not diminished the presumptive role played by *Geopolitik* in ideologically justifying and legitimating both National Socialist rule and *its* ideological deployment of geo-spatial political terminology. Haushofer's articulation of one particular geopolitical conception and his efforts to nurture an institutional base on behalf of the emerging discipline of geopolitics, bespeaks and intervenes on behalf of a conservative nationalist disposition that appears constitutive of German academic geography more broadly in the early twentieth century. Traumatized by the outcome of World War I, conservative nationalism in and outside geography was predisposed to either support or not oppose the resolution of "the crises" of German national identity promised by National Socialism (Natter, 1999).

Haushofer and World War I

Without discounting the obvious influence of Haushofer's impressions of Japan (1908–10) in generating a geographic imaginary (similar to the effect travels in the USA had on Friedrich Ratzel), World War I, particularly its outcome, was a defining point in Haushofer's life and subsequent world view. While the former additionally provided the regional basis for the writing of both his dissertation and second dissertation, his perceptions regarding the conduct of the war made palpable for him the necessity of articulating geopolitical thought for a nation sorely in need of it. The career officer fought both on the eastern and western fronts, commanding a unit of some 3000 soldiers. Haushofer strongly affirmed the soldierly virtues in evidence on the front lines. He believed that the war had brought out the best in his fellow soldiers, had enabled the development of true comradeship, idealism, and a new national community. He was deeply distressed when the war ended in defeat. Wartime letters addressed to his wife evidence a catalogue of attitudes that not only offer insight to his thinking during the war, but are symptomatic of the emergence of a broadly based, disenchanted, conservative nationalism which would play an important role between the wars (Natter, 1999). It is in these contexts that the *Zeitschrift für Geopolitik* which he co-founded and edited for 20 years, may be properly situated. Within this view, the front had given birth to a new sense of national community. The German soldier had unflinchingly accomplished heroic deeds, and reasons for the nearly inexplicable defeat would need to be sought elsewhere. "It is at the front that one finds true freedom, and also humanity," he wrote in January

1916, but in that same letter noted that "that spirit already begins to fade beyond the border of the zone where shots are fired, in divisional headquarters and beyond." With respect to those at the front, Haushofer wrote as late as August, 1918, "I continue to believe that all those who survive the war and its terrible events will have become better human beings. I see that in myself." In other letters, he contrasted this "spirit of the trenches" with what he saw in evidence at the "home front," and its liberal and socialist representatives. In a letter of August, 1917, he railed against those of the left, the bureaucrat, and other civilians who threatened to "rob we at the front of the fruit of our exertions" (Jacobsen, 1987, pp. 122, 124, 136). Most of the elements of what would become a widely articulated platform of the German right during the 1920s – the "stab in the back" thesis – thus already find expression in his war-time correspondence (Natter, 1999).

Also in evidence in this correspondence is Haushofer's own personally drawn consequence from his war experience, to resign from the military and pursue an academic career. Much impressed by his study at the front of Kjellen's work, and building upon prior familiarity with Ratzel, he resolved to develop a geopolitics that would provide the insights necessary to avoid the mistaken strategic conceptions of the present war. By the end of the Weimar Republic, Haushofer offered a "classic" formulation of geopolitical teaching in his *Weltpolitik von Heute* [*Contemporary World Politics*], dedicated to his friend Rudolf Hess, with, as so often before, explicit reference to the lost war's outcome (Haushofer, 1932a). In it, geopolitics was explicitly offered as a weapon to combat the geographical error produced by Versailles: the division of Europe into colony-possessing powers in the West, space-possessing powers in the East, and strangulated states in the center. The primary political task that followed from this insight was the need to restore the space of the German Reich, in all its dimensions. These dimensions included (1) military space, which in 1934 was even smaller than the territory of the extant Reich, (2) the territory of the Second German Reich, and (3) the compact mass of the German "folk" soil, which Haushofer extended to the Polish Corridor, the Südetenland, Upper Silesia, Teschnia, Austria, Alsace-Lorraine and southern Denmark. He also extended the sphere of influence of German space – not necessarily specifying appropriate administrative-state or economic structures – to all territories where German language and culture was in evidence, and to the independent Dutch–Flemish spaces as well. As will be detailed further below, throughout the 1920s and 1930s Haushofer persistently argued for the necessity of Germany foreign policy to think globally and in terms of building either pan-regionalisms or a "continental, Indo-European block" (Haushofer, 1931). This was necessary, he argued, in order to protect the development of Germany's sovereignty and hegemony in Central Europe and as a counterbalance to the influence of the quintessential sea power, England. These geopolitical considerations, moreover, ought have ramifications for German military planning (Haushofer, 1932b).

The Institutional Geopolitics of the *Zeitschrift für Geopolitik*

Picking up on his prior association with Munich professor of Geography Erich von Drygalski, Haushofer completed his second academic thesis in 1919 and became an honorary professor in Munich. From this social position, he conversed with more or less like-minded individuals, including a newly minted publisher, Kurt Vowinckel, who had decided to foreground geographic and geopolitical publications in his new enterprise, including a journal of geopolitics. Haushofer and Vowinckel were able to interest the political geographers Erich Obst, Kurt Lautensach, and soon after, Otto Maull, who had just authored a political geography textbook (Maull, 1925), to serve as the members of an editorial collective, along with the geographer and journalist Fritz Termer. Beginning in January, 1924, the *Zeitschrift für Geopolitik* appeared monthly. Between 1924 and 1944, the journal published a total of 1269 essays and position papers, authored by 619 people. During its first ten years, the journal attracted contributions from a majority of Germany's established political geographers, along with others from a variety of fields.

The history of the journal reveals numerous evolutions and, broadly speaking, three clearly defined phases, which are reflected in its content, list of contributors, and editorial board. The temporal markers of these phases are (i) 1924–31, (ii) 1931–40, and (iii)1940–44. Whilst Haushofer's continuous editorial involvement may suggest that the journal served as an organ continuously reflecting his views between 1924 and 1944, the situation is far more diffuse. In the journal's first phase, concluded in 1931 when Haushofer became the editor-in-chief, each of the principal co-editors was autonomously responsible for publications within a geographically defined area: Haushofer was responsible for the Indo-Pacific region, Obst for Europe and North Africa, and Termer for the New World and the rest of Africa. Lautensach was responsible for global and systematic questions. When Termer resigned, Maull replaced him and became responsible for the region of the New World.

Given coherence by the goals of providing knowledge for policy decision-making and enhancing scientific recognition of the developing field of geopolitics, the editors and publisher of the journal soon ascertained substantial differences of orientation among themselves. Already in 1926, the correspondence editor Fritz Hesse, who had authored the lead article of the very first issue, complained to his fellow editors about the lack of a common theoretical perspective uniting the journal's autonomous fields as well as a divergence of opinion regarding the prioritization of tasks to be pursued to influence the current climate (Harbeck, 1963, p. 20). Hesse's departure resolved that crisis, but Vowinckel's amalgamation of a contemporaneous journal, *Weltpolitik und Weltwirtschaft* [*World Politics and Economy*] led to new tensions. The influence of its former editors, now part of the *Journal of Geopolitics* (Alfred and then Arthur Ball, Kurt Wiedenfeld), led the core group of geographers to decry the "babble" (e.g. Obst speaking of von Rheinbaben) now to be found in the journal's pages. Obst and Lautensach threatened to resign in 1928 (Lautensach did so a year later), with Lautensach lambasting the damage to their scholarly reputations because of the influence of these newcomers. Vowinckel, whose overall influence on the journal behind the scenes was at least as great as Haushofer's throughout the journal's existence, and whose political commitments were aligned relatively early on with the Nazi party, repeatedly locked horns with Maull and Obst, complaining in particular of the former's lack of "organic" orientation (a code word which for Vowinckel meant a racialized orientation to geopolitics), and his unwillingness to integrate geopolitics with political science (the *World Politics and Economy* impulse). In a letter from October, 1931, to Haushofer, in which he first proposed that Haushofer assume sole editorship, Vowinckel complained that while fascism had elevated an organic theory of the state to official doctrine and while this state theory was also becoming the majority opinion of the populace, "we (the journal) are limping along on the side of these developments" (Harbeck, op. cit., p. 39). In his capacity as editor responsible for selecting each issue's lead article, Vowinckel invited others sympathetic to his views (e.g. three articles by Hans Zehrer, a leader of the "conservative revolution" movement, in 1930). As recorded in correspondence, Haushofer himself contemplated resigning several times between 1928 and 1930.

These substantial disputes shed considerable light on how to read both the panoply of perspectives represented in the journal during these years, as well as the general definition of geopolitics finally offered four years after its founding:

> Geopolitics is the science of the conditioning of spatial processes by the earth. It is based on the broad foundations of geography, especially political geography, as the science of political space organisms and their structure. The essence of regions as comprehended from the geographical point of view provides the framework for geopolitics within which the course of political processes must proceed if they are to succeed in the long term. Though political leaders will occasionally reach beyond this frame, the earth dependency will always eventually exert its determining influence. As thus conceived, geopolitics aims to be equipment for political action and a guidepost in political life" (translation by Heske in O'Loughlin, 1994).

The wording of the statement of principle was agreed upon only after considerable correspondence, disagreement, and a specially convened editor's meeting (in which Haushofer participated *in abstentia*) and its points of emphasis bespoke considerable compromise on the part of all involved. Thus, instead of being read, after the fact, as it were, as a guiding manifesto of the journal's activity, as often inferred in secondary literature, I wish to stress the need to read it as a compromise formation in a particularly contentious phase of the journal's existence. The journal's organizational arrangements (separate, autonomous editorial spheres, Vowinckel's responsibilities as correspondence and later, lead article editor), also help account for the relative diversity of positions, regional understandings, and emphases within the general spectrum of the German right and center-right.

Every new discipline seemingly needs to identify and represent canonical figures from whom guiding orientation and legitimation can be drawn. For the journal editors, the question of whom, precisely, to accord that role, was a matter of debate. Some nuance is therefore required in reading Maull's editorial in 1928 asserting that Friedrich Ratzel, not Kjellen, is the father of geopolitics. His editorial implicitly addresses a number of "behind the scenes" maneuverings: Vowinckel's primary theoretical orientation on Kjellen, his primacy of *political science*, and his organicist *and* racial understanding of the state (all in contrast to Ratzel), the severe dissatisfaction on the part of Obst, Lautensach and Maull with Vowinckel's activist influence, and from their perspective, the journal's regrettable "digression" into nongeographical orientations. With all due respect to Kjellen, Maull offered, Ratzel's work and thought (including various emphases not carried forth by Kjellen) contain virtually everything of substance given the name geo-politic by Kjellen. "The development of geopolitics is unthinkable without Ratzel. No one else, therefore, not even Kjellen, as is occasionally done from ignorance, can be characterized as the father of geopolitics. It is Ratzel." Thus, political geography was the basis of geopolitics, not political science, as for the former *World Politics and Economy* editors, not the entwinement of folkish with so-called "racial science" perspectives favored by Vowinckel, but the anthropogeographical orientation contained within Ratzel's political geography. To the extent that Maull would have wished to emphasize this, he could also have mentioned (but did not) certain other expressed differences between Ratzel and both Kjellen and the orientation of Vowinckel. While Ratzel, more than any other German geographer, had identified the trend toward *Großraumformen* (forms of large space – not identical to nation-states) in the contemporaneous phase of globalization, he offered numerous implications from this insight that contradict Kjellen. Maull could have explicitly stressed that Ratzel was never an adherent of racist theories of statehood, that Ratzel did not follow the conviction that all members of an ethnic group rightfully belong within a single state structure, that Ratzel never argued for national self-determination on the basis of unifying the settlements of ethnic Germans who had migrated to Eastern Europe – these were instead citizens of other states – or finally, that Ratzel's regionalization of Germany within Central Europe advocated the establishment of an economic union of existing states (Natter, forthcoming). Furthermore, Maull might have stressed, but did not, that Ratzel was a sharp critic of contemporaneous efforts by Gobineau and Chamberlin to explicate human identity on the basis of biological race and that Ratzel found laughable the notion that contemporary Germans were the descendents of Aryans. Yet in the same year, in an essay which appeared in a book co-edited by Haushofer and his fellow journal editors, Haushofer would insist, this time apparently siding with Vowinckel, that "Kjellen's book was the work... in which the theory of geopolitics is most clearly developed" (Haushofer, 1928). Tellingly, Haushofer's own selection from and commentary on parts of Ratzel's copious work similarly mostly eliminated precisely these points of difference between Ratzel and Kjellen in rendering the former more fully as the great canonical progenitor of 1930s *Geopolitik* albeit in a direction completed by Kjellen (Haushofer, 1940).

In these years, the diversity of geopolitical orientation on the part of the editors is accompanied by a broadening of application to various subfields, topics, and methodologies. Essays appeared on communication and news technology (Fritz Runkel, 1930), on the geopolitics of aviation (Hans Hochholzer, 1930), on geo-military studies (Haushofer and Banse), on geopolitics and film (Erich Maschke, 1928), on the geopolitics of German industry (L. Hamp, 1930), and on geomedicine and demographic analysis. Essays by Manfred Langhans-Ratzburg reported on his efforts to systematize a new subdiscipline of geographical legal studies. Recurring reports throughout the 1920s by Haushofer on the Indo-Pacific region consistently stressed the need to adjust recognition to the fundamental importance of "the heartland" and consequently an accomodation with the Soviet Union and Japan, as the centers of emerging pan-regions (along with the Americas). Alternatively, these two were viewed as co-developers in the construction of a continental block (landpower) intended to offset the sea power of England. Reports equal in number by Erich Obst regularly argued Germany's primary need to work towards its central emplacement within an Euro-North African sphere of influence and the development of the pan-idea of Eurafrica. In a different direction, an article published in 1924 by the geographer Otto Schlütter argued for the need to recognize that it is not only a people that creates a state, but the state a people. Perpetuating an understanding derived from Ratzel's anthropogeography, Schlütter detailed the mutual interdependence of culture, nature (not race), and the state. Schlütter also restated Ratzel's insight that all peoples have been advanced by including initially foreign elements within it, and further, that a people is defined by a commonality of inner experience brought about by cultural means. Appearing in a journal whose general orientation was unmistakably the revanchistic re-territorialization of pre-Versailles borders, Schlütter proposed that the goal for national thought in the present should not be a drive towards rapid expansion in distant space, but instead the consolidation and deepening of spiritual community (Schlütter, 1924).

Books published in the journal's occasional series, included Josef Cohn's *England and Palestine* (1930), which concluded that Dr. Weizmann's efforts to secure a national homeland would succeed, "that Zionism would overcome all political difficulties, because behind these efforts stand the unbeatable power of an idealistic and revolutionary movement"(Cohn, 1930, p. 196). Another book in the series contained essays offering completely contradictory opinions about whether Germany was under- or overpopulated (Haushofer thought the latter), and what was to be done under the circumstances (Harmson and von Loesch, 1929). Equally typical are essays decrying the French effort, on the basis of Vidal's *France de L'Est*, to declare the Alsace region as a French (not German) cultural and political space, an entire issue amplifying reasons for the demand that Germany should be given back its colonies (1926), a geopolitical attack on the Versailles Treaty (Tiessen, 1924), and editor Fritz Hesse's lead article in the very first issue of the journal, which set out, purportedly on the basis of Ratzel's thought and Kjellen's deepening of it, to define the "law of expansive space" (1924). A textbook example of geopolitical analysis represented during the period is the book-length study by Hesse regarding the Mossul region. The study argued that the political tension between France and England there would result in England's further support of France's ambitions against Germany in Europe in order to be allowed by France to have its way in the Orient (Hesse, 1926).

In summarizing this phase of the journal's existence with respect to Haushofer's own disposition regarding geopolitics, a letter from him in March, 1930, to an author (Marten) who had published an essay elsewhere in a liberal periodical critical of the geopolitics represented in the journal, is indicative. In his letter, Haushofer replied that the best proof he could offer regarding the diversity of the journal – contrary to Marten's characterization of them as "a group of geopolitically oriented nationalists" – is that he personally would have gladly published Marten's essay in the journal as a lead article ("allowing for the need for some revisions"). After all, he continued, the journal had generally published a range of opinion, including essays by the Swiss social democrat Reinhard, arguing from a pacifist perspective. Did not the value attributed geopolitics by those on the left permit him, Haushofer, to pursue the development of the thoughts of Kjellen and Ratzel "for us" (Jacobsen, 1987, vol. II, p.102)? Finally, wishing to see geopolitics continue to develop required that it needed to be open "to every side" in order to achieve a level it had not yet , and could not yet attain – which is why Haushofer noted to Marten, the book published by the journal's editors in 1928, refered to earlier, was titled "building blocks" (*Bausteine*), and not the "science" (*Lehre*), of geopolitics (Haushofer et al., 1928).

In 1931, the journal's ongoing crises were "resolved" by the resignation/ouster of the other editors, the transfer of sole editorship to Haushofer but with Vowinckel's continued activist participation, and the appointment of Albrecht Haushofer as contributing editor to essentially act as an intermediary between his father and Vowinckel. This second phase came to an end with the declaration of war against the Soviet Union, and Albrecht Haushofer's resignation from his journal duties. During the early 1930s, Karl Haushofer remained interested, as previously, in offering some diversity of opinion within the journal, whose goals, he thought, should remain equally those of providing guidance to policy makers *and* the consolidation of geopolitics as a recognized academic discipline. It is this later consideration, which above all seems to have led to his willingness both to compromise with his publisher in those instances after 1933 when the promise of state affirmation seemed likely to follow, and to continue to attract "serious" political geographers and their geopolitical opinions (Hennig, Hassinger) even when they conflicted with both Vowinckel, his *Arbeitskreis für Geopolitik* (Geopolitical Working Group or *AFG*), and a racialized understanding of geopolitics. It is noteworthy that shortly after his sole editorship began, Haushofer published the Marxist theoretician Karl Wittfogel in the journal (1932) and that he continued to at least tenuously support Hennig even after Hennig was severely criticized by Vowinckel and party functionaries in the *AFG* for his "over emphasis" on spatial issues, i.e. his underemphasis on the priority of race in establishing political space. Further reflective of this dimension of his editorship, is publication in the journal of Wilhelm Volz's work on "Industry in the East," even though it contained an opinion antithetical to his own regarding the development of industrialization (not agriculture exclusively) in the east as a response to the problem of "excess population" (1933). He also published Hugo Hassinger on the topic of the state as a creator of landscape (1932) whose aim was to re-situate geopolitics as a subfield within political geography (on the basis of the study of the mutual constitutions of landscapes and states), and a book on Bulgaria in the journal's book series that advised that Bulgaria's borders should neither expand nor be reduced in size (Gellert, 1933).

Equally telling, however, is his reluctance to tackle the racial question head on, as Hennig recommended, in order to draw a border line between "racial science" (*Rassenkunde*) and geopolitics. Haushofer responded that doing so would merely lead to further misunderstandings and antipathy (letter correspondence in Jacobsen, 1987). With other correspondents, however, he offered a severe criticism of any effort to reduce geopolitics to a subdiscipline of a racialized folkish science (*Völkerkunde*). Yet at the same time, such a racialized understanding of geopolitics – though of course not in the form of its reduction to a subdiscipline of another field – was precisely the one advocated by his publisher and it permeated the *AFG*'s platform as well. Ever more explicitly, Vowinckel sought for *Geopolitik* nothing less than the role of its elevation into the central ideological platform for the foreign policy of the Nazi state. A pre-condition for its becoming this, as he frequently told Haushofer, was *Geopolitik*'s unconditional embrace of a *Blood and Earth* ideology. After the mid-1930s, such an understanding increasingly, though not exclusively, found entry into the journal's pages. But as reported by Vowinckel, a meeting of the *AFG* in November, 1937, had revealed serious complaints with the Haushofer line on the part of various influential members of the working group. This list included the journal's presumed overemphasis on the influence of space (against Hennig and Knieper), its undermining of racial thinking, and its presumptive pro-Soviet position (directed against Haushofer himself). Charges were also leveled that geopolitics was not really a scientific discipline (Jacobsen, op. cit., pp. 328–9). As doubtlessly interested as Vowinckel was in giving this version of the facts to Haushofer, it nonetheless makes the point that Haushofer's version of geopolitics was under considerable attack from within.

In the final five years of the journal's existence, however, virtually all depictions of geopolitical space are informed by a fundamental correlation between race (not culture) and the earth. Tellingly, these depictions are offered without explicit "theorization" of the linkage – rather, the linkage has become a presupposed discursive fact. Even so, Vowinckel's (and Haushofer's) aspirations for *Geopolitik* to become the guiding political theory of the Nazi state failed to materialize. For too many Nazi leaders, the journal and its science had not proven itself reliable enough. Indeed, the journal, like others in Germany, was subjected to ever sharper pre-censorship restrictions starting in the late 1930s, a policy Haushofer unsuccessfully contested. The *AFG* was dissolved in 1941. Within institutional life, Haushofer's own active, "political" role had effectively ended by 1937 when he was unable to realize, despite the support of Hess, his organizational conception of an office to coordinate German policy toward ethnic Germans abroad, and was ousted from the presidency of the German Academy. Deemed important for the war effort, the journal continued until 1944. In the last years, the articles contain little overt sign of engagement with the realities of a losing war, whether for reasons of preventive censorship or not, nor with the brutal realities of genocide or "ethnic cleansing," but do increasingly take recourse to a language of fate (*Schicksal*) – in offering a prognosis on the war's likely results. A rare "academic" article by professor of history Johannes Kühn, who made this last point in 1943, countered that the power of belief is what finally would inscribe itself in history: should Germany and its allies not "fully" succeed in securing their "separate space" (*Sonderraumansprüche*) against the powers seeking world domination (not Germany, apparently), he predicted the war's conclusion would leave only two global powers, the Soviet Union and the United States, who, with their block of subservient allies, would in short order become deadly rivals for global control of "space, resources, power and live-style" (Kühn, 1943, p. 254). The Jews would profit in either case, for Kühn, since "the Jews" incarnated the quest for world control, whether on the side of "super"-capitalism or communism.

In its final phase, a general judgement often made about the journal as a whole does apply in full force. Most articles demonstrate no effort to mount even modest academic arguments, but instead expand upon popularized party slogans and the offerings of a censored news media, illustrated by dramatic and simplifying black and white maps with arrows pointing to presumptive flows of space (on these maps, see Herb, 1996). In the final war years, the journal's articles legitimated – after the fact – whatever policy had most recently been announced by the Nazi state using a language that simplified events into an overarching narrative of *Blood and Earth*, and the legitimate hunger for living space (*Lebensraum*), if not *Großraum* (large orders of space). The more "modest" designs for German hegemony expressed between 1931 and prior to 1939, which at their most expansive were grounded on the basis of ethnic German concentration and prior geopolitical territorialization and sometimes leaving open the appropriate form of state-administrative structuring of this *Großraum*, gave way to a type of *Großraum* argumentation which simply presupposes the requirement and legitimacy of the Reich's adjudication of all conquered territories. In this regard, Carl Schmitt's articulation of the idea of *Großraum* – the subsumption of existing state structures into expansive territorial blocks (or sea powers) under the hegemony of a few individual centers (e. g. Germany) – dovetails with opinion expressed in the journal (Schmitt, 1940). The journal had finally become in all senses an officially sanctioned propaganda instrument explicating some circumspect aspects of German foreign policy – after the fact – for a readership now increasingly made up primarily of high school teachers and their students. And yet, unlike his son Albrecht, Karl Haushofer never took the step of actively joining opposition groups and his public statements and writings gave hardly a clue regarding the severe doubts he apparently expressed in private (Jacobsen, 1987, p. 454). Indeed, Haushofer does not seem to have been familiar with the full extent of Nazi policy, particularly regarding its policies of annihilation, until the summer of 1945. Furthermore, Haushofer at various times, particularly until 1938, also presented geopolitics as the search for a "just" partitioning of the world, a form of conflict resolution to head off war. However, just partitioning as it affected Germany, meant, as with Schmitt, an embrace of the idea of a pan-German *Großraum*, and thus the need to re-shape state alignments and relations. As with Mackinder, Euro–Asia received particular attention for him as the pivot of history, and as a consequence, "Eastern Europe" as a field of German hegemony. In positing the understanding for more than a decade before the outbreak of World War II, as did Schmitt beginning in the late 1930s, that contemporaneous globalization had fostered the development of a few hegemonic blocs, which Germany was positioned to either command or be subsumed under, Haushofer had in principle articulated a geopolitics that presupposed a social Darwinistic battle of the survival of the fittest, whose outcome would not permit Germany from shying away, "if necessary," from a total mobilization of resources. In World War I, he wrote in the 1930s, one had learned the heavy burden of duty to send forty men in a row to undertake a task that would cause their deaths because of the justified presupposition that the forty-first would succeed. This presupposition also fully marks the geopolitics Haushofer developed.

Conclusion: The State of German *Geopolitik*

Generally speaking, German geopolitics, in the writings of the majority of those associated with the journal, particularly during its first two phases of existence, is an effort to think through a phase of globalization with reference to Germany's presumptive beleaguered position following the "unhappy outcome" of World War I. Most of the *Geopolitik* published in the journal stressed the primacy of physical space in understanding space for the politics of the state. That is, all developmental tendencies of space ensue from its primary identification with the earth and the ground it offers for cultural and political developments. As noted earlier, other dimensions of space, including those that ensue from technological developments, also received considerable attention, extending geopolitics well beyond objects of analysis of earlier traditions of political geography, but generally an effort is made to link these developments to the primary, physical space from which they emanate. To this extent, the critique made by various Nazi functionaries that *Geopolitik* was a form of geo-materialism is doubtless correct.

At the same time, German geopolitics, though national-conservative and German-centered, is only in a limited sense primarily explicable with reference to the structures of the nation-state (see also Sprenger, 1996). Indeed, German geopolitics of the 1920s and 1930s, like the reflection on *Großraum* expressed by Carl Schmitt, demonstrates an engagement with the belief that the age of the nation-state was ebbing and that an epoch of global spaces was reconstituting multiple cores and peripheries. Morever, in contrast to much present-day globalization theory, the primacy of the political, not the economic, was foregrounded. German geopolitics emphasized, first, the political as the lever of global change, and second, stressed its operation as occurring on the earth and as a contestation over the earth. In Haushofer's famous equation, "perhaps 25%" of the political equation was determined. In this sense, German geopolitics combined both a geographical materialism and the idea of a teleological possibilism. As argued by Haushofer's incisive critical contemporary Franz Neumann, as well as in more recent literature, this pre-eminence of politics and the contest over space was at the same time linked to a devaluation of the state as a determinate of space-maintenance (Neumann, 1944). That the territorial arrangements fixed by the Versailles Treaty were likewise to be viewed as a mere momentary fixation of space, goes to the core of much of the animus that inspired this generation of geopoliticians to begin with. The state as found in the 1920s and 1930s was not seen to signify the mediated assemblage of legal, ethical, political, educational, and social imperatives posited more or less by and for those who live under it. In the journal's publications of the 1920s, this devaluation of existing state arrangements, as well as of the state itself as a central arbitrator of socio-political-ethical values, was also implicitly applied to Weimar Germany's state, which was seen, like all states, as artificial and contingent. Recalling Lefebvre's presumption regarding the city as the place where the social space of the state (but also where an alternative geopolitics) is produced, Haushofer's pejorative treatment of *the state* – whose tendency needed to be overcome – can be analysed as being part of the same understanding. Space "itself," often seemingly posited ontologically as a result, had the tendency to override existing state-localized territorial arrangements within an inexorable contemporanious tendency towards large orders of space. The unhappy need to grapple with the lessons to be learned from the outcome of World War I, however, put a limit on any overly emphatic spatial determinism attached to such an ontology. For surely, one of the most obvious deterministic "lessons" from 1918 could have been that geography itself precluded Germany's attainment of the status of a world power (see also Sprengel, 1996 and Murphy, 1997). Instead, the remarkably contingent and constantly re-makable character of political space became a main presupposition.

Under the umbrella of certain shared assumptions, the exposition of a relatively heterogenous panoply of sought after global-regional affiliations and regional concepts was explicated under the sign, variously of *Lebensraum* and *Großraum*. That Germany should seek to attain the status of a core center was a given for writers in the journal, but understandings regarding which were the necessary paths towards the realization of this desideratum varied considerably. One recurring assumption, nonetheless, was that a simply national framing of Germany's presence was insufficient unto itself. The intra-German regionalism partially overcome in 1871 was one thing, the loss of territory in the wake of the Versailles Treaty another, but the overwhelming supposition iterated in the journal was the belief that Germany, not to say Europe, required a supplement, both demographically and economically, in order to survive the contemporaneous global restructuring whose tendency was the consolidation of regionalisms into a limited number of large orders of space. The outcome of this restructuring was anything but guaranteed – so much for geographical determinism. What seemed certain in the journal was simply that restructuring was proceeding apace, that a new spatial imaginary was required, and that present endeavors – a matter of politics – would affect the ultimate outcome. For Germany, the stakes were high, with either the attainment of its hegemony in one *Großraum* an imaginable result, or its consignment to a subordinate regional presence equally imaginable. But Germany could not, in this new spatial imaginary, be content to stand still as one of the many states of varied sizes while political, economic and demographic processes recast regional identifications within global space. For all of these reasons, the logic of German *Geopolitik* portended an embrace of whatever expansive foreign policy the Nazi state pursued.

Despite an identification with substantial aspects of National Socialist ideology, however, the aspirations articulated for *Geopolitik* during the 1920s and early 1930s to become both a science and a signpost to policy makers generally failed on both counts. Far from being a foreign policy mastermind of the Third Reich, Haushofer's "career" with respect to foreign policy after 1933 is marked by a trajectory of increasing irrelevance. Like Carl Schmitt, who on occasion quoted Ratzel and Haushofer, but above all Mackinder, Haushofer refused his identification with the policies of Hitler after the war's conclusion. The comparison is instructive. Schmitt, who re-cast international law under the sign of *Großraum* in order to stress that international law ought merely to protect dominant ethnic groups (their industry, agriculture and trade) from intervention by other dominant ethnic groups ensconced in other *Großraum* formations, and who in 1938 had offered the view that Jews were not an enemy in political space, but the enemy of all political space, because of "their" presumptive non-earth connectedness, professed in 1946 that Hitler "had not pursued a politics of *Großraum* in the sense of the theory, but rather had pursued a politics of conquest inimical to principles or ideas, which one could only then label the politics of *Großraum* if one empties it of its specific meaning and substitutes for it an empty slogan for any kind of expansion" (quoted by Schmoeckel, 1995). This may – or may not – be the case, but with – to my mind – an even greater identity to be drawn between Schmitt and actual Nazi policies undertaken, the careers of both still point to the limits of the intentionality fallacy when reflecting on the topic of Nazi *Geopolitik*.

In terms of contemporaneous effects, the importance of Haushofer and the *Zeitschrift für Geopolitik* are most firmly locatable in their having generated intellectual arguments and linguistic formulations that buttressed the general legitimacy of a revanchistic Nazi foreign policy, the full dimensions of which Haushofer could neither control nor, increasingly, even influence. Franz Neumann's judgement in *Behemoth* (1944) is both correct and offers a correct implicit judgement of Haushofer and German *Geopolitik*: "It would be fatally wrong to assume that National Socialist leadership has pre-determined the final limit to German domination over

Europe or the eventual form of its empire. The boundaries are being determined by the political situation – by military success, by strategic motives, by economic considerations, which may or may not coincide" (Neumann, 1944, p. 171). In this regard, too, critical geopolitics cannot but work through *Geopolitik*, despite repulsion on substantive grounds, for what it tells us about the demarcation of geopolitics more generally, the development of the subdiscipline in relation to geography more broadly, and perhaps above all, as a lesson in the enabling limits of the aspiration to guide policy makers and the state. To the extent that reflection on Nazi Geopolitics can offer guidance to these vexing issues, it also remains a matter of interest both to geography's past and its future.

BIBLIOGRAPHY

Bassin, M. 1987. Race contra space: the conflict between German *Geopolitik* and National Socialism. *Progress in Human Geography*, 11, 473–95.
Boehm, H. 2000. Magie eines Konstruktes. Anmerkungen zu M. Fahlbusch: Wissenschaft im Dienst der nationalsozialistischen Politik? Die "Volksdeutschen Forschungsgemeinschaften" von 1931–1945. Baden-Baden 1999. *Geographische Zeitschrift*, 88, 177–96.
Cohn, J. 1931. *England und Palaestina*. Berlin-Grünewald: Kurt Vowinckel.
Eberlin, F. 1994. *Geopolitik*. Berlin: Akademie Verlag.
Edney, M. 1999. *Mapping an Empire*. Chicago: University of Chicago Press.
Godlewska, A. and Smith, N. (eds.). 1994. *Geography and Empire*. Oxford: Blackwell.
Harbeck, K.-H. 1963. *Die "Zeitschrift für Geopolitik" 1924–1944*. Kiel: Universität Kiel, unpublished doctoral dissertation.
Haushofer, K. (ed.). 1928. *Bausteine zur Geopolitik*. Berlin: Kurt Vowinckel Verlag.
Haushofer, K. 1931. *Geopolitik der Pan-Idee*. Berlin: Zentralverlag.
Haushofer, K. 1932a. *Weltpolitik von Heute*. Berlin: Zeitgeschichte.
Haushofer, K. 1932b. *Wehrgeopolitik*. Berlin: Junker und Duennhaupt.
Haushofer, K. 1940. *Erdenmacht und Völkerschicksal*. Berlin and Stuttgart: Krönerverlag.
Herb, G. 1996. *Under the Map of Germany*. London: Routledge.
Heske, H. 1987. Karl Haushofer: his role in German geopolitics and in Nazi politics. *Political Geography Quarterly*, 6, 135–44.
Heske, H. Karl Haushofer. In J. O'Loughlin (ed.) *Dictionary of Geopolitics*. Westport, CT: Greenwood Press, 112–13.
Hipler, B. 1996. *Hitlers Lehrmeister*. St. Ottilien: EOS Verlag.
Hooson, D. (ed.). 1994. *Geography and National Identity*. Oxford: Blackwell.
Kost, K. 1988. *Die Einflüsse der Geopolitik auf Forschung und Theorie der Poltischen Geographie von ihren Anfaengen bis 1945*. Bonn: Ferd. Duemmlers.
Kühn, J. 1943. Der Sinn des gegenwaertigen Krieges. *Zeitschrift für Geopolitik*, 20, 255–6.
Jacobsen, H. A. 1987. *Karl Haushofer. Leben und Werk*, vols I and II. Schriften des Bundesarchives 24. Boppard am Rhein: Harald Boldt.
Jaeckel, E. 1969. *Hitlers Weltanschauung*. Tübingen.
Lefebvre, H. 1990. *The Production of Space*. Oxford: Blackwell.
Livingstone, D. 1992. *The Geographical Tradition*. Oxford: Blackwell.
Maull, O. 1925. *Politische Geographie*. Berlin: Borntraeger.
Maull, O. 1928. Fredrich Ratzel zum Gedaechtnis. *Zeitschrift für Geopolitik*, 5.
Murphy, D. T. 1997. *The Heroic Earth*. Kent, OH: Kent State University.
Natter, W. 1999. *Literature at War, 1914–1940. Representing "the Time of Greatness" in Germany*. New Haven, CT:Yale University Press.

Natter, W. 2000. Hyphenated practices: what put the hyphen in geopolitics? *Political Geography*, 19, 353–60.
Neumann, F. 1944. *Behemoth: The Structure and Practice of National Socialism*. New York: Oxford University Press
O'Loughlin, J. 1994. *Dictionary of Geopolitics*. Westport, CT: Greenwood Press.
Ó Tuathail, G. 1996. *Critical Geopolitics*. Minneapolis, MN: University of Minnesota Press.
Roessler, M. 1990. *Wissenschaft und Lebensraum. Geographische Ostforschung im Nationalsozialismus. Ein Beitrag zur Disziplinsgeschichte der Geographie*. Hamburg.
Schmitt, C. 1940. Raum und Grossraum in Völkerrecht. *Zeitschrift fuer Voelkerrecht*, 22, 145–79.
Schmoeckel, M. 1995. *Die Großraumtheorie*. Berlin: Dunkler & Humblot.
Schoeller, P. 1957. Wege und Irrwege der Politischen Geographie und Geopolitik. *Erdkunde*, 11, 3–48.
Schultz, H.-D. 2001. Geopolitik "avant la lettre" in der deutschsprachigen Geographie bis zum Ersten Weltkrieg. *Geopolitik und Kritische Geographie*, 14, 29–50.
Smith, W. 1986. *The Ideological Origins of Nazi Imperialism*. New York: Oxford University Press.
Sprengel, R. 1996. *Kritik der Geopolitik. Ein deutscher Diskurs 1914–1940*. Berlin: Akademie Verlag.
Troll, C. 1947. Die geographische Wissenschaft in Deutschland in den Jahren 1933 bis 1945. Eine Kritik und Rechtfertigung. *Erdkunde*, I, 3–48.
Walsh, E. 1948. *Total Power. A Footnote to History*. New York: Doubleday.
Wardenga, U. 2001. *Zur Konstruktion von "Raum" und "Politik" in der Geographie des 20.ten Jahrhunderts*. Heidelberg: Selbstverlag des Geographischen Instituts der Universität Heidelberg.

Chapter 14
Cold War Geopolitics

Klaus Dodds

Introduction

In August 2000, a terrible disaster befell the Russian Navy when the nuclear-powered submarine *Kursk* was reported missing 60 miles north of the Russian port of Severomorsk, close to the Eastern Norwegian coastline. Within hours of the first media reports, it became apparent that a catastrophic accident had occurred, which left the vessel stranded on the seabed. Onboard, 118 sailors were trapped at a depth of 350 feet. Notwithstanding the human dimension to this tragedy, the *Kursk* represented a potential nuclear disaster. Senior Russian naval officers, when asked to explain how the accident might have occurred, suggested that the submarine had been involved in an accident with another vessel, possibly a NATO patrol ship or submarine. NATO and Western journalists swiftly denied this accusation while their governments speculated that the cause of the accident could have been an explosion onboard the submarine. Over the following week, the Russian armed forces' attempted rescue operations were foiled by inadequate equipment, strong undercurrents, and a damaged escape hatch. Initially declining offers of help from the British and American governments able to supply specialist rescue equipment, the Russian government finally accepted help from Norwegian and British experts. After Norwegian divers gained access to the stricken vessel, their worst fears were confirmed; the submarine was flooded and all the crew had perished.

Recriminations followed and criticism that the Russian Navy failed to seek help earlier mounted inside and outside of the Russian Federation. It was suggested that a misplaced sense of pride in combination with a fear that foreign navies would use the disaster to gather intelligence, were blamed for the Russian reluctance to invite outside help. The British newspaper, the *Independent* in an editorial dated August 16, 2000 drew a sobering conclusion:

> The plight of the *Kursk* is not just a terrible human drama. It is a metaphor for the decline of a superpower, and for the decay of Russia's armed forces since the collapse.... By accepting American or British help in rescuing its stricken submarine, Russia could demonstrate that Cold War bygones are truly bygones. In return the US should begin to do likewise without

waiting for a Start III agreement. If it does, a disaster in the Barents Sea may actually help to make the world a little safer.

A decade after the collapse of the Berlin Wall and the break-up of the Soviet Union, the Russian Navy, in a manner reminiscent of the Cold War, still operates at such high levels of secrecy that families of serving sailors do not know their exact whereabouts. After sustained pressures from the relatives of the sailors, the Russian President Vladimir Putin was forced to make a humiliating apology on behalf of the armed forces.

For many senior officers in the Russian armed forces, the Cold War was a powerful frame of reference which determined not only the strategic significance of regions such as the Arctic, but also the responses to rival NATO armed forces such as the UK and the USA. During the Cold War, areas such as the Barents Sea were highly sensitive and were extensively patrolled by the Russian Northern Fleet. Beneath the waters, a formidable fleet (over 120 nuclear- and electric-powered submarines) played a dangerous game of close encounters with rival NATO vessels (and this continues albeit on a smaller scale). Over a period of 40 years, virtually every square kilometer of the sea floor in the high Arctic had been surveyed and together with an extensive network of communication stations helped to monitor events above and below the icy waters of the Barents Sea (Chaturvedi, 1996; Jalonen, 1988). In comparison with the 1970s and 1980s, the present state of the Northern Fleet is poor and many vessels are deteriorating, including the nuclear reactors on board its submarines. While the *Kursk* was subsequently recovered from the sea bed, the continued nuclear-related pollution of the Barents Sea remains an unwelcome legacy of the Cold War (see, more generally, Stokke, 2000).

While few would deny that much has changed since the collapse of the Soviet Union in 1991, the Russian armed forces demonstrated that the political practices and interpretative dispositions associated with the Cold War remain. This incident indicates that geography remains significant, as the Barents Sea and the Arctic more generally were some of the most militarized spaces in the Cold War era. Despite the creation of the Barents Euro-Arctic region of cooperation in 1993, ocean basins, continental shelves, fjords and mountains are still perceived to be strategically significant for individual states (Chaturvedi, 1996; Stokke and Tunander, 1994; Tunander, 1989). The development of long-range missile technology in combination with the advent of the nuclear-powered submarine, have transformed the political geography of the Arctic. The voyage of the USS *Nautilus* under the Arctic ice cap in 1957 was crucial in this geopolitical metamorphism. The icy waters of the Arctic offered no defense against modern technology, which led both superpowers to radically reappraise their views on relative proximity. NATO and Soviet submarine fleets bolstered by early warning systems and remote military airfields continued to patrol the Arctic Ocean. By the late 1950s, the Cold War had become truly icy as the uneasy co-existence predominated.

This chapter is concerned with the Cold War and the political practices and interpretative schemas associated with the period between 1945 and 1990. Initial attention will be paid to the literature of critical geopolitics and by using the categories of analysis developed there, aspects of practical, formal, and popular forms of Cold War geopolitical reasoning will be investigated. While the focus is mainly on the USA, there are Soviet equivalents of Cold War geopolitical reasoning (see Hough, 1988). For much of the 50 years of the Cold War, the Americans represented the Soviet Union as an expansionist "threat" which challenged the USA sense of mission based on the promotion of liberal democracy and global capitalism. Communism had to be confronted regardless of location and the Cold War was not only an ideological struggle between two rival superpowers, but also an intensely physical struggle, which transformed even the high Arctic into a zone of confrontation. The final part of the chapter questions the assumption that the Cold War has come to an end and instead suggests that America's cold war against Russia and China continues.

Critical Geopolitics

Over the last ten years, a growing body of literature called critical geopolitics has questioned the language and practices of geopolitics (Ó Tuathail, 1996; Ó Tuathail and Dalby, 1998). Geopolitics can be described as a problem-solving approach to international politics with due emphasis given to the territorial dimensions of diplomacy and foreign policy (Dodds and Atkinson, 2000). Within the Anglophone literature, geopolitics had been perceived as a repository for ideas and advice to policy makers; the notion of global balance of power and "geography" in particular tended to be conceptualized as a permanent and unchanging backdrop to world politics (see Agnew, 1998). This intellectual disposition encouraged a restricted view where undue emphasis was placed on the fixed geographic location of continents and oceans. Geopolitical thinkers employed terms such as heartland, lifeline, choke point, and domino effect to convey how these geographic features decisively shaped world politics (Agnew and Corbridge, 1989, p. 267). Writing in the 1940s, the American geographer Nicholas Spykman articulated the dominant conception of geopolitics:

> Geography is the most fundamental factor in the foreign policy of states because it is the most permanent. Ministers come and go, even dictators die, but mountain ranges stand unperturbed (Spykman, 1942, p. 41).

In contrast, critical geopolitical writers have argued that the world political order is actively constituted through particular modes of geopolitical reasoning. Mountain ranges and oceans are not naturally significant but they tend to be labeled as "strategic." In other words, critical geopolitics investigates the ways in which geopolitical forms of reasoning have interpreted the "world political map." Thus, the creation of geographic knowledge is closely bound up with power relations. During the Cold War, for example, the geopolitical reasoning of American administrations contributed to a dangerous simplification of politics as global areas were divided into "friendly" and "hostile" spaces. The subsequent investment in military forces and weapons programs in Europe and the wider world was justified on the basis of conflicting geographic and ideological blocs.

The representation of global political space is not confined to the policy-making circles of national governments and their official advisers. Geopolitics can also be explained as a broader social and cultural phenomenon, which permeates throughout societies. Critical geopolitics recognizes that geopolitical reasoning can be investigated at three distinct if inter-linking levels of analysis: practical, formal and popular. *Practical* geopolitical reasoning referred to the depictions and rationales produced by national governments and their supporting armed forces and bureaucracies. *Formal* geopolitical reasoning describes the research ideas and descriptions produced by academics working in universities and so-called "think-tanks." *Popular* geopolitics refers to the geographic representations found within the popular media whether it be mass-market magazines, movies and/or cartoons (Dodds, 2000; Ó Tuathail and Dalby, 1998). As Ó Tuathail and Dalby have proposed:

> Its [i.e. geopolitics] sites of production are multiple and pervasive, both "high" (like a national security memorandum) and "low" (like the headline of a tabloid newspaper), visual (like the images that move states to act) and discursive (like the speeches that justify military actions)... these practices and much more mundane practices... are constituted, sustained and given meaning by multifarious representational practices throughout cultures (Ó Tuathail and Dalby, 1998, p. 5).

Hence the contest between the USA and the Soviet Union was unquestionably an ideological, territorial, and cultural struggle, which permeated the everyday life of citizens and nations alike (Billig, 1995).

Practical Geopolitical Reasoning and the Cold War

The Cold War dominated the second half of the twentieth century as American and Russian governments committed huge amounts of resources and personnel to testing one another's convictions. Coined in 1947 by the American journalist, Walter Lippman the "Cold War" referred to a period of forty years of Soviet–American geopolitical and ideological competition (see Stephanson, 1998). For former American President Ronald Reagan, the geopolitical stakes were enormous as he urged his fellow Americans in 1982 to fight for freedom and resist the "Evil Empire" of the Soviet Union (see O'Loughlin and Grant, 1990). In a speech to the British Parliament entitled "The crusade for freedom," Reagan drew a sharp distinction between the USA and its adversary, the Soviet Union:

> Must civilisation perish in the hail of fiery atoms? Must freedom wither in a quiet, deadening accommodation with totalitarian evil?... We in America now intend to take additional steps.... What I am describing now is a plan and a hope for the long term – the march of freedom and democracy will leave Marxism-Leninism on the ash heap of history as it has left other tyrannies which stifle the freedom and muzzle the self-expression of the people. The British people know that, given strong leadership, time and a little bit of hope, the forces of good ultimately rally and triumph over evil.... Well the emergency is upon us. Let us be shy no longer. Let us show our strength. Let us offer hope. Let us tell the world that a new age is not only possible but probable (Reagan, 1982).

While we may raise our collective eyebrows at such a hyperbolic statement in 2002, these were not simply hollow words. Under the Carter administration, the USA boycotted the 1980 Moscow Olympics because it disapproved of the Soviet invasion of Afghanistan the previous year. This was the first time that a major sporting nation had declined to participate in this hugely popular global event. On his election to the presidency in 1981, Reagan had implemented further increases in defense spending and even approved funding for the ill-fated Strategic Defence Initiative (commonly called Star Wars) which purported to offer protection against any incoming Soviet intercontinental missiles (see Hough, 1988, p. 238 for details on Soviet reactions). Billions of dollars were invested in defence programs and the Reagan administration even approved funding to military groups fighting Soviet-backed governments and organizations in Central America and southern Africa (see Chomsky, 1982; Halliday, 1983, 1989). In 1984, the Soviet Union returned the compliment and boycotted the Los Angeles Olympics.

Forty years earlier as wartime allies, a rather different sort of relationship existed between the USA and the Soviet Union. After the end of the World War II, however, these two countries became embroiled in a "war of words" as crises in Berlin and the Korean peninsula ended their wartime co-existence. This collapse in relations demanded a new mode of interpretation. For Joseph Stalin, geopolitical interest in proximate Eastern European states was understandable given the sacrifices made by the Soviet Union during last war. As he noted in 1946:

> In other words the Soviet Union's loss of life has been several times greater than that of Britain and the USA put together [subsequently estimated at 20 million people].... And so what can there be surprising about the fact that the Soviet Union, anxious for its future safety, is trying to see to it that governments loyal in their attitude to the Soviet Union should exist in these countries? How can anyone who has not taken leave of his senses, describe these peaceful aspirations of the Soviet Union as *expansionist* tendencies on the part of our state (Stalin, 1946, cited in Lane, 1985, pp. 105–6, emphasis added).

While the Soviet Union set about creating a "sphere of influence" in Eastern Europe, Western observers began to warn their governments of Soviet expansionism and Bolshevik ideology. An American official, George Kennan stationed in Moscow composed one of the most formative documents (called the "Long Telegram") of the early Cold War period. Writing in February 1946, Kennan observed that the Soviet Union was a very different society in comparison to the USA and as a consequence there could be little prospect of long-term rapprochement. As he noted:

> The Kremlin's neurotic view of world affairs is traditional and instinctive. They have always feared foreign penetration, feared direct contact between the Western world and their own, feared what would happen if Russians learned the truth about the world.... And they have learned to seek security only in patient but deadly struggle for [the] total destruction [of a] rival power. We must see that our public is educated to the realities of the Russian situation. Press cannot do this alone. It must be done by [national] government. World communism is like a malignant parasite, which feeds only on diseased tissue (cited in Kennan, 1967, pp. 549–57).

This form of analysis depicted the Soviet Union as an authoritarian and anti-democratic state, which was inherently expansionist and uncompromising. The "evidence" for such an interpretation (of the Soviet Union) was to be found in the historical and geographical evolution of Russia (see also Kennan, 1947). As the largest country in the world, Russia appeared to offer considerable evidence of territorial colonialism and occupation and to Western writers such as Kennan this despotic nature of the modern Soviet Union owed much to the legacy of Czarist rule. Fifty years later, he reiterated this belief in an interview with the editor of *US News and World Report*, "There have been two great periods of Russian expansion into Western Europe. One was under Catherine the Great, and it lasted until the First World War... and after the World War II there was this other great expansion of Russia into areas of Eastern and Central Europe. If none of my previous literary efforts had seemed to evoke even the faintest tinkle from the bell at which they were aimed, this one [the Long Telegram] to my astonishment struck it squarely and set it vibrating" (Kennan, 1996).

While the Long Telegram appeared more reminiscent of a religious sermon than a policy memorandum, Kennan proposed that the USA had to be prepared to "contain" the expansionist tendencies of the Soviet Union. As a national security memorandum NSC 68 noted in 1950:

> The United States, as the principal center of power in the non-Soviet world and the bulwark of opposition to Soviet expansion, is the principal enemy whose integrity and vitality must be subverted or destroyed by one means or another if the Kremlin is to achieve its fundamental design (cited in Gaddis, 1982, p. 104).

Interestingly, as Susan Sontag noted, communism was either compared to an illness (such as cancer) or disease (such as a parasite), which raised the prospect as to whether there could ever be a "cure" (Sontag, 1980). Geographical proximity to the Soviet Union was considered to be the prime source of infection and as such neighboring states would not only have to be vigilant but also prepared to fight the spread of communism. Shortly after George Kennan had sent the "Long Telegram," Operation Rollback was initiated by the USA promoting espionage, subversion, and sabotage behind the communist Iron Curtain. These attempts to undermine the authority of the Soviet Union were applied in Eastern Europe (and then later the wider world) in the belief that once one state was infected, the disease would spread to other states. It has been mooted that American involvement in Vietnam in the 1960s was precipitated by a fear of the "domino effect" in South East Asia (Sullivan, 1986). If South Vietnam was to fall to communism then, so it was argued, other states such as Thailand and Burma would be vulnerable to Soviet-inspired action too. According to Rear Admiral Arthur Redford, an American nuclear strike on North Vietnam would have been justified in 1953 [more than ten years before the large scale involvement of American forces] because of fears that neighboring countries would be "like a row of falling dominoes" (cited in Sharp, 1999, p. 184).

As with social metaphors, analogies often reduce phenomena to simple assertions rather than a complex system of ideas and values. More disturbingly, these simple analogies can escalate and provide a basis for justifying aggressive policies designed to counteract communism or other apparent threats to national security. Between 1965 and 1973, for example, over 600,000 people died in Vietnam and Laos as a direct consequence of the conflict between the USA and North Vietnam. Bombs in combination with defoliants such as Agent Orange devastated the human and physical ecology of the region. Even after their humiliating withdrawal in 1973, the Nixon administration argued that the conflict had been justified because of the threat of communism to South East Asia.

American geopolitical reasoning also had profound implications for many parts of the Third World as well as for the social and political life within the USA. This representation of the Soviet Union as an expansionist threat effectively obscured the violent interventions of the USA. For much of the Cold War era, American administrations firmly believed that the Soviets could only be contained by a series of military and economic alliances throughout the world. For many conservative writers, the responsibility for the Cold War lay with the Soviets and their behavior in Eastern Europe and the Third World. This view was further reinforced by a "crisis here, a Soviet move there, and an analysis of the protagonists which insisted that Moscow was compelled to expand and that only the United States could prevent it from achieving world domination" (Cox, 1990, p. 30). Global political space was effectively divided into friendly and unfriendly spaces and successive American Presidents argued that conflicts in the Third World were an integral part of this struggle to contain the Soviet Union (Dodds, 1999). Cold War geopolitical space was also conceptualized as a "three fold partition of the world that relied on the old distinction between traditional and modern and a new one between ideological and free. Actual places only became meaningful as they were slotted into these geopolitical categories, regardless of their particular qualities" (Agnew, 1998, p. 112).

In this simplified geopolitical setting, the ideological conflict with the Soviet Union was used by the USA to justify armed intervention in Latin America and elsewhere on the basis that these countries had to be saved from communism. Direct intervention in the affairs of Third World states was epitomized by the involvement of the US Central Intelligence Agency in the overthrow of socialist or allegedly radical governments in Guatemala (1954), Dominican Republic (1965), Chile (1973) and Grenada (1983). In the Dominican Republic, for example, 20,000 US Marines landed in the country on the orders of President Johnson who feared a "communist coup." After the Cuban revolution of 1959 and the emergence of President Fidel Castro, American paranoia of communism in its geopolitical "back yard" became ever more intense and led to new economic programs such as *Alliance for Progress* which provided commercial and military assistance to Latin American countries. When the American troops were not invading, successive governments were prepared to support nasty military regimes in Latin America provided they did not flirt with communism or the Soviet Union. For example, in 1973 the Central Intelligence Agency (CIA) supported the violent overthrow of the democratically-elected Chilean government of Salvador Allende. The new President, General Augusto Pinochet, initiated a murderous campaign (between 1973 and 1990) against so-called left wing subversives. Human rights protection was not a priority for Cold War US administrations who overwhelmingly favored stable political relations with (if necessary) military regimes in Latin America and apartheid South Africa (see McMahon, 1984).

Formal Geopolitical Reasoning and the Cold War

For most of the forty years after the ending of World War II, geopolitics as an intellectual term and form of academic analysis was in dispute (Hepple, 1986,

p. S21). When the late American geographer, Richard Hartshorne, exclaimed in 1954 that geopolitics was an "intellectual poison," he confirmed the widespread opinion that geopolitical reasoning was synonymous with Nazi spatial expansionism (Dodds and Atkinson, 2000; Hartshorne, 1954). Moreover, he argued that theories and approaches gathered under the label "geopolitics" were little more than a bogus "pseudo-science" whose political contamination brought shame on academic geography. Nearly ten years after this act of condemnation, another geographer reiterated this hostility towards "geopolitics:" "revival of the term geopolitics is probably premature and may remain so as long as most people associate the term with the inhuman policies of Hitler's Third Reich" (Pounds, 1963, p. 410). The association between National Socialism and German *Geopolitik* was almost fatal to geopolitics in the aftermath of World War II. Given the enormous human cost of the conflict with Nazi Germany, Soviet geographers too were hostile to the use of the term "geopolitics" (Hepple, 1986, p. S22). Unsurprisingly, only a few Anglophone geographers were willing to be associated with the language of geopolitics (see Chubb, 1954) while the rest chose to ignore it (Crone, 1967; Golbet, 1955; Jackson, 1964).

As a consequence of this intellectual rejection, the political geography of the Cold War attracted scant analysis by English-speaking geographers. Although some geographers analysed global views on containment and geographical areas such as the Middle East and Eastern Europe, their research was described as "political geography" (East and Moodie, 1956; Jones, 1955). Ironically, the practical geopolitical reasoning of Cold War officials such as George Kennan was inspired by a geographical view of world politics. As he noted in his "Mr X" article published in 1947, "Soviet pressure against the free institutions of the Western world is something that can be contained by the adroit and vigilant application of counter force at a series of constantly shifting geographical and political points" (cited in Grose, 2000, p. 6). While Kennan did not make an explicit reference to earlier geopolitical writers such as Sir Halford Mackinder or Nicholas Spykman, his view of the world nevertheless appeared to echo their concerns for the geo-power of the Soviet Union. In essence it was argued that the USA had to be prepared to intervene in geographical regions proximate to the Soviet Union such as the Middle East, Eastern Europe, East Asia and South and South-East Asia. Hence, American administrations were not only prepared to fight in places such as Vietnam and Korea, but also rearmed Asian allies such as Japan, South Korea, and Taiwan. All these actions were motivated by the sole desire to "contain" the Soviet Union.

One of the very few exceptions to this formal academic rejection of "geopolitics" was the American geographer, Saul Cohen. As a Jewish–American scholar, Cohen could negotiate (better perhaps than others) the alleged connection of geopolitics with Nazism. In 1963, Cohen published an analysis of the postwar international system, which actively engaged the earlier geopolitical ideas of Halford Mackinder and Nicholas Spykman (Cohen, 1963, 1973). His basic purpose was to question the policy of containment (something which had been fiercely rejected by Walter Lippman as unworkable and over-ambitious in 1948) and to demonstrate that US Cold War strategy had already misinterpreted the Soviet Union as a "land power." The policy of containment was exposed as geographically flawed because as Cohen's regional world order identified, the areas proximate to the Soviet Union are highly disparate (Cohen, 1963). Two of the regions, the Middle East and South-East Asia, were identified as "shatterbelts," characterized by intense geopolitical rivalry between the Soviet Union and the USA. In subsequent analyses of the Cold War, Cohen (1973, 1982) not only identified sub-Saharan Africa as a third "shatterbelt," but also regional centers of power such as China, Japan, and emerging Third World states such as Brazil and India. As his geopolitical models became more complex, Cohen's research became increasingly divorced from the earliest practical geopolitical reasoning of George Kennan.

Former Secretary of State (and German–Jewish émigré) Henry Kissinger was credited with reviving the term "geopolitics" at a time when Saul Cohen's second edition of *Geography and Politics in a Divided World* appeared (Cohen, 1973). As part of his review of US foreign policy in the midst of the Vietnam debacle, Kissinger invoked the term geopolitics to consider the global geopolitical equilibrium (Hepple, 1986, p. S25). As Leslie Hepple explained, "Kissinger's use of the term is thus part of an attempt to turn American foreign policy towards a *Realpolitik* (though Kissinger's only use of this term is ironical) to address the balance-of-power perspective. He is concerned to thwart Soviet expansionism, but sees US containment policy as [an] excessively ideological... concept of the balance of power" (Hepple, 1986, p. S26). In the early 1970s, Kissinger (who served in the disgraced Nixon administration) was troubled that the Soviets (and their allies) were establishing a strategic presence in southern Africa and the Middle East. Geopolitics appeared to provide a vocabulary for Kissinger to describe this potentially new and dangerous situation.

Kissinger's pronouncements encouraged other academics and commentators to use the term and language of geopolitics. Some writers such as Colin Gray (1977) returned to the ideas of Mackinder and Spykman and argued that American foreign policy needed to recognize the geopolitical realities of Soviet expansionism. In conjunction with Kissinger, Gray postulated that an obsession with nuclear strategy had "blinded" American administrations to the fact that the Soviet Union remained an expansionist territorial power, seeking to establish communist allies in Africa, Asia, Europe, and Latin America. Other commentators, such as the geographers Pepper and Jenkins (1984), were beginning to develop a more critical evaluation of the geopolitics of the Cold War (see also O'Sullivan, 1986). At the same time, a new literature concerned with international diffusion of war and conflict complemented this renewed interest in geopolitics and new theories such as world-systems analysis (O'Loughlin, 1986; Taylor, 1985). In contrast to the previous forty years of minimal geopolitical commentary, the practical geopolitical reasoning of the Cold War was beginning to attract critical appraisals. French-speaking geographers such as Yves Lacoste (and his journal *Hérodote*) developed rigorous critiques of American geopolitical power and Euro-American nuclear strategy (for recent reviews see Claval, 2000; Hepple, 2000).

By the late 1980s, both critical geopolitics and *Hérodote* had established a framework of analysis and critique (see Dalby, 1990; Ó Tuathail, 1986; Ó Tuathail and Agnew, 1992). And as Leslie Hepple warned over fifteen years ago, "Geopolitics must come to terms with its past, and examine the nature of its discourse" (Hepple, 1986, p. S34). The re-evaluation of Cold War geopolitical reasoning has established an important link in the investigation of post-1945 American "security intellectuals." Underpinning this literature has been the recognition that the geographies of global politics (either in the Cold War or the post-Cold War era) were neither immutable nor inevitable, but were constructed culturally and sustained politically by the discourses and practices of American foreign policy.

Popular Culture and the Cold War

By the early 1950s, the Cold War had become a self-fulfilling prophecy for successive administrations as they simply considered the USA to be responding to the persistent threat of the Soviet Union. Armed interventions in Latin America, Vietnam, and Korea were conceived as defensive reactions rather than as a series of violent acts designed to perpetuate American political and commercial hegemony. This persistent assumption about the relatively benign quality of these foreign adventures contributed to a cultural politics of Cold War identity (Campbell, 1992; Slater, 1999). Doctrines such as containment contributed to the construction of "America" as a "bastion" or "fortress" of democracy and liberty. This sense of the USA as a state entrusted with a moral mission had implications for domestic political life too. By 1947, the US House of Representatives Un-American Activities Committee had already descended upon Hollywood armed with the names of suspected communists and socialists. As a consequence of these "witch-hunts," some people were arrested and imprisoned for their alleged connections to communism and hundreds of actors, writers and directors were placed on an unofficial industry blacklist. In the 1950s, the US Senate created a committee headed by Joe McCarthy to investigate anyone judged to be sympathetic to communism or the Soviet Union. Suspected of being "un-American" led some of the condemned to commit suicide. The discourses and practices of the Cold War strategies were linked to the domestic life of the USA. Slogans such as "reds under the bed" in conjunction with television and cinema propaganda helped to legitimize the internalization of danger.

In the aftermath of World War II, Hollywood film studios maintained close relations with the US Department of Defense, as successive administrations recognized that film could play an important role in the ideological struggle against the Soviet Union. The decision to produce a film-length version of *Animal Farm* by the English novelist, George Orwell provided one such example (another would be *I Married a Communist* 1950). Published in 1945, *Animal Farm* was widely interpreted as a parody of the 1917 Russian Revolution and the subsequent descent into totalitarianism. Orwell's writings were also profoundly affected by his experiences during the Spanish Civil War in the 1930s and remained wedded to the reasonable belief that Stalin had cynically failed to support socialist struggle in the Iberian Peninsula. The main characters, Napoleon and Snowball (two farm pigs) organize an animal revolt against the oppressive farm owner, Mr Jones. Despite their lofty intentions they quarrel violently over how to manage the farm and when Napoleon (representing Stalin) emerges as the supreme leader, the principles behind the animal revolution are shown to have been totally betrayed as one form of oppression was simply replaced by another. The final chapter of Orwell's novel warned of the emergence of communist–capitalist rivalry, which was seized upon and used by the American and British secret services as anti-Soviet propaganda (Shaw, 2000).

In June 1948, the US National Security Council created a new agency called the Office of Special Projects (OSP) to conduct deniable political, economic and psychological operations. Housed within the Central Intelligence Agency, the OSP (later the Office of Policy Co-ordination) launched an ambitious psychological warfare program and *Animal Farm* was identified as a valuable cultural medium for transmitting anti-Soviet propaganda to American and wider audiences. In 1951, after months of negotiation and planning, the animation rights to *Animal Farm* were sold to Louis de Rochement Associates who were linked to the CIA-funded Campaign for Cultural Freedom. Under the direction of the British animation team of John Halas and Joy Batchelor, work began in 1951 and filming was completed in 1954, after more than 250,000 drawings and 1,000 colored backgrounds had been created by a team of eight animators (Shaw, 2000). Interestingly, when the production process was reviewed by the US Psychological Strategy Board (PSB), it recommended that the ending of the film be changed. In the original story by George Orwell, the triumphant animals (mainly Napoleon and his dogs) form the new elite and behave in a manner reminiscent of their human masters. The book ends on a pessimistic note, with the other animals meekly accepting the new enforced conditions. In essence, Orwell suggested that there was effectively no difference between communist and capitalist tyrannies. The revised film version, however, depicts other animals such as the horses and chickens overthrowing the dictatorial leaders due to their corrupt way of governance. This shift in the plot was considered crucial by the PSB and the CIA because it demonstrated that communist oppression could be overthrown by a counter-revolution. The film was eventually released in New York in December 1954 and London in January 1955.

Most critics applauded the film for its technical competence and many reviewers recognized its thinly veiled critique of Soviet totalitarianism. While the film was not a box office success either in the USA or the UK, it provided evidence of the close association between Hollywood and US governments. By the time *Animal Farm* became standard reading in Anglo-American schools and colleges, Hollywood had embarked on collaboration with the Defense Department in order to produce the World War II film *The Longest Day* (1961) and the Vietnam film, *The Green Berets* (1968). The latter, starring John Wayne, was considered a morale-boosting movie for US audiences in the wake of the controversial *My Lai* massacre, which led to the killing of scores of unarmed Vietnamese civilians by American troops led by Lieutenant William Calley (Dodds, 2000, p. 76). By that stage, large-scale anti-war protests in the USA and the UK were further heightened by the widespread television coverage of death and destruction in Vietnam.

Other films such as *The Manchurian Candidate* (1962) encouraged the American government and the CIA to actively investigate whether they could control secret agents without their knowledge (Sharp, 1999, p. 186) thereby creating the perfect spy (one who simply carries out instructions without critical reflection). A year later, *Dr Strangelove* (1963) featured a deranged Cold War warrior in command of a US air base who launched a pre-emptive nuclear strike against the Soviet Union. The film parodies the feverish negotiations between the American and Soviet political leaders as they attempted to prevent a nuclear Armageddon. Despite their endeavors, diplomacy failed to avert the explosion of bombs over the two states. Where American governments judged films to have positive geopolitical potential, they were prepared to lend men and equipment to help the filming process. The most famous example of Cold War collaboration was the movie, *Top Gun* (1986) starring

Tom Cruise as a US Navy pilot with the codename Maverick. The American Navy supplied a number of Tomcat fighter planes as well as the carrier *Enterprise* for the duration of filming. By the end of 1986, the film had earned $130 million and eventually grossed $350 million in worldwide cinema sales (Kellner, 1995, p. 80). The media critic, Douglas Kellner argued that *Top Gun* was indicative of a particular period of American Cold War adventurism:

> Aggressive military intervention in the Third World, with the invasion of Grenada, the US-directed and financed Contra war against Nicaragua, the bombing of Libya, and many other secret wars and covert operations around the world. Hollywood films nurtured this militarist mindset and thus provided cultural representations that mobilised support for such aggressive policy (Kellner, 1995, p. 75).

In conjunction with other films such as *Rambo: First Blood Part II* (1985), it has been argued that Hollywood reflected the Reagan administration's willingness to intervene in Latin America and elsewhere in the name of defending the "free world" from the Soviet Union. Perhaps as a former Hollywood actor, President Reagan was particularly sensitive to the cultural power of film (Sharp, 1998).

Successive Soviet leaders have also appreciated the power of film to mould political ideas and shape national identities. Under Joseph Stalin, for example, Soviet cinema flourished in the 1920s and 1930s. In the aftermath of World War II, however, cultural repression became the norm and only 124 feature films were made between 1946 and 1954 (Leyda, 1983). For much of this austere period, film viewing was restricted to subject matter such as the heroic role of the Soviet Union during World War II. Anti-American films such as Abran Room's *Court of Honour* (1949) were intended to warn viewers that Soviet citizens (in this case scientists) should not be seduced by US society and its material wealth. Intense political repression (including deporting many Soviet citizens to their death in Siberian labor camps – *gulags*) went hand in hand with cultural suffocation. After Stalin's death in 1953, Soviet cinema emerged once more and film-makers tackled new subjects such as Soviet youth and the coming of age in conjunction with Cold War propaganda movies (Kenez, 1992, 1997). Despite some greater cultural freedoms under Khrushchev (1953–69), Soviet cinema was tightly controlled and regulated by the Communist Party of the Soviet Union.

While film was a facet of the cultural politics of the Cold War, popular American magazines such as the *Reader's Digest* and *Time* were also formative sources of geopolitical representation (Sharp, 2000). These magazines frequently commented on the expansive power of the Soviet Union and warned readers of the perils should the USA fail to remain ever vigilant. Images and words played their part in fighting the Cold War. In some cases, they could even swing presidential elections as the famous "Daisy Girl" party political commercial was once used by the Johnston administration to suggest that his presidential rival Senator Barry Goldwater would not be sufficiently judicious in his response to the Soviet Union. The commercial, shown on primetime television in 1964, depicted a young girl picking flowers whilst a countdown to a simulated nuclear attack occurs behind her back. Viewers were then left to imagine the consequences for her and others caught in the midst of a possible nuclear explosion in the USA. It was considered particularly shocking as it appeared shortly after the 1962 Cuban Missile Crisis when the USA and the Soviet Union were brought to the brink of nuclear war. Amazingly, a recent anti-China group in the USA re-released this advert in order to warn American viewers that they still could not trust the communist government of China.

Conclusions

While the chapter opened with a discussion of the recent Russian submarine tragedy, most of the attention has been directed towards the USA and the geopolitics of the Cold War. Both the USA and the Soviet Union (since 1991 effectively Russia) have had to come to terms with the effects of the ending of fifty years of ideological and territorial struggle. David Campbell has argued that the Cold War manifested itself as a series of boundaries between civilization and barbarism and consequently helped to render "America" as a place and form of secure cultural identity (Campbell, 1992). Strategies such as "containment" helped to secure the "United States" within a complex geography of evil and danger. Likewise, the cultural politics of the Soviet Union were unquestionably shaped by a long-standing belief that the USA threatened the security of territorial borders. In contemporary Russia, there remains considerable suspicion and mistrust of the USA, eager to expand membership of the Cold War security organization NATO to their former members of the rival Warsaw Pact (O'Loughlin, 2000). In the USA, new threats in the form of "rogue states" such as Iraq, Libya, North Korea, and Serbia co-exist uneasily with enduring suspicions of Russia and China. Recent American plans to develop a National Missile Defence (NMD) programme at the cost of some $60 billion will contribute to fraught relations with both Russia and China (for a critical review, see Gottfried, 2000). The Cold War may be over but as the *Kursk* disaster illustrated, both protagonists (USA and Russia) are struggling to develop new post-Cold War national identities.

Acknowledgements

I would like to thank Gearóid Ó Tuathail for his helpful comments on this chapter.

BIBLIOGRAPHY

Agnew, J. 1998. *Geopolitics*. London: Routledge.
Agnew, J. and Corbridge, S. 1989. The new geopolitics: the dynamics of geopolitical disorder. In R. Johnston and P. Taylor (eds.) *A World in Crisis?* Oxford: Blackwell, 266–88.
Billig, M. 1995 *Banal Nationalism*. London: Sage.
Campbell, D. 1992. *Writing Security: United States Foreign Policy and the Politics of Identity*. Manchester: Manchester University Press.
Chaturvedi, S. 1996. *The Polar Regions: A Political Geography*. Chichester: Wiley.
Chomsky, N. 1982. *Towards a New Cold War*. New York: Pantheon Books.
Chubb, B. 1954. Geopolitics. *Irish Geography*, 3, 15–25.
Claval, P. 2000. Hérodote and the French Left. In K. Dodds and D. Atkinson (eds.) *Geopolitical Traditions: A Century of Geopolitical Thought*. London: Routledge, 239–67.

Cohen, S. 1963. *Geography and Politics in a Divided World*. New York: Random House.
Cohen, S. 1973. *Geography and Politics in a Divided World*, 2nd edn. New York: Oxford University Press.
Cohen, S. 1982. A new map of the global geopolitical equilibrium: a developmental approach. *Political Geography Quarterly*, 1, 223–42.
Cox, M. 1990. From the Truman doctrine to the second superpower détente: the rise and fall of the Cold War. *Journal of Peace Research*, 27, 25–41.
Crone, G. 1967. *Background to Political Geography*. London: Museum Press.
Dalby, S. 1990. *Creating the Second Cold War*. London: Pinter.
Dodds, K. 1999. Taking the Cold War to the Third World. In D. Slater and P. Taylor (eds.) *The American Century*. Oxford: Blackwell, 166–80.
Dodds, K. 2000. *Geopolitics in a Changing World*. Harlow: Longman.
Dodds, K. and Atkinson, D. (eds.). 2000. *Geopolitical Traditions: A Century of Geopolitical Thought*. London: Routledge.
East, G. and Moodie, A. (eds.). 1956. *The Changing World*. London: Harrap.
Gaddis, L. 1982. *Strategies of Containment*. New York: Oxford University Press.
Golbet, Y. 1955. *Political Geography and the World*. London: George Philip.
Gottfried, K. 2000. A missile defence certain to backfire. *Financial Times*, August 7, 2000.
Gray, C. 1977. *The Geopolitics of the Nuclear Era*. New York: Crane and Russak.
Grose, P. 2000. *Operation Rollback: America's Secret War Behind the Iron Curtain*. New York: Houghton Mifflin.
Halliday, F. 1983. *The Making of the Second Cold War*. London: Verso.
Halliday, F. 1989. *Cold War, Third World*. London: Hutchinson.
Hartshorne, R. 1954. Political geography. In P. James and C. Jones (eds.) *American Geography: Inventory and Prospect*. Syracuse, NY: Syracuse University Press, 211–14.
Hepple, L. 1986. The revival of geopolitics. *Political Geography Quarterly*, 5 (supplement issue), S21–S36.
Hepple, L. 2000. Geopolitique de Gauche: Yves Lacoste, Hérodote and French radical geopolitics. In K. Dodds and D. Atkinson (eds.) *Geopolitical Traditions: A Century of Geopolitical Thought*. London: Routledge, 268–301.
Hough, J. 1988. *Russia and the West: Gorbachev and the Politics of Reform*. New York: Simon and Schuster.
Jackson, W. 1964. *Politics and Geographic Relationships*. Englewood Cliffs, NJ: Prentice Hall.
Jalonen, O. 1988. The strategic significance of the Arctic. In K. Mottola (ed.) *The Arctic Challenge: Nordic and Canadian Approaches to Security and Co-operation in an Emerging International Region*. Boulder, CO: Westview Press, 157–82.
Jones, S. 1955. Global strategic views. *Geographical Review*, 45, 492–508.
Kellner, D. 1995. *Media, Culture and Society*. London: Routledge.
Kenez, P. 1992. *Cinema and Soviet Society 1917–1953*. Cambridge, Cambridge University Press.
Kenez, P. 1997. Soviet film under Stalin. In G. Nowell-Smith (ed.) *The Oxford History of World Cinema*. Oxford: Oxford University Press, 389–98.
Kennan, G. 1947. The sources of Soviet conduct. *Foreign Affairs*, 25, 566–82.
Kennan, G. 1967. *Memoirs 1925–1950*. New York: Pantheon.
Kennan, G. 1996 *Transcript of an interview with George Kennan and David Gergen, 18 April 1996*. Accessed from pbs.org/newshour/gergen/kennan.html (September 4, 2000).
Lane, D. 1985. *State and Politics in the USSR*. Oxford: Blackwell.
Leyda, J. 1983. *Kino: A History of Russian and Soviet Film*. Princeton, NJ: Princeton University Press.

McMahon, J. 1984. *Reagan and the World: Imperial Policy in the New Cold War.* London: Pluto.

O'Loughlin, J. 1986. Spatial models of international conflict: extending current theories of war behaviour. *Annals of the Association of American Geographers*, 76, 63–80.

O'Loughlin, J. 2000. Ordering the crush zone: geopolitical games in post-Cold War Europe. In N. Kliot and D. Newman (eds.) *Geopolitics at the end of the 20th Century.* London: Frank Cass, 34–56.

O'Loughlin, J. and Grant, R. 1990. The political geography of presidential speeches, 1946–1987. *Annals of the Association of American Geographers*, 80, 504–30.

O'Sullivan, P. 1986. *Geopolitics.* London: Croom Helm.

Ó Tuathail, G. 1986. The language and nature of the "new geopolitics" – the case of US-El Salvador relations. *Political Geography Quarterly*, 5, 73–85.

Ó Tuathail, G. 1996. *Critical Geopolitics.* London: Routledge.

Ó Tuathail, G. and Agnew, J. 1992. Geopolitics and discourse: practical geopolitical reasoning in American foreign policy. *Political Geography*, 11, 190–204.

Ó Tuathail, G. and Dalby, S. 1998. Introduction: rethinking geopolitics: towards a critical geopolitics. In G. Ó Tuathail and S. Dalby (eds.) *Rethinking Geopolitics.* London: Routledge, 1–15.

Pepper, D. and Jenkins, A. 1984. Reversing the nuclear arms race: geopolitical bases for pessimism. *Professional Geographer*, 36, 419–27.

Pound, N. 1963. *Political Geography.* New York: McGraw-Hill.

Reagan, R. 1982. "The crusade for freedom" President Ronald Reagan's Speech to the House of Commons, June 8, 1982. Accessed from reagan.webteamone.com/speeches/empire.html (September 4, 2000).

Sharp, J. 1998. Reel geographies of the new world order: patriotism, masculinity, and geopolitics in post-Cold War American movies. In G. Ó Tuathail and S. Dalby (eds.) *Rethinking Geopolitics.* London: Routledge, 152–69.

Sharp, J. 1999. Critical geopolitics. In P. Cloke et al. (eds.) *Introducing Human Geographies.* London: Arnold, 181–8.

Sharp, J. 2000. *Condensing Communism: the Reader's Digest and American Identity, 1922–1994.* Minneapolis, MN: University of Minnesota Press.

Shaw, T. 2000. *British Cinema and the Cold War.* London: I B Tauris.

Slater, D. 1999. Locating the American century: themes for a post-colonial perspective. In D. Slater and P. Taylor (eds.) *The American Century.* Oxford: Blackwell, 17–34.

Sontag, S. 1980. *Illness as Metaphor.* Harmondsworth: Penguin.

Spykman, N. 1942. *America's Strategy in World Politics.* New York: Harcourt Brace.

Stephanson, A. 1998. Fourteen notes on the very concept of the Cold War. In G. Ó Tuathail and S. Dalby (eds.) *Rethinking Geopolitics.* London: Routledge, 62–85.

Stokke, O. 2000. Radioactive waste in the Barents and Kara Seas: Russian implementation of the global dumping regime. In D. Vidas (ed.) *Protecting the Polar Marine Environment.* Cambridge: Cambridge University Press, 200–20.

Stokke, O. and Tunander, O. (eds.). 1994. *The Barents Sea Region: Co-operation in Arctic Europe.* London: Sage.

Taylor, P. 1985. *Political Geography.* Harlow: Longman.

Tunander, O. 1989 *Cold War Politics: The Maritime Strategy and Geopolitics of the Northern Front.* London: Sage.

Chapter 15

Postmodern Geopolitics: The Case of the 9.11 Terrorist Attacks

Timothy W. Luke

At this juncture, there is no commonly agreed understanding of "postmodern geopolitics." Nevertheless, there are a series of tendencies that political geographers, international relations scholars, and many others characterize as the condition of postmodernity. In this chapter, I explore three of these tendencies – risky vulnerabilities of living amidst complex technoscientific infrastructures, the cultural conflicts of existing societies and virtual networks, and globalization after the end of the Cold War. These tendencies, in turn, are discussed with reference to the shocking 9.11.01 terrorist attacks in the USA on the World Trade Center (WTC) and Pentagon as events that dramatically illustrate how these tendencies now are developing.

Before addressing "postmodernity," however, what is "modernity?" The project of modernity, as the general condition or basic state produced by the workings of modernization, has been tied conventionally to an increasing latitude of choice in human affairs (Apter, 1967; Mills, 1959; Onuf, 1989). Opposed to "tradition," which often suggests the systematic constriction of individual choice, most recent social scientific and philosophical accounts of modernity see it as that moment at which human beings gain control over their lives by coming to manage their economy, society, and technology more rationally (Kern, 1983). As one influential account of modernization from the 1960s asserted, modernization takes hold "when a culture embodies an attitude of inquiry and questioning about how men make choices – moral (or normative), social (or structural), and personal (or behavioral). The problem of choice is central for modern man.... To be modern means to see life as alternatives, preferences, and choices" (Apter, 1967, p. 10).

Making choices implies greater levels of instrumental rationality, disputation over alternatives, command over complex technologies, and management of natural resources (Luke, 1996; Foucault, 1980, 1991; Marx and Engels, 1978). Economy, government, and society are reimagined in technical terms, which highlights the importance of varying systems of choice, and then more choices between differing systems (Fukuyama, 1992; Jameson, 1992). Yet, modernization also was a militant struggle against resistant traditions, and their many embedded feudal, religious or communal legacies of real choicelessness. Once the ambit of choice is expanded, and the confining constraints of aristocratic authority, enduring poverty, dynastic privilege or repressive religion are lessened, then modernity "wins," and tradition "loses" (Poggi, 1978). Once modernity is left working by itself, the terrains of everyday life shift significantly to the much less definite dimensions of postmodernity (Harvey, 1989).

Here the collective purposes of modernization, which pitted the promise of democracy, affluence, equality, and reason against the varied realities of aristocracy, poverty, privilege, and religion, become more diffuse once tradition is overcome by modernity (Poster, 1995). Evincing individual choice against collective predestination is a heroic struggle; yet, it arguably has won in many places around the world by the end of the twentieth century (Taylor, 1996). Defining and determining which rational choices should be made over and above other rational choices is much more difficult; and, as early as 1959, C. Wright Mills saw these more indefinite ambiguities beyond modernity as the stuff of "the post-modern" (1959, pp. 178–94).

With the triumph of technology over nature, the secular over the sacred, and affluence over poverty, science is believed to have improved life (McNeill, 2000). Still, science "it turns out, is not a technological Second Coming. That its techniques and its rationality are given a central place in a society does not mean men live reasonably and without myth, fraud, and superstition" (Mills, 1959, p. 168). So at "the post-modern climax" of modernity, for Mills, change bogs down, or even collapses. This is what is most characteristic of contemporary life as the prerogatives for decision-making, culture-creation, and choice-elaboration are centralized, and then granted mostly to professional–technical elites (Beck, 1997; Bourdieu, 1998; Virilio, 1995). Postmodernity is "the collapse of the expectations of the Enlightenment, that reason and freedom would come to prevail as paramount forces in human history" (Mills, 1959, p. 183).

Like Mills, Lyotard also no longer believes in modernity's grand narratives from the Enlightenment, which have clad most of Western capitalist society's economic, political, and social practices in fables of reason and freedom. A ceaseless search for performance and profit, then, is the real essence of today's postmodern conditions (Kennedy, 1992; Reich, 1991). As Lyotard claims, a growth-driven capitalist agenda "continues to take place without leading to the realization of any of these dreams of emancipation" (1984, p. 39). With little trust in any metanarratives, or canonical narratives of truth, enlightenment or progress, the science and technology behind big business, Lyotard argues, are slipping into the register of "another language game, in which the goal is no longer truth, but performativity – that is, the best possible input/output equation" (1984, p. 46). On another level, as Jameson claims, these persistent advances toward greater performativity are spinning up "a new social system beyond classical capitalism," proliferating through "the world space of multinational capital" (1992, pp. 54 and 59).

Rather than being a "break," or "crisis," or "rupture" in modernity, postmodernization perhaps is merely a "turn" in the existing routines of modernized being (Ohmae, 1990; Poster, 1995; Reich, 1991). In accord with making the consumption of commodities a way of everyday modern life, postmodernity essentially mimics the "fast capitalism" (Agger, 1989) of markets: it rejects closed structures, fixed meaning, and rigid order in favor of chaos, incompleteness, and uncertainty (Ó Tuathail, 1999; Rosenau, 1990). Its politics often repudiate fixed territories, sacred spaces, and hard boundaries in favor of unstable flows, secularized practices, and permeable borders. Even so, postmodernity is not a wholly new social order; rather it is instead a systemic adaptation of culture to capitalism itself as the production and reproduction of an almost totally commercialized way of life becomes generalized on a transnational scale (Beck, 1992; Bourdieu, 1998; Luke, 1999).

For geopolitics, this incredulity before the workings of metanarrative calls into question the meaning and purpose of nation-states, fixed territoriality, common governance, and scientific–technological progress within a stable international order. If enlightenment, liberation, and progress are not to be believed, then their historically most common geographical containments and political practices also come into doubt (Ó Tuathail and Dalby, 1998). Endless choices between many different empowering, enriching or edifying alternatives bring people into a postmodern condition in which "a risk society" emerges from the outlines of societies centered upon the production and distribution of material satisfactions (Beck, 1992). Almost all of these choices are defined and controlled by networks of professional–technical elites who work and live in a few sheltered places even though the effects of their efforts are mostly felt in many particular places by those who cannot easily affect how the choices are identified or made (Martin and Schumann, 1998).

The Vulnerabilities of Contemporary Life

Some believe that destroying the WTC and damaging the Pentagon were futile efforts to destroy the global economy and American military power. And, in some ways, they are right. World trade really has no single center, and the armed forces of the USA can be controlled from many different points scattered all around the nation. Nonetheless, buildings are signs, as well as sites, of wealth, power, and culture. So destroying or damaging any significant building becomes a successful first strike in a sign war against the nation that still dominates the means of communication and relations of signification at the dawn of the twenty-first century. Such acts are an ultimate propaganda of the deed, and those who committed them know that the relations of signification will replay the images of them in deadly success over and over again in accord with the contemporary media's prime directive: "if it bleeds, it leads."

Contemporary life depends upon a network of complex, interlinked technostructures (Beck, 1992). Whether it is communication, nutrition, and transportation or finance, housing, and medicine, ordinary technical artifacts and processes will always afford terrorists innumerable embedded assets that can be used for destructive purposes. Lethal capabilities can be created simply by contrafunctioning the everyday uses of many technics. Resourceful resistance fighters must create weapons from what is at hand, and the Internet, 24/7 finance markets, global airlines, agricultural fertilizers, rental trucks, and tourist industries readily provide the organizations, intelligence, weapons, and/or target needed for a terrorist act. Combining a full-fueled wide-body airliner with a kamikaze pilot clearly can create a strange new type of cruise missile whose kinetic energy, chemical fuel, and symbolic impact can forever alter the world's air transport system, New York's skyline, and the exceptionalist myths of invulnerability that once flew over the USA. Yet, this capability remains deployed as long as airliners fly, and gritty geopolitical conflicts produce more suicide plots.

Protecting against any future attacks, however, becomes a nightmarish defense problem once the generic liberal assumptions of rational, life-enhancing utility presumed by all modern technics are pushed outside in the daily equations of technological use. Many large technical systems become highly problematic, threatening, and uncontrollable dangers if one repurposes their applications to cause havoc or harm rather than generate power or profit. The most relevant case in point is the American air transport system. Every day prior to 9.11.01 (hereafter 9.11), 35,000 to 40,000 airplanes took off and landed, which included 4,000 commercial flights, at 460 FAA-controlled airports to serve almost 2 million passengers (*Washington Post*, September 12, 2001, p. A5). A forewarning of 9.11 was uncovered in 1995 in the plot to hijack and/or bomb twelve US airliners in Asia and Oceania. Nevertheless, on any given day, finding terrorist suicide pilots among nearly 2 million passengers on 40,000 planes and 4,000 commercial flights is nearly impossible; and each of these flights can become a terrorist-guided missile.

The *modus operandi* of the Al Qaeda networks, allegedly behind many acts of domestic and international terrorism over the past decade, displays a measure of versatility and adaptability that probes for such possibilities in many places. Consequently, 9.11 is most likely not going to be repeated in exactly the same way. Instead, the next major strike will undoubtedly leverage another embedded asset in some existing technostructures to raise havoc at home or abroad. Those who cling to common liberal assumptions about life, liberty, and the pursuit of property as "happiness" cannot easily accept that illiberal assumptions sparked in other degraded lives, following from degraded liberties, and spawning from dangerous properties, will lead to darker ends of equal import to those who embrace them. Even so, these illogical logics are real, so banning fingernail clippers, tweezers, and pocket knives on airliners will not stop hijackings by anyone who is determined enough to pit everyday technologies against their original intended purposes.

Organized stateless violence emerged with great fervor during the Cold War in wars of national liberation, narcocapitalist crime syndicates, ethnic secessions, and shadowy counterintelligence units. Tolerated by the superpowers from the 1940s to the 1990s, these entities often proved to be reliable tools in the border conflicts between the capitalist and socialist zone-regimes that once were arrayed around Washington and Moscow. In the political vacuums created in far too many states after 1989–91, however, these entities have acquired quasi-sovereign powers in far too many territorial areas across Africa, Asia, Latin America, and even parts of the former Soviet Union. Consequently, one finds small organized war machines with varying levels of capability, but no real closure over entire territories and populations, demodernizing many places around the world in pursuit of their contrasovereign illegitimate power (Bowden, 2000). From the Congo, Somalia, Liberia, and Sierra Leone to Afghanistan, Iraq, Chechnya, and Palestine, there are new demodernized wildzones in which these stateless formations for organized violence play out their stateless institutional quest for power on both a local and global level (Rashid, 2000).

Mostly dismissed as minor turmoil when their first effects were registered on 2.26.93 at the WTC, they now are regarded as sources of major chaos after the destruction at the WTC, Pentagon, and rural Pennsylvania on 9.11. Indeed, the USA has now entered into "a state of war" with "stateless warriors" – a situation that has not prevailed in the republic since its "civilizing campaigns" against Native Americans, the Barbary pirates, and Caribbean buccaneers in the eighteenth and nineteenth centuries. Instead of considering this condition a historical oddity, however, America needs to ask what strategic failures, inconsistencies or discontinuities so plague its global roles as the world's last superpower, that such demodernizing tendencies are now becoming much more endemic. Over 2,800 people died at the WTC, and they came from 86 different countries. The biggest single-day death toll of Americans from violent terrorism in US history also has proved equally true for Great Britain, Canada, Japan, Chile, Colombia, Australia, Pakistan, and many other countries in their respective histories with terror.

Just as the extent of Washington's collaboration with Saddam Hussein from 1978 to 1990 has never been fully disclosed, because of Baghdad's former roles in the West's resistance against Islamic revolution in Iran, the full measure of American support for "the Afghan" Arabs who now are "Al Qaeda" during the Soviet invasion of Afghanistan probably cannot be known (Cooley, 2000). After being left high and dry by the USA once the Soviet Union began to fragment, the many Algerian, Egyptian, Gulf, and Saudi Arabs who answered America's call to defeat communist invaders in Afghanistan became enraged in 1990–1 by the massive American military build-up in Saudi Arabia. Seeing these moves as a Western attempt to occupy the holiest places of the faith, Osama bin Laden and his confederates have apparently spent the past decade infiltrating at least thirty-five countries to strike back against the USA in particular and the advanced industrial West in general. Yet, the USA does not really know how much of this may be true, and its counterterrorist efforts to date have been total failures (Gerecht, 2001).

Al Qaeda, if it is indeed behind 9.11, is tied to the puritanical Wahhabi Islam followed by the Saudi dynasty, which has flogged this rigid dogmatic code in the face of its luxuriant excesses to maintain Riyadh's flagging authority at home and abroad. The radicals of Al Qaeda in Algeria, Egypt and Saudi Arabia tried to oust their home countries' rulers in the early 1990s, but they consistently failed. Hence, they turned to assailing the USA as the patron of those more conservative regimes to disrupt the stability of the Middle East. Many found their mission first as mujahaddin in Afghanistan fighting against the Soviet Union. Once Moscow withdrew, they wandered, causing trouble in Algeria, Sudan, Egypt, Yemen, Saudi Arabia, and Palestine. Al Qaeda, or "the base," first operated in Sudan where bin Laden had many private investments, but after President Clinton's ill-conceived response to Khartoum, which expelled bin Laden and his group, bin Laden threw in with the Taliban in Afghanistan.

Like many underground guerrilla or terrorist movements, however, Al Qaeda's networks are very discontinuous and decentralized in the ways they go about developing their tactics, raising their support, and finding their followers. 9.11 was quite shocking in its scope and intensity, but it really is not all that surprising. The Bush administration, of course, denied having any forewarnings of 9.11 for months after those tragic events, but several FBI and CIA documents were leaked out during spring 2002 that indicated a few field agents were extremely concerned in August 2001 about a major terrorist attack being launched by Al Qaeda within the

USA. After all, the WTC was the target of a terrorist truck bomb on 2.26.93, which failed to collapse the complex, even though it killed six and injured thousands. This is when America's "new war" really began, and there have already been many other clashes with radical Islamic terrorist groups over the past eight years: in Somalia during 1993 Al Qaeda-trained guerrillas stymied American efforts to capture Somali warlord Hussein Aideed during the American intervention in Mogadishu; in the Philippines during 1994 an airliner was bombed killing one and injuring 10 as part of a plot to destroy 12 US jumbo jets all around the Pacific; during 1995 an Islamic plot to assassinate Pope John Paul II was uncovered in Manila; in 1996 a truck bomb killed 19 US service personnel in Saudi Arabia; in Egypt during 1997, 58 foreign tourists were killed at ancient ruins along the Nile; during 1998 the American embassies in Kenya and Tanzania were bombed killing 224; during 1999 plots were foiled in Jordan and the USA that aimed to disrupt millennium celebrations; during 2000 the U.S.S. *Cole* was bombed in Yemen's Aden harbor killing 17 American naval personnel; and, in 2001, the events of 9.11 have killed thousands in New York, Pennsylvania, and Virginia. Of course, these incidents are only those that are believed to be connected somehow to Osama bin Laden, Al Qaeda, and its many allies (Griffin, 2001). Other radical Islamic groups with ties in Palestine, Iran, China, Algeria, Uzbekistan, Tajikistan, Azerbaijan, Kyrgizistan, Albania, Georgia, and Russia also can be tied to violence elsewhere from 1991 to 2001. Many see these incidents as a neo-medieval jihad that aims to topple the highly modernized Western nations for abusing and/or exploiting the nations of Islam, but they also are a response to the modernity of failure brought by corporate globalism to the poor and powerless.

The geopolitical underpinnings of 9.11 are not new: they are ragged contours cut by a unipolar correlation of forces that has emerged after the Gulf War and the collapse of the Soviet Union (Campbell, 1992). The New World Order of 1991, however, soon devolved into carpet bagging, fiscal skullduggery or benign neglect as many individuals and firms in the USA looked inward to seek El Dorado on the World Wide Web instead of dealing with the disintegration of the communist bloc. As a result, large swaths of the old "Second" and "Third World" decayed, disconnected or devolved into demodernized chaos on a scale not seen since the seventeenth century as the Asian financial crisis of 1997, the Russian market crash of 1998, and the global slump of 1999–2001 deflated even the once robust economies of the Pacific Rim countries. While the American economy boomed throughout the 1990s, the Arab economies in the Middle East grew only 0.7 percent annually and the Islamic states in one-time Soviet Central Asia actually contracted without big subsidies from Moscow (*Business Week*, October 1, 2001, p. 47).

On one level, the terrorist networks behind 9.11 may represent a failure of modernity, which rarely has been acknowledged in the triumphalism of the past decade. In 1991, the USA oversaw the successful recapture of Kuwait from Saddam Hussein; and, then a few months later, it watched in awe as the Soviet Union totally unraveled. During the intervening years, the USA quickly washed its hands of many Cold War alliances and policies, which often had been connected to authoritarian allies relying upon using violent means. What had seemed necessary to resist the Soviet Union was no longer required. At the same time, the USA slowly turned away from many larger internationalist responsibilities that befell it as the world's sole remaining superpower. Instead of continuing to stand resolutely for unshakeable modern ideals, like democracy, equality, and freedom, the USA left tyrants like Saddam Hussein in place after Kuwait's oil was once again secure, permitted gangster capitalism to establish itself securely in places as varied as Russia, Columbia, Romania, Congo, and Ukraine, and temporized as horrendous civil strife racked East Timor, Sri Lanka, Rwanda, Bosnia, Congo, Iraq, and most of former Soviet Central Asia as well as Afghanistan. At the same time, Washington ineffectively brokered a fragile peace process between Israel and the Palestinians that only increased tensions between Jews and Arabs as more militant groups on both sides pushed more extreme measures to attain their goals after the Oslo peace process (Masalha, 2000; Usher, 1995).

The difficult detail that most overlook in the putative triumph of "the West" over "the Rest" in the 1990s, then, is how fully a modernity of failure can coexist beneath, behind or beside the modernizing successes brought on by globalization through transnational corporate commerce (Kaplan, 1996). For every Hong Kong, Singapore, Frankfurt, or San Jose in the 1990s, there were five Groznys, Kabuls, Sarajevos, or Kinshashas (Gourevitch, 1999; Power, 2001). As the twenty-first century dawned in many places, several others slipped back into sixteenth or seventeenth century conditions of demodernizing disintegration (Luke and Ó Tuathail, 1997). Large parts of the world now do not have effective territorial governance by modern nation-state institutions (Anderson, 1991). Many regions of the world have slipped back into early modern relations of trade in which black markets for gems, oil, weapons, drugs, timber or even people clearly eclipse the open exchange for legitimate goods and services. And, in this chaotic flux of change, the modernity of failure suffered by many is easily blamed upon a modernity of success enjoyed by the few with the USA at the top of that small pile of highly modernized nation-states.

Culture War and Postmodern Geopolitics

In many ways, this current wave of terror is so terrific because it remains thusfar anonymous. While the White House believes Osama bin Laden and Al Qaeda are its authors, those forces have disclaimed responsibility for 9.11. Terror in the past typically was committed in the name of a nation seeking nationhood, a revolution pushing for realization, or a faith expressing frustration. Even the madness in Bosnia, Kosovo, and Macedonia fit this model, and the USA could intervene in those circumstances, albeit often at great cost and with considerable loss of military efficacy, in a "policing" or "stabilization" action (Langewiesche, 2001). All of these forces might be involved in 9.11, but no credit for them has been claimed. Instead these acts are now a strange species of stealth bombing in which the bomb is known, but the bombers remain cloaked in obscurity to prevent counterstrikes, raise the rhetorical heat, and preserve their freedom to maneuver. Because the USA is the world's sole superpower, it makes eminent sense to strike on its territory against all people who work and live there in support of the global economy. There are tremendous embedded assets waiting to be artfully abused, the media system willingly will serve as the PR office of the terror by putting its most terrible moments on continuous replay, and the familiar "day late, and a dollar short" defensive reaction of the US government can be counted upon to create as much, if not more, collateral damage as the bombing did merely by preparing to stave off the unthinkable acts that already have passed. Remaining unnamed, unknown, and undetected simply supercharges this propaganda of the deed with even more energy.

A culture war also rests at the core of 9.11, but it is not one between Islam and Christianity, even it can be tied to the incommensurability of secularism and devotion in many respects of everyday life. Liberal ideologies rest at the core of modern consumer society. Without the codes of conduct that manage everyday human behaviors through codes for autonomous rational agency, the technics that underlie market exchange, instrumental action, and personal happiness would grind to a halt. To live is to consume, and to consume is to live (Davidson, 1997). By these lights, few, if any, modern individuals even can imagine rationally and choosing freely not to consume or to die. Consequently, the common sensibilities of the American public have been shocked from Guadalcanal in World War II to Hue during the 1968 Tet offensive in Vietnam to the WTC bombings in 2001 by dedicated violence committed under illiberal visions of existence that readily will put other collective goals far ahead of individuals choosing to consume, work or acquire property. Believing God, History or Nature is on their side, these "others" willingly can sacrifice themselves, their family, and their wealth to attaining long-term strategic goals. While destroying the Pentagon or WTC might not seem to offer many strategic benefits, the audacious devotion to such violent goals always can, first, confuse, and, then, shock liberal understandings of the self and society down to their core.

Therefore, any defense of ordinary liberal capitalist ways of life always will require an uncomfortable on-going effort to comprehend the radical indifference to its codes of conduct that illiberal ways of acting and thinking can generate. All too often it takes a final phone call to loved ones who relay the latest CNN updates about terrorist attacks elsewhere to awaken ordinary consumers to such foundational threats to their existence. Then some readily rise to the call, even if it is too late for them. Radical Islamism obviously fits these shoes as its advocates allege a new world order tied to liberal capitalist values, and the American society and state that stand behind them, are threatening Islam as a whole. Moreover, the Americans in Iraq, Serbs in Bosnia, Hindus in Kashmir, Russians in Chechnya, the French in North Africa, or the Israelis in Palestine are all working to destroy the faith. Hence, its dispossessed radical followers can swear allegiance to "defeat the mightiest military power of modern times" by trusting, as bin Laden maintains, how fully "your lives are in the hands of God" (*Newsweek*, September 24, 2001, p. 44). This absolute profession of religious faith keeps radical Islam disciplined and resourceful, but its origins also highlight how easily everything can dissolve in the poorer, less developed regions of the world from Morocco to Indonesia as the peaceful followers of the faith struggle to coexist with a fundamentalistic Islam. Moreover, the generic forms of liberal capitalist life brought to millions by transnational firms now compete on the same terrain with Al Qaeda not only in Egypt or Sudan, but also in Russia or Bosnia as well as Ontario or Florida.

In many ways, it is clear that this normalized generic liberalism at the core of modern markets, technics, and societies is what radical Islamicists reject. The "Occidentalosis" that Islamic critics and clerics have been decrying since the 1970s is not focused especially upon the "Disneyfication" or "McDonaldsization" of everyday life that America has represented in many anti-globalization campaigns. Actually, the police investigation thus far into 9.11 shows the alleged terrorists' lifestyles were entwined in a remarkably normal level of typical suburban consumerism – cars, credit cards, kids, gym workouts, wives, technical schooling, vacations, and rowdy bar-hopping. Those aspects of American life seemed acceptable, if only to maintain deep cover as sleeper agents, but they also were important in sustaining the terrorists' everyday life.

What these radical Islamicists appear more dead-set against are older liberal principles in American life that the Right and Left in America are still contesting: the separation of religion and government, basic natural rights to life, liberty and property, the emancipation of women, even scientific reason. This perceived threat in the prophet's homeland sparked bin Laden's *jihad* against the USA, and these precepts are what globalization often portends for "the Rest" as they confront "the West." If they wish to resist this new opposition to modern life, then both the Left and the Right need to move past their current cultural warring over small beer, and decide which foundational practices in Western modernity are worth reaffirming. Yet, they must also be aware of how much those principles aggravate the anxieties of outsiders who see their values smothered by a civilization of cultural clash that constantly is inundating them with unpalatable changes. Of course, Islam can coexist – and has done so in the past – with scientific scepticism, the freedom of women, basic natural rights, and a separation of the faith and the state; it is mostly religious fundamentalists in Islam who assail these principles. Still, as the Reverend Jerry Falwell so artlessly illustrated when he interpreted 9.11 as God's retribution against America for being a nation of sinners, these illiberal tendencies also plague Christianity.

These strikes on the Pentagon and WTC are hardly apocalyptic events, even though the American media drone on as if the destruction wrought upon these iconic buildings will mark a sea change in Western civilization. Of course, if such assertions are repeated often enough, then many people might come to regard them as true in a cycle of self-fulfilling prophecies. This outcome may well follow from 9.11. There is a frightening insularity shared by most average American consumers as they buy more and more of the world's products at their local Walmarts, while remaining utterly clueless about how those goods are so abundant, cheap, and endless; why their credit is so steady, sound, and bottomless; or whose welfare elsewhere in the world is not so solid, certain, and strifeless. And, as long as major media outlets continue to replay the explosive impacts of airliner crashes on the WTC, the apocalyptic strains of 9.11 will continue to shake consumer confidence and public security, which is precisely what the terrorists had hoped to achieve with this propaganda of the deed.

Apocalypse should be understood as a world-ending catastrophe; and, in a sense, one world has ended – the world of exceptionalist American smugness about being safe from foreign threats, secure in the hands of complex technological systems, and stable before the onslaughts of any foreseeable threat. Rather than apocalyptic, these events are instead instances of a low-tech iconoclasm, landing blows on a few key icons of American power and wealth. The iconoclastics who succeeded at crushing these symbols did not materially lessen that power or wealth, but they have cracked some important icons of American invincibility, industry, and invention simply by putting ordinary inventions and industries to other insidious uses for which they were not designed.

Globalization and the End of the Cold War

For many, globalization is the key trope tying together neo-liberal capitalist rationalization, informational technics, mass consumption culture, and integrated world markets of a postmodern geopolitics (Rodrik, 1997). The globalizing impetus of postmodern geopolitics is considerably different from that which prevailed during the Cold War. As Friedman suggests:

> If the defining perspective of the Cold War world was "division," the defining perspective of globalization is "integration." The symbol of the Cold War system was a wall, which divided everyone. The symbol of the globalization system is a World Wide Web, which unites everyone. The defining document of the Cold War system was "The Treaty." The defining document of the globalization system is "The Deal".... While the defining measurement of the Cold War was weight – particularly the throw weight of missiles – the defining document of the globalization system is speed – speed of commerce, travel, communication, and innovation. Globalization is about Moore's law, which states that the computing power of silicon chips will double every eighteen to twenty-four months. In the Cold War, the most frequently asked question was: "How big is your missile?" In globalization, the most frequently asked question is: "How fast is your modem?".... If the defining anxiety of the Cold War was fear of annihilation from an enemy you knew all too well in a world struggle that was fixed and stable, the defining anxiety in globalization is fear of rapid change from an enemy you can't see, touch, or feel – a sense that your job, community or workplace can be changed at any moment by anonymous economic and technological forces that are anything but stable (Friedman, 1999, pp. 8, 9 and 11).

This extended explication of alikenesses and differences in globalization remediates the postmodern world's meaning in the measures of increasing speed, instability, and collaboration all tied to remaking geopolitics around 1s and 0s.

The Al Qaeda terrorist networks also are the epitome of contemporary globalization. Whether it is the easy facility of these cadres with global finance, world travel or international communication, one must not mistake the cultural traditionalism of Al Qaeda's mujahededdin for some sort of technological unsophistication or political obtuseness. Indeed, the loosely articulated cellular structures of these networks are highly specialized "virtual organizations," pulling people, money, resources, and tactics from different places at different times into a single cohesive task performance team without necessarily following any overly centralized strategy. Whether it is truck bombs, hijacked airliners, ship bombings, individual murders, or seizing symbolic buildings, global means of communication, organization, and transportation make it easier to refunction embedded technical assets as tools of terror in such loosely coupled, flexible means of destructive organization.

Ó Tuathail (1998, pp. 27–8) suggests that the problematics of geopolitics, as practiced in world politics by the dominant states, can be reconsidered by asking:

1. How is global space imagined and represented?
2. How is global space divided into essential blocs or zones of identity and difference?

3. How is global power conceptualized?
4. How are global threats spatialized and strategies of response conceptualized?
5. How are the major actors shaping geopolitics identified and conceptualized?

Given these criteria, how does a postmodern geopolitics stack up in comparison to more traditional visions of geopolitical activity?

In postmodern geopolitics, space can be first imagined, and then broadly represented, in both statalized and nonstatalized terms (Luke, 1993; Ó Tuathail and Dalby, 1998). While the USA continues to recognize a modernized map of nation-states, its campaigns in the 1990s and 2000s against criminal narcocapitalists, shadowy infowarriors, and Islamic terrorists acknowledged global threats which are transnational yet stateless as well. After the terrorist attacks of 9.11, global space is sharply divided between zones of identity and difference that stand or fall on supporting an international coalition of antiterrorist countries or cooperating with a nebulous network of radical Islamic fundamentalists who exploit postmodern spaces for antimodern purposes. Global power, in turn, is being recast in informatic terms as well. American "soft power" (Nye, 1990) is what threatens radical Islamicists and surviving state socialists. Likewise, international threats are now imagined in destatalized, dematerialized, and deformalized network terms. Whether it is the unknown hacker, a faceless narcocapitalist or an Islamic underground terrorist cell, the threat to the USA is presented in terms of deterritorialized and decentered network assaults, and not a certain secure statal authority.

Modern geographies rewrote the earth in terms of visible spatial formations created by modern industrial capitalism (Agnew, 1994). Landscapes of cities and farms drawn and defended by nation-states anchored identities scripted out by a print capitalist press or national broadcasting systems. The modernization project is tied to concrete production, and national traditions of cultural, economic, and geophysical geography propound their mappings of railway lines, telegraph systems, road networks, urban settlements, electrical grids, agricultural outputs or linguistic zones in writing the geopolitics of mastered modern space (Agnew and Corbridge, 1995).

A postmodern geopolitics does not break with these realities for they clearly persist in much of the world's collective social practices. Yet, it must turn on, and from, them, in realizing a postmodern geography of the earth that also must account for less visible and tangible spatial flows in contemporary transnational life (Appadurai, 1996). Discontinuous and diverse global webs of informatic exchange throw forth new mediascapes, infoscapes or cyberscapes behind, beneath or beside landscapes (Barber, 1995). Global cities pull together their exchanges transnationally as they pull apart from national engagements (Greider, 1996; Harvey, 1996). Transnational diasporas of dispossessed peoples build little Kabuls, Kashmirs, or Kurdistans all over the world, but without a statalized national homeland (Appadurai, 1996; Doty, 1996). Alliances of telematic systems, corporate enterprises, global laborers, and local sales in a "soft capitalism" (Thrift, 1998) capture world market share in a mappable manner, but without being definitively based in any one physical site. This "unruly world" (Herod et al., 1998) poses a problem for geography, governance, and globalization, because one finds many "unmastered" postmodern spaces coevolving with the mastered measures of modernist spatiality (Debrix, 1999). What once were discrete "solid state" circuitries for geopolitical power in closed hierarchical systems must face open-sourced architectures of power in which capital and authority work at nodes in networks (Diebert, 1997; Luke, 1999; Luke and Ó Tuathail, 1997). Thus, one finds odd contradictions at work today: on the one hand, Osama bin Laden and his Al Qaeda terrorist network are believed to be behind the bombings of the Khobar Towers in Saudi Arabia to disrupt the hold of the Saudi establishment; but, on the other hand, the Saudi establishment is populated by many members of the larger bin Laden family, and the family's construction company has been awarded contracts to help rebuild the Khobar Towers after the 1995 attacks there (Mayer, 2001, p. 65).

The "coming anarchy" foretold by Robert Kaplan in the 1990s cannot be easily disentangled from the triumphalism of the USA under Presidents George H. W. Bush and Bill Clinton (Kaplan, 1996). President George W. Bush was elected in 2000 hoping to continue riding on those same waves of exceptionalist neglect, but 9.11 has brought him, his administration, and the nation back to earth. The "world" of the WTC with thousands of people working from scores of countries under America's aegis has been shattered by scores of people from "another world" shut off from world trade centers by thousands of grievances rooted in ethnic, ideological or religious complaints about perceived American arrogance. Today's nascent world civilization carries the workings of both "worlds" within its civilized practices. This cannot be reduced to a "clash of civilizations" (Huntington, 1998). Religious leaders from both Islam and Christendom have roundly condemned the acts of 9.11. This day is instead one more outcome of a failed modernity resting upon a commercial civilization rooted in cultural, economic, and technological clashes. Today, myths of exceptionalism, Edenic isolation, and a ceaseless quest for growth are colliding with the quiddities of material limits, global villages, and common tragedies. These new geopolitical realities need to be faced, and then responded to, rather than evaded in self-centered excess (Walker, 1993).

For the most part, however, an extraordinary moment for a new world order during the 1980s has been squandered. Rather than accepting the immense responsibilities of world superpower, the USA shrank from them under both Republican President George H.W. Bush and Democratic President Bill Clinton (Halberstam, 2001). For all the talk about human rights, this empty political project has been pursued either weakly, as South Africa, Bosnia, and Kuwait suggest, or not at all, as Rwanda, Chechnya, and East Timor illustrate. While touting the merits of a modern civil society, Washington often looked the other way as Serbs butchered non-Serbs in racial pogroms, the Congo erupted in murderous tribal warfare, the Taliban "Islamicized" most of Afghanistan, and autocratic gangsters reasserted themselves in Byelorussia, Romania, and Ukraine. Even though the USA believes itself to be the world's pre-eminent power, it has shown little leadership over the past decade on global warming, the AIDS pandemic, world poverty, nuclear proliferation, and economic instability beyond its propensities for fighting antiseptic air wars from B-2 bombers or fomenting technological upheaval with "dot com" capitalism.

The geopolitical balance in this new war is unusual. A $10 trillion economy in the USA is pitted in the first round against a complete economic basket case – Afghanistan – whose main source of income is opium farming, and where nearly thirty percent of the population depend upon international relief agencies for food, most of which comes from the USA. Almost 300 million Americans face about 30 million Afghans, and Washington's Star Wars-era military forces are well equipped to cope with the Taliban's leftover Cold War-era Soviet arms. The NATO countries, Japan, Australia, New Zealand, and tens of other nations are lining up to aid the USA, including Russia, China, and India. No nations recognized the Taliban except Pakistan, Saudi Arabia, and the United Arab Emirates, and no nations admit to aiding bin Laden and Al Qaeda. The mujahededdin of the Taliban resisted fiercely, but this is what President Bush needs. Before 9.11, the world was ignoring Washington's efforts at resisting the Kyoto global warming accords, dismantling the 1972 ABM treaties, and touting more NAFTA-like trade pacts. After 9.11, a new enemy is producing a measure of cohesion and compliance not seen since 1947 at the dawn of a Cold War with the Soviet Union.

An elusive bin Laden, with ever-present signs of Al Qaeda subversion at home and abroad, is precisely the tonic that the times demand when President Bush and his national security team assess today's threat environments. Nonetheless, the first battle in this new war was a catastrophic defeat for the USA. The whole terrorist budget for 9.11 could have been as little as $200,000, which is an astounding geoeconomic reality pitted against the estimated $60 billion in direct and indirect losses from the attacks of 9.11, the $140 billion in federal stimulus packages to rev up the US economy, the $1.4 trillion in stock declines the week after the bombings, and the 150,000 jobs cut in the immediate wake of these incidents (*Newsweek*, October 1, 2001, pp. 29 and 55).

9.11 at the WTC cannot be forgotten as easily as the 2.26.93 attack on the WTC was. The ordinary architectures of modern life make such calamities possible; it only takes slight efforts by a few clever zealots to turn common conveniences into tools for utterly uncommon outrages. In the aftermath, one hears of the FBI discovering persons with Arab names checking out crop dusting planes in Florida, getting HazMat licenses in Pennsylvania, and lurking about chemical plants in Texas. This will continue, a globalist economy simultaneously creates possible weapons and angers possible enemies. The ongoing processes of globalization, as such, and the persistent application of neoliberal ideologies of globalism *per se*, will almost certainly continue creating new resentments against the USA and the West. Furthermore, a feckless Republican administration whose agenda had been tied to ill-conceived tax cuts and slap-doodle economic restructuring now has resurrected Cold War era logics of enemy-definition, threat-containment, and defense-mobilization on a scale that Al Qaeda could not have foreseen before 9.11.

Summary

What was rooted in the old realities of geopolitics as they rolled forward into 9.10.01 now has found fresh narratives, projects, and energies in the new geopolitical realities of 9.12.01. Promising to go anywhere anytime to fight any terrorist force, the new Bush administration is reimagining global terrorism as a foe worthy of permanent war. This search for a "new enemy" in the register of grand Cold War-era struggles has been an ongoing project since 1991. Saddam Hussein, Communist China, Latin narcocapitalists, and post-1991 Russia have all failed the screen test, but radical Islamic terrorism has been forced into playing this lead (Johnson, 1999). Where the war will be fought, by whom, and at what cost remain to be seen, but it begins with Bush having the highest approval ratings of any US President ever, Congress declaring an undeclarable war with a very generous initial $40 billion credit line, NATO invoking its collective defense charter for only the first time in its history, scores of nations from Great Britain to Kazakhstan pledging full support to Washington for the fight, and young Americans flocking to armed services recruiting stations at rates not seen since December 1941. For those who attacked the USA on 9.11.01 as the homeland of "a rogue superpower," these new geopolitical realities will not easily be ignored. Yet, for those rallying to the flag in the USA, the costs of the last Cold War in people and resources also should not be forgotten. And, rather than trusting entirely in the US marines or stealth bombers, America needs to ask how much its globalist system of political economy and cultural production are behind the conflicts it currently faces.

This chapter takes the premise of postmodernity quite seriously, and it has evaluated how space is being refashioned and reimagined at the local level and on a global scale after the end of the Cold War (Crang and Thrift, 1999). The 1990s and 2000s have proven to be a decade of rapid widespread change in most of the cultural, diplomatic, economic, political, and strategic structures that have been in place since the end of World War II. Moreover, 9.11 suggests that boundaries, practices, and territories that many thought were almost permanent features of everyday life have broken apart (Agnew, 1998). And, in many instances, there are no clear means of replacement, succession, or reorganization effectively following them at this time. Consequently, the most basic principle for organizing geographical space in the twentieth century – namely, that physical territory, collective identity, and political power are all integrated by, for, and in autonomous nation-states – is now eroding. While their operational unity always was precarious, the alignment of these once stable forces is being disrupted, and new forms of geography are developing on many fronts in ethnic diasporas, informatic networks, religious blocs, terrorist communities, and underground economies (Appadurai, 1996; Barber, 1995).

Consequently, one must rethink how these new structures for organizing space are being reimagined as the world system tumbles into the twenty-first century. Most importantly, the political maps of the world must be re-evaluated at every level – the local, regional, national, and global. Here several questions must be asked: How are geographical borders constructed and dismantled? Why do the old boundaries between activities, ideologies, and peoples no longer hold? Where are territorial struggles succeeding and failing, and what is at stake when states, corporations, nations, and organizations try to rethink the workings of space today? How does this change tranversalize political practices? In what ways are intellectuals and politicians discussing these changes, and what implications do their conceptual, ideological, or moral debates have for the global community? Most importantly, postmodern geopolitical analysis at this time, requires crossing disciplinary boundaries, and including issues covered by cultural studies, communications, economics, international studies, political science, rhetoric, and sociology as well as the subfields of geography.

BIBLIOGRAPHY

Agger, B. 1989. *Fast Capitalism*. Urbana, IL: University of Illinois Press.

Agnew, J. 1994. The territorial trap: the geographical assumptions of international relations theory. *Review of International Political Economy*, 1, 53–80.

Agnew, J. 1998. *Geopolitics: Re-Visioning World Politics*. London: Routledge.
Agnew, J. and Corbridge, S. 1995. *Mastering Space*. London: Routledge.
Anderson, B. 1991. *Imagined Communities*, rev. edn. London: Verso.
Appadurai, A. 1996. *Modernity at Large: Cultural Dimensions of Globalization*. Minneapolis, MN: University of Minnesota Press.
Apter, D. E. 1967. *The Politics of Modernization*. Chicago: University of Chicago Press.
Barber, B. 1995. *Jihad vs. McWorld*. New York: Times Books.
Beck, U. 1992. *The Risk Society: Towards a New Modernity*. London: Sage.
Beck, U. 1997. *The Reinvention of Politics*. Oxford: Polity.
Bourdieu, P. 1998. *Acts of Resistance: Against the Tyranny of the Market*. New York: The New Press.
Bowden, M. 2000. *Black Hawk Down: A Story of Modern War*. New York: Penguin.
Campbell, D. 1992. *Writing Security*. Minneapolis, MN: University of Minnesota Press.
Cooley, J. K. 2000. *Unholy Wars: Afghanistan, America, and International Terrorism*, 2nd edn. Cherndon, VA: Pluto.
Crang, P. and Thrift, N. 1999. *Thinking Space*. London: Routledge.
Davidson, J. D. and Lord Rees-Mogg, W. 1997. *The Sovereign Individual: How to Survive and Thrive During the Collapse of the Welfare State*. New York: Simon & Schuster.
Debrix, F. 1999. *Re-Envisioning Peacekeeping*. Minneapolis, MN: University of Minnesota Press.
Diebert, R. 1997. *Parchment, Printing and Hypermedia: Communication in World Order Transformation*. New York: Columbia University Press.
Doty, R. 1996. *Imperial Encounters*. Minneapolis, MN: University of Minnesota Press.
Foucault, M. 1980. *The History of Sexuality, Vol. I: An Introduction*. New York: Vintage.
Foucault, M. 1991. Governmentality. In G. Burchell et al. (eds.) *The Foucault Effect: Studies in Governmentality*. Chicago: University of Chicago Press, 87–104.
Friedman, T. 1999. *The Lexus and the Olive Tree*. New York: Knopf.
Fukuyama, F. 1992. *The End of History and the Last Man*. New York: Free Press.
Gerecht, R. M. 2001. The counterterrorist myth. *The Atlantic Monthly*, 288(1), 38–42.
Gourevitch, P. 1999. *Tomorrow We Wish to Inform You that We Will be Killed With Our Families: Stories from Rwanda*. New York: Picador.
Greider, W. 1996. *One World, Ready or Not: The Manic Logic of Global Capitalism*. New York: Simon & Schuster.
Griffin, M. 2001. *Reaping the Whirlwind: The Taliban Movement in Afghanistan*. Herndon, VA: Pluto.
Halberstam, D. 2001. *War in a Time of Peace: Bush, Clinton, and the Generals*. New York: Scribner.
Harvey, D. 1989. *The Condition of Postmodernity*. Oxford: Blackwell.
Harvey, D. 1996. *Justice, Nature and the Geography of Difference*. Oxford: Blackwell.
Herod, A., Ó Tuathail, G., and Roberts, S. M. (eds.). 1998. *Unruly World? Globalization, Governance, and Geography*. London: Routledge.
Huntington, S. P. 1998. *The Clash of Civilizations and the Remaking of World Order*. New York: Simon & Schuster.
Jameson, F. 1992. *Postmodernism, or the Cultural Logic of Late Capitalism*. Durham, CT: Duke University Press.
Johnson, C. 1999. In search of a New Cold War. *The Bulletin of the Atomic Scientists*, 55(5), 44–51.
Kaplan, R. D. 1996. *The Ends of the Earth: A Journey at the Dawn of the 21st Century*. New York: Random House.
Kennedy, P. 1992. *Preparing for the Twenty-First Century*. New York: Random House.

Kern, S. 1983. *The Culture of Time and Space: 1880–1918*. Cambridge, MA: Harvard University Press.

Langewiesche, W. 2001. Peace is Hell. *The Atlantic Monthly*, 288(3), 51–80.

Luke, T. W. 1993. Discourses of disintegration, texts of transformation: re-reading realism in the New World Order. *Alternatives*, 18, 229–58.

Luke, T. W. 1996. Governmentality and contra-governmentality: rethinking sovereignty and territoriality after the Cold War. *Political Geography*, 15(6/7), 491–507.

Luke, T. W. 1999. *Capitalism, Democracy, and Ecology: Departing from Marx*. Urbana, IL: University of Illinois Press.

Luke, T. W. and Ó Tuathail, G.. 1997. On videocameralistics: the geopolitics of failed states, the CNN International, and (UN) governmentality. *Review of International Political Economy*, 4, 709–33.

Lyotard, J-F. 1984. *The Postmodern Condition: A Report on Knowledge*. Minneapolis, MN: University of Minnesota Press.

Martin, H-P. and Schumann, H. 1998. *The Global Trap: Globalization & the Assault on Democracy & Prosperity*. London: Zed.

Marx, K. and Engels, F. 1978. The Communist Manifesto. In R. C. Tucker (ed.) *The Marx-Engels Reader*. New York: Norton, 469–500.

Masalha, N. 2000. *Imperial Israel and the Palestinians: The Politics of Expansion*. Herndon, VA: Pluto.

Mayer, J. 2001. The House of Bin Laden. *The New Yorker*, November 12, 54–65.

McNeill, J. R. 2000. *Something New Under the Sun: An Environmental History of the Twentieth-Century World*. New York: Norton.

Mills, C. W. 1959. *The Sociological Imagination*. Oxford: Oxford University Press.

Nye, J. 1990. *Bound to Lead: The Changing Nature of American Power*. New York: Basic Books.

Ó Tuathail, G. 1998. Postmodern Geopolitics: the modern geopolitical imagination and beyond. In G. Ó Tuathail and S. Dalby (eds.) *Rethinking Geopolitics*. London: Routledge, 16–38.

Ó Tuathail, G. 1999. *Critical Geopolitics*. Minneapolis, MN: University of Minnesota Press.

Ó Tuathail, G. and Dalby, S. 1998. *Rethinking Geopolitics*. London: Routledge.

Ohmae, K. 1990. *The Borderless World: Power and Strategy in an Interlocked Economy*. New York: Harper and Row.

Onuf, N. 1989. *World of Our Making: Rules and Rule in Social Theory and International Relations*. Columbia, SC: University of South Carolina Press.

Poggi, G. 1978. *The Development of the Modern State: A Sociological Introduction*. Stanford, CA: Stanford University Press.

Poster, M. 1995. *The Second Media Age*. Cambridge: Polity.

Power, S. 2001. Bystanders to genocide. *The Atlantic Monthly*, 288(2), 84–108.

Rashid, A. 2000. *Taliban: Militant Islam, Oil and Fundamentalism in Central Asia*. New Haven, CT: Yale University Press.

Reich, R. 1991. *The Work of Nations: Preparing Ourselves for 21st Century Capitalism*. New York: Knopf.

Rodrik, D. 1997. *Has Globalization Gone Too Far?* Washington, DC: Institute for International Economics.

Rosenau, J. 1990. *Turbulence in World Politics: A Theory of Change and Continuity*. Princeton, NJ: Princeton University Press.

Taylor, P. 1996. *The Way the Modern World Works*. New York: Wiley.

Thrift, N. 1998. The Rise of Soft Capitalism. In A. Herod et al. (eds.) *Unruly World? Globalization, Governance and Geography*. London: Routledge, 25–71.

Usher, G. 1995. *Palestine in Crisis: The Struggle for Peace and Political Independence after Oslo*. London: Pluto.
Virilio, P. 1995. *The Art of the Motor.* Minneapolis, MN: University of Minnesota Press.
Walker, R. B. J. 1993. *Inside/Outside: International Relations as Political Theory.* Cambridge: Cambridge University Press.

Chapter 16

Anti-Geopolitics

Paul Routledge

Geopolitical knowledge tends to be constructed from positions and locations of political, economic, and cultural power and privilege. Hence, the histories of geopolitics have tended to focus upon the actions of states and their elites, understating rebellion and overemphasizing statemanship. However, the geopolitical policies enacted by states, and the discourses articulated by their policy-makers have rarely gone without some form of contestation by those who have faced various forms of domination, exploitation, and/or subjection which result from such practices. As Foucault has noted, "there are no relations of power without resistances . . . like power, resistance is multiple and can be integrated in global strategies" (1980, p. 142).

Indeed, myriad alternative stories can be recounted which frame history from the perspective of those who have engaged in resistance to the state and the practices of geopolitics. These histories keep alive the memory of people's resistances, and in doing so suggest new definitions of power that are not predicated upon military strength, wealth, command of official ideology, and cultural control (Zinn, 1980). These histories of resistance can be characterized as a "geopolitics from below" emanating from subaltern (i.e. dominated) positions within society that challenge the military, political, economic, and cultural hegemony of the state and its elites. These challenges are counter-hegemonic struggles in that they articulate resistance to the coercive force of the state – in both domestic and foreign policy – as well as withdrawing popular consent to be ruled "from above." They are expressions of what I would term "anti-geopolitics."

Drawing upon Konrad's (1984) notion of antipolitics, anti-geopolitics can be conceived as an ethical, political, and cultural force within civil society – i.e. those institutions and organizations which are neither part of the processes of material production in the economy, nor part of state-funded or state-controlled organizations (e.g. religious institutions, the media, voluntary organizations, educational institutions, and trades unions) – that challenges the notion that the interests of the state's political class are identical to the community's interests. Anti-geopolitics represents an assertion of permanent independence from the state *whomever is in power*, and articulates two interrelated forms of counter-hegemonic struggle. First, it challenges the *material* (economic and military) geopolitical power of states and global institutions; and second, it challenges the *representations* imposed by political and economic elites upon the world and its different peoples, that are deployed to serve their geopolitical interests.

Anti-geopolitics can take myriad forms, from the oppositional discourses of dissident intellectuals to the strategies and tactics of social movements (although the former may frequently be speaking on behalf of the latter). While anti-geopolitical practices are usually located within the political boundaries of a state, with the state frequently being the principal opponent, this is not to suggest that anti-geopolitics is necessarily localized. For example, with the intensity of the processes of globalization, social movements are increasingly operating across regional, national and international scales, integrating resistance into global strategies, as they challenge elite international institutions and global structures of domination.[1]

Colonial and Cold War Anti-Geopolitics

Historically, anti-geopolitics has been articulated against both colonialism and the Cold War. Resistances posed to colonialism took two forms. First, there were challenges to the (mis)representation of other cultures and places as primitive, savage, and uneducated, in need of Western civilization and enlightenment. For example, in the classic text *Orientalism*, Edward Said (1978) shows how such representations were "imaginative geographies" or fictional realities, that shaped the West's perception and experience of other places and cultures. Through an analysis of various texts written by Westerners during colonial times, Said shows how these representations were constructed around essentialist conceptions of (non-Western) *others* that equated difference with inferiority, and served to inform and legitimate geopolitical strategies of control and colonization by the Western countries, as they subjected other territories to military conquest and commerical exploitation. Second, there were material challenges to colonialism, through violent and nonviolent struggle waged by national liberation movements (e.g. in India and Kenya against the British, and in Vietnam and Algeria against the French). Writing about the anti-colonial struggle in Algeria, Fanon (1963, 1965) has argued that decolonization entails both the physical removal of the occupier from one's territory and a decolonization of the mind (see also Ngugi, 1986). This involves opposition to Western ways of representing and organizing the world and the peoples in it – a struggle over who decides and controls how different cultures are interpreted and represented. Moreover, as Fanon (1963) notes, the decolonization process was a global phenomenon, influenced by both other anti-colonial struggles and the geopolitics of the Cold War. Both the US and the USSR attempted to support and control independence movements as part of broader geopolitical strategies against one another.

Concerning the Cold War, there were challenges to the domination by the USA and the Soviet Union of their respective "spheres of influence" – Latin America and Western Europe, in the case of the USA, and Eastern Europe in the case of the Soviet Union – and to their military interventions in the Third World (e.g. US intervention in Vietnam, and Soviet intervention in Afghanistan). Such challenges took intellectual and material forms. A prominent intellectual challenge to US interventions was conducted by a group of scholars known as "dependency theorists," who variously sought to analyse the extent to which the political economy of developing countries was influenced by a global economy dominated by the advanced capitalist countries (e.g. see Amin, 1976; Cardoso and Faletto, 1979; Frank, 1967).

Material challenges took the form of numerous peasant guerrilla movements which emerged throughout Central and Latin America that attempted to challenge authoritarian regimes and allieviate poverty – such as the Cuban Revolution of 1959, and the Nicaraguan Revolution of 1979 (see Armstrong and Shenk, 1982; Dixon and Jonas, 1984; Pearce, 1981) – and national liberation struggles against US intervention such as in Vietnam (e.g. see Chailand, 1977, 1982). These struggles were often supported by anti-war and solidarity movements in the USA and Western Europe.

Despite Soviet military dominance within Eastern Europe, popular uprisings against Soviet occupation and control periodically surfaced within its "satellites" – in the German Democratic Republic in 1953, in Hungary in 1956, in Czechoslovakia in 1968, and in Poland in 1981. Although these expressions of opposition proved unsuccessful, they were indicative of broader counter-hegemonic currents within the Soviet bloc. What first came to the notice of the West as "dissent" – articulated by dissidents such as Andrei Sakharov (in the USSR), and Vaclav Havel (1985) in Czechoslovakia – was symptomatic of the development within the Soviet Union and the Warsaw Pact countries of various independent initiatives that emerged "from below." These sought to extend the space available within society for autonomous action out of the control and discipline of state political culture, articulating a "second culture." Moreover, such dissent set up parallel – and frequently underground – organizational forms that challenged the state's claims to truth and sought to strengthen the development of an independent civil society (e.g. the Czechoslovakia-based human rights group Charter 77).

The dissident movements in Eastern Europe also forged links with what proved to be the largest popular resistance against the Cold War itself, the Peace Movement, which opposed the deployment of Cruise and Pershing missiles in Europe by NATO and SS20s by the Soviet Union. The movement comprised a variety of anti-nuclear and anti-militarist groups, including the Nuclear Freeze in the USA, the Campaign for Nuclear Disarmament (CND) in Britain, and the European Nuclear Disarmament (END) movement. This movement posed both representational and material challenges to the Cold War. Material challenges took the form of a variety of nonviolent direct actions, demonstrations, and peace camps throughout Western Europe and the USA (e.g. see Harford and Hopkins, 1984. McAllister, 1988). Representational challenges articulated a theoretical critique of the Cold War, voicing opposition to the superpower arms race and the division of Europe into ideological and militarized blocs. For example, E.P. Thompson (1985) argued that the expansionist ideologies of the USA and the Soviet Union were the driving force of the Cold War, each legitimated through the threat of a demonized *other* (communism and capitalism respectively). Thompson also argued that in reality the principle threat of the Cold War was not the demonized other, but rather was within each of the superpower blocs, i.e. the peace movements of the Western bloc, and the dissident movements of the Eastern bloc. These movements articulated both material challenges to superpower militarism – through direct action, underground organizations, etc. – and an intellectual challenge to the geopolitical othering that the Cold War was predicated upon. Their calls for international solidarity, rather than antagonism, were seen as a threat to the power of political elites within each bloc to determine geopolitical spheres of influence. Moreover, by attempting to revitalize spaces of public autonomy, these movements challenged each superpower's ability to control public opinion (see Albert and Dellinger, 1983; Smith and Thompson, 1987).

Anti-Geopolitics in the New World Order

With the revolutions of 1989, the demise of the Soviet Union, and the Gulf War of 1990–91, the geopolitical discourse of the USA in particular, and the West in general, has shifted from that of the Cold War to that of the so-called "new world order." The Gulf War provided the rationale for this new discourse, which has geopolitical and geoeconomic dimensions. The geopolitical dimension involves the maintenance of the US national security state, and the legitimation of (continued) US military and economic intervention around the world in order to ensure "freedom" and "democracy."

The geoeconomic dimension of the new world order involves the doctrine of transnational liberalism or, as it is also called, neoliberalism. The fundamental principal of this doctrine is "economic liberty" for the powerful: that is, that an economy must be free from the social and political "impediments," "fetters," and "restrictions" placed upon it by states trying to regulate in the name of the public interest. These "impediments" – which include national economic regulations, social programs, and class compromises (i.e. national bargaining agreements between employers and trade unions, assuming these are allowed) – are considered barriers to the free flow of trade and capital, and the freedom of transnational corporations to exploit labor and the environment in their best interests. Hence, the doctrine argues that national economies should be deregulated (e.g. through the privatization of state enterprises) in order to promote the allocation of resources by "the market" which, in practice, means by the most powerful. As a result of the power of international organizations like the International Monetary Fund (IMF), the World Bank, and the World Trade Organization (WTO) to enforce the doctrine of neoliberalism upon developing states desperately in need of the finances under their control, there has been a drastic reduction in government spending on health, education, welfare, and environmental protection across the world. This has occurred as states strive to reduce inflation and satisify demands to open their markets to transnational corporations and capital inflows from abroad. Transnational liberalism celebrates capital mobility and "fast capitalism," the decentralization of production away from developed states and the centralization of control of the world economy in the hands of transnational corporations and their allies in key government agencies (particularly those of the seven most powerful countries, the G-7), large international banks, and institutions like the World Bank, the IMF and the WTO. As transnational corporations have striven to become "leaner and meaner" in this highly competitive global environment, they have engaged in massive cost-cutting and "downsizing," reducing the costs of wages, health care provisions, and environmental protections in order to make production more competitive.

Transnational liberalism has been institutionalized through various international free trade agreements, such as the North American Free Trade Agreement (NAFTA) between the USA, Canada, and Mexico. These agreements are based upon the doctrine that each country and region should produce goods and services in which they have a competitive advantage, and that barriers to trade between countries (such as tariffs) should be reduced. However, such agreements are more concerned with removing the barriers to the movement of capital, to enable transnational corporations to operate without government interference or regulation, and to exploit the "competitive advantage" in cheap labor, lax environmental regulations, and natural resources.

The resulting global competition for jobs and investment has resulted in the pauperization and marginalization of indigenous peoples, women, peasant farmers, and industrial workers, and a reduction in labor, social, and environmental conditions – what Brecher and Costello (1994) term "the race to the bottom" or "downward levelling." However, such processes have not occurred without challenges by their victims. Anti-geopolitical struggles in the new world order challenge the power of the state, transnational corporations, and global institutions in order to protect and improve people's livelihoods, culture, and environment. A multiplicity of groups including social movements, squatter organizations, neighborhood groups, human rights organizations, women's associations, indigenous rights groups, self-help movements amongst the poor and unemployed, youth groups, educational and health associations and artist's movements are involved in various types of anti-geopolitical struggle. Many of these struggles take place within the realm of civil society, i.e. those areas of society that are neither part of the processes of material production in the economy nor part of state-funded organizations.

The Realms of Anti-Geopolitical Struggle

Social movements articulate anti-geopolitics on a number of interrelated realms within society, including the economic, cultural, political and environmental. Economic struggles may also contain political dimensions, political struggles may also contain cultural elements and so on. Moreover, the responses of state authorities to social movements vary, according to the type of movement resistance, and the character of the government involved. When faced with social movement challenges, governmental responses include repression, cooption, cooperation, and accommodation. Repression can range from harassment and physical beatings, to imprisonment, torture, and the killing of activists.

In the economic realm, social movements articulate conflicts over access to productive natural resources such as forests and water that are under threat of exploitation by states and transnational corporations. The economic demands of social movements are not only concerned with a more equitable distribution of resources between competing groups, but are also involved in the creation of new services such as health and education in rural areas (Guha, 1989). In the cultural realm, social movement identities and solidarities are formed, for example, around issues of class, kinship, neighborhood, and the social networks of everyday life. Movement struggles are frequently cultural struggles over material conditions and needs, and over the practices and meanings of everyday life (Escobar, 1992). In the political realm, social movements challenge the state-centered character of the political process, articulating critiques of neoliberal development ideology and of the role of the state. By articulating concerns of justice and "quality of life," these movements enlarge the conception of politics to include issues of gender, ethnicity, and the autonomy and dignity of diverse individuals and groups (Guha, 1989). In the environmental realm, social movements are involved in struggles to protect local ecological niches – e.g. forests, rivers, and ocean shorelines – from the threats to their environmental integrity through such processes as deforestation (e.g. for logging or cattle grazing purposes) and pollution (e.g. from industrial enterprises). Many of these social movements are also multidimensional, simultaneously addressing, for example, issues of poverty, ecology, gender, and culture. Also, whereas environmental struggles in the developed countries tend to concentrate upon "quality of life" issues, in developing countries, movements have often focused upon access to economic resources. Such groups articulate an "environmentalism of the poor" (Martinez-Allier, 1990), whose fundamental concerns are with the defence of livelihoods and of communal access to resources threatened by commodification, state take-overs, and private appropriation (e.g. by national or transnational corporations), and with emancipation from material want and domination by others.

Two examples of contemporary social movements exemplify these interrelated realms of anti-geopolitical struggle: the *Ejercito Zapatista Liberacion National* – the EZLN or the Zapatistas and the *Narmada Bachao Andolan* (Save the Narmada Movement, NBA). The Zapatistas in Chiapas, Mexico, have articulated resistance to the NAFTA and the Mexican state. The Zapatistas, a predominantly indigenous (Mayan) guerrilla movement, have emerged in Chiapas due to several factors. First, the state of Chiapas is rich in petroleum and lumber resources that have been ruthlessly exploited causing deforestation and pollution. Second, the increasing orientation of capital-intensive agriculture for the international market has led to the creation of a class of elite wealthy farmers, and forced Indian communities to become peasant labor for the extraction and exploitation of resources, the wealth of which accrues to others. In addition, large landowners and ranchers control private armies who are used to force peasants off their land, and to terrorize those with the temerity to resist. Third, although it is resource-rich, Chiapas is amongst the poorest states in Mexico with 30 per cent of the population illiterate, and 75 per cent of the population malnourished. Fourth, the production of two of the main crops from which *campesinos* (peasants) earn a living in Chiapas – coffee and corn – have undergone severe economic problems in recent years, and will be further damaged by NAFTA. Finally, government reforms in 1991 enabled previously protected individual and communal peasant land-holdings to be put up for sale to powerful cattle ranching, logging, mining, and petroleum interests.

The Zapatistas initially engaged in a guerrilla insurgency by occupying the capital of Chiapas and several other prominent towns in the state. However, they staged their uprising in a spectacular manner to ensure maximum media coverage, and thus gain the attention of a variety of audiences, including civil society, the state, the national and international media, and international finance markets. For, although their guerrilla bases were in the Lacandon jungle in Chiapas, the Zapatistas were particularly concerned to globalize their resistance. The appearance of an armed insurgency, at a moment when the Mexican economy was entering into a free trade agreement, enabled the Zapatistas to attract national and international media attention. Through their spokesperson, Subcommandante Marcos, the Zapatistas engaged in a "war of words" with the Mexican government, fought primarily with rebel communiques (via newspapers and the Internet), rather than bullets. Through their guerrilla insurgency and this war of words the Zapatistas have attempted to raise awareness concerning the unequal distribution of land, and economic and political power in Chiapas; challenge the neoliberal economic policies of the Mexican government; articulate an indigenous worldview which promotes Indian political autonomy; and articulate a call for the democratization of civil society. They have also been able to forge an international solidarity network of groups and organizations. They have thus posed both material and representational challenges. The success of the Zapatista struggle has lain in its ability, with limited resources and personnel, to disrupt international financial markets, and their investments within Mexico, while exposing the inequities on which development and transnational liberalism are predicated (Harvey, 1995; Ross, 1995).

However, despite certain successes, the movement has been faced with repression from the Mexican government. Over 15,000 army personnel have been deployed in Chiapas; villages suspected of being sympathetic to the Zapatistas have been bombed; and peasants suspected of being Zapatistas have been arrested and tortured. At present an uneasy cease-fire is in place between the Zapatistas and the government and peace-talks between them are stalled. Since its emergence in 1994, the Zapatistas have attempted to pose a political challenge to the Mexican state. In their demands for equitable distribution of land, their calls for indigenous rights and ecological preservation (i.e. an end to logging, a program of reforestation, an end to water contamination of the jungle, preservation of remaining virgin forest), they also articulate an economic, ecological, and cultural struggle.

Another example of anti-geopolitical struggle is that of the resistance against the Narmada river valley project in India. This river, which is regarded as sacred by the Hindu and tribal populations of India, spans the states of Madhya Pradesh, Maharashtra, and Gujarat, and provides water resources for thousands of communities. The project envisages the construction of 30 major dams along the Narmada and its tributaries, as well as an additional 135 medium-sized and 3,000 minor dams. When completed, the project is expected to flood 33,947 acres of forest land, and submerge an estimated 248 towns and villages (Sangvai, 2000). According to independent estimates, up to fifteen million people will be affected by the project – either by being forcibly evicted from their homes and lands as they are submerged, or by having their livelihoods seriously damaged (Roy, 1999). With six of the dams already built, opposition to the project has been focused on the Sardar Sarovar and Maheshwar dams – the former funded by the World Bank until 1993, when, after an independent review, the Bank withdrew its funding, and the latter partly funded by transnational corporations such as Siemens and Asea Brown Boveri. The resistance to the project has been coordinated by the Narmada Bachao Andolan (NBA) – a network of groups, organizations, and individuals from various parts of India who have demanded the curtailment of the scheme.

The movement's repertoire of protest has included material struggles such as mass demonstrations, road blockades, fasts, public meetings, and disruption of construction activities. While localized protests have occurred along the entire Narmada valley, wider public attention has been drawn to spectacular events such as mass rallies and protest marches. Moreover, representational challenges have also been posed, such as the critical analysis of the impacts of large dams upon local economy and ecology conducted by the NBA and allied non-government organizations (NGOs), and an actively articulated discourse of sustainable irrigation and development alternatives by the movement. While the movement has been almost completely nonviolent, its leaders and participants have been harassed, assaulted, and jailed by police. In one tragic event, police opened fire on a demonstration in the Dhule district of Maharashtra, killing a 15-year-old boy. The movement has attracted widespread national support – through participating in the National Alliance of People's Movements (a coalition of different social movements in India) – and global support from various environmental groups and NGOs such as the London-based Survival International and the International Narmada Campaign. However, despite the resistance, construction of the dams continues. In representing a threat to the ecology of the area surrounding the Narmada river, the construction of the dams also threatens the economic survival of the tribal and peasant peoples who will be evicted from their homes and lands – from which they earn their livelihoods – when the land is submerged. Moreover, these inhabitants have a profound religious connection to the landscape around the Narmada river. This spiritual connection to place – which eviction threatens to sever – intimately informs their customs and practices of everyday life. Hence, opposition to the dam also articulates the inhabitants desire for cultural survival. In addition, many of the villages that border the Narmada are demanding a level of regional autonomy, seeking "our rule in our villages," thereby articulating political demands as well (Gadgil and Guha, 1995).

Globalizing Anti-Geopolitics

The use of the Internet to organize and inform resistance has enabled activists and place-based social movements such as the NBA and the Zapatistas to effect a strategic mobility in their struggles enabling the organizing and coordination of resistance across national and international spaces. This is because such resistances are frequently responses to local conditions that are in part the product of global forces. In contrast to official political discourse about the global economy, these challenges articulate a globalizing anti-geopolitics: an evolving international network of groups, organizations, and social movements.

The recent creation of the WTO and its use by corporations has hastened the extension of previously local struggles to the international level. While the WTO is serving to increase the centralization of global economic policy-making, it has also provided a central object of protest. Local conflicts between citizens, governments, and transnational institutions and corporations have begun to globalize as a result of the increased uniformity of policies and international agreements among governments to implement global sets of rules. This has resulted in the perception of common interests amongst resistance formations to challenge these rules. In addition, the common ground that has begun to be articulated against neoliberalism and its agents, has manifested itself in myriad local points of protest (Cleaver, 1999). Ironically, such a globalizing anti-geopolitics has been facilitated by the discourses of globalization, since it has allowed resistances to engage in critical analysis of the present in which no theoretical or political privilege is given, *a priori*, to experience or analysis of any social group or actor about their vision of the future. In addition, globalization has enabled the forging of new political alliances, as different social movements representing different terrains of struggle experience the negative consequences of neoliberalism (Wallgren, 1998).

Writing in a recent issue of *Race and Class*, Robinson (1998) has argued that effective anti-geopolitical struggle requires: (i) a political force and a broader vision of social transformation that can link different place-based social movements; (ii) the creation of viable socioeconomic alternatives to neoliberalism which can emerge out of ongoing political, economic, environmental and cultural struggles; and (iii) the need for social movements to transnationalize their struggles. Such recommendations are vague, and raise serious questions about whose vision of social transformation is to be mobilized. This is because resistances attempt to privilege the powers of everyday existence over incursion and exploitation by states, and national and transnational corporations. As resistance globalizes, attempting to organize and coalesce across space, there is a simultaneous attachment of particular resistances (e.g. in the form of social movements) to particular places – whereby geographies of difference are articulated and asserted – and a critical engagement with the global, through both the targeting of opponents (e.g. the World Bank) and through emerging globalizing networks and alliances. Whereas globalism tends to involve "transmogrifying places into homogenous space ready to be quantified and commodified" (Apffel-Marglin and Parajuli, 1998, p. 17), globalizing resistance articulates myriad claims to difference within its changing networks and places of struggle.

The recent emergence of People's Global Action (PGA) provides a fascinating example of such a process. The PGA owes its genesis to an encounter between international activists and intellectuals that was organized by the Zapatistas in Chiapas in 1996. In a meeting in Spain the following year, that sought to build upon the Zapatista encounter, the idea of a network between different resistance formations was launched by ten social movements including *Movimento Sem Terra* (Landless Peasants Movement) of Brazil, and the Karnataka State Farmer's Union of India. The official "birth" of the PGA was in February 1998; its stated goal was to facilitate the sharing of information between grassroots social movements without the mediation of established NGOs. At the 1998 Ministerial Conference of GATT/WTO in Geneva, an alternative conference of groups from Asia, Africa, and Latin America was held under the PGA banner and convened by such groups as *Movimento Sem Terra* (Brazil), Karnataka State Farmer's Association (India), Movement for the Survival of the Ogoni People (Nigeria), the Peasant Movement (Philippines), the *Central Sandinista de Trabajadores* (Nicaragua) and the Indigenous Women's Network (North America and the Pacific). From this meeting, a manifesto was established which called for direct confrontation with transnational corporations and an end to globalization.

However, the PGA is not an organization. Rather, it represents a *convergence space* of social movements, resistance groups, and individuals from across the world. Its main objectives are: (i) inspiring the greatest number of persons, movements, and organizations to act against corporate domination through nonviolent civil disobedience and people-oriented constructive actions; (ii) offering an instrument for coordination and mutual support at the global level for those resisting corporate rule and the capitalistic development paradigm; and (iii) giving more international projection to the struggles against economic liberalization and global capitalism.[2] It has been the PGA network that has put out the calls for the recent global days of action against capitalism, such as the event actions of June 18, 1999 (when protests occurred in 100 cities in 40 countries across the planet), and the protests against the WTO in Seattle in November 1999.

Globalizing anti-geopolitical resistance is all about creating networks: of communication, solidarity, information sharing, and mutual support. The core function of networks is the production, exchange, and strategic use of information. The speed, density, and complexity of such international linkages has grown dramatically in the past twenty years. Cheaper air travel, and new electronic communication technologies have speeded up information flows and simplified personal contact among activists (Keck and Sikkink, 1998). Indeed, information-age activism is creating what Cleaver (1999, p. 3) terms a "global electronic fabric of struggle" whereby local and national movements are consciously seeking ways to make their efforts complement those of other organized struggles around similar issues. Certainly, the use of telecommunications has the potential to alter the balance of power in social struggles. This is in part effected by the refusal of social movements to accept the boundaries of communication taken for granted by established systems of domination (e.g. states). Through their use of media vectors social movements can escape the social confines of territorial space, upon which much of the legitimacy of the state is predicated. Indeed, the globalization of communications provides new opportunities for decentralized political practices as many social movements increasingly locate their strategies within local and translocal spaces as well as national and transnational spaces (Adams, 1996, p. 419).

Moreover, it is usually in the interests of governments to restrict the bounds of a conflict in order to effect containment (if not total control) of events, whereas social movements frequently wish to publicize their struggles in order to attract the attention of as wide an audience as possible to their aims and grievances. The uses of various media by social movements can be construed as "going globile" (Routledge, 1998): effecting strategic mobility within the increasingly global space of contemporary media. As Wark (1994) has argued, the occupation of time in the information network is an important aspect of contemporary struggle, but the occupation of space in the symbolic and physical landscape remains an important means to that end.

Different issues get framed to target a variety of audiences (e.g. international media, states, civil societies), to attract attention, and encourage action. Such networks, greatly facilitated by the Internet, enable fluid and open relationships that are more flexible than traditional hierarchies. They have the ability to generate information quickly and deploy it effectively. Indeed this is central to their identity. Participation in networks has become an essential component of collective identities of the activists involved, networking forming part of their common repertoire. Information enhances the resources available to domestic actors in social/political struggles and leads to action. Many information exchanges are informal such as by telephone, e-mail, fax, and the circulation of newsletters, pamphlets, and bulletins through a variety of means including by hand, by post, and via the Internet. They provide information not otherwise available, from sources that might not otherwise be heard, and they make this information comprehensible and useful to activists and publics who might be geographically and/or socially distant (Keck and Sikkink, 1998).

Through the use of informational and symbolic politics, social movements are forging alliances through globalizing networks. The Internet and e-mail are being used to obtain and share information and also to serve as an important means of organizing resistance initiatives. For example, the June 18, 1999 global day of action was organized "globally" over electronic vectors by the PGA, via its website and through various activist electronic mailing lists. Such information included the date of the action, its purpose, and the call for mobilization. However, the actual material mobilizations "on the ground" were organized in particular (primarily urban) localities. The circulation of information has the ability to by-pass the elite-controlled mass media, and also involves the circulation of interpretation and evaluation. Hence virtual conferences and mailing lists provide forums for discussion and debate of actions, tactics and strategies. Indeed certain tactics known in other areas have been adapted to the electronic environment, such as protest letter-writing campaigns, graffitti and billboard art modifications on the World Wide Web, and "hactivism" (computer hacker activism) (Cleaver, 1999).

What characterizes convergence spaces such as the PGA is a fragmented geography, one that is heterogeneous, fluid, and discontinuous, where the virtual geography of the Internet and other media vectors becomes entangled with the materiality of place, local knowledge, and concrete action. It is comprised of myriad grounded material struggles in particular places as well as a globalizing network of alliances that are attempting to share information, support one another and coordinate various struggles. Hence, both the Zapatistas and the Narmada Bachao Andolan engage with the convergence space of the PGA network.[3]

Some of the globalizing forms of anti-geopolitical struggle may be characterized as (i) *globalized local actions*, which are political initiatives which take place either at the same or different times, in different locations across the globe, in support of a particular localized struggles (such as the various solidarity actions that have taken place around the world in support of the Zapatistas and the Narmada dam struggles) or against particular targets (such as the global day of action on June 18, 1999) and (ii) *localized global actions* whereby different social movements and resistance groups coordinate around a particular issue or event in a particular place, such as the recent global days of action against the WTO in Seattle. Although only in their nascent stage, such globalizing anti-geopolitics represent potentially potent material and representational challenges to the exploitation and domination of the new world order.

ENDNOTES

1. Of course, solidarities across borders are not a new phenomenon. During the nineteenth century international alliances were established in the antislavery movement and the Internationals of the Communist and Anarchist parties, while in the twentieth century, internationalism has been present in the campaign for women's suffrage, the International brigades (in Spain in the 1930s, Cuba since the 1960s and Nicaragua in the 1980s), and in

the anti-nuclear movement. What is different in present international alliances is the means, speed, and intensity of communication between the various groups involved.

2. The specific hallmarks of the PGA are:
   (i) A very clear rejection of the WTO and other trade liberalization agreements (like APEC, the EU, NAFTA, etc.) as active promoters of a socially and environmentally destructive globalization.
   (ii) A rejection of all forms and systems of domination and discrimination including, but not limited to, patriarchy, racism, and religious fundamentalism of all creeds. PGA embraces the full dignity of all human beings.
   (iii) A confrontational attitude, since PGA does not think that lobbying can have a major impact in such biased and undemocratic organizations, in which transnational capital is the only real policy-maker.
   (iv) A call to nonviolent civil disobedience and the construction of local alternatives by local people, as answers to the action of governments and corporations.
   (v) An organizational philosophy based on decentralization and autonomy.
3. For further information on particular resistances mentioned in this paper, see the following websites:
   Zapatistas: www.eco.utexas.edu/faculty/Cleaver/zapsincyber.html
   Narmada Bachao Andolan: www.Narmada.org
   People's Global Action: www.agp.org

BIBLIOGRAPHY

Adams, P. C. 1996. Protest and the scale politics of telecommunications. *Political Geography*, 15(5), 419–41.

Albert, M. and Dellinger, D. (eds.). 1983. *Beyond Survival: New Directions for the Disarmament Movement*. Boston: South End Press.

Amin, S. 1976. *Unequal Development: An Essay on the Social Formations of Peripheral Capitalism*. New York: Monthly Review Press.

Appfel-Marglin, F. and Parajuli, P. 1998. Geographies of difference and the resilience of ecological ethnicities: "Place" and the global motion of capital. *Development*, 41, 14–21.

Armstrong, R. and Shenk, J. 1982. *El Salvador: The Face of Revolution*. Boston: South End Press.

Brecher, J. and Costello, T. 1994. *Global Village or Global Pillage*. Boston: South End Press.

Cardoso, F. H. and Faletto, E. 1979. *Dependency and Development*. Berkeley, CA: University of California Press.

Chailand, G. 1977. *Revolution in the Third World*. London: Penguin.

Chailand, G. 1982. *Guerrilla Strategies*. London: Penguin.

Cleaver, H. 1999. Computer-linked social movements and the global threat to capitalism. Accessed from polnet.html at www.eco.utexas.edu.

Dixon, M. and Jonas, S. (eds.). 1984. *Nicaragua Under Siege*. San Francisco: Synthesis.

Escobar, A. 1992. Culture, Economics, and Politics in Latin American Social Movements: theory and research. In A. Escobar and S. E. Alvarez (eds.) *The Making of Social Movements in Latin America*. Boulder, CO: Westview, 62–85.

Fanon, F. 1963. *The Wretched of the Earth*. New York: Grove.

Fanon, F. 1965. *A Dying Colonialism*. New York: Grove.

Foucault, M. 1980. *Power/Knowledge*. New York: Pantheon.

Frank, A. G. 1967. *Capitalism and Underdevelopment in Latin America*. New York: Monthly Review Press.

Gadgil, M. and Guha, R. 1995. *Ecology and Equity.* London: Routledge.
Guha, R. 1989. The Problem. *Seminar,* March, 12–15.
Harford, B. and Hopkins, S. 1984. *Greenham Common: Women at the Wire.* London: The Women's Press.
Harvey, N. 1995. Rebellion in Chiapas: rural reforms and popular struggle. *Third World Quarterly,* 16(1), 39–72.
Havel, V. 1985. *The Power of the Powerless.* Armink, NY: ME Sharpe.
Keck, M. E. and Sikkink, K. 1998. *Activists Beyond Borders.* Ithaca, NY: Cornell University Press.
Konrad, G. 1984. *Antipolitics.* New York: Henry Holt
Martinez-Allier, J. 1990. Ecology and the poor: a neglected dimension of Latin American history. *Journal of Latin American Studies,* 23, 621–39.
McAllister, P. 1988. *You Can't Kill The Spirit.* Santa Cruz, CA: New Society.
Ngugi, wa Thiong'o. 1986. *Decolonising the Mind.* London: Heinemann.
Pearce, J. 1981. *Under the Eagle.* London: Latin American Bureau.
Robinson, W. I. 1998. Latin America and global capitalism. *Race and Class,* 40(2/3), 111–32.
Ross, J. 1995. *Rebellion from the Roots.* Monroe, ME: Common Courage Press.
Routledge, P. 1998. Going Globile: Spatiality, embodiment and mediation in the Zapatista insurgency. In S. Dalby, and G. Ó Tuathail (eds.). *Rethinking Geopolitics.* London. Routledge, pp. 240–60.
Roy, A. 1999 *The Greater Common Good.* Bombay: India Book Distributors.
Said, E. 1978. *Orientalism.* New York: Vintage.
Sangvai, S. 2000. *The River and Life.* Mumbai: Earthcare.
Smith, D. and Thompson, E. P. (eds.). 1987. *Prospectus for a Habitable Planet.* London: Penguin.
Thompson, E. P. 1985. *The Heavy Dancers.* New York: Pantheon Books.
Wallgren, T. 1998. Political semantics of "globalization": a brief note. *Development,* 41(2), 30–2.
Wark, M. 1994. *Virtual Geography.* Bloomington. IN: Indiana University Press
Zinn, H. 1980. *A People's History of the U.S.* New York: Harper & Row.

Part IV States, Territory, and Identity

Chapter 17

After Empire: Identities and Territorialities in the Post-Soviet Space*

Vladimir Kolossov

Introduction

There are a lot of theoretical models explaining the outburst of nationalism in the post-Soviet space since the collapse of the Soviet Union. Some authors stress that nationalism in Eastern Europe is in principle different from the relatively liberal and "inclusive" nationalist traditions of Western Europe (Anderson, 1996), where the membership of an individual in a nation became a function of his civil behavior, and of countries of immigration, where tolerance towards an ethnic "other" has been a natural feature of state-building. It is argued that ethnic nationalism in Eastern Europe was based on traditions of local life in communities, which meant that group membership was grounded in feelings of kinship, collectivism and solidarity, and this prevented the formation of a civil nationalism (Greenfeld, 1993).

This hypothesis is not confirmed by empirical studies, especially in Russia, which show that many citizens identify not only with their ethnic group, but also the state and the region in which they live, even during the period of painful transition, and that self-identification is influenced to an increasing extent by concerns about civil rights and political freedom (Chernysh, 1995). There is evidence that the frequency and the acuity of ethnic conflicts, or their likelihood, do not depend strictly on the ethnic, social and demographic structure of populations. For instance; the ethnic compositions of Latvia, Kirghizstan, and Kazakhstan are similar, but the processes of nation- and state-building and the fate of national minorities in these newly independent countries differ markedly. Areas with the most worrying ethnic situation do not coincide with the border zones between "civilizations" (i.e. between the ethnic groups which are most culturally dissimilar), as predicted by Huntington (1996). Nor do they coincide with spaces shared by different peoples. Obviously, this issue depends on the complex interplay of ethnic, religious, political, regional,

*An earlier version of this chapter first appeared in *GeoJournal*, 48, 7141, 1999. The present version appears with the kind permission of Kluwer Academic Publishers, Netherlands.

and other identities. The assessment of ethnopolitical risks cannot be based on only one group of factors. "Objective," material factors, such as economic gaps between neighboring areas, need to be combined with "subjective" trends, especially the evolution of identity as the main attribute of ethnicity.

According to the view on ethnicity which is dominant since the appearance of works by Barth (1969), communities called "nations" are social constructs which appear as a result of the purposeful efforts of political elites and the political institutions that they create, especially the state. These communities are based on representations about belonging to a community, or on an identity, and on the ideas of solidarity shared by its members. Ethnic identity is a matter of choice. Its limits, formed on the basis of selected cultural characteristics, are fluid and change over time, depending upon historical (and geographical) situations. Ethnic identity becomes more salient or less salient for members of a community as circumstances change (Hobsbawm and Ranger, 1983; Thompson, 1989; Tishkov, 1997). It is only one of several possible focuses for self-identification, such as sex, age, education and social status, etc. According to numerous representative polls in different regions of the Russian Federation, ethnic identity is usually only the fourth or even the fifth most important among all human identities: each individual identifies primarily with the people of the same sex, then with their social groups. If ethnic identity becomes more important, it clearly indicates that the situation in the region or in the community in question is not normal (Drobizheva et al., 1996). Identities are multiple and "negotiable," and the same individual or the same group may privilege one identity over another according to the situation and the moment (Burke, 1996). *Ethnic identity* in this chapter is defined in self-ascriptive terms: it is the self-identification of an individual with the *ethnic group*(s) that he or she believes they belong to. *National identity*, or state-political identity in this chapter refers to the self-identification of an individual with their *state* of origin and/or of residence.

In the post-Soviet space, identities are changing rapidly. Newly independent countries have to struggle for the loyalty of their citizens as they face the problem of state- and nation-building, i.e. they have to forge or strengthen both *national and ethnic identities*. They inherited borders from the Soviet Union that were usually drawn more or less arbitrarily under Stalin and often do not correspond with ethnic, linguistic and cultural borders. Many ethnic groups shared the same territory for centuries. Moreover, the ethnic heterogeneity of the post-Soviet space increased dramatically in the Soviet years because of the industrialization of peripheral areas, which involved the import of labor, mainly Slav and particularly Russian, so that the major cities in all the republics came to have a higher percentage of Russians than other regions (Kolossov, 1993; Kolossov et al., 1992). The ratio of "Europeans" in the Asian part of the former Soviet Union increased during the last century, as did the share of "Asian" peoples in the population of almost all regions in the European part (Kolossov, 1993). Most of the countries that emerged from the collapse of the Soviet Empire have large Russian minorities.

The political elites in these newly independent countries quickly appreciated the advantages of independence and are now establishing social structures aimed at perpetuating cultural distinctions between "titular" groups (those nominally recognized as belonging to a certain region or state, even if they are not the majority) and "others" (especially Russians), and between citizens of the new states and the inhabitants of neighboring countries. In heterogeneous post-Soviet societies, this increases inequalities and the competition among ethnic groups for jobs, incomes, education, and other resources. The rise of nationalism amongst "titular" ethnic groups and the counter-reaction of the minorities in the new independent countries (as well as in "autonomous" regions in the Russian Federation), has often led to latent and open ethnic conflict. Ted Gurr argues that the greater the salience of ethnic identities, the greater the likelihood of open conflict, and the longer open conflict persists, and the more intense it becomes, the stronger and more exclusive do these group identities become (Gurr, 1994, p. 350).

The objective of this chapter is two-fold. First it examines the complex landscape of ethnic and political identities in post-Soviet space and secondly analyses those factors influencing the evolution of identity regimes amidst nation- and state-building projects in post-Soviet space. Throughout I will pay particular attention to the Ukraine, a country where I have conducted extensive research. In the first section I consider the post-Soviet changes in ethnic and political identities; the second section considers the role of the different means used by the political elites and the state to transform these identities.

Multiple, Nested and Shifting Ethnic and State-Political Identities in Post-Soviet Space

An attachment to territory is as old as human society, and there is little to suggest that the powerful ideological bonds that link identity, politics, and territory will be loosened in the future (Brubaker, 1992; Murphy, 1996, p. 109). It is already well demonstrated that the relationship between identity and territory is becoming more complex (Newman and Paasi, 1998; Paasi, 1996) and that the "deterritorialization" of the state leads to the creation of multilayered and mixed identities. Pirie (1996) distinguished at least four types of ethnic self-identification: (i) strong identification with only one ethnic group; (ii) strong, stable identification with two groups simultaneously; (iii) marginal identification – weak or unstable identification with two or more ethnic groups, and vacillation between them. This may lead to a complete rejection of ethnic identity, resulting in "ethnic nihilism" and (iv) "pan-ethnic" identification – strong identification with a group that encompasses several ethnic groups (for example, East Slav, or Pan-Russian, or the former "Soviet people"). Moreover, national identity, though still occupying the central place in the hierarchy of human territorialities, is gradually losing its hegemony. In the contemporary interrelated and interdependent world, more and more individuals have a mixed ethnic background, move between regions and countries as a result of urbanization and globalization, or are forced to leave their homes because of civil wars, ethnic conflicts or environmental disasters (Kolossov and O'Loughlin, 1998).

Mixed, blurred and "hierarchical" identities

"Hierarchical" multiple identities are common in many parts of the post-Soviet space, reflecting the multiethnic nature of the Russian state since the sixteenth century. The vast territories incorporated into the Russian Empire were colonized by ethnic Russians over centuries. In the Soviet era, ethnic heterogeneity in many republics and regions was so high that the share of mixed marriages reached substantial proportions. For instance, in the 1980s in Kazakhstan each fifth couple was ethnically mixed, and in Donbass (East Ukraine) 55% of all children were born from mixed marriages. Not surprisingly, the role of the territorial factor was clearly salient, as the content of the ethnic "cocktail" varied strongly between towns and the countryside, urbanized and rural areas, transitive cultural zones and "internal" areas with more homogeneous population. Finally, many territorially autonomous regions across Soviet space included smaller national political–territorial units or specially designated ethnic areas. Specific regional territorial identities developed according to local conditions.

The constantly changing hierarchy of territorial identities consequently consisted of ethnic, state-political ("all-Union" and Republics), regional, and local components. It was related to the mobility of the borders of main territorial units, which have become relatively stable only since the late 1950s. The concept of "matrioshka" nationalism – from the Russian dolls that are hidden inside each other – describes adequately the post-1989 political developments in former Eastern Europe (Taras, 1993). For example, in East Ukraine political geographers identified up to six identities – Soviet, Ukrainian national, Ukrainian ethnic, Russian national and ethnic, and then regional and local ones layered within each like the dolls (Holdar, 1995; Pirie, 1996).

Overlapping of political and national identities

This extremely complicated hierarchy of territorial identities logically results in the coincidence over time and space of the processes of state- and nation-building. Indeed, in the new countries the state is not fully established yet, because:

- a large part of the population, especially in particular regions, does not identify yet with the political nation-state, which presupposes a consensus about the values of common citizenship between all social, ethnic, and regional groups;
- as societies are deeply split between "winners" and "losers" in the economic reforms (Kitschelt, 1995), the state has not achieved full legitimacy, i.e. the recognition by all the citizens of the authority of the central power. To be truly legitimate, a state needs recognition not only by the majority of the titular ethnic group, but also by the majority of each of its minorities;
- political participation is not satisfactory, as most often it does not ensure territorial integrity, national unity and political stability;
- political distribution, i.e. equal accessibility to material resources, values and privileges, is still far from being recognized as fair by the main political actors. For instance, regions do not all have the same status and may not agree with the existing system of financial redistribution (though, of course, a system which would completely satisfy each region can hardly be created);
- the state has not succeeded in penetrating all ethnic, territorial and social segments of the society to the same degree: the existence of self-proclaimed republics in Russia (Chechnia before 1999), Moldova (Transdnestria), Georgia (Abkhazia and South Osetia), and Azerbaijan (Nagorno–Karabakh) provides evidence of this.

Ethnic identities are often unstable as well, because different regional groups of the titular and/or dominant ethnic group do not yet share common representations about themselves, their past and future, their relations with ethnic "others" and with neighbors. In early 1996, 31% of the adult population of Ukraine would have liked their country to reunite with Russia (Khmelko, 1996). A survey in 1991 showed that only 55% of the 977,000 who were classed in the 1989 census as local "Ukrainians" in Transcarpathia thought of themselves as such, whereas 27% considered themselves to be Rusyn or other ethnic groups. In 1992, they established their own political party, the Subcarparthian Republican Party, which has demanded cultural and even political autonomy for the region. Kiev has granted the local ex-communist elite considerable economic freedom to discourage it from making a common cause with the Rusyns (Wilson, 1997). In Russia, seven years after the dismantling of the Soviet Union, 12.4% of the population still identified with the whole of the former Soviet territory, and almost a quarter of the population still felt embarrassed and have an ambiguous idea about their territoriality and state affiliation (table 17. 1). In other words, more than 40% of the Russian population was not able or did not wish to identify with the country in which they were living. At the same time, in 1997, only 34% of the Russian population believed that the Russian Federation should remain an independent state within her contemporary borders, without reuniting with another country.

As a result of the complicated history (of differing dates and circumstances of incorporation into Russian Empire/Soviet Union, remainders of tribal and clan ties etc.), strong ethnic identities have yet to be created among titular peoples, *except in the Baltic states*. The state has to, at the same time, "glue" together the titular people and the nation consisting of many ethnic groups. Thus, it has to identify its priorities and strategy and to decide whether it will consider as its main support the nation as whole, or the titular people as a "core group." Long and intense conflicts, between political elites who attempted to reform their countries along foreign lines and those who fought to preserve their traditional local cultures from modernization, are typical for many states. In this case the conflict between center and periphery and the conflict between dominant and subordinate classes virtually merged into one (Burke, 1996; Burns, 1980).

Each strategy has its advantages and its shortcomings. The exclusive option, meaning the choice in favor of the nineteenth-century concept of the nation-state, makes the political mobilization of the core group easier and allows the titular elite to redistribute and to retain property and power in its own hands, but it runs the risk of exacerbated nationalism, provoking a counter-reaction from minorities and condemnation by the international community. The alternative, contemporary, inclusive option is often perceived by the ruling elites as threatening the integrity of the state territory and the real independence of the young state (Chinn and Kaiser, 1996). In

Table 17.1 Self-identification of Russian citizens in 1997

What country do you feel a citizen of?	1993	1995	1997
Russia	45.6	53.1	58.2
USSR	12.7	15.5	12.4
Citizen of the world	8.8	7.3	5.1
Don't know	32.9	24.1	24.3
Total	100	100	100

Source: Gorshkov, 1997.

most cases, the political elites cannot unambiguously select either of these options and so they have to try to combine both of them. They declare their adherence to the concept of the political nation and, in necessary and unavoidable cases, even take real steps and make concessions in order to build it. But, in practice, they never hesitate to insist on the privileges of the titular nations and their tongues and unremittingly work on diffusing and enrooting of old and new national myths in order to impose them on all populations.

Quite clearly, under these conditions, the concept of double citizenship cannot be adopted because it contradicts the basic postulates of state-building: it is assumed that each individual can select only one motherland. Only Turkmenistan, Tajikistan and the Russian Federation, preoccupied (rhetorically at least) with the rights of the large Russian minorities in neighboring countries, permits its citizens to have another citizenship. But in areas with especially complicated identities, there are lots of people without any definite identity at all or without citizenship, as in Latvia and Estonia. As often happens in such transitive zones (Sahlins, 1989), the people in such areas manipulate their identities/passports. In the self-proclaimed Transdnestrian Moldovan Republic (TMR), whose citizenship is not recognized, many people illegally have two or three passports: the Transdnestrian, the Russian and the Moldovian ones. In Latvia and Estonia, as well as in TMR, a considerable percentage of the permanent residents have Russian citizenship.

With the partial exceptions of Turkmenistan, and probably of Lithuania, all post-Soviet states are consequently experiencing a crisis of identity, which can be defined as a period when ethnic or other region-specific subnational segments of a society create obstacles to national unification and the identification with a certain political community. As a result, a considerable part of the population does not recognize the boundaries of the territorial state as a legitimate political unit. This crisis of identity is due not only to the multiethnic character of all the successor states of the Soviet Union, but also to the variety and the heterogeneity of the ethnic identity of the titular peoples.

The continuing existence and even strengthening of supranational identities

As different parts of the territory, with a few exceptions, coexisted within the same state for at least a century, and more often for a much longer time, numerous "transnational," "transbordal" or "supranational" identities appeared and continue to exist in the post-Soviet space. There is a lot of evidence that the "Soviet" identity continues to exist, although in most cases it has no ideological character, and most of its bearers would not like a return to the Soviet, or to a communist power. The territorial identity of these people embraces the whole or most of the Soviet territory. According to the polls conducted by the All-Union (now All-Russian) Center of Public Opinion Studies (YTsIOM) in 1989, 30% of Russians perceived themselves as "Soviet." In the cosmopolitan capital cities, Moscow and Leningrad (Petersburg), the share of "Soviet people" among ethnic Russians was even greater at 38%. Today the spread of the "Soviet" identity significantly varies over the territory of Russia. Research in June 1998 by myself and others in the city of Stavropol (in southern Russia, part of the North Caucasus region) and a surrounding multiethnic district showed that about 25% of the respondents identified themselves as "Soviet people," especially those who belonged to the small "nonterritorial" minorities in these areas, those who were born in other former Union republics, and those who recently moved into the area, independent of their ethnic background (Kolossov et al., 2001). In other regions, for instance, in the Urals, the share of "Soviet" people does not exceed 10%. In the city of Moscow, by far the richest subject of Russian Federation, only 6% of people thought of themselves firstly as Soviet people.[1] It is a common phenomenon in all the new independent states, but, of course, especially in Russia and amongst Russians who formerly felt comfortable everywhere (although with time to a lesser and lesser extent, as indicated by the flow of Russian migrants from many republics, which began as far back as the 1970s or even the late 1960s).

The group with a "Soviet" identity in many republics includes mostly Russian speakers of different ethnic origins, similar to one another in terms of status, occupation, education, and culture. Most of them are urban dwellers engaged in industry, health care, education, science, and other activities requiring a relatively high level of education and skill. They usually do not speak the language of the titular people. Opposition to the transition to a new "state language" is the strongest factor forging a common identity among them. The evolution of the "Soviet" identity is crucially important for relations between Russia and other former Soviet republics. It can be peacefully incorporated into new political identities and/or become a kind of regional identity or occupy the leading place in the hierarchy of territorial identities that will perpetuate old conflicts and provoke new ones.

Tishkov (1996, 1997) is correct when he argues that the only possible way to accommodate ethnic tensions in post-Soviet space is to cultivate a double identity – ethnic and political (civic). Such a double non-mutually exclusive identity began to emerge in the Soviet era, and despite all obstacles this process partly continues today: it is the reason for the wide use of the politically correct term "Rossiisky," i.e. those who belong to Russia, or the Russian Federation as a state, instead of "Russky" (ethnically Russian). Polls show that a stable all-Russian political (civic) identity was already well established for some time on the basis of the use of Russian language and of common cultural traditions. It is strong not only among ethnic Russians, but also among numerous representatives of other peoples living outside their titular political units and even among a considerable number of people living in their ethnic homelands. According to Tishkov, the implementation of the principle of a double identity requires a gradual "de-ethnization" of the state at all levels and "de-politicized" ethnicity, without of course putting in question the existence of contemporary republics. In other words, power in the federal center should not be the preserve only of Russians, while ethnic elites should not monopolize power in their titular political units.

The Politics of Identity Maintenance and Manipulation

According to the old myth of liberal nationalism, the natural course of history for a nation culminates in the establishment of a state (Gellner, 1983; Lind, 1994), and, therefore, political identity develops from ethnic identity. But more often states, rather than nations, build both political and ethnic identities. The states use all the means they possess to select and to diffuse the cultural markers which allow each individual to distinguish himself from an ethnic "other," as social groups tend to define themselves not by reference to their own characteristics, but by exclusion, that is by comparison with "strangers" (Lanternari, 1986). Language policy and the education system, the construction or the revival of historical myths and social representation, the creation of the national informational space and, finally, powerful economic leverages at the disposal of the state are the means most often used in nation- and state-building.

Language policy

Language is often considered the most powerful mark of identity. The rise of nationalism and the nation-state were generally associated with attempts to impose a standardized form of the national language on speakers of dialects, or, in some cases, of other languages. According to Gellner, during the transition from the agricultural to the industrial society culture needed to be standardized through the system of general education in the national tongue, which can be ensured only by the nation-state (Gellner, 1983). The educational system and the nation-state are strictly interdependent: one cannot exist without another. In the nineteenth century French was taught in Brittany, and English in Ireland and Wales (Wall, 1969, pp. 81–90). Teachers punished pupils in Breton schools for using the Breton language instead of French (Burke, 1996). In the former Soviet Union, Russian was considered as the language of the upwardly socially mobile and, because it was so widespread, it became the main marker of the "Soviet" and Russian political identities.

The political elites in all the new independent states realize the importance of primary socialization and of elementary education in state-building and, consequently, in the creation of new political identities. The use of Russian is interpreted as a major threat to national identity and as a tool of "Russian imperialism." With the exception of Latvia and Estonia, where the nationalist elites do not even disguise their intentions, the new republics usually adopt "good" democratic laws but their practical implementation, based on ministerial instructions, pursues other objectives.

Unlike the two Baltic countries, where the state refused from the very beginning to fund higher education in Russian, a Law on National Minorities adopted in the Ukraine in late 1991 (just before independence) included the right to be educated in one's native tongue in a state educational institution. But the ministry of education proclaimed the principle of the "optimal concordance" of the language of education with the ethnic composition of population according to passport nationality. Russian began to be ascribed the status of one of the "languages of the peoples of Ukraine," appearing in alphabetical order after Polish (0.4%) and Romanian (0.2%).

Yet Russian is the mother tongue of many ethnic Ukrainians. Numerous survey data show that the 1989 census indicator "mother tongue" severely underestimated the actual use of Russian. Only 48.4% of ethnic Ukrainians speak mostly or exclusively Ukrainian. The situation varies strongly by regions (table 17.2) and by settlements of different size. In cities with more than 200,000 dwellers, 63% use Russian, while Ukrainian largely dominates in the countryside (73%) (Khmelko, 1999; Shulga, 1999). The principle of "optimal concordance" means that, for instance, in Donetsk oblast 40% of the 1,100-odd schools would need to change their language of instruction. Since 1993, entrance examinations to higher educational institutions are taken in Ukrainian, and, as a rule, first-year classes should

Table 17.2 The percentage of population speaking in Ukrainian and/or Russian by regions in July 1999

Regions	Only or preferably Ukrainian	Ukrainian and Russian	Only or preferably Russian
West	89.7	4.3	5.6
West-Center	61.7	19.2	18.7
East-Center	31.4	26.5	42.1
South	11.2	27.6	59.4
East	2.9	24.3	72.0
Ukraine as a whole	40.3	20.5	39.2

Source: Survey of the Center of Political Science and Conflict Studies in Kiev under the leadership of Prof. Valery Khmelko, 1999.

now be taught in Ukrainian. This conforms with the language law which foresaw the Ukrainization of the higher educational network.

Fortunately, the zeal of local authorities in the implementation of this principle varies from one region to another and fluctuates in time. As for the Ukrainization of higher education, there are objective problems with regard to textbooks, of the fluency in Ukrainian amongst the teaching staff, and even with regard to the Ukrainian equivalents of professional terms. As a result, the Ukrainization of education, at least, in East Ukraine is taking place much slower than the national "revivalists" would like. Enrollment in Ukrainian-language schools grew in the Donbas and southern regions, including Crimea, slower than in other regions (Arel, 1995; Okhrimenko, 1998). But the situation in the capital is striking and provides an "example" to all the country. Since 1997, in Kiev there are no Russian kindergartens anymore and no groups studying in Russian in the National University. In 1991, there were 129 Russian schools in the Ukrainian capital but in 1999 they were only 10 and it was planned to reduce the number to four. In 1998, only 9% of first class pupils were enrolled to Russian classes, while even according to the 1989 census, the ratio of ethnic Russians was more than 22% (Krasniakov, 1999). In Ukraine as a whole, in 1992–7 the ratio of pupils in Russian schools and classes had decreased from 50% to 36.4%. Only seven out of 300 Russian-language schools remain in West Ukraine (*Nezavisimaya Gazeta*, 26.05.1999, p. 10).

This policy clearly originates from the nationalist perception that Ukrainians who do not speak the Ukrainian language are traitors. The discourse of Ukrainian nationalists is the following: "Those who refrain from using Ukrainian are not supporting Ukrainian statehood, do not believe it, and are somehow hoping for a return of Ukraine to a new colonial slavery." Even in the parliament, Russian-speaking deputies were severely accused by their nationalist colleagues: "The Russian speech of deputies morally and psychologically legitimizes a lack of respect toward Ukrainian as a language of official use and a contempt for Ukrainian culture and its people in general" (nationalist deputy quoted in Okhrimenko, 1998, p. 99). Nationalists introduced the concept of "indigenous" and "non-indigenous" national minorities. The first ones have their homeland in Ukraine and share their territory with Ukrainians (Crimean Tatars, for instance) and, therefore, should be allowed to speak their mother tongue. The second ones, including Russians, are "migrants," "newcomers" on the Ukrainian territory, and their languages are no more than "languages of other peoples of Ukraine."

In Moldova, Uzbekistan, Estonia, and Georgia too, language was the main basis of identity for 80–90% of the people belonging to titular groups in the 1970s–1980s, and for 35–45% for those who belonged to titular peoples in the autonomous republics of the Russian Federation (Drobizheva et al., 1996). In Kazakhstan, Russian is still the language of preference for 70% of the population (Ogneva, 1996). The use of Russian as the second state language is the main ethnopolitical issue. As the Russian-speaking population is underrepresented in the Kazakh parliament, after long debates Kazakh has been proclaimed the only state language of the country (half of the population of which does not belong to the titular group). Not only the Russian speakers but also most Kazakhs in Northern Kazakhstan support the idea of the second state language – from 49% of Kazakhs in Kustanai oblast to 60–5% in other northern oblasts (Susarov, 1997). Kazakhstan runs the risk of splitting along regional, rather than ethnic dividing lines. This was one of the reasons why President Nazarbaev moved his capital to the center of the country from Alma-Aty. The former capital is situated on the territory of the so-called Elder Zhuz (horde) of Kazakhs, who are more nationalist, Asia-oriented and traditionally dominant in Kazakh politics.

With the sole exceptions of Belarus and Kirghizia, all of the newly independent states firmly refuse to make Russian the second state language, which would obviously contradict their objectives in state- and nation-building. Titular languages will certainly increase their role, and require state support at the first stage of independence. But the processes in this delicate field are very gradual. By trying to speed up the natural course of events, nationalist elites run the risk of strengthening the "Soviet" identity of Russian speakers and of slowing down the formation of more inclusive state-political identities in their countries. Language laws are used to discriminate against Russians and other non-titular dwellers in the post-Soviet republics: to close them off from access to higher and often even to secondary education in their tongue, to eliminate their participation in privatization, and to ignore their cultural needs and traditions.

The creation of the national informational space

The policy in the fields of language and education is closely related to the formation of the national informational space. Nation- and state-building are now only possible using the mass media and communications, which allow the state bureaucracy and urban activists and intellectuals to "teach" identity and to mobilize adherents, especially amongst the rural population (Deutsch, 1953; Gellner, 1983). Nationalist elites attempt to protect the informational space under their control from "foreign" influence, i.e. from the impact of Russian media and, in general, to limit the diffusion of newspapers and broadcasting in "non-state" languages (i.e Russian). The modern territorial state plays a dominant role in the generation and dissemination of information. It is difficult to imagine media that would not be state-based (Murphy, 1996, p. 103). They strongly accelerate the development of different identities among the same ethnic group separated by political boundaries. Media create a different informational context on both sides of a border, which deprives an individual of the ability to catch all the meaning of the texts even in his/her mother tongue (Deutsch, 1953).

Russian speakers in the former Union republics cannot now watch Russian TV, except for a few entertainment programs which still are broadcast outside the Russian borders. Even in the predominantly Russian-speaking East Ukraine and in Crimea, the only national channel broadcasting Russian programs – the so-called Ukrainian International Television – broadcasts mostly in Ukrainian and can interrupt the only news program from Moscow if it does not fit the scheduling planned in Kiev. This is officially explained by economic reasons, because Russian TV companies and the state do not want to pay for translation abroad. Likewise, programs from new independent countries are not broadcast in Russia. "Foreign" (i.e. Russian) newspapers are rather expensive, as are transportation and post services.[2]

The invention of national myths and stereotypes shaping territorial identities

The most effective argument in nation-building is that of a shared historical destiny closely linked with a myth regarding the common origin of the ethnic group, and having an exclusive right to control its homeland, the area marked by its heroic past. There is a tendency on the part of post-Soviet states to present their titular nations as ancient European peoples who can be proud of the glorious periods in their history, mainly in the Middle Ages, but who are also the innocent victims of primordially hostile "other" nations. More recent periods are described exclusively as a dark era of foreign occupation, and the whole of history is depicted as a sequence of heroic and tragic battles for national independence (von Hagen, 1995, pp. 658–9) in much the same way as Marxist–Leninist historians reduced history to the class struggle. History really is transformed into a true battleground (Kohut, 1994). Ballads and archeological remains become the objects of conflicting claims to ownership by competing nations. These claims are important because they reinforce images of present-day identity: who "we" are depends on who "we" were (Burke, 1996, p. 297).

State- and nation-building processes in the post-Soviet space have been accelerated at all the levels. Ukraine provides a brilliant example of the effective use of old and new historical myths in the construction of a new national and political identity and also social representations of traditional enemies and friends and the new geopolitical codes. Ukrainian ideologists have already inundated the book market with an abundant literature in the attempt to enroot in the mass consciousness a new set of primordial myths about Ukrainian history on which a Ukrainian identity can be built (Keenan, 1994). The main myth concerns the origin of Ukrainians as a nation and the origin of Ukrainian statehood. The officially sanctioned Ukrainian historical school promotes the chronicles of the Galicia–Volhynian princedom and views this polity and even the Grand Duchy of Lithuania (before the union with Poland in 1569) as the only heirs of Kievan Rus' and embodiments of Ukrainian statehood. This view is clearly opposed to the traditional Russian historiography and to Russian national historical myths, according to which, Vladimir–Suzdal and, later, Muscovy were the direct heirs to Kievan Rus'. Kievan Rus' is viewed as the common motherland of all East Slavic peoples – Russians, Ukrainians, and Belorussians. Ukrainian ideologists also argue that only the Ukrainians are a truly ancient European people, whose historical destinies were linked with Europe, while the Russian ethnos was formed only in the fourteenth century and is not truly Slavic.

According to the second myth, the polity of Zaporozhian Cossacks in the seventeenth and early eighteenth centuries was a true independent state, the most democratic in Europe (it is claimed that Cossacks had the first constitution in the world), that the Cossacks were all ethnic Ukrainians, their state incarnated the traditions of Kievan Rus', and their area of settlement embraced all of Southern and Eastern Ukraine. The Cossack myth is probably the most important in grounding the present-day all-Ukrainian identity and territoriality (the boundaries of the independent Ukrainian state, including even Crimea and Sevastopol itself). In fact, the Cossack polity, the hetmanate, included only a relatively small part of the Ukrainian territory; the Cossacks were irreconcilable enemies of the Uniate Church in the then Polish part of Ukraine, approximating Galicia; and the Brest Union with the Catholic Church was regarded as a betrayal. But, paradoxically, it was the Galician intelligentsia who not only initially adopted the Cossack myth in the nineteenth century, but also – owing to the works of the national Ukrainian poet Taras Shevchenko who extolled the Cossack love of freedom and their struggle against Polish domination – transformed it into a nationwide mythology. Special sub-myths about the migration of people from Galicia to Dnieper Ukraine were invented in order to justify all-Ukrainian unity and territoriality (Plokhy, 1996). The Cossack myth is also clearly opposed to the previously dominant interpretation of the history of the whole south of the European part of the former Soviet Union, which was usually associated with the colonization efforts of Catherine II and of the Russian Empire.

The third Ukrainian historical myth is that Russian and Soviet (i.e. also Russian) dominance was imposed on Ukrainians by force, it had a colonial character and it caused irreversible demographic, economic and environmental damage to the Ukrainian people and threatened their very existence (Bilinsky, 1994; Luk'ianenko, 1992, etc.). Ukrainian nationalists claim that the sufferings of Ukrainians under Russian oppression have no precedents in history: when B. Khmelnitsky took the decision in 1654 to agree to a union with Russia, there were supposedly more Ukrainians than Russians (Wilson, 1997).

The fourth myth concerns the history of Ukrainian republic of 1918–21, which is regarded as a truly democratic state defeated by the Russian Bolshevik–chauvinist intervention in the absence of its own well-organized army, resulting in a new forcible incorporation and colonization, and the internal weakness of this republic (or, better to say, these republics, because several very different regimes replaced one another in Kiev during a short time, so vividly described in the novels of Mikhail Bulgakov) and the lack of a mutual understanding between Kiev and Galicia. In fact, there were many Ukrainians in the Red Army.

The fifth myth, which played a very important role in mobilizing support for independence on December 1, 1991, is the representation of Ukraine as the breadbasket of Europe, a rich country which suffered from the colonial exploitation by Russia and which would thrive if not subjugated to Russia. The difficult years immediately after independence, the absence of the anticipated large-scale support from the West, and the relatively more prosperous economic situation in Russia came as a surprise to many Ukrainians.

No national myth in the post-Soviet countries, including, of course, Russia, can be accepted without criticism. For instance, the Kievan state was neither Ukrainian nor Russian, just as Charlemagne was neither French nor German (Kappeler, 1995, p. 698; Wilson, 1997). As for Russian historical myths, in a similar way, it was not an awareness of the Kievan inheritance, and the motive of reconstituting the erstwhile Kievan unity (the "gathering of the Russian lands"), that explained Muscovy's earliest expansions into neighboring Russian principalities; rather its great princes simply behaved as pragmatic opportunists and tried to exploit weaknesses of their neighbors (Riasanovsky, 1994). But in the newly independent states, these myths form the basis of school and university textbooks, of new state symbols and ceremonies, result in the creation of sacred places, and efficiently contributed to the formation of an ethnic identity.

At their own level, quite in the spirit of Hobsbawm's concept (Hobsbawm and Ranger, 1983), the political elites in the Russian republics are also intensively inventing traditions and strengthening identities. For example, in Yakutia (Sakha), ideas are promoted regarding a distinctive northern civilization and of national self-respect on the basis of high economic potential, the supposedly high level of local culture and high education levels by international standards (according to official statistical data, the per capita number of persons with higher education, especially of PhDs and professors, among Yakuts is considerably higher than among Russians). The purpose of Yakutian nationalism is political: economic sovereignty, meaning increased autonomy from Moscow in the control and the use of natural resources by the titular elite.

In North Osetia, it is a glorious history which is mostly used in the construction of ethnic stereotypes. Most people of the titular nationality are perfectly bilingual and do not desire to drop Russian as one of two state languages. Religion cannot be used as the basis of identity either, because although most Osetians are Orthodox Christians, many are Muslims. So, the titular intelligentsia diffuses representations about the powerful state of Alania (the predecessor of Osetia), which survived the invasion of Mongols, adopted Christianity and controlled vast territories in the Caucasus. The republic is now officially called North Osetia–Alania. Territoriality is a major national issue in this case as well, as Ingushes claim a part of the Osetian territory (the so-called Suburban district, whose territory used to be settled by Osetians and Ingushes before the 1992 bloody conflict), and as South Osetia remains an unrecognized republic within Georgia. A strengthening of identity is needed to continue the struggle for territory. According to the polls of the Institute of Ethnology of the Russian Academy of Sciences, in the mid-1990s Osetians had the highest share of the ethnic component in all individual self-identifications. Osetian nationalism can be defined as "defensive" (Drobizheva et al., 1996). In Tuva, the memory of the recent state independence[3] and also writings about a more remote past serve to ground nationalist representations.

Against this background of the efficient manipulation of ethnic identities in the new countries and in the republics of Russia, the failure of Russia as a whole to respond to this process with its own national idea is striking. The Yeltsin administration created special commissions whose mission was to "compile" a concept of the national idea. But, of course, such an idea cannot be established by order "beginning next Monday." The problem of national iconography combining different values is especially difficult. As A. Miller noticed, there are no national symbols that could belong to the Russian political nation, but would not be Russian in ethnic terms (Miller, 1997). The Putin administration adopted state symbols from different historical epochs: it left the old tsarist coat of arms and returned the Soviet national anthem. The self-respect of Russians as citizens of their country slightly increased, partly as a result of a certain improvement of the economic situation: in late 2001, 21% supposed that Russia was at the very bottom of the list of world countries, against 28% in 1998, while 10% believed that it was one of the top ten countries, against 4% three years before. But still there is no consensus in society about the future of the nation. Polls show the growing ambiguity of Russian citizens' self-images, i.e. to an increasing extent they choose conflicting options.

Russian public opinion remains deeply split, and the self-images and identities of Russian citizens and of Russians proper varies from region to region. For instance, the affiliation with the region in Stavropol is stronger than it is in Voronezh, and in Voronezh it is stronger than in Moscow. The identity with Russia is stronger in Moscow than in both Voronezh and Stavropol, while the "Soviet" identity is strongest in Voronezh (the largest oblast of the "red belt") (Survey INTAS, 1996). In Tatarstan, the urban Russian population supports the republican administration to a higher extent than it does even the federal (properly "Russian") authorities; likewise, in Yakutia to almost the same extent. This is a very favorable factor for reaching a national consensus. But in Tuva the situation is just the opposite (Drobizheva et al., 1996).

Thus, the role of the territorial factor in the formation of self-images is clearly salient. As the history of the peoples settling the post-Soviet space is deeply interrelated, the "battle for history" not only intensifies ethnic distinctions, but potentially aggravates the risk of ethnic conflicts. In cases where the elites manipulate historical representations in an instrumentalist fashion to promote ethnic or national identity, spatial cultural and political borders tend to become more solidified.

The increasing role of economic leverages and factors

Nairn (1977) and Anderson (1991) point to economic factors as clearly modern underpinnings of nationalism. Nationalist forces are usually supported in the capital cities, where political elites and intellectuals are concentrated, and by people in the more rural, less urbanized and industrialized regions, where the ethnic structure is more homogeneous, and the titular group has a dominant position. Because of their relative backwardness, and as a consequence of their loyalty to nationalist ideals, these areas become the recipients of financial transfers from the central government, while more urbanized and less nationalist regions become the donors to the central budget.

This situation is typical for Ukraine, Kazakhstan, and Kirghizstan and is widely used in the nationalist discourse. Besides, areas with different proportions between the titular group and ethnic minorities often have a historically established economic specialization caused in particular by natural factors. For example, in Kazakhstan there exists, on the one hand, a cultural division of labor between Kazakhs and the non-Kazakh ("European") population, and, on the other, between Kazakhs living in northern, central and southern regions. The socioeconomic divisions therefore do not strictly follow ethnic watersheds (Susarov, 1997).

In most of the new independent states in the post-Soviet space, the ethnic composition is clearly different between urban and rural areas: Russians and other Russian speakers are usually engaged in declining manufacturing industry in towns, while the titular population lives in the countryside which is now being rapidly urbanized. This increases the competition between the titular group and ethnic "others" for jobs, housing, status and privileges, strengthens ethnic identities, and contributes to political mobilization on both sides. The distinction in ethnic structure between towns and the countryside is exploited in the nationalist propaganda: "Russians can leave for their country, but for us, this land is the only motherland." Experience shows that in such situations, the departure of the Russian speaking population is inevitable.

In the Russian Federation, nationalist arguments and territorial identities are a convenient tool for local elites in bargaining with the federal center – according to the "classical" model of "instrumental" nationalism (Smith, 1996). Relations between the federal center and the regions were transformed into a competition between ruling oligarchic groups. Regional legislatures turned into meeting points for functionaries and local businesses, creating cooperative alliances or trying to counteract the strong Moscow banks and companies (Afanasiev, 1998). Despite the new wave of centralization required to balance the consequences of the direct elections of governors, and despite an attempt by Moscow to deprive republican leaders of privileges they obtained in the early 1990s, the strengthening of an extreme regionalism or even separatism remains a potential threat against the federal center. The most economically powerful republics, Tatarstan and Bashkiria, tried to establish their own citizenship and independent republican courts, which contradicts the federal Constitution, and claimed control over land, state property and natural resources. In the 1990s, local authorities often received unlimited political and economic power within the borders of their republic in exchange for loyalty towards Moscow. This created "local tyrannies" federal laws were ignored, electoral procedures were openly adapted to the needs of the leader, political rivals were pursued, and outcomes of elections were falsified. The greater the degree of autonomy of a republic from Moscow, the fewer civil freedoms and human rights are respected. Leaders of many republics, as for instance, in Bashkortostan (Bashkiria) and Kalmykia, have established autocratic regimes and manipulate elections eliminating all serious rivals with the use of sophisticated bureaucratic procedures (Politicheskii almanakh Rossii, 1998). It is the reason why President Putin reformed in 2000–1 the Council of Federation of the Federal Assembly, which diminished the political role of governors, and launched the campaign aimed to put in conformity regional laws with the federal Constitution and legislation.

Almost everywhere the post-Soviet ethnic bureaucracy firmly keeps power and is the main protagonist of nationalism. In the Russian republics, incumbent leaders try to present themselves as moderate mediators between the center, the Russian-speaking population and radical nationalists, especially the titular intelligentsia. But in reality, the pressure of nationalism allows them to come to profitable compromises with the federal center and to control material resources. Interestingly and not accidentally, the per capita number of regional functionaries in the small "national" republics is much higher than in the larger Russian provinces. In "Russian" regions, it varies from 1:1,200 to 1:1,500, while in Kabardino–Balkaria, Yakutia, Bashkiria, and in Tatarstan it reaches 1:600–1:1,000. In the relatively small northern city of Yakutsk, the capital of the Republic of Yakutia-Sakha, there are 14 ministries and 13 state committees. The national political elites now controlling higher education successfully reproduce themselves and do not have any will to lose their dominant positions (Afanasiev, 1998). They created quasi-feudal hierarchies of their vassals at the lower levels of administration, most of whom belong to the titular nationality. For example, in Buriatia, Buriats in the mid-1990s made up 28% of the total population, but 40% of deputies of the People's Khural (republican parliament), 35.6% of deputies of local governments, and 50% of the heads of district administration, including five districts with more than 50% of the Russian population (Kamyshev, 1997, p. 168).

This development has already practically transformed (at least, in North Caucasus) the titular ethnic groups, and not the respective republics, into real subjects of the Russian Federation (Khoperskaya, 1997). This can create a basis for future conflicts. Indeed, why should this privileged status be confined to peoples possessing territorial autonomy? What about others? What about the ousting of the Russian population? These questions remain open.

Conclusion

Political control over a geographic space is the main opportunity to realize nationalist aspirations as a political program. The nationalist perception of a territorial identity with the soil of one's ancestors as the place which belongs only to the members of the nation, and which is the only place where its historical destiny can be fulfilled, is being transformed into a feeling of national exclusiveness. Nationalism has become the principal equivalent of national territoriality, identified as a political strategy that nationalists can use to control the fate of the nation (Kaiser, 1994). National stereotypes, as an element of ethnic identity, necessarily include images of space: regions incorporated into the state territory by the national consciousness get their codes, and many of them became national symbols (like Sevastopol for Russia). Negative stereotypes are purposefully cultivated, especially when the national elites feel a threat to their national integrity and culture; these representations become the key elements in human territoriality, as in the case of Ukraine. The hierarchy of identities and their very content depends upon the regional context. The disintegration of the Soviet Union provoked obvious disjunctions between the political organization of territory and ethnic and other territorial identities, and the geography of economic, social, and political life.

Mass ethnic hysteria is created by "activists" among the same group, or by the state authorities, in order to achieve concrete political objectives. Therefore, "the voice of the people" against or for something becomes a widespread myth. Fortunately, a very large percentage of the population is absolutely indifferent to national issues: in Russia, this percentage varies from 35 to 43% among titular peoples in the republics and from 45 to 53% among Russians. Our survey in the territory of Stavropol showed that for more than 80% of respondents, the ethnic background of their friends and colleagues does not matter and for about 30–50% of respondents, even the ethnic origin of their closest family, plays no role (Kolossov et al., 2001). In most regions of the former Soviet Union and, in particular, of Russia (though, unfortunately, not in all of them), other sociological surveys also indicate a relatively high degree of ethnic tolerance and demonstrate that most people are not ready to sacrifice their well-being in order to achieve "superior" ethnic or political goals. This inspires cautious optimism about the future of nation- and state-building in this part of the world.

ENDNOTES

1. This survey was conducted in March 2000 as part of a joint project with the Institute of Behavioral Science, University of Colorado at Boulder, sponsored by the National Science Foundation (authors: John O'Loughlin, Vladimir Kolossov, James Bell and Olga Vendina). About 3500 Moscovites were interviewed door-to-door in 15 carefully selected representative districts.
2. Partly because of bureaucratic reasons: for example, in Ukraine a special tax was established for kiosks selling "foreign" newspapers in 1996, which made the diffusion of Russian media financially exhorbitant. As a result, the main Russian newspapers are sold with the same content (except advertising and weather), and under the same titles, but with two additional words – "in Ukraine" ("Izvestia in Ukraine," etc.) because they are registered in Kiev as Ukrainian periodicals.
3. Tuva profited from a high degree of autonomy within the Russian Empire, which it entered as a protectorate only in 1914; it was independent between 1921 and 1944.

BIBLIOGRAPHY

Afanasiev, M. 1998. *Suverenitet khorosh, kogda on ogranichen* (Sovereignty is good when it is limited). Izvestia, June 2.

Anderson, B. 1991. *Imagined Communities*. London: Verso.

Anderson, M. 1996. *Territory and State Formation in the Modern World*. Cambridge: Polity.

Arel, D. 1995. Ukraine: The temptation of the nationalizing state. In: V. Tismaneanu (ed.) *Political Culture and Civil Society in Russia and the New States of Eurasia*. Armonk, NY: M.E. Sharp, 157–82.

Barth, F. (ed.). 1969. *Ethnic Groups and Boundaries: The Social Organization of Cultural Difference*. Boston: Little, Brown and Co.

Bilinsky, Ya. 1994. Basic factors in the foreign policy of Ukraine. The impact of the Soviet experience. In F. Stair (ed.) *The Legacy of History in Russia and the New States of Eurasia*. Armonk, NY: M.E. Sharp, 171–93.

Brubaker, R. 1992. *Citizenship and Nationhood*. Cambridge, MA: Harvard University Press.

Burns, E. B. 1980. *The Poverty in Progress: Latin America in the Nineteenth Century*. Berkeley, CA: University of California Press.

Burke, P. 1996. We, the people: popular culture and popular identity in modern Europe. In: S. Lash and J. Friedman (eds.). *Modernity and Identity*. Oxford: Blackwell, 293–308.

Chernysh, M. F. 1995. Natsionalnaya identichnost': osobennosti evolutsii (National identity: features of evolution). *Sociological Journal*, 2, 110–14.

Chinn, J. and Kaiser, R. 1996. *Russians as the New Minority*. Boulder, CO: Westview Press.

Deutsch, K. 1953. *Nationalism and Social Communication: An Inquiry into the Foundation of Nationalities*. Cambridge, MA: Harvard University Press.

Drobizheva, L. M., Aklaev, A. R., Koroteeva, V. V., and Soldatova, G. U. 1996. *Demokratizatsia i obrazy natsionalizma v Rossiiskoi Federatsii v 90-kh gg* (Democratization and images of nationalism in Russian Federation in the 90s). Moscow: Mysl.

Gellner, E. 1983. *Nations and Nationalism*. Blackwell, Oxford. (Russian edition 1991, Moscow: Progress.)

Gorshkov, M. 1997. Shto proiskhodit s nami? (What is happening with us?) *Nezavisimaya Gazeta*, May 15.

Greenfeld, L. 1993. *Nationalism: Five Roads to Modernity*. Cambridge: Cambridge University Press.

Gurr, T. R. 1994. Peoples against states: Ethnopolitical conflict and the changing world system. *International Studies Quarterly*, 38, 347–77.

von Hagen, M. 1995. Does Ukraine have a history? *Slavic Review*, 54(3), 658–90.

Hobsbawm, E. and Ranger, T. (eds.). 1993. *The Invention of Tradition*. Cambridge: Cambridge University Press.

Holdar, S. 1995. Torn between East and West: the regional factor in Ukrainian politics. *Post-Soviet Geography*, 36(2), 112–32.

Huntington, S. 1996. *The Clash of Civilizations and the Remaking of World Order*. New York: Simon and Schuster.

Kaiser, R. J. 1994. *The Geography of Nationalism in Russia and the USSR*. Princeton, NJ: Princeton University Press.

Kamyshev, A. D. 1997. *Mezhetnicheskie vzaimodeistvia v Buriatii: sotsialnaya psikhologia, istoria i politika* (Interethnic interactions in Buriatia: Social psychology, history, and politics). Ulan Ude: University of Buriatia Press.

Kappeler, A. 1995. Ukrainian history from a German perspective. *Slavic Review*, 54(3), 691–701.

Keenan, E. L. 1994. On certain mythical beliefs and Russian behaviors. In F. S. Starr (ed.) *The Legacy of History in Russia and the New States of Eurasia*. Armonk, NY: M.E. Sharp, 19–40.

Khmelko, V. 1996. Otnoshenie k Rossii (The attitude to Russia). *Nezavissimaya Gazeta*, 19.12.96.

Khoperskaya, L. L. 1997. *Contemporary Ethnopolitical Processes in North Caucasus: A concept of ethnic subjectivity*. Rostov: SKAGS.

Kitschelt, H. 1995. Formation of party cleavages in post-communist democracies. *Party Politics*, 1, 447–72.

Kohut, Z. 1994. History as a battleground: Russian-Ukrainian relations and historical consciousness in contemporary Ukraine. In S. F. Stan (ed.) *The Legacy of History in Russia and the New States of Eurasia*. Armonk, NY: M.E. Sharp, 123–46.

Kolossov, V. 1993. La Russie dans le monde. In J. Lèvy et al. (eds.) *Le monde: espace et systèmes*, 2nd edn. Paris: Dalloz.

Kolossov, V., Galkina, T. and Krindatch, A.. 2001. Territorialnaya identichnost' i mezhetnicheskie otnoshenia (na primere vostochnykh raionov Stavropolskogo krya) (Territorial identity and ethnic relations (the case of eastern districts of Stavropol region)). *Polis (Political Studies)*, 2(61), 67–77.

Kolossov, V. and O'Loughlin, J. 1998. New borders for new world orders: territorialities at the fin-de-siècle. *Geojournal*, 44(4), 259–73.

Kolossov, V., Glezer, O., and Petrov, N. 1992. *Ethnoterritorial Conflicts in the Former USSR*. Durham, UK: IBRU Press.

Krasniakov, E. 1999. *The Right to Study on the Mother Tongue. The Dialogue Between the Ukrainian and the Russian Cultures in Ukraine*. Kiev: The Russian Foundation, the Committee on the Freedom of Information of the Supreme Rada, the Institute of Sociology of the Ukrainian National Academy of Sciences, 84–91.

Lanternari, V. 1994. *Identità e differenza: Percorsi storico-antropologici*. Naples: Liguori.

Lind, M. 1994. In defense of liberal nationalism. *Foreign Affairs*, 73(3), 325–46.

Luk'ianenko, L. 1992. Veruyu v Boga i v Ukrainu (I trust in God and in Ukraine). In O. Gonchar (ed.) *Ukraine as the Heir of the Ukrainian People's Republic. Rozbudova derzhavy*, 5.

Miller, A. 1997. *O diskursivnoi ptirode natsionalizma* (About the discursive nature of nationalism). *Pro et Contra*, 2(14), 141–52.

Murphy, A. B. 1996. The sovereign state as political-territorial ideal. In: T. J. Biersteker and C. Weber (eds.) *State Sovereignty as Social Construct*. Cambridge: Cambridge University Press, pp. 81–119.

Nairn, T. 1977. *The Break-up of Britain: Crisis and Neo-Nationalism*, 2nd edn. London: New Left.

Newman, D. and Paasi, A. 1998. Fences and neighbors in a postmodern world: boundary narratives in political geography. *Progress in Human Geography*, 22(2).

Ogneva, V. A., 1996. *Vkhodia v demokratiu. Opyt politicheskogo analiza problem perekhodnogo perioda v Kazakhstane* (Entering in democracy. An experience of a political analysis of the problems of the transitive period in Kazakhstan). Alma-Ata.

Okhrimenko, V. S. 1998. *Russkoe naselenie Ukrainy i Belorussii: rasselenie, migratsii, orientatsii* (Russian population in Ukraine and Betorussia: settlement, migrations, orientations). Doctoral thesis, Institute of Geography of the Russian Academy of Sciences, Moscow.

Paasi, A. 1996. *Territories, Boundaries and Consciousness: The changing geographies of the Finnish-Russian border.* Chichester: Wiley.

Pirie, P. S. 1996. National identity and politics in Southern and Eastern Ukraine. *Europe Asia Studies*, 48(7), 1079–104.

Plokhy, S. P, 1995. The history of a "non-historical" nation: Notes on the nature and current problems of Ukrainian historiography. *Slavic Review*, 54(3), 702–19.

Politicheskii almanakh Rossii, 1998. *Political Almanac of Russia* 1997. Moscow: Carnegie Endowment for International Peace.

Riasanovsky, N. V. 1994. *A History of Russia*. Oxford: Oxford University Press.

Sahlins, P, 1989. *Boundaries: The Making of France and Spain in the Pyrenees*. Berkeley, CA: University of California Press.

Shulga, N. 1999. *The Situation of Russian Culture in Ukraine: Myths and Reality. The Dialogue Between the Ukrainian and the Russian Cultures in Ukraine*. Kiev: The Russian Foundation, the Commitee on the Freedom of Information of the Supreme Rada, the Institute of Sociology of the Ukrainian National Academy of Sciences, pp. 7–17.

Smith, G. 1996. Russia, ethnoregionalism and the politics of federation. *Ethnic and Racial Studies*, 19, 403–4.

Susarov, A. 1997. Priroda kulturnykh distantsii i granits v Kazakhstane (The nature of cultural distances and borders in Kazakhstan). *Bulletin of the Center of Ethnopolitical and Regional Studies*, 7(63), 5–54.

Survey INTAS "Prometee." 1996. *VTsIOM, 12.1995–01.1996*. Moscow: INTAS.

Taras, R. 1993. Making sense of matrioshka nationalism. In I. Bremmer and R. Taras (eds.) *Nations and Politics in the Soviet Successor States*. Cambridge: Cambridge University Press, pp. 513–38.

Thompson, R. H. 1989. *Theories of Ethnicity: A Critical Appraisal.* Westport, CT: Greenwood.

Tishkov, V. A. 1996. What is Russia? Prospects of nation building. *Russian Politics and Law*, March–April, 5–27.

Tishkov, V. A. 1997. *Ethnic Conflicts in the Post-Soviet Space*. New York: Sage.

Wall, M. 1969. *Peasants and Frenchmen*. London: Chatto and Windus.

Wilson, A. 1997. *Ukrainian Nationalism in the 1990s. A Minority Faith*. Cambridge: Cambridge University Press.

Chapter 18

Nation-states

Michael J. Shapiro

Introduction: A Peruvian Prelude

Behind patriotism and nationalism there always burns the malignant fiction of collectivist identity, that ontological barbed wire which attempts to congregate "Peruvians," "Spaniards," "French," "Chinese," etc., in inescapable and unmistakable fraternity. You and I know that these categories are simply abject lies that throw a mantle of oblivion over countless diversities and incompatibilities... (Vargas Llosa, 1999, p. 170).

These lines, by a Peruvian writer and former presidential candidate (and former radical turned neo-liberal) reflect more than a desire to demythologize patriotism. The eloquent and hyperbolic statement, in a novel focused primarily on a series of familial love triangles, conveys an appreciation of the work of fiction, which is integral to the contemporary nation-state's maintenance of its existence. The style of Vargas Llosa's novel – "a rich confusion of art and fact, fiction and reality" (Kendrick, 1999) – mimics the contemporary state's performance of nationhood, which also bundles "art and fact, fiction and reality," in a variety of genres of expression, not as in Vargas Llosa's novel, to disclose the ways in which fantasy mediates erotic encounters, but to perpetuate itself as the container of a coherent and unified national culture. Yet, as Vargas Llosa clearly appreciates, the tasks, though disparate in their aims, produce similar, hybrid modes of expression.

Although I have begun this treatment of the nation-state with a literary voice, situated within the global literary culture in general and the Peruvian politico-intellectual culture in particular, there is a multitude of *locations* (loci of enunciation) from which diverse kinds of voices emerge and bear on the status of nation-states. There is also a variety of *texts* in diverse genres (modes of enunciation), which articulate the issues surrounding the status of nation-states. In this chapter I invoke a variety of loci and modes of enunciation for two reasons, one theoretical and the other historical. Theoretically, multiple positions are required because of the essential contestability of the meaning and status of the nation-state. Historically, the political, geographic, and economic prerogatives of the contemporary nation-state have been and are increasingly contested. In response, states have been required, now more than ever, to perform their identities, to maintain their practical as well as ontological statuses as "nation-states." Although state control over the meaning of territory and bodies is increasingly challenged by globalizing forces, "the concept-metaphor 'nation-state'" manifests a "ferocious re-coding power" (Spivak, 1992, p. 101). While the initial aggregations forming the dominant model for subsequent states – the European state system established by the Treaty of Westphalia in 1648 – involved military and fiscal initiatives, coercive and economic aspects of control have been supplemented by a progressively intense cultural governance, a management of the dispositions and meanings of citizen bodies, aimed at making territorial and national/cultural boundaries coextensive.

This chapter has two basic aims. At a simple level, the aim is to provide an understanding of the emergence and persistence of nation-states (for some traditional approaches, see Bendix, 1964; Emerson, 1960; Smith, 1986). At another, more conceptual level – and in keeping with Vargas Llosa's critically disrupting, mixed metaphor – I want to cut the "ontological barbed wire" surrounding the metaphysical nation-state in order to disclose the dense set of interpretive and material practices through which the (misleading) nation segment of the hyphenated term, nation-state achieves its standing. "Nations" – at least those that are arguably contained within states – should be regarded as dynamic and contentious domains of practice. Instead of treating them as autonomous entities and as static objects of analysis, they should be regarded as "categories of practice" (Brubaker, 1996, p. 7). At a symbolic level, they are imaginaries (abstract domains of collective coherence and attachment), which persist through a complex set of institutionalized modes of inclusion and exclusion.

However, to simply refer to "the nation" as an imaginary is insufficient for supplying a critical understanding of nation-states. Adding a historical perspective on the media through which the imaginary is created and sustained, which will occupy much of this chapter, moves in the direction of a historically and politically perspicuous purchase on the contemporary nation-state. The "ontological barbed wire" to which Vargas Llosa refers has been installed by the largely symbolic, self-inscribing practices through which modern states claim nationhood. But the historical emergence of a nationalizing "statecraft," a term for a complicated territory- and people-managing mode of governmental practice, must be understood in the context of the variety of specific, aggregating and disaggregating, material as well as symbolic, conditions shaping the contemporary political entities recognized (in varying degrees) as nation-states.

While the historical trajectory and social dispersion of the practices supporting state claims to containing coherent national communities receives considerable conceptual emphasis in this chapter, to supply a sufficient context for Vargas Llosa's scepticism, I turn first to the functioning of statecraft during its most crucial nation-building stage in his "homeland," Peru. As a polity that emerged from colonial rule, the boundaries of the Peruvian state resulted initially from Spanish imperialism, then from contests among Creole functionaries to manage a fairly amorphous territory, and then from conflicts between Peru and other states. Forging a *nation*-state, however, has been more complicated.

It cannot be claimed that the Peruvian case is exemplary because there are different ways in which states have sought to become nation-states. Nevertheless,

Peru's nationalizing process warrants scrutiny because conceptualizations (as opposed to generalizations) are best developed in the context of specific historical episodes. In the Peruvian case, state government began its nationalizing initiatives in the face of regional power configurations that were a legacy of the colonial period. Peru's relatively late independence (1824) came about as the result of a bourgeois revolution that displaced the precapitalist, colonial-imposed oligarchy (of planters, mine owners, and wool merchants: Guardino and Walker, 1992, p. 12). Inasmuch as virtually all interdependencies – material, social, political and religious – functioned within regions, the process of nationalizing the Peruvian population necessitated a wresting of control from various regional elites. The state sought to effectively re-terriorialize attachments and structures of exchange through the "annihilation of regional space by state power" (Nugent, 1994, p. 338).

In the nineteenth century, the Peruvian state had maintained its bureaucratic and fiscal controls through a set of clientele relationships. The ruling elites were required to manage such state administrative functions as revenue collection and peace keeping in exchange for being allowed to control their regional affairs. At the end of the century, however, "the massive expansion of North Atlantic capital into Peru... initiated a process of social transformation and national consolidation that brought to an end the fragmented conditions that had dominated the 19th century" (Nugent, 1994, p. 337). The subsequent process of centralization, in which the state expanded fiscal, bureaucratic, and cultural controls, involved a complex set of shifting alliances among different classes and different locales. In the center, the Lima ruling class has controlled credit and commercial structures, and in the periphery, different regions, villages, and groupings have had a history of unstable partisanship with respect to local interdependencies and varying degrees of cooperation and resistance with respect to state centralization (Nugent, 1994, p. 333).[1]

While the historical process of centralization in Peru, as in other cases, speaks to the emergence of the modern fiscal-military state (a more or less continually contested "emergence" in Peru's case, for as late as the 1990s the state military fought a Maoist insurgency, instituted a period of terror against suspected collaborators, and wiped out much of the black economy, based on coca production), the story of the coherence of the nation segment of the Peruvian nation-state is case-specific and has been historically unstable. The process of integrating the diverse population of Creoles, "Indians," and *mestizos* into a coherent citizen body has been contested at both elite and mass levels at every stage. At the elite level, the Creole nation-building strategy has been inconsistent. State-oriented intellectual discourses constructed the problem of indigenous citizenship (*indigenismo*) in various ways during the mid 1940s and early 1970s, as the state attempted to achieve nationhood through "revolutionary social and educational reforms" (Devine, 1999, p. 64).

In the 1940s, a conception of a dual citizen body was perpetuated: "The Ministry of Education... implemented an educational policy infused with a purist, anti-*mestizo* ideology that sought to forge a State out of two physically and culturally separate nations" (Devine, 1999, p. 68). Some twenty years later, however, the biopolitical dimension of the state's nation-building initiative (the state's warranting of politically eligible versus ineligible identities) was radically reversed.[2] Whereas before, educational policy attempted to reinstall a "pure" Indianness as a separate cultural existence within the state, under the revolutionary regime of Juan Velasco

Alvarado, "Indian" was abolished as an official term and replaced with the term peasant (*compesino*). Although this displacement was effected primarily through the national educational curriculum, it was also implemented through a variety of other genres with which states have historically scripted the characteristics of their nationhood, for example national *fetes*. In Peru, June 24 which had previously been "the day of the Indian," was now to be "the day of the peasant" (Devine, 1999, p. 69).

Yet state scripting – the imposition of meanings on citizen bodies as well as space – has been frequently resisted by counter-scripts. In the case of Peru since independence, spatially, the state geographic imaginary, the imposition of a coincidence between state and nation, has been countered by, for example, "cartographic practices by indigenous groups" who privilege local rather than national "geographies of identity" (Radcliffe, 1998, p. 275). And temporally, the "Creole nationalist 'discursive frameworks' of the liberal republican state," which have lent the Peruvian state a macro-level, "republican history," have been opposed by an alternative "history of indigenous rights and property," which can be discovered when one descends "from elite texts" (for example the successive drafts of the Peruvian Constitution beginning in 1822) to a genre that mixes elite and vernacular idioms: "the petty archives of local courts and notaries where peasant voices were registered" (Thurner, 1995, pp. 291–3).

In short, the existence of a distinctive postcolonial Peruvian nationhood has been contested and negotiated in a clash between alternative writing performances (or modes of enunciation) and between alternative loci of enunciation, those issuing from either official/national or local space. Official republicanism, defined in legal texts ranging from the founding Constitution to a historical trajectory of tax codes, has been opposed by an indigenous version of "Indian republicanism" (*indios repubicanos*), which, though invented in part by a state taxing authority, took on local meanings, registered within local ledgers (Thurner, 1995, p. 302). Despite historical attempts by Creole nationalists to negate "the historical agency of republican Indians" (for example constructing them as "pre-political beings" [Thurner, 1995, p. 318] during the early twentieth century *indigenismo* educational initiative treated above), the local ledgers of the nineteenth century nation-building period stand as a challenge to "the teleological historicity of Creole nation-building" (Thurner, 1995, p. 291).

The unstable nation-building dynamic evident in the case of postcolonial Peru, where "political keywords like *republic* and *nacion* resisted univocal definition" (Turner, 1995, p. 295), reflects a situation of incomplete national integration, which the ever sceptical Vargas Llosa has attributed to his subcontinent as a whole: "We in Latin America do not yet constitute real nations . . . our countries are in a deep sense more a fiction than a reality" (Vargas Llosa, 1990, p. 51). But, as Vargas Llosa himself implies elsewhere (in the epigraph above), this incompleteness is not peculiar to the newer states of Latin America. It remains the case in "the 'old' nations of Europe as well," where "national culture is very much an ongoing construction and live issue," (Foster, 1991, p. 238) and is also true of the USA, whose process of nation building remains live and continually contested. In order to emphasize the persistence of the contestation over national cultures, I turn to the USA and to a different scripting and counter-scripting genre, music.

Consolidating Euro-America: Rogers and Hammerstein's *Oklahoma!*

Opening in New York in 1943, the Broadway musical, *Oklahoma!*, reflected the Euro-American fantasy of an attunement between domestic and national life. Although focused on the Oklahoma territory just before statehood in the early part of the century, its geographic imaginary is the "American" nation-state as a whole:

> The apparently modest and homely social world that *Oklahoma!* dramatically produces and enacts has a more fundamental, symbolic reality as the realization of the American dream. The domestic world of *Oklahoma!* is also the hoped and promised "land of the free" which is to be wrought in the USA by civilizing nature. This is to be accomplished democratically by the individual and collective labor of its inhabitants, and will make the USA the earthly Garden of Eden that is both "God's own Country" and the world's first, modern nation-state (Filmer et al., 1999, p. 383).

Although as a commercial venture, *Oklahoma!*'s complicity with national legitimation appears to be gratuitous, its production can be in part attributed to the state's cultural governance policies. President Franklin D. Roosevelt's New Deal policy included a "Federal Arts Project" (1935–43), "which encouraged a form of democratic realism in art as a means of rhetoricising the economic and social transformations needed to relocate capitalism as a morally defensible economic system for a democratic society" (Filmer et al., 1999, p. 385). The Arts Project created an accepting climate for such productions as *Oklahoma!*, which, like the other funded projects, aimed at developing a "vernacular tradition of representation" in response to "the perceived dominance of European art forms" (ibid.). *Oklahoma!* exemplifies this version of an American vernacular, and it thematizes the power of an American commonality of values to retard threats to the new order.

From its opening, lyrical celebration of daily life with "Oh What a Beautiful Morning" to its closing celebration of conjugal fulfilment with a communal dance routine, *Oklahoma!* suppresses a history of usurpation, displacement, and violence. What was to become "Oklahoma" in the early twentieth century was peopled largely by Western and displaced Eastern Native American nations by the mid-nineteenth century (and was in fact officially designated as "the Indian Territories"). The wholly white Oklahoma depicted in the Broadway musical (and subsequent film version) reflects a sudden white population surge – the 1889 "land rush" provoked when President Benjamin Harrison "announced [in violation of the treaties with the displaced Eastern nations] that the "Unassigned Lands" would be opened for public homesteading at noon on April 22" (Day, 1989, p. 193).

Ignoring this violent historical process and the still-mixed ethnoscape – of Eastern nations, Western nations and a considerable population of "Negro cowboys"[3] – the drama of *Oklahoma!* unfolds as a containment of threats to a unified and integrated white national culture. The dangers to an accord between person and nation include a rapidly developing modernity (in such places as the "up-ter date" Kansas City, which must be carefully and slowly integrated with the simple pastoral life on the plains), the intrusions of urban and foreign seducers of gullible women (e.g., the crafty Ali Hakim), other outsiders (e.g. Jud Frye who is neither a cowboy nor a farmer, and whose "primordial instinctuality" is a threat to a communal libidinal order), and, generally, any representative of extrafamilial sexuality (e.g. the temptations of prostitutes and dance hall women, shown in a choreographed dance – Filmer et al., 1999, p. 386).

While the immediate thematic context of *Oklahoma!* involves a peaceful accord between cowmen and farmers, the music and choreography manifest a broader theme. They are aimed at integrating all potentially centrifugal elements of the social order, at maintaining coherent community in the face of change. Thus, the choral song style celebrates the transition from territory to state (in "Many a New Day..."). And as people arrive by train from such places as Kansas City, the dance routines present alternative dance idioms side-by-side, which ultimately become integrated routines. In each staging of music and dance, the implicit narrative is a movement from separate idiom to integrated genre, a movement toward an organic whole, representing "the prevailing ideological commitment in the USA to an integration of its constituent states into a unified society, able to both acknowledge and reconcile its differences" (Filmer et al., 1999, p. 387).

Resistant Genres: Centrifugal National Attachments

The cultural governance articulated through Rogers and Hammerstein's *Oklahoma!* has been subjected to a telling reinflection in one instance. While in its original context, one of *Oklahoma!*'s songs, "The Surrey with the Fringe on Top," serves as a romantic ballad to express a desire for a communal witnessing of an inchoate marriage bond, when the black musician, Sonny Rollins, does a jazz version of the song, it takes on significance in connection with his parody of a singularly white narrative of the settling of the West. Rollins' riff on "The Surrey with the Fringe on Top" is of a piece with his *Way Out West* album, which contains "aberrant readings of 'I'm an Old Cowhand' and 'Wagon Wheels'" and features him on the cover, a "Negro cowboy" enacting an "appropriation of the iconography of the American West" (Jarrett, 1994, p. 233).

Rollins' musical gestures are typical of the ways in which African-American jazz musicians have resignified the nation-integrating orientation of the music of Tin Pan Alley composers. They reflect not only an alternative musical idiom, but also a different political sensibility. In this case the political agenda is a critique of *Oklahoma!*'s evacuation of the dark bodies from the ethnoscape of the "Indian territory," which, as a result of the white land grab, was incorporated into a Euro-American-dominated USA. But Rollins' musical critique is a very small part of an African-American musical politics. Recognition of an "America" that emerges when the jazz tradition is in focus requires attention first to a domestic diaspora, a movement of African-Americans from the South and South-west to Northern and Midwestern cities. This movement tells a different story than the integrative nation-building story that supplies *Oklahoma!*'s implicit narrative.

Secondly, African-American music articulates a musical geography that reaches beyond its domestic history and issues a challenge to the dominant Euro-American model of US nation building, which emphasizes the white bodies that moved from Europe across the Atlantic. The musical idioms articulate an alternative, supranational imaginary, for they are associated not only with a domestic, US diaspora but also with a geographically extensive movement of black bodies (and consequently musical genres) throughout the Atlantic region. Despite the diaspora, however, there remains a coherent dispersion of "structures of feeling," constituting what Paul Gilroy calls a "Black Atlantic" with cultural forms that are "stereophonic, bilingual," and "bifocal" (Gilroy, 1993, p. 3).

Achieving an Appropriate Grammar: Nationhood as State Practice

The cases of Peru and the USA point to the pervasive incompleteness of state nationalizing initiatives, an incompleteness that has challenged the traditional grammar of questions posed about nation-states. In recognition of the limitations of traditional queries, Walker Connor has changed the nation-state question from "what is a nation?" to "when is a nation?," because, as he puts it, "it is problematic whether nationhood has even yet been achieved" (Connor, 1990, p. 99).[4] Connor's alteration of the question of nationhood provides a threshold for a critical analysis of nation-states (for more on the critical tradition see Bhabha, 1990). His displacement of the "is" with a temporal trajectory challenges the ontological status of nations by shifting attention from the ideal of the nation to the vagaries of nation-building. But while historicizing rather than reifying is an important step in resisting ideological complicity with "the state effect" (Mitchell, 1999) – the representational practices through which nation-states appear as autonomous, self-contained agents, separate from the activities that constitute them – the "when" question is an analytic dead-end unless another query is added. An inquiry-empowering approach also requires a "how" question. "We should not," in the words of Rogers Brubaker, "ask 'what is a nation' but rather: how is nationhood as a political and cultural form institutionalized within and among states?" (Brubaker, 1996, p. 16).

Once attention is shifted to the "how" question, there remains a need to generate conceptions with which to elaborate state nation-building-as-practice. A preface to the conceptual task is supplied in Ana Maria Alonso's reflections on the "politics of space and time" involved in the nationalizing aspects of state formation. Referring to "the cultural inscription of the idea of the state," she notes that it "has in part been secured through the spatialization of time" (Alonso, 1994, p. 381). Alonso's use of the writing metaphor ("inscription") provides a powerful frame for treating state practices. At a concrete historical level, it accords with Benedict Anderson's insights into the role of print technology and its various genres of expression – especially novels and newspapers – in creating and consolidating national consciousness (Anderson, 1991). At a conceptual level, it articulates well with Gilles Deleuze's and Felix Guattari's way of construing the state as a scripting machine that "overcodes" (Deleuze and Guattari, 1977, pp. 139–53) alternative ways of encoding bodies and territories. Manifesting a "dread of uncoded flows" (ibid., p. 140), states have historically constituted themselves through the production of increasingly dense systems of inscription, not only for creating and policing boundaries but also for coding the movement of bodies. Effectively, nationalizing states translate biological bodies into social bodies.[5] The Deleuzian state is therefore a machine of capture (Deleuze and Guattari, 1987, pp. 424–73). In Alonso's terms, it spatializes time; it turns a dynamic of self-production into a reified, unproblematic existence. In order to capture its "people," the state must also capture time; it must monopolize the temporal trajectory through which its existence is made natural and coherent.

Pursuing the temporal aspect of state nation building, we can note that state attempts to control temporality have produced not only the familiar "invention of tradition" that Hobsbawm has analysed in his historical treatment of nationalism (Hobsbawm, 1990), but also attempts to control interpretations of the future. States have put considerable energy into managing anticipation as well as historical memory. For example, as Reinhart Koselleck has pointed out, European states, after the establishment of the Westphalian system in the seventeenth century, instituted controls over astrological readings. They sought to maintain their people's focus on national rather than apocalyptic futures (Koselleck, 1985, pp. 10–12).

By heeding the temporality associated with nation building, we can therefore discern a disjuncture between the state and nation portion of the hyphenated nation-state. The spatial discourse that identifies the state system is uneasily articulated with a temporal one. The state, in its contemporary realization, is understood as a territorial entity, even though it has a history of emergence, having gradually or rapidly, as the case may be, expanded its political, legal, and administrative control by monopolizing violence and incorporating – by statute, by force, and/or by other means – various subunits into a legal and administrative entity with definitive boundaries. The primary understanding of the modern "nation" segment of the hyphenated term is that a nation embodies a coherent culture, united on the basis of shared descent or, at least, incorporating a "people" with a historically stable coherence.

Inasmuch as few if any states contain coherent historically stable communities of shared descent, the symbolic maintenance of the nation-state requires a contentious management of historical narratives as well as territorial space. Effectively, state aspirations to nation-state existence are realized in various modalities of collective autobiography. The nation-state is scripted – in official documents, histories and journalistic commentaries, among other texts – in ways that impose coherence on what is instead a series of fragmentary and arbitrary conditions of historical assemblage. At the same time, other modalities of writing – e.g. journals, diaries, novels, and counter-historical narratives – challenge the state's coherence-producing writing performances.

To resist the metaphysical hypostatization of the nation-state and, at the same time, to appreciate the dynamic of scripting and counter-scripting, as the state has attempted to achieve nation-state coherence and has faced challenges to its self-production, we must again look at concrete historical instances. And, as was noted in the treatments of the Peruvian case, where state constitutions and tax codes have met a challenge at the level of the local ledger, and the US case, where alternative musical genres scripted both dominant and resistant modes of national attachment, we must pay attention to the diverse writing genres within which the ongoing contest over the nation-state political form has been conducted. The significance of writing, as both a technology for the production of national consolidation and as a potentially subversive form, became apparent to the nationalizing state quite early in the European nation-building process. For example, in England, in the early sixteenth century, the state sought to repress oppositional cultures by extending "the scope of treason . . . to encompass treason by word as well as overt deed" in addition to making other "sustained attempts . . . to suppress oppositional writing" (Corrigan and Sayer, 1985, p. 49).

Scripting National Cultures: State Initiatives and Oppositional Forms

The contemporary nation-state's departure from the pre-modern state consisted of both its establishment of clearly marked borders rather than ambiguous frontiers and its development as a political form that became increasingly associated with territorial rather than dynastic markers. Among the analyses which emphasize the shift from the pre-modern to the modern state as one involving a frontier versus border practice respectively is Anthony Giddens' (1983). By the Renaissance, for example, the geopolitics scripted by European cartographers began to reflect a diminution of the significance of emperors and monarchs. In the case of Britain, the spaces on official maps, which had previously been assigned to "insignia of royal power," decreased as the maps increasingly emphasized the markers of land configuration and the boundaries of national territory (Helgerson, 1986, p. 56). By the seventeenth century, the succession of images on maps reflected a historical sequence "from universal Christendom, to dynastic state, to land centered nation" (ibid., p. 62).

French national territory was also inscribed cartographically, but key episodes in the genealogy of French cartography reflect a significant difference in the mode of French nation-building, which stemmed from a revolutionary change in the class basis of national legitimation. Rather than a shift from monarchical inscription to a land-based territory, the key changes in French cartographic practices were aimed at displacing the spaces of aristocratic privilege with a uniform space in keeping with a republican ideal: "the uniform application of law and administration" (Konvitz, 1990, p. 5). Accordingly, the new national cartographic surveys repressed regional differences in order to depersonalize the formerly hierarchically-oriented social order (ibid., p. 4.). In effect, the "redrawing of France's administrative boundaries in the late eighteenth century was a moral act inspired by and symbolizing the highest political ideals;" it reflected a state in the process of producing a modern politico-moral territoriality, one in which the "subordination of the bourgeoisie to the nobility and church" would be overturned (ibid., p. 5).

Map making has been only one of the early genres of state nation building. In subsequent periods, as I have noted, the state's cultural governance has been a major aspect of its modern form of management and legitimation. In Europe, the cultural production aspect of its nation-building practices intensified in the nineteenth century when states found it necessary to produce homogeneous national cultures, not only to legitimate and celebrate their "essential" national characteristics, but also to mobilize their population for work and military service. In the case of an industrializing Europe, the state's scripting was often aided and abetted by the development of national arts programs – for example national theater initiatives in both England and France.

But in both states, "theatrical nationhood" (national theater developments beginning in the mid-nineteenth century) was contested. The inauguration of national theaters took place amidst a struggle "between advocates of a centralized national theater that might reconcile the nation from above and rival, perhaps antagonistic, 'popular' cultures on the social and geographic periphery which resisted this reconciliation under duress" (Kruger, 1992, p. 3). Both movements drew their legitimation from the developing ideology of nineteenth-century nationalism, the idea that the state contains a coherent nation, a "people" that provides the basis of its authority. But, as Loren Kruger has pointed out, "notions of the 'people' offer no stable ground or ruling principle on which to erect the nation or the nation's theater, but rather a battleground of intersecting *fields* on which the legitimacy of national popular representation is publicly contested" (ibid., p. 6). In the French case, the legitimations associated with the struggle over the national theater reflect the same impetus that is evident in the postrevolutionary cartographic genre. Advocates of a national theater saw it the way they saw the *fetes publiques*, as a way to "invite the people into the nation" (ibid., p. 33).

As in the case of national theater initiatives, the contention between state nationalizing and modes of resistance can be observed in a wide variety of genres.[6] To make the issue manageable, however, I focus here on the role of the novel in a variety of nationalizing venues, beginning with the case of the UK in the nineteenth century, because the novel's role in British nation building provides the basis for some telling comparisons with its role in other nation-building contexts. With respect to the relationship between the novel and the nation-state, the novel's "literary geography," the ways in which it encodes territory, is, as Franco Moretti's (1988) investigation has shown, of particular significance.

In his treatment of the British case, Moretti begins with some observations on the remarkable exclusions in the novelistic geography of Jane Austen's plots. Her Britain has no Ireland, Scotland or Wales, and her England is missing its industrial North. The sentimental novel, the genre to which Austen's works belong, is focused on social rather than national issues. As a result, the "small homogeneous England of Austen's novels" (ibid., p. 14) represents a marriage market for the families, ranging from local gentry to national elite, living in England's most well-to-do and populated area. Given Austen's "ideology of space," there is little concern with British nation building as whole, much less with Britain's management of its colonial possessions. When the men in an Austen novel travel abroad, the writer's purpose is not to map global space or to elucidate the political economy of colonialism but either to remove the character from the plot or to allow the character to acquire wealth. And wealth in Austen's plots has nothing to do with economic imperialism and everything to do with funding a character in order to enable him to participate effectively in the social relations of his class.

In contrast to the geography of the sentimental novel, which is focused primarily on social rather than nation-state space, are nineteenth-century historical novels, for example those of Sir Walter Scott, which offer "a veritable phenomenology of the border" (ibid., p. 35). Scott's novels evoke a complex bordered world at a time in which the nation-state system is taking shape, as some borders are hardened and others contested. His novelistic geography treats most of Britain and is concerned with the delineation of its internal, anthropological boundaries. Moreover the historical novels of Scott and other nineteenth-century writers do not simply tell static stories of borders; their plots are associated with a nation-building, geographic dynamic, a process of the erasure of borders "and of the incorporation of the internal periphery into the larger unit of the state" (ibid., p. 40); they reflect and reinforce the geopolitical preoccupations of the nineteenth-century nation-state. The historical novel therefore exceeds Jane Austen's "middle sized-world" (ibid., p. 22);

it articulates a highly politicized geography in comparison with the depoliticized, social geography of the sentimental novel. And, the enlarged space it references enunciates the political consolidation of the state system.

The nineteenth-century English historical novel therefore aided and abetted two major dimensions of the British nation-building project. In their general preoccupation with boundaries, Scott's novels mapped state territorial space, and with their erasure of "internal" anthropological divisions, they complied with the state's domestic biopolitical project, its attempt to fashion a homogenous national culture.[7] However, Scott's literary geography was not uncontested both within Britain's "Celtic fringe" and in her colonies. Pointing toward an alternative, "bardic nationalism," which was developed in a nativist, antiquarian literature, Kate Trumpener describes the nature of the literary opposition, in nineteenth-century romantic novels, to Scott's implied narrative of a progressive British cultural nationhood. The antiquarian revival in the works of such writers as James Macpherson, Sydney Owenson, John Galt, and Charles Maturin valorized disparate folk communities. Their novels were aimed at preserving the separate cultural spaces that Scott's novels sought to displace with a unified British national culture (Trumpener, 1997).

Yet despite significant opposition by writers in Britain's colonies (as well as within the Isles), Scott's novelistic approach to national history – his placement of fictional characters in actual historical situations – has been appropriated to other nation-building contexts in former colonial states. For example, in twentieth-century "Spanish America," the project of consolidating unitary "American" national cultures "propelled the formation of national literatures and histories... seeking to display each country's uniqueness" (Gerassi-Navarro, 1999, p. 109), and many Spanish-American authors saw Scott's historical novels as a model. However, as Nina Gerassi-Navarro points out, "unlike England or France, where the distant past had the ability to awaken a consonant patriotic fervor (as Scott's Waverley novels did), the colonial past in Spanish America was a much more controversial period." As a result, she reads twentieth-century pirate novels "as metaphors for the process of nation building in Spanish America" (ibid., p. 7).

Gerassi-Navarro found a nationalist ideology evinced not as a play of identity/difference within the state, but as an ambiguous play of association and disassociation between local and European cultural attachments. Inasmuch as "Spanish American republics were seen as being contingent on existing European models," the novels tended to articulate the problem of the political and cultural gap between the former colonial powers and the new Spanish American states (ibid., p. 8). Characteristically, then, twentieth-century pirate novels used history not to find national origins but to define their national orders *vis-à-vis* "other identities" (ibid., p. 119).[8] Nevertheless, in Spanish America as well as in many other newer states, which have extracted themselves from colonial rule, literary culture in general and the novel in particular has often contested the state's nation-building project, as for example in the case of some African states. Effectively, as Rhonda Cobham puts it, "the transformation of the anti-imperialist struggle in Africa into a nationalist movement exacerbated a crisis of individual and collective identity that is staged in the African novel" (1992, p. 43).

In the case of Somalia, the novels of Nuruddin Farah are exemplary in this respect, for they have provided a counter-script to that developed during Somalia's most intense nation-building period under General Siyad Barre in the post-independence period (1969–91). Faced with a population of people with a disjunctive set of attachments – a social order composed of lineage clans and sub clans, different classes with, among others, a huge noble-commoner divide, as well as a north–south divide as a colonial legacy – Barre employed educative as well as coercive strategies. While mounting a military irredentist campaign to recapture the Ogaden territory under Ethiopian rule, he also mounted a literacy campaign, seeking to displace tribal attachments with national ones by, among other things, having everyone learn to read and write the Amharic language (Bestemen, 1996).

In his novels, especially *Sweet and Sour Milk* and *Maps*, Farah contests the imposition of a uniform cultural space in Barre's nationalizing project. In *Maps*, for example, the main character, Askar is the posthumous offspring of two patriotic martyrs with Somali nationalist attachments. At the same time, however, he is intimately connected with a foster mother and uncle whose backgrounds do not fit the new Somali national profile in terms of heritage or language.[9] And most significantly there is a disjuncture between the novel's structures of intimacy and the new imposed cultural/political map. For example, speaking to his surrogate mother, Misra, Askar notes that on the one hand, he has intense bodily dependence on her, but at the same time there are "the maps which give me the distance between you and me" and that as a result "we are a million minutes apart, your 'anatomy' and mine" (Farah, 1986, p. 18). And, in turn, Misra refers to the divisive forces of the nationalizing campaign, telling Askar "when you do well, the credit is not mine but your people's that is your [Somali] nation whose identity I do not share" (ibid., pp. 40–1).

Ultimately, the interacting characters in Farah's *Maps* reflect the instabilities of the identities of the people contained within the nationalizing state. Through the characters' queries about their attachments and interrelationships and their confusions about shifting designations of place – for example Askar's Uncle Hilaal sees a map in Askar's room in which the word Ogaden has been erased and replaced by the term "Western Somalia" (ibid., p. 216) – Farah challenges the presumption of a coherent Somali nation and resists complicity with the project of imposing an unambiguous nation-state frame on a complex African inter-cultural order.

In some cases, the novel has been resistant to the state's culture governance in the USA as well. Recalling the above-noted challenge of African-American musical forms – through both their alternative loci of enunciation and their alternative structures of intelligibility (or modes of enunciation)[10] – a similar challenge is apparent in African-American literary forms, for example in the novels of Toni Morrison. Morrison's political challenge to a unitary national literary culture is explicit. She has pointed to a paradox inherent in her participation as a novelist in a culture of literacy. Although she "participates in the public sphere constituted by print literacy, . . . her fiction strains to constitute itself as anti-literature and to address a type of racial community that she herself recognizes to be unavailable to the novelist" (Dubey, 1999, p. 188). That "racial community," moreover, is not simply another American. Like many black writers, artists, and musicians, especially those who have preceded her in the twentieth century (and most notably those associated with the Harlem Renaissance), Morrison's audience/constituency takes on its coherence as a protean transnational black culture, forged as much through structures of exclusion and episodes of displacement as through practices of solidarity. And much of the cultural imaginary, which forms the implied readership of her novels, is "preliterate" (ibid.). For Morrison, as for many black intellectuals and artists, there is no unitary national culture within the USA. Like the resistant subcultures of Native Americans, who in films, novels and other genres have attempted to resist their inscription as racial minorities and reconstitute themselves as nations, many African-American artists and intellectuals have seen their contributions to cultural expression as a form of extra-state nationalizing against a history of the state's culture governance.[11]

The Nation-State and its Others: The State as a Territorial and Biopolitical Enterprise

In a telling response to Jürgen Habermas' recent gloss on the "European Nation-State" (Habermas, 1998), Timothy Mitchell points out that contrary to Habermas' implication, imperialism, rather than being incidental to state nationalisms, was the context shaping them (Mitchell, 1998, p. 413). As my analysis thus far has implied, this is also true of a major dimension of imperialism's legacy – racism. Rather than merely a policy problem for the nation-state (as implied in much of official and academic policy writing) racism has been a fundamental shaping force, although in different ways in different state nation-building venues. In Ecuador, for example, the institutionalized "ideology of national identity results in a racial map of national territory: urban centers... are associated with modernity, while rural areas are viewed as places of racial inferiority, violence, backwardness, savagery, and cultural deprivation" (Rahier, 1999, p. 106).

The Ecuadorian "racial/spatial order," which is reflected in a regional cultural genre, a beauty contest that is significantly racially coded, contrasts with the way the US racial/spatial order is often addressed – through erasure – as in the mid-twentieth century production of *Oklahoma!* But as the Ecuadorian "beauty contest" example suggests, there remains an additional shaping context to be addressed in this analysis of the state's nationalizing project – its management of the meaning and role of women, particularly with respect to the relationship between "nationalism and sexualities."[12] For example, treating "nations" as "elaborate social *practices* enacted through time," and expressed in diverse media and print genres, Anne McClintock, in her analysis of the case of South Africa, points out the ways in which women have been confined within the state's national story to the role of "biological producers of National groups... transmitters and producers of cultural narratives,... and reproducers of the boundaries of the nation by accepting or refusing sexual intercourse or marriage with prescribed groups of men" (McClintock, 1991, p. 104).

Biopolitically, then, state nationalizing often constructs good versus bad women on the basis of their relationship to a racial, ethnic, or class-legitimated familial sexuality. The nation-building impetus of this construction was evident in the musical, *Oklahoma!*, at the point where a choreographed dance routine has "bad women" (prostitutes and dance hall girls in skimpy outfits and mesh stockings), being displaced by farmers' wives in long calico dresses. It is a construction also discerned in the case of Argentina, where the state's nationalizing discourse was shaped by locating women's appropriate versus inappropriate sexuality: "The issue of 'bad women' triggered the discourse on female nationality" at a historically crucial moment in Argentina's nation-building process, a time in the nineteenth century when "thousands of women left their European homelands . . . in search of a better life in the Americas." Argentina's "imagined community," therefore owed much of its contour to the regulation of women's sexual identities: "the Honor of the nation required a particular practice of sexual virtue" (Guy, 1992, p. 202).

Because the kind of regulation of women in state nation-building and maintenance varies considerably, I want to conclude by contrasting two disparate cases, the USA and India. In the USA, the economic exploitation of women increased when, in the 1970s changes in global economic forces, and accompanying movements of persons and a resulting fiscal austerity led to a "crumbling" of what Nancy Fraser refers to as "the gender order . . . centered on the ideal of the family wage." Within the modern, industrialized welfare state, she asserts, an imposed "normative picture of the family" meant that "people were supposed to be organized into heterosexual, male-headed nuclear families, which lived principally from the man's labor market earnings" (Fraser, 1994, p. 591). Among the consequences of the crumbling of the gender order and the insufficiency of the model of the "family wage" (because in a period of post-industrial capitalism fewer families could survive on a single income), "needy women" were increasingly subjected to an "exploitable dependency" (ibid., p. 597).

In an industrializing state such as India, where there is no state-sponsored "welfare system" the situation of "exploitable dependency" is far more pervasive, uninhibited, and violent. The exploitation is especially severe in the case of "tribal" women who, as the writer and tribal activist, Mahashweta Devi, points out, have not been a part of "the decolonization of India" (Devi, 1995, p. xi). Many tribal women end up as prostitutes in "the bonded labor system," handed over by their families to pay off a debt, which can never be fully paid because the accounts carry compounded interest (ibid., p. xix). While Indian nation-building is valorized in official texts as a case of successful development, Mahashweta has described her counter-hegemonic task as, among other things, resistance to "development" because of the state's complicity in the devastating effects of the capitalist market on tribal women and children (ibid., p. xxii).

In addition to her strenuous political organizing among "tribals," Mahashweta's writing constitutes an exemplary counter-script to that of the nationalizing state, which has exploited indigenous bodies, while, at the same time, engaging in a biopolitical, nationalizing discourse that erases their presence. In her story, *Douloti the Bountiful*, Mahashweta displaces the state's national cartography with "the socially invested cartography of bonded labor," (Spivak, 1992, p. 98), and she challenges the state's biopolitical discourse by showing how women's bodies are casualties of "national industrial, and transnational global capital" (ibid., pp. 101–2). Her Douloti, the daughter of a tribal bonded worker, is sold into prostitution. She dies on a journey homeward, after having given up on being taken in by a hospital to be treated for venereal disease.

Ultimately, therefore, to describe nationhood as the mere aggregation of diverse groups, or to use such cleansing language as "the forging of a national culture," is to launder some of nation-building's catastrophic effects. Mahashweta supplies a powerful recognition of the violence of a nationalizing state. Her Douloti, exploited and ravaged by illness, lies down to die on a large map of India that a teacher,

Mohan, has inscribed for his students in a clay courtyard to celebrate Independence Day. As Mahashweta puts it (in the last lines of the story):

Today on the fifteenth of *August*, Douloti has left no room at all in the India of people like Mohan for planting the standard of the Independence flag. What will Mohan do now? Douloti is all over India (Devi, 1992, p. 93).

I will give the last words to Gayatri Spivak, whose characterization of Douloti's final gesture eloquently captures this chapter's theme: the cartographic and biopolitical scripting agon involved in the struggle between nation-states and those who resist them. Douloti, Spivak writes, "reinscribes this official map of the nation by the zoograph of the unaccommodated female body restored to the economy of nature" (1992, p. 112).

ENDNOTES

1. For example, as Nugent points out, in one region, the Cachapoyas in the Northern Sierra, there were two different periods with respect to reactions to state centralizing initiatives. In the 1930s, the state was regarded as a "protector and potential liberator of a self-defined 'moral community'" so that the local, marginalized groups used the state to resist the local power brokers and assisted the state in the process of integrating the region and nationalizing the regional population but in the second phase, of centralization in the 1970s, the same "moral community" regarded the state's expansionist policies as immoral and mobilized to thwart its attempt to impose greater control.
2. In referring to state biopolitical initiatives, I am influenced by Giorgio Agamben's analysis of the topology of sovereignty. According to Agamben, sovereignty exists in "the intersection between the juridico-legal and biopolitical models of power." In addition to its legal supports and legitimations, it is situated in a complex topology of lives, both inside and outside its jurisdiction (Agamben, 1998, p. 6).
3. The so-called "civilized tribes" from the East brought black slaves and black relatives with them ("thousands of Negroes were neighbors or slaves of the Five Civilized Tribes"), and many became cowboys (Durham and Jones, 1965, pp. 18–19).
4. It should be noted in addition that the recognition of an incomplete nationalizing process is often part of a state's nation-building strategy. As David Lloyd and Paul Thomas have pointed out, the state's justification for the need to continue the forging of a harmonious national order encourages it to locate itself in a narrative of an "asymptotically deferred harmony" (Lloyd and Thomas, 1998, p. 33).
5. This way of formulating Deleuzes' and Guattari's approach to the state's coding of bodies belongs to Paul Patton (2000, pp. 140–1).
6. For example, in the case of France the contestation over creating a national culture is manifested in the development of the French Grand Opera, a story well told in Fulcher (1987).
7. Georg Lukacs, perhaps the best-known critical commentator on Scott's novels, also recognized the role of the novels, particularly the historical novel, in Britain's nation-building project. He saw in Scott's novels an inscription of "the complex and intricate path which led to England's national greatness and to the formation of the national character" (1965, p. 54).

8. Ibid., p. 119. The identity practices of the former colonial states have in effect reciprocated the uses of the other reflected in these novels. As Michael Kearney points out, these states intensified their biopolitical practices to supplement its territorial management. They became "social, cultural, and political form[s]" with "absolute geopolitical and social boundaries inscribed on territories and persons, demarcating space and those who are members from those who are not" (Kearney, 1991, p. 54).
9. My analysis here is influenced by the discussion in Wright (1997).
10. As Houston A. Baker, Jr. has put it "A Nation's emergence is always predicated on the construction of a field of meaningful sounds." And, he adds, the "*national* enterprise of black artists and spokespersons, since the beginning of the century, has been involved in a mode of *sounding* reality, that resists the Euro-American state's of a unitary nationalizing enterprise" (1987, p. 71).
11. For a treatment of the racialization of Native Americans during the US nation-building process, see Borneman (1995). There is a variety of genres within which Native Americans have countered the historical identity lent them in the Euro-American nation-building story. Among the contemporary genres of resistance are the many novels and stories of Sherman Alexie (as well as his feature film, *Smoke Signals*), the public installations of Edgar Heap of Birds (e.g. 1986), and activist Leonard Peltier's writings (1999).
12. For a general treatment of this relationship, see Parker et al. (1992).

BIBLIOGRAPHY

Agamben, G. 1998. *Homo Sacer: Sovereign Power and Bare Life*, Transl. Daniel Heller-Roazen. Stanford, CA: Stanford University Press.

Alonso, A. M. 1994 The politics of space, time and substance: state formation, nationalism, and ethnicity. *Annual Review of Anthropology*, 23, 379–405.

Anderson, B. 1991. *Imagined Communities*, extended and revised edn. New York: Verso.

Baker, H. A., Jr. 1987. *Modernism and the Harlem Renaissance*. Chicago: University of Chicago Press.

Bendix, R. 1964. *Nation-Building and Citizenship*. New York: Wiley.

Bestemen, C. 1996. Violent politics and the politics of violence: the dissolution of the Somali nation-state. *American Ethnologist*, 23(3), 579–96.

Bhabha, H. K. (ed.). 1990. *Nation and Narration*. New York: Routledge.

Borneman, J. 1995. American anthropology as foreign policy. *American Anthropologist*, 97(4), 663–72.

Brubaker, R. 1996. *Nationalism Reframed*. New York: Cambridge University Press.

Cobham, R. 1992. Misgendering the nation: African nationalist fictions and Nuhruddin Farah's *Maps*. In M. R. Parker et al. (eds.) *Nationalism and Sexualities*. New York: Routledge, 42–59.

Connor, W. 1990. When is a Nation. *Ethnic and Racial Studies*, 13(1), 92–103.

Corrigan, P. and Sayer, D. 1985. *The Great Arch*. Oxford: Basil Blackwell.

Day, R. 1989. "Sooners" or "Goners," they were hellbent on grabbing free land. *Smithsonian*, 20(8), 192–206.

Deleuze, G. and Guattari, F. 1977. *Anti-Oedipus: Capitalism and Schizophrenia*, Transl. R. Hurley et al. New York: Viking.

Deleuze, G. and Guattari, F. 1987. *A Thousand Plateaus: Capitalism and Schizophrenia*, Transl. B. Massumi. Minneapolis, MN: University of Minnesota Press.

Devi, M. 1995. The author in conversation (with Gayatri Chakravorty Spivak). In G. C. Spivak (transl.) *Imaginary Maps*. New York: Routledge, ix–xxii.

Devine, T. L. 1999. Indigenous identity and identification in Peru: *Indigenismo*, education and contradictions in state discourses. *Journal of Latin American Cultural Studies*, 8(1), 63–74.
Dubey, M. 1999. The politics of genre in *Beloved. Novel: A Forum on Fiction*, 32(2), 187–206.
Durham, P. and Jones, E. L. 1965. *The Negro Cowboys*. New York: Dodd, Mead & Co.
Emerson, R. 1960. *From Empire to Nation*. Cambridge, MA: Harvard University Press.
Farah, N. 1986. *Maps*. NY: Pantheon.
Filmer, P., Rimmer, V., and Walsh, D. 1999. *Oklahoma!*: ideology and politics in the vernacular tradition of the American musical. *Popular Music*, 18(3), 381–95.
Foster, R. J. 1991. Making national cultures in the global ecumene. *Annual Review of Anthropology*, 20, 235–60.
Fraser, N. 1994. After the family wage: gender equity and the welfare state. *Political Theory*, 22 (4), 591–618.
Fulcher, J. 1987. *The Nation's Image: French Grand Opera as Politics and Politicized Art*. New York: Cambridge University Press.
Gerassi-Navarro, N. 1999. *Pirate Novels: Fictions of Nation Building in Spanish America*. Durham, NC: Duke University Press.
Giddens, A. 1983. *The Nation State and Violence*. Cambridge: Basil Blackwell.
Gilroy, P. 1993. *The Black Atlantic: Modernity and Double Consciousness*. Cambridge, MA: Harvard University Press.
Guardino, P. and Walker, C. 1992. The state, society, and politics in Peru and Mexico in the colonial and early Republican periods. *Latin American Perspectives*, 19 (2), 10–43.
Guy, D. J. 1992. "White slavery," citizenship and nationality in Argentina. In M. R. Parker et al. (eds.) *Nationalism and Sexualities*. New York: Routledge. 201–17.
Habermas, J. 1998. The European Nation-State: on the past and future of sovereignty and citizenship. *Public Culture* 10(2), 397–416.
Heap of Birds, E. 1986. *Sharp Rocks*. Buffalo, NY: CEPA.
Helgerson, R. 1986. the land speaks: cartography, chorography, and subversion in Renaissance England. *Representations*, 16(4), 51–85.
Hobsbawm, E. J. 1990. *Nations and Nationalism Since 1780*. New York: Cambridge University Press.
Jarrett, M. 1994. The tenor's vehicle: reading *Way Out West. LIT*, 5, 1–12.
Kearney, M. 1991. Borders and boundaries of State and Self at the end of Empire. *Journal of Historical Sociology*, 4(1), 52–74.
Kendrick, W. 1999. Review quotation on the back of Mario Vargas Llosa 1999.
Konvitz, J. W. 1990. The nation-state, Paris and cartography in eighteenth- and nineteenth century France. *Journal of Historical Geography*, 16(1), 3–16.
Koselleck, R. 1985. *Futures Past: On the Semantics of Historical Time*, Transl. K. Tribe. Cambridge, MA: MIT Press.
Kruger, L. 1992. *The National Stage: Theater and Cultural Legitimation in England, France, and America*. Chicago: University of Chicago Press.
Lloyd, D. and Thomas, P. 1998. *Culture and the State*. New York: Routledge.
Lukacs, G. 1965. *The Historical Novel*, Transl. H. and S. Mitchell. New York: Humanities Press.
McClintock, A. 1991. "No Longer a Future Heaven": women and nationalism in South Africa. *Transition*, 51, 104–23.
Mitchell, T. 1998. Nationalism, imperialism, economism: a comment on Habermas. *Public Culture*, 10(2), 417–24.
Mitchell, T. 1999. Society, economy, and the state effect. In G. Steinmetz (ed.) *State/Culture*. Ithaca, NY: Cornell University Press, 76–97.
Moretti, F. 1998. *Atlas of the European Novel: 1800–1900*. New York: Verso.

Nugent, D. 1994. Building the state, making the nation: the bases of limits of state centralization in "modern" Peru. *American Anthropologist*, 96(2), 333–69.
Parker, M. R., Sommer, D., and Yaeger, P. (eds.). 1992. *Nationalisms and Sexualities*. New York: Routledge.
Patton, P. 2000. *Deleuze and the Political*. London: Routledge.
Peltier, L. 1999. *Prison Writings*. New York: St. Martin's Press.
Radcliffe, S. A. 1998. Frontiers and popular nationhood: geographies of identity in the 1995 Ecuador-Peru border dispute. *Political Geography*, 17(3), 273–93.
Rahier, J. M. 1999. Body politics in Black and White: *Senoras, Mujeres, Blanqueamiento* and Miss Emeraldas 1997–1998, Ecuador. *Women & Performance*, 11(1), 103–19.
Smith, A. D. 1986. *The Ethnic Origins of Nations*. Oxford: Blackwell.
Spivak, G. C. 1992. Woman in Difference: Mahashweta Devi's "*Douloti the Bountiful*. In M. R. Parker et al. (eds.) *Nationalism and Sexualities*. New York: Routledge. 96–117.
Thurner, M. 1995. "*Republicanos*" and "*la Comunidad de Peruanos*": unimagined political communities in postcolonial Andean Peru. *Journal of Latin American Studies* 27(2), 291–318.
Trumpener, K. 1997. *Bardic Nationalism: The Romantic Novel and the British Empire*. Princeton, NJ: Princeton University Press.
Vargas Llosa, M. 1990. Questions of conquest. *Harper's*, 281 (1687), 45–53.
Vargas Llosa, M. 1999. *The Notebooks of Don Rigoberto*, Transl. E. Grossman. New York: Penguin.
Wright, D. 1997. Nations as fictions: postmodernism in the novels of Nuhruddin Farah. *Critique*, 38(3), 193–204.

Chapter 19

Places of Memory

Karen E. Till

In 1989, shortly after being elected the new leader of Yugoslavia, Slobodan Milosevic rescinded political autonomy in the province of Kosovo and established a monumental memorial. The memorial was erected at a historic battlefield known as Kossovo Polje – or field of the black birds – a place long considered to be the heart of "Old Serbia" (Kaplan, 1993). According to Serbian legend, it was at Kossovo Polje that Prince Lazar gave up his worldly kingdom to fight for god's glory in the battle between the Turks and Serbs in 1389. The words Lazar is believed to have spoken before his defeat are now inscribed at the memorial's base:

> Whoever is a Serb and of Serbian birth
> And who does not come to Kossovo Polje
> To do battle against the Turks
> Let him have neither a male
> Nor a female offspring
> Let him have no crop
> (quoted in Kaplan, 1993, p. 39).

Establishing a sacred place of honor in a predominantly Albanian Muslim province was a symbolic and political act. The memorial's mythic location, monumental form, and associated ceremonial rituals were intended to create a sense of continuity between past and present, and to demarcate a Serbian presence in the landscape. By making a particular rendering of an ethnic national past concrete, the new Milosevic regime attempted to (re)claim political control over space.

The construction of the Kossovo Polje memorial illustrates the more general process whereby a group represents the past through place in an attempt to claim territory, establish social boundaries, and justify political actions. Historically, officials of the state have sanctified places of memory at prominent locations in symbolic settings as part of the process of nation building (Anderson, 1991; Hobsbawm and Ranger, 1983).[1] Elites in many societies have constructed statuary, memorials, museums, grand boulevards, public squares, and ornate buildings to function as "theaters of memory" where selective histories about the state could be ritually enacted (Boyer, 1994).[2] However, the material and symbolic presence of monumental landscapes and sacred topographies does not necessarily indicate a coherent set of political agendas (after Agnew, 1998). Nor are such places mere "stages" that reflect state power or a regime's historical interpretations. Rather, places of memory and the processes associated with their establishment, demonstrate the complex ways that nationalist imaginations, power relations, and social identities are spatially produced (after Johnson, 1995).

This chapter provides an overview of the geographical literature about place and memory with a focus on national commemorations.[3] I have organized the literature according to three themes: the sacralization of national imaginaries, changing political regimes, and conflict over national places. Before discussing this literature, I first introduce key theoretical works about social memory in the following section. By social memory, I mean the ongoing process whereby social groups "map" their myths of self onto and through a place and time (Till, 1999).[4] When these cultural practices take place in public domains, the results may be quite politicized. Groups and individuals may struggle with one another to gain authority to represent their version of the past in the built environment, the media, and legal arenas. The "politics of memory" associated with such places, then, refers to the spaces and processes of negotiation about whose conception of the past should prevail in the public realm (ibid.). Because the meanings of these places are not stable in time or space, the politics of memory also refers to the ways and reasons groups attempt to "fix" time and identity through the material and symbolic qualities of place.

Topographies and Sites of Social Memory

Authors writing about social memory in recent years have analysed national commemorations (Gillis, 1994), representations of traumatic pasts (Dowler, 1998; Sturken, 1997; Young, 1993), and tourism and the heritage industry (DeLyser, 1999; Hoelscher, 1998; Lowenthal, 1996). In this wide-ranging interdisciplinary discussion, the works of Maurice Halbwachs (a French philosopher and sociologist) and Pierre Nora (a French social historian), have been particularly significant. For Halbwachs, personal memory is not "stored" in the unconscious, as Sigmund Freud suggested, but is always constructed and located in the social environments of the present (Halbwachs, 1992 [1952, 1941]). By belonging to social groups, including family, class, and religious organizations, individuals learn narratives about their world that make social life intelligible. They also engage in repetitive cultural performances that provide continuity between past and present. Individuals learn to "remember" the past through written chronicles, traditions, and commemorative activities, including collecting relics, naming places, and observing anniversaries. These social narratives and cultural practices (that constitute "collective frameworks of memory") are infused with moral meaning and continuously change according to the needs of the present (Halbwachs, 1925).

Because stories about the past are always in flux, groups often create "topographies" of memory to make the connection between past and present seem permanent and tangible. Halbwachs argued that group remembrances endure when they have a "double focus – a physical object, a material reality such as a statue, a monument, a place in space, and also a symbol, or something of spiritual significance, something shared by the group that adheres to and is superimposed upon this physical reality" (Halbwachs, 1992 [1952, 1941], p. 204). The dense experiential and social qualities of place and landscape therefore not only frame social memory, they also situate and spatially constitute group remembrances (Till, 1996). Although Halbwachs has been criticized for his implicit assumption of a Durkheimian collective conscious (Withers, 1996), his work is important because he emphasized the role of language and landscape in social recall.

Like Halbwachs, Pierra Nora also calls attention to the material and symbolic nature of memory. Nora's (1997) ambitious project documents the diversity of French national sites of memory, including architecture, public festivals, books, and monuments. Nora (1989) argues that because memory has a history, the "where" of memory also changes through time. We no longer live in environments of memory (*milieux de mémoire*) according to Nora, but instead create self-reflexive "sites" (*lieux de mémoire*). Sites of memory result from the "deritualization" of the world: "we must deliberately create archives, maintain anniversaries, organize celebrations... because such activities no longer occur naturally" (Nora, 1989, p. 12). Modern memory, as a self-conscious representation of the past that has been transfigured by history, is archival and relies on the materiality of the trace and the visibility of the image. In contrast, "true memory" is "concrete, [and located in] spaces, gestures, images, and objects" (ibid, p. 9). Although the distinction between "true" and "modern" memory practices and geographies is problematic, Nora's work demonstrates how national and elite memory is fragmented and opposed by alternative forms of social memory (Withers, 1996).

Inspired by Halbwachs and Nora, scholars from many disciplines now pay attention to the material landscapes and cultural performances of social memory. Yet while the memory literature is replete with spatial metaphors, most scholars neither acknowledge the politically contestable and contradictory nature of space, place, and scale (Alderman, 1996), nor examine the ways that social memory is spatially constituted. Some scholars fetishize geographical concepts or classify the complex space–time formations of memory according to temporal categories only. Implicit to Nora's work, for example, is an underlying nostalgia for the supposed loss of "real" environments in which experience and its "true" recollection take place "naturally." Nuala Johnson (1999) argues that the temporal framework of "Traditional vs. Modern" subsumes the geographies of remembrance under the histories of memory "in ways which treat space as epiphenomenal to the historical process" (p. 39).

Recent work by geographers demonstrates that places of memory are more than monumental stages or sites of important national events. They also constitute historical meanings, social relations, and power relations. Places are spatial and social contexts of events, activities, and peoples (Agnew and Duncan, 1989; Entrikin, 1991; see also Staeheli, chapter 11, this volume); they are centers of meaning, memory, and experience for individuals and groups (Tuan, 1974). Far from being "rooted" or stable, places are porous networks of social relations that continuously change because of the particular ways they are interconnected to (and in turn shape) other places and peoples (Massey, 1997). As I describe below, political struggles over who controls the past in the public arena are often intimately linked to competing interests over the production of power relations and political–economic space. The process of place-making, in other words, is central to social memory and the formation of cultural and political communities (Adams et al., 2001; Agnew and Duncan, 1989; Keith and Pile, 1993).

Place and the Sacralization of National Imaginaries: Monuments, Memorials, and Museums

During the period of nation building in Europe, official places of memory were created to establish a topography of "a people" and to maintain social stability, existing power relations, and institutional continuity. Selective elite interpretations of the past (by predominantly white males) tended to be abstract and normative (Bodnar, 1992), and understandings of the nation as timeless and sacred were represented through the relative locations, designs, and functions of places like monuments, memorials, and museums. Historical narratives and representations of empire, nation, and state were also naturalized through gender relations, in particular through the adulation of male, heroic bodies in public spaces. These memorial landscapes defined sacred centers and political power; they drew from or competed with previously existing topographies of social recollection (Edensor, 1997; Johnson, 1995; Lowenthal, 1985).

In late nineteenth and early twentieth century Italy, for example, officials attempted to define a united nation through the Vittoriano Emanuele II Monument in Rome. According to David Atkinson and Denis Cosgrove (1998), a timeless, imperial Rome was evoked through the veneration of dead heroes, grandiose and ornate designs, and the monument's location. As initially conceived in 1878–82, the Vittoriano was built as a sacred altar honoring a dead king. Later, in 1921 under Mussolini's fascist regime, the site also celebrated the Cult of the Unknown Soldier. The white marble monument – reaching from earth (symbolized by dead bodies) to the heavens (represented by the Beaux-Arts, neoclassical design) – materialized a mystical understanding of Italy as a transcendental, classical empire. Located on the north-facing slope of the Capitoline Hill, adjacent to the Imperial and Roman Fora, the Vittoriano also linked the new political state to the mythical acropolis of ancient Rome and the ideal of an "eternal city." These spectacular productions, however, led to "affectionate popular dismissal" rather than awe. Furthermore, John Agnew (1998) suggests that the aesthetic "overproduction" of places like the Vittoriano reflected the ideological incoherence, rather than popularity, of nationalistic agendas, especially under Fascism.

Through monuments, officials evoked myths of a timeless nation (see also Azaryahu and Kellerman, 1999) as well as attempted to naturalize gendered representations of the nation. Gendered symbols of social identities and power relations in the landscape implicitly or explicitly construct a national "norm" that becomes an unquestioned experience of everyday life (Nash, 1994; Sharp, 1996). The establishment and unveiling of the George Etienne Cartier monument in Montreal, Canada in 1919, for example, symbolized the "drama, romance, and emotion of several dimensions of patriotism" (Osborne, 1998, p. 445) through the embodiment of a heroic male citizen. Although Cartier's role in history was ambiguous (leading to various social interpretations of the statue), cast in stone he stood with arms outstretched, surrounded by nine female figures representing the Canadian Provinces. Twenty meters overhead, the bronze-winged female figure Renown bestowed a laurel wreath upon his head. Typical of national statuary and commemorative practices elsewhere, the "great man" metonymically represented the nation, whereas female figures symbolized the "motherland" (Johnson, 1995; Monk, 1992; Sharp, 1996; Warner, 1985).

Gendered national imaginaries are also reified through war memorials. Although women have been active participants in warfare, they are often represented as mothers only (and not also warriors) in social memory practices of war and are thereby excluded from public political landscapes (Dowler, 1988). War memorials are masculine spaces and include monuments for generals, tombs to the Unknown Soldier, mass or military cemeteries, commemorative fields, historic battlefields, prisons, and their associated ceremonies (Mosse, 1990; Raivo, 1998). At war memorials, soldiers – represented by sacred relics of dead male bodies – are commemorated as national martyrs who died protecting their homeland and "vulnerable" citizens. Although these places of memory reinforce gendered understandings of the nation as a fraternal brotherhood, the meanings and social identities of "war dead," and of victims, perpetrators, and heroes change through time. Further, public commemorations of war are far from straightforward and vary in different national, local, and political contexts. The "Great War" (World War I), for example, was difficult to commemorate and remember socially because the experience of that war was unintelligible for the living (Johnson, 1999; Winter, 1995). Different cultural practices resulted from social interpretations of already existing local and national traditions in particular postwar political contexts. France paid for the return home of the soldiers' dead bodies (Sherman, 1994), whereas Britain resisted attempts by families to repatriate bodies from the Western Front (Heffernan, 1995). In Ireland, the population did not respond to peace parades and annual remembrance-day celebrations uniformly because such commemorations were taking place in the context of a civil war (Johnson, 1999).

The sacred time–space of death and the veneration of relics also structure and define the museum (cf. Fabian, 1983). The "bodies" on display at such "memory palaces" symbolize a possessive understanding of the nation (Handler, 1988). As a type of place, museums represent the nation through the cultural objects that have been collected, classified, sorted, and exhibited. During the nation-building period, museum exhibitions institutionalized classification schemes based upon the natural sciences, resulting in knowledges biased toward vision and categories defined by temporal, racial, and gendered hierarchies (Bennett, 1995; Haraway, 1988; Kulik, 1989). Exhibitions typically demonstrated the progress of "civilized" national cultures and their distinction from more "primative" ones through the spatial relationships of artefacts on display (Clifford, 1988).

Museum scholars have examined the institutional histories of museums, their collecting practices, architectural forms, representational strategies, and systems of display, as well as the social subjectivities produced through such discursive apparatuses (after Rose, 2001). According to Rose (2001), however, the literature neglects the archives, laboratories, service areas, and offices where museum professionals produce their knowledges. In addition, few scholars have discussed the popular reception of exhibitions or the controversies surrounding museums and their practices (for an exception see Karp et al., 1992). Recent work has begun to examine visitors' interpretations of exhibitions and the work of museum experts through ethnographic approaches. For example, the "Chapters of Life, 1900–1993" exhibition that opened at the German Historical Museum in Berlin shortly after unification had a strong visitor response (Till, 2001). In this exhibit, the West German state was represented as the inevitable precursor to the unified Germany, whereas the present past of the German Democratic Republic (GDR) was displayed as belonging to a more distant past, one that also included National Socialism. Although many visitors appear to have (re)framed their personal experiences and historical understandings according to the exhibition's interpretative spaces, others – in particular former East Germans – were critical of the exhibition. Through visitor books, surveys, and informal conversations, they challenged the exhibition authors' interpretations of the past and attempted to "rewrite" the narrative of "Germany" and the GDR. Exhibition authors, impressed by these responses, considered ways to incorporate visitors' voices into future exhibits about contemporary German history. This case study suggests that while museum experts work within institutional constraints, they view the museum as a dialogic, rather than authoritative, social space. It also demonstrates that popular interpretations of official narratives are far from uniform. Indeed, citizens' everyday knowledges may challenge dominant cultural practices and representations associated with national places of memory.

Changing Political Regimes and Cityscapes

When everyday routines, political regimes, economic structures, and symbolic systems are in flux, the constructed "normality" of places – and their associated identities, power relations, and social practices – may be questioned. Localized struggles over the meanings, forms, and locations of places of memory are often tied to larger political disputes about who has the authority to represent the past in a society. Renaming streets and urban districts, for example, is one way that officials have attempted to canonize a version of the past in the urban landscape to support a particular political order (Alderman, 1996; Azaryahu, 1997; De Soto, 1996; Pred, 1992). Helga Leitner and Peter Kang (1999) describe how Chinese Nationalists renamed streets, schools, theaters, and other public buildings, squares, and spaces in Taipei, Taiwan after 1949. Nationalists attempted to remap "the geography of the Chinese homeland onto the cityscapes of Taipei" (Leitner and Kang, 1999, p. 221). To promote their territorial aspirations, they used place names from the asserted living space of the Chinese nation-state, Chinese nationalistic slogans, and names of national heroes. Because street names create symbolic and routine landscapes, individuals and groups may establish group identities and make sense of their city through them. But they may also contest them. Indeed, recent opposition movements in Taiwan challenged the Kuomintang (KMT) by demanding that the city street names reflecting authoritarian Chinese Nationalism be changed (ibid.).

During times of political transition, officials may invest a lot of money and time through more dramatic landscape inscriptions to accumulate "symbolic capital" and project a particular worldview (Forest and Johnson, forthcoming, after Bourdieu, 1990 and Verdery, 1999). The postcolonial Tamil Dravida Munnetra Kazhagam party, for example, erected statues, temples, and buildings in Madras after 1949

(Lewandowski, 1984). Similarly, statues of British political figures were removed and replaced by those of Ceylonese (later labeled Sri Lankan) nationalist leaders after independence in Kandy (Duncan, 1989). Yet such "symbolic decolonizations" of urban space do not necessarily indicate coherent nationalist movements. As James Duncan (1989) described, this process reaffirmed the hegemonic political position of only one party (the United National Party) rather than a working alliance of parties after independence in Kandy. Furthermore, the replacement of one nation-state or political regime by another does not necessarily entail the simple erasure of memory in the material landscape. Although new regimes may add symbols, alter the physical context, or relocate monuments to articulate power relations, they also incorporate imagery and features from previous regimes or states to make their "new" memorials.

The (re)establishment of national places of memory in capital and other symbolic cities thus provides evidence about continuities between past and present states and regimes. Current research about memorials and monuments in former authoritarian societies describes a multifaceted process of commemoration through which political change and continuity, and formations of civil society can be analysed (Foote et al., 2000; Legacies of Authoritarianism, 2001; Till, 1999). Benjamin Forest and Juliet Johnson (2001), for example, describe public monuments in Moscow to examine domestic political struggles at various scales in post-Soviet Russia. Their study of the Victory Park monument at Poklonnaya Gora (commemorating the Soviet victory over Nazism), Lenin's Mausoleum, the Exhibition of Economic Achievements of the Soviet Union, and the "Park of Totalitarian Art" suggest that while Russian elites may be uncomfortable with the Soviet legacy, they would rather reinterpret than erase this past. Visitor surveys conducted at these places also indicated the limited popular appeal of civic nationalism in Russia and the associated difficulties of creating new (i.e. post-Soviet) symbolic capital (ibid.).

Conflict over National Places of Memory

National places of memory are not simply imposed onto an empty landscape by a seemingly coherent elite. Different political parties, factions, and "publics" negotiate understandings of the past (and of social identity) at multiple scales through place. David Harvey's (1979) description of the history of the Basilica of Sacré Coeur in Paris, for example, demonstrates that deep fissures existed in nineteenth-century Parisian and national politics. The erection of the building resulted from an uneasy alliance between conservative Catholics and monarchists to commemorate their martyrs. Yet communards also laid symbolic claim to Montmartre and commemorated their lost heroes at the site. Groups and individuals evoke geographic imaginations in very different, and often competing, ways in the construction of place and memory to realize their goals in the present and future. Even internationally hallowed places like Auschwitz – the symbolic site of the Holocaust – have been claimed by various groups to achieve various political agendas (Charlesworth, 1994). Conflicting Catholic and former Communist claims to the site as a place of Polish martyrdom resulted in contested memorializations and meanings of Auschwitz, interpretations that underplayed the importance of the site for Jewish memory.

Social groups have contested official representations of the past and offered alternatives to claim a political "voice" in the public realm. During the centenary celebrations of the 1798 rebellion in Ireland, for example, Irish nationalists erected statutes at historic sites to confront British unionists' interpretations of the past (Johnson, 1994). At the same time, however, the unveilings of these figures were often controversial as a result of the existing divisions within the nationalist opposition. Although other such historical examples exist, John Gillis (1994) argues that the public scrutiny of national history increased after the 1960s and resulted in new national memorial practices and landscapes by the 1980s. He cites the Vietnam War Memorial in Washington, D.C. as exemplary of this shift to "postnational" Western commemorative activities. Maya Lin's abstract 493-foot black granite wall sinks ten feet below the white marble Washington Mall, inverting traditional, national representational spaces of the dead and aesthetics of commemoration. The design, which lists the names of soldiers who died or went missing, encourages individualized relationships to a traumatic national past. While grandiose ritual ceremonies still take place, the memorial represents the absence of life through the presence of individuals leaving gifts (photos, teddy bears, shoes, dogtags) or making tombstone rubbings rather than through monumental form (Richardson, 2001; Sturken, 1997; Wagner-Pacifici and Schwartz, 1991). A more recent example of a reconfigured national memorial is the Little Bighorn Battlefield National Monument in southern Montana (Foote, 1997). Previously established as the Custer Battlefield National Monument in 1946, the field now honors Custer's soldiers as well as Native American warriors who died in 1876. Although this place of memory still works within a masculinst tradition of commemorating heroes, Kenneth Foote argues that it challenges traditional racial and ethnic understandings of patriotism through a new national culture of memory defined by "equal honor on the battlefield" (p. 327).

Citizen groups that wish to give voice and presence to peoples, pasts, and places forgotten in national narratives often establish alternative places of memory, such as the Lower East Side Tenement Museum in New York or the District Six Museum in central Cape Town, South Africa. The latter is located in an old church that stands in a landscape that was razed under apartheid (Lewis, 2001). Museum visitors challenge the official national violence writ in the surrounding landscape by creating a socially vibrant memory of their home(land). They may weave part of a memory fabric or sketch in the location of their (former) homes onto a giant laminated area street map on the museum's floor. Other interactive, grassroots places of memory include history workshops, neighborhood tours, mappings, historical exhibitions, community centers, public art, and gardening projects. For example, brightly painted Puerto Rican "casistas" in abandoned spaces of dilapidated tenements in New York "evoke a memory of the homeland for immigrants" (Hayden, 1995, pp. 35–6) as well as house political meetings and community activities. In former West Berlin, citizen groups "discovered" an abandoned field that was the former headquarters of the Gestapo, SS, and Reich Security Service. After many years of protest, activism, and even excavations, they established a new type of educational and historical center called the "Topography of Terror" (Till, 1996; Young, 1993).

Although alternative places of memory call attention to official amnesias, many derive their meaning from dominant social and discursive memory practices and may reinforce problematic identity categories (Lewis, 2001; Till, 1996). Furthermore, Owen Dwyer (2000) argues that tourism economies and practices may result in simplified narratives about the past at politically progressive memorials. For example, Civil Rights monuments, plaques, parks, museums, buildings, and street names have significantly challenged formerly segregated commemorative topographies in the American South. Nonetheless, Dwyer demonstrates how their locations and forms are limited by the political economic logic of the tourism and heritage industries. Because these places are created for a mass audience, the complexity of civil rights activism is presented as a straightforward history, as a battle won on the streets by charismatic "great men." The significance of women and working class people is downplayed, the national and local tensions in the *multiple* civil rights movements are ignored, and the complex social relations of racism are left unaddressed (ibid.; see also Alderman, 1996 and Savage, 1994). Clearly, existing memory traditions as well as the political economies of place at various scales influence what pasts are to be remembered by whom, where, and in what form.

Concluding Comments

Places of memory include museums, monuments, cemeteries, statuary, public buildings and squares, streets, historic preservation projects, plaques, and memorials, as well as the rituals, images, and practices associated with them. They punctuate and create symbolic space, and function as nodes of collective politics at and through which notions of identity (such as race, class, gender, and the nation) are performed and contested (after Johnson, 1995). Many places of memory are built as overtly political projects intended to justify existing power relations or to disrupt old ones. Certainly this would, in part, explain why so much time, money, and symbolic capital is invested in the construction of monumental buildings and their topographies. Nonetheless, while officials have historically attempted to legitimate their contemporary political acts through such places, simply because they are built does not mean that they inevitably serve to sacralize state politics (Agnew, 1998). Nor does their establishment indicate a coherent ideological basis among officials of a state or regime.

Places are not fixed in absolute or Cartesian space; their dominant social images and understandings – what Linda McDowell (1999) calls "regimes of place" – are open to question and change, especially during periods of political, economic, and social instability and transformation. Although elites have had more control over the establishment of places of memory in public settings, they cannot control how they are perceived, understood, and interpreted by individuals and various social groups. Furthermore, the recent popularity of places of memory as tourist sites does not indicate that visitors accept the narratives exhibited, or that they share uniform interpretations and experiences of these places. Places of memory in a national capital, for example, are probably experienced differently by international visitors, by national citizens from various regions and social groups, and by residents living in that city. Finally, social groups establish places of memory for different reasons, such as to challenge dominant power relations or to contribute to their "politics of everyday life." Because different understandings of the past influence individual and social experiences of the present and future, studies about place and memory provide material and symbolic evidence about how these complex social processes may lead to political action.

ENDNOTES

1. "Officials" refer to the community of state bureaucrats, leaders, and experts that attempt to legitimize a particular worldview through abstract and selective representations of space and place (after Ó Tuathail and Agnew, 1992).
2. Monuments are often associated with the plastic arts (Young, 1993) and have a dramatic presence in material and symbolic landscapes. Memorials are commemorative places often associated with historic sites. Depending on the context, a monument may also be a memorial.
3. Although I mainly discuss Western case studies, post-colonial theorists have significantly challenged Western models of the "nation." See chapters 18 and 23, this volume.
4. In this chapter, social memory is used broadly to include cultural, public, and collective memory (cf. Bal et al., 1999; Ben-Amos and Weissberg, 1999; Fentress and Wickham, 1992; Sturken, 1997; Till, 1999).

BIBLIOGRAPHY

Adams, P., Hoelscher, S., and Till, K. 2001. Place in Context: rethinking humanist geographies. In P. Adams et al. (eds.) *Textures of Place*. Minneapolis, MO: University of Minnesota Press.

Agnew, J. 1998. The impossible capital: monumental Rome under Liberal and Fascist regimes, 1870–1943, *Geografiska Annaler*, 80B, 229–40.

Agnew, J. and Duncan, J. 1989. Introduction. In J. Agnew and J. Duncan (eds.) *The Power of Place: Bring Together Geographical and Sociological Imaginations*. Boston: Unwin Hyman, 1–8.

Alderman, D. 1996. Creating a new geography of memory in the South: (re)naming the streets in honor of Martin Luther King, Jr. *Southeastern Geographer*, 36, 51–69.

Anderson, B. 1991. *Imagined Communities*. New York: Verso.

Atkinson, D. and Cosgrove, D. 1998. Urban rhetoric and embodied identities: city, nation, and empire at the Vittorio Emanuele II Monument in Rome, 1870–1945. *Annals of the Association of American Geographers*, 88, 28–49.

Azaryahu, M. 1997. German reunification and the politics of street names: the case of East Berlin. *Political Geography*, 16, 479–93.

Azaryahu, M. and Kellerman, A. 1999. Symbolic places of national history and revival: a study in Zionist mythical geography. *Transactions of the Institute of British Geographers N.S.*, 24, 109–23.

Bal, M., Crewe, J., and Spitzer, L. (eds.). 1999. *Acts of Memory: Cultural Recall in the Present*. Hanover, NH: University Press of New England.

Ben-Amos, D. and Weissberg, L. (eds.). 1999. *Cultural Memory and the Construction of Identity*. Detroit: Wayne State University Press.

Bennett, T. 1995. *The Birth of the Museum: History, Theory, Politics*, London: Routledge.

Bodnar, J. 1992. *Remaking America: Public Memory, Commemoration, and Patriotism in the Twentieth Century*. Princeton, NJ: Princeton University Press.

Bourdieu, P. 1990. *The Logic of Practice*, Stanford, CA: Stanford University Press.

Boyer, M. C. 1994. *The City of Collective Memory: Its Historical Imagery and Architectural Entertainments*. Cambridge, MA: MIT Press.

Charlesworth, A. 1994. Contesting places of memory: the case of Auschwitz. *Environment and Planning D: Society and Space*, 12, 579–93.

Clifford, J. 1988. *The Predicament of Culture: Twentieth Century Ethnography, Literature, and Art*. Cambridge, MA: Harvard University Press.
De Soto, H. G. 1996. (Re)inventing Berlin: Dialectics of Power, Symbols and Pasts, 1990–1995. *City and Society*, 1, 29–49.
DeLyser, D. 1999. Authenticity on the Ground: Engaging the Past in a California Ghost Town. *Annals of the Association of American Geographers*, 89, 602–32.
Dowler, L. 1998. And they think I'm just a nice old lady: women and war in Belfast, Northern Ireland. *Gender, Place and Culture*, 5, 159–76.
Duncan, J. 1989. The Power of Place in Kandy, Sri Lanka: 1780–1980. In J. Agnew and J. Duncan (eds.) *The Power of Place: Bringing Together Geographical and Sociological Imaginations*. London, Unwin Hyman, 185–201.
Dwyer, O. 2000. Interpreting the Civil Rights Movement: place, memory, and conflict. *Professional Geographer*, 52, 660–71.
Edensor, T. 1997. National identity and the politics of memory: remembering Bruce and Wallace in symbolic space. *Environment and Planning D: Society and Space*, 29, 175–94.
Entrikin, J. N. 1991. *The Betweenness of Place: Towards a Geography of Modernity*. Baltimore, MD: Johns Hopkins University Press.
Fabian, J. 1983. *Time and the Other: How Anthropology Makes its Object*, New York: Columbia University Press.
Fentress, J. and Wickham, C. 1992. *Social Memory*. Oxford: Blackwell.
Foote, K. 1997. *Shadowed Grounds: America's Landscapes of Violence and Tragedy*. Austin, TX: University of Texas Press.
Foote, K., Tóth, A., and Árvay, A. 2000. Hungary after 1989: inscribing a new past on place. *Geographical Review*, 90, 301–34.
Forest, B. 1995. West Hollywood as symbol; the significance of place in the construction of gay identity. *Environment and Planning D: Society and Space*, 13, 133–57.
Forest, B. and Johnson, J. 2001. Unraveling the threads of history: monuments and post-Soviet national identity in Moscow. *Annals of the Association of Geographers*, in press.
Gillis, J. (ed.). 1994. *Commemorations: The Politics of National Identity*. Princeton, NJ: Princeton University Press.
Gillis, J. 1994. Memory and Identity: the history of a relationship. In J. Gillis (ed.) *Commemorations*. Princeton, NJ: Princeton University Press, 3–26.
Halbwachs, M. 1925. *Les cadres sociaux de la memoire*. Paris: F. Alcan.
Halbwachs, M. 1992 [1941, 1952]. *On Collective Memory*, Chicago: University of Chicago Press.
Handler, R. 1988. *Nationalism and the Politics of Culture in Quebec*. Madison, WI: University of Wisconsin Press.
Haraway, D. 1989. *Primate Visions: Gender, Race, and Nature in the World of Modern Science*. New York: Routledge.
Harvey, D. 1979. Monument and myth. *Annals of the Association of American Geographers*, 69, 362–81.
Hayden, D. 1995. *The Power of Place: Urban Landscapes as Public History*. Cambridge, MA: MIT Press.
Heffernan, M. 1995. For ever England: the Western Front and the politics of remembrance in Britain. *Ecumene*, 2, 293–324.
Hobsbawm, E. and Ranger, T. (eds.). 1983. *The Invention of Tradition*. Cambridge: Cambridge University Press.
Hoelscher, S. 1998. *Heritage on Stage: The Invention of Ethnic Place in America's Little Switzerland*, Madison, WI: University of Wisconsin Press.
Johnson, N. 1994. Sculpting heroic histories: celebrating the Centenary of the 1798 Rebellion in Ireland. *Transactions of the Institute of British Geographers*, 19, 78–93.

Johnson, N. 1995. Cast in stone: monuments, geography, and nationalism. *Environment and Planning D: Society and Space*, 13, 51–65.

Johnson, N. 1999. The spectacle of memory: Ireland's remembrance of the Great War, 1919. *Journal of Historical Geography*, 25, 36–56.

Kaplan, R. 1993. *Balkan Ghosts: A Journey Through History.* New York: Vintage Books.

Karp, I., Mullen Kreamer, C., and Lavine, S. (eds.). 1992. *Museums and Communities: The Politics of Public Culture*. Washington, DC: Smithsonian Institution Press.

Keith, M. and Pile, S. (eds.). 1993. *Place and the Politics of Identity.* London: Routledge.

Legacies of Authoritarianism. 2001. *Cultural Productions, Collective Trauma, Global Justice.* University of Wisconsin-Madison, available at http://wiscinfo.doit.wisc.edu/globalstudies/LOA/.

Leitner, H. and Kang, P. 1999. Contested urban landscapes of nationalism: the case of Taipei. *Ecumene*, 6, 214–33.

Lewandowski, S. 1984. The built environment and cultural symbolism in post-colonial Madras. In J. Agnew et al. (eds.) *The City in Cultural Context*. London: Allen and Unwin, 237–54.

Lewis, M. 2001. *Crossroads to District Six: Reclaiming Official History/Restaging Popular Memory.* Paper delivered at Space, Place, and Memory Symposium, Humanities Institute, University of Minnesota, April 2001 (available from the Department of Theatre Arts and Dance, UMN, Minneapolis, MN 55414).

Lowenthal, D. 1985. *The Past is a Foreign Country.* Cambridge: Cambridge University Press.

Lowenthal, D. 1991. *Possessed by the Past: The Heritage Crusade and the Spoils of History.* New York: Free Press.

Massey, D. 1997. A global sense of place. In T. Barnes and D. Gregory (eds.) *Reading Human Geography*. London: Arnold.

McDowell, L. 1999. *Gender, Identity and Place: Understanding Feminist Geographies*. Minneapolis, MN: University of Minnesota Press.

Monk, J. 1992. Gender in the landscape: expressions of power and meaning. In K. Anderson and F. Gale (eds.) *Inventing Places: Studies in Cultural Geography.* Melbourne: Longman Cheshire/ New York: Wiley, 123–38.

Mosse, G. 1990. *Fallen Soldiers: Reshaping the Memory of the World Wars*. New York, Oxford University Press.

Nash, C. 1994. Remapping the Body/Land: new cartographies of identity, gender, and landscape in Ireland. In A. Blunt and G. Rose (eds.) *Writing Women and Space: Colonial and Postcolonial Geographies*. New York: Guilford Press, 227–50.

Nora, P. 1989. Between memory and history: Les Lieux de Memoire. *Representations*, 26, 7–25.

Nora, P. 1997. *Realms of Memory.* New York: Columbia University Press.

Osborne, B. 1998. Constructing landscapes of power: the George Etienne Cartier Monument, Montreal. *Journal of Historical Geography*, 24, 431–58.

Ó Tuathail, G. and Agnew, J. 1992. Geopolitics and discourse: Practical geopolitical reasoning in American foreign policy. *Political Geography*, 11, 190–204.

Pred, A. 1992. Capitalisms, Crises, and Cultures II: notes on local transformation and everyday struggles. In A. Pred and M. Watts (eds.) *Reworking Modernity.* New Brunswick, NJ: Rutgers University Press, 139–41.

Raivo, P. 1998. Politics of Memory: Historical Battlefields and Sense of Place. *Nordia Geographical Publications*, 27, 59–66.

Richardson, M. 2001. The Gift of Presence: the act of leaving artifacts at shrines, memorials and other tragedies. In P. Adams et al. (eds.) *Textures of Place: Rethinking Humanist Geographies*. Minneapolis, MN: University of Minnesota Press.

Rose, G. 2001. Discourse Analysis II: Institutions and Ways of Seeing. In *Visual Methodologies*. London: Sage, 164–86.
Savage, K. 1994. The Politics of Memory: black emancipation and the Civil War Monument. In J. Gillis (ed.) *Commemorations: The Politics of National Identity*. Princeton, NJ: Princeton University Press, 127–49.
Sharp, J. 1996. Gendering Nationhood: a feminist engagement with national identity. In N. Duncan (ed.) *Body Space*. London: Routledge, 97–108.
Sherman, D. 1994. Art, commerce, and the production of memory in France after World War I. In J. Gillis, (ed.) *Commemorations: The Politics of National Identity*. Princeton, NJ, Princeton University Press, 186–211.
Sturken, M. 1997. *Tangled Memories: The Vietnam War, the AIDS Epidemic, and the Politics of Remembering*. Berkeley, CA: University of California Press.
Till, K. 1996. *Place and the Politics of Memory: A Geo-Ethnography of Museums and Memorials in Berlin*. unpubl. dissertation, Department of Geography, University of Madison-Wisconsin.
Till, K. 1999. Staging the past: landscape designs, cultural identity and Erinnerungspolitik at Berlin's Neue Wache. *Ecumene*, 6, 251–83.
Till, K. 2001. Reimagining National Identity: "Chapters of Life" at the German Historical Museum. In P. Adams et al. (eds.) *Textures of Place: Rethinking Humanist Geographies*. Minneapolis, MN: University of Minnesota Press, 273–99.
Tuan, Y.-F. 1974. Space and place: humanistic perspective. *Progress in Human Geography*, 8, 213–52.
Verdery, K. 1999. *The Political Lives of Dead Bodies: Reburial and Postsocialist Change*. New York: Columbia University Press.
Wagner-Pacifici, R. and Schwartz, B. 1991. The Vietnam Veterans Memorial: commemorating a difficult past. *American Journal of Sociology*, 97, 376–420.
Warner, M. 1985. *Monuments and Maidens: The Allegory of the Female Form*. New York: Atheneum.
Winter, J. 1995. *Sites of Memory, Sites of Mourning: The Great War in European Cultural History*. Cambridge: Cambridge University Press.
Withers, C. W. 1996. Place, memory, monument: memorializing the past in contemporary Highland Scotland. *Ecumene*, 3, 325–44.
Young, J. 1993. *The Texture of Memory: Holocaust Memorials and their Meaning*. New Haven, CT: Yale University Press.

Chapter 20

Boundaries in Question

Sankaran Krishna

...perhaps it is because our sense of what is the case is constructed from such inadequate materials that we defend it so fiercely, even to the death...(Rushdie, 1991, p. 13)

Introduction

Two of the more insightful social commentators of recent times remind us that the atlas of the modern world of nation-states is an arbitrary product of contingency and conjuncture. Immanuel Wallerstein's delightful counterfactual exercise runs thus:

> Suppose in the period 1750–1850 what had happened was that the British colonized primarily the old Mughal Empire, calling it Hindustan, and the French had simultaneously colonized the southern (largely Dravidian) zones of the present-day Republic of India, giving it the name of Dravidia. Would we today think that Madras was "historically" a part of India? Would we even use the word "India"? I do not think so. Instead, probably, scholars from around the world would have written learned tomes, demonstrating that from time immemorial, "Hindustan" and "Dravidia" were two different cultures, peoples, civilizations, nations, or whatever. There might be in this case some "Hindustan" irredentists who occasionally laid claim to "Dravidia" in the name of "India", but most sensible people would have called them "irresponsible extremists" (Wallerstein, 1991, p. 130).[1]

In similar vein, Benedict Anderson (1991) muses on the uncanny propensity for the boundaries of the newly-emergent nations of the twentieth century to be "isomorphic" with the administrative divisions of the erstwhile colonial empire. That postcolonial boundaries have their genesis in historically recent colonial conquests, and the balance of power politics of the nineteenth century, explains the enormous energy and intensity with which nationalists in these societies have sought to endow them with timeless sanctity and forms of geographic sacrality. Furthermore, the proximate and mimetic character of their origins has imparted an anxious and obsessive character to most discussions about boundaries in postcolonial societies.

This essay focuses on postcolonial variations on a modernist geopolitical imaginaire[2] that rules the contemporary world. It should be clear that the anxieties and obsessions described here are not confined to a Third World out there but are central to the formation of the nation-states everywhere, including those of the Western world. Indeed, one might even say that the First World has been more thoroughly colonized by this geopolitical imaginaire that it no longer sees this way of being as remarkable.[3] Besides delineating postcolonial understandings of space, this essay attempts to accentuate certain nuances that distinguish it from postmodern representations. In the concluding section, by focusing on certain aspects of Indian public culture, it examines the impact of globalization on the protean character of postcolonial space.

Political Boundaries, Difference, and Violence

Somewhere during the last few centuries, a certain understanding of global space has come to be dominant. This understanding avers that the world is divided into enclosed territories called nation-states. Each of these nation-states is an immanent destiny that has traveled intact through time and, to varying degrees across this planet, has achieved a fulsome presence today. Each of these bordered or enclosed units is regarded, ideally, as a container of a singular identity – call it a nation, a culture, a civilization, a people, a genius, a race, or an ethnic group. If multiple identities are stridently jostling for supremacy within a bounded space, it is marked as an unstable and transitional zone still on the way to that pacific and unambiguous plateau of national arrival. Contemporary nations are arrayed from the most developed to the most severely under-developed "quasi-states" in terms of their degrees of separation from such a desired plateau.[4]

Given such a world-view, the overwhelming desire of state-makers becomes one of aligning "territory" with "identity," of ensuring that each geopolitical unit of the modern imaginaire is populated by a singular sense of identity. Implicitly or explicitly, this desire to re-make the world in conformity with this vision underlies much of our thinking about nations, international relations, and space in general[5]. Indeed, much of what passes for "history" in modern times is the story of the ongoing, violent, and largely unsuccessful effort to pulverize various recalcitrant identities into the univocal code of national citizenship. "Nation-building," to use the sanitized phrase of choice, is the never-ending effort to colonize all other ways of being into conformity with a simulacrum called the nation – an idealized copy based on an original that does not exist anywhere. Political boundaries are deemed to mark the separation of inside from outside, self from other, identity from difference, safety from danger, community from anarchy, home from world, and so on. The "boundary," then, becomes the quintessential site for the production and enactment of identity and difference, for the deployment of a rhetoric of self-making. Belying its location at the edge or the periphery of an enclosed space, the boundary moves to the very center of being – it is where the most fundamental contestations over identity occur. Predictably too, the boundary and its defense occasion the most egregious forms of violence in our world, violence seen as necessary, unavoidable, and, indeed, an annealing and strengthening force in constructing the nation. It is important to note that here violence is neither a regrettable by-product nor an aberration of the state–sovereign system. Rather, it is encrypted within the narrative, it is a chronicle of deaths foretold, for the boundary is no mere line on the ground but the demarcation between different ontologies – the point at which we end and they are.

There are other stories one could tell about boundaries – stories that see them as bridges connecting landscapes marked by distinctive languages or ways of being; as sites where the movement of peoples, ideas, religions, languages, fashions, and commerce intersect and interact; stories that draw the boundary not territorially but through lineage or kinship, or the song-lines of nomadic peoples; stories that see the boundary as a therapeutic space where one might go in order to de-familiarize oneself from what one has become; and, ultimately, stories that see the boundary not as a line that separates the already-constituted identities of "us" and "them" but rather as one that allows for the very formation of an "us" and "them." These alternative stories of boundary spaces are not legible within the sober realms of statecraft, diplomacy, military strategy, geopolitical discourse, and international relations. Unsurprisingly, they have enjoyed a vibrant life in fiction, poetry, cinema, humor and, I would like to think, the secret underworld of every sentient human being's imagination.[6] Nevertheless, my task here is to register the hegemonic story told about boundaries – it is overwhelmingly encoded as a discourse of danger. It is seen as the entry point for unwelcome immigrants, terrorist guerrillas, dangerous ideologies, fatal diseases, and illegal narcotics. Its protection, sanctity and successful maintenance becomes the ultimate yardstick of a state's efficacy and legitimacy. The boundary becomes the definitive site for the production of otherness, and thereby for the self.

The story thus far has been a general meditation on modern spatiality and violence as an entailment of that narrative. There is nothing distinctly postcolonial about this story – it is one that traduces all of us. The enactment of this story, as it proceeds in postcolonial space, has often resembled the movement from tragedy to farce. Boundary discourse in postcolonial space is inflected with an anxiety and an obsessiveness that marks it as a special case. As the introduction suggested, here the artifices and arbitrariness associated with national boundaries are more recent. Lacking the patina of credibility that comes with time, boundary conflicts have consequently been more endemic and violent in this space. Secondly, the boundaries of the new nation-states were dictated to coincide with the boundaries of the erstwhile colonial administrative units. They were often violently negotiated between imperial states and had little or nothing to do with "ethnic" fault lines, linguistic demarcations, religious affiliations, geographical landmarks or other such "natural" lines of cleavage between territories. While the desired fit between national boundaries and cultural or linguistic or "ethnic" identity was always fictitious and forced in the so-called original Western nations, the absurd character of these presuppositions stood out more starkly in postcolonial space. Thirdly, the very fact that postcolonial boundaries are relics of colonial rule makes them particularly appropriate sites for attempts at re-negotiation in current times. The struggle to restore an imagined pre-lapsarian boundary for the nation-state becomes alloyed with a form of anti-colonial nationalism, or anti-neocolonialism, which retains considerable appeal decades after decolonization.

Finally, the career of the nation-state is informed by some of the most retrogressive aspects of modernity. It has been steadily anthropomorphized since the early nineteenth century. Critical to this are pseudo-scientific theories of race and of social Darwinism; romantic reactions against the burgeoning industrialization of life and the destruction of pre-urban communities; the latent mercantilism that energized the quest for colonies in the late nineteenth century; a destructively competitive mode of production, capitalism, that has encouraged a self-understanding among both individuals and societies that they are each eviscerated, utility-maximizing entities locked in zero-sum competition within an anarchic milieu; and the mediation of such ideas to produce ever-larger communities of identity through the technologies of print and electronics. Together, these forces have interacted to move us decisively away from thinking about each civilization as a miracle in itself to viewing the world as a hierarchy of differentially endowed, competitive spaces ranging from advanced and developed nations to underdeveloped or backward nations.

As Agnew and Corbridge (1995, p. 49) note, we arrive at the unprecedented and "singular trait of modern geopolitical discourse:" namely, its propensity to regard "... others as 'backward' or potentially disadvantaged if they remain as they are."

The desire on the part of the postcolonial societies to "catch up" with advanced nations (a desire central to the discourse of "developmentalism") results in fetishizing certain attributes that supposedly mark the latter, notably, inviolate borders. As one commentator notes with ill-concealed despair about the fuzzy contours of India: "According to political scientists one of the prerequisites for the ordered growth of a modern nation state is settled boundaries. Once a country has well defined borders, the planned development of the various sectors of the economy becomes easier and predictable. Also, as a member of the comity of nations it will have more credibility, if not confidence, in its relations with other nations... After almost 40 years as an independent country, India still has undefined borders with two of its neighbors" (Menon, 1987, p. 14). Development, credibility, and self-confidence await the acquisition of "well defined borders," a proposition certified by no less than political scientists. The border becomes the bellwether of the state of the nation.

Postcolonial and Postmodern Understandings

In articulating a difference between postcolonial and postmodern standpoints on questions of space, I adumbrate an argument I have made elsewhere.[7] There is much that is common to these discourses. They are critical of the modernist tendency to fetishize national space and boundaries, and they are against the ontopological belief that seeks to align territory with identity. They share an anti-essentialism that sees identity and difference as mutually constitutive of each other, with no Archimedean points serving as their foundations. They are sceptical of grand narratives (liberalism, nationalism, socialism, capitalism, etc.) that have inscribed social and political life with a telos, and instead see them as constituting a nexus of power/knowledge that imposes interpretations on a reality that is not complicit with our efforts to understand it.

Yet, one can discern a difference between these two discourses on the way their deconstructions of modernist notions of space play themselves out. Specifically, the difference between the postcolonial and the postmodern sensibility is that while the latter seems more attuned to the deconstructive moment than it is toward the articulation of alternative spatializations based on contingent notions of identity, justice, fairness, or equality, the former tries to be/ is forced to be equally attentive to both concerns. I define a postcolonial standpoint as one that is engaged in the simultaneous task of deconstructing, historicizing, and denaturalizing all identities even as it envisions and struggles for justice and fairness in the worlds we do inhabit – struggles in which one might have to strategically essentialize identity in order to achieve these aims. In other words, it is a standpoint that is simultaneously deconstructive *and* oriented towards the achievement of specific political goals and visions. Appiah, for instance, makes a related distinction between the two on grounds that postcolonial discourse, while sharing the deconstructive and anti-essentialist positions of postmodernism, makes an explicit commitment towards a contingent ethical universal:

> Postcoloniality is *after* all this: and its *post*, like postmodernism's, is also a *post* that challenges earlier legitimating narratives... But it challenges them in the name of the ethical universal; in the name of *humanism*... And on that ground it is not an ally for Western postmodernism but an agonist, from which I believe postmodernism may have something to learn... For what I am calling humanism can be provisional, historically contingent, anti-essentialist (in other words, postmodern) and still be demanding. We can surely maintain a powerful engagement with the concern to avoid cruelty and pain while nevertheless recognizing the contingency of that concern. (Appiah, 1992, p. 155, emphases as original)

A postcolonial perspective, then, entails acknowledging the socially constructed, anti-essentialist, and open-ended character of our narratives even as it accepts responsibility to affect the trajectories along which those narratives seem headed, a responsibility not so much denied by the postmodernist as left implicit or unelaborated by her.

Perhaps the difference between the two discourses on questions of space might become clearer by considering two instances. In a piece written after a trip through the violent, ethno-nationalist cauldron that is northern Myanmar today, Amitav Ghosh observes,

> Burma's borders are undeniably arbitrary, the product of a capricious colonial history. But colonial officials cannot reasonably be blamed for the arbitrariness of the lines they drew. All boundaries are arbitrary: there is no such thing as a "natural" nation, which has journeyed through history with its boundaries and ethnic composition intact. In a region as heterogeneous as Southeast Asia, any boundary is sure to be arbitrary (Ghosh, 1996, p. 49).

So far, so po-mo, one might say. What makes Ghosh's work distinctively postcolonial is that he proceeds to say: "On balance, Burma's best hopes for peace lie in maintaining intact the larger and more inclusive entity that history, albeit absent-mindedly, bequeathed to its population almost half a century ago" (ibid). In other words, Ghosh does not (cannot?) stop at the deconstructive moment – that is, with displaying the artifices and caprices that inform Burma's boundaries – but he pragmatically ("on balance") considers the question of what is to be gained by renegotiating those boundaries in today's Burma. He is interested in assessing what kinds of spatial orders further pluralism, ethnic reconciliation, nonviolence, individual rights, freedom of speech, and the emergence of a multiparty liberal democracy in contemporary Burma. Notions such as multiparty systems, individual rights, liberal democracy, and fair elections, sound rather odd as explicit political commitments within a postmodern sensibility more attuned to deconstructing the modern condition; yet these are often inescapable questions for a postcolonial. Even as one exposes the artifices of colonial boundaries, one is faced with the fact that most alternatives to that artifice, represented by the various ethno-nationalist movements, see Myanmar as irretrievably fragmented into ethnic enclaves.[8] I am not suggesting that postmodernists are not for peace, liberal democracy, and fair elections, or that they are not interested in ethnic reconciliation or individual rights. I am, however, suggesting that the responsibility that Ghosh feels in making an explicit statement in support of the pluralist possibilities contained in the inherited boundaries of postcolonial Burma is a responsibility that a postmodernist does not feel in the same way. (I explicate the reasons for this difference below.)

My second instance is from Sri Lanka. Faced with the onslaught of the Sinhalese majority, the minority Tamils of that country have embarked on a struggle for self-determination. As part of this struggle they have, inevitably, "invented" the argument of their "traditional" homelands and in various ways have mythologized and fetishized a desired space called "Eelam." To deconstruct their claims, to show the artifices and the primordial claims of Tamil narratives in this regard, would be an easy task. Yet, in the political, economic and cultural context of the marginalization of the Sri Lankan Tamils by the Sinhalese majority, it is politically and normatively insufficient to stop with such a deconstruction. One has to ask the further question: *what sort of a political gesture must this deconstruction of the Tamil claim to a traditional homeland be accompanied by in order for it to escape complicity with the unitary nation-building project of the majoritarian (Sinhala) community and its state?* It is only in the context of a genuinely pluralist, multiethnic, egalitarian and nondiscriminatory Sri Lankan (not Sinhalese) nation-state that one can stop with demonstrating the historical vacuity of the Tamil claim regarding a traditional homeland. In other words, the deconstruction of the Tamil claim to a traditional homeland has to be accompanied by an explicit commitment to a pluralist Sri Lanka that recognizes itself as a multiethnic space, otherwise that deconstruction could have disastrous consequences for the Tamils in that country.

In large part, this difference in perceived political responsibility that I outline between the postcolonial and postmodern viewpoint results from their different loci of enunciation. While Arif Dirlik (1997) and Aijaz Ahmad (1992) are undoubtedly correct that both species are perhaps most populous in the metropolitan academy, the fact is that postcolonial scholarship is practiced by many who are either originally from the "Third World," or still live there, or are scholars whose subject matter happens to be the formerly colonized countries. The consequences of their political choices and their academic analyses, while rarely a matter of life or death for their own selves, often have a direct and unmediated impact on the lives and fortunes of people in the societies they study.

There are consequential and inescapable differences between the "First" and "Third" worlds. Significant among these differences, surely, is the fact that the rule of law, individual rights, freedom of expression, minority rights, and ethnic pluralism are comparatively more well entrenched in the former than the latter. As the two examples outlined above show, a deconstruction of postcolonial spaces (arbitrary boundaries inherited from the colonial era in the case of Myanmar, and the Tamil claim for a traditional homeland in the case of Sri Lanka), conducted in the absence of the rest of the accoutrements of a full-fledged liberal democracy, could have political consequences that no progressive could support. In the first instance it could further the collapse of Myanmar into warring ethnic enclaves, and in the second would wind up being complicit with the ethno-nationalist project of the Sinhala majority. Deconstructions of identities and spaces can often have different imports in First and Third World settings – and this difference has its entailments for postmodern versus postcolonial understandings of national space.[9]

Globalization and Postcolonial India: the Killer Instinct of a Soft State

Recent years have seen the ascendance of neo-liberal ideas on the centrality of "free" markets to spur economic growth, and the decisive demise of planning or state-directed strategies of development. We are also increasingly inhabiting a single, global, synchronic time-zone, as the speed of communications, and the movement of ideas, fashions, capital, commodities, images, and (very selectively) of labor, has accelerated greatly in the last two decades. One might, provisionally, regard these as two central aspects of a process called "globalization." What has been the impact of globalization on ideas of postcolonial space? This concluding section of the essay offers some thoughts on this question, by a detour through the public culture of contemporary India.

The spatialization of the world into a hierarchy of nation-states ranging from "advanced" to "backward" has been greatly accentuated in recent years with the spread of a global media: satellite television has brought home the extraordinary disparities in standards of living in different parts of this world. Faced with glitzy, if atypical, images of rampant consumerism in spaces such as South Korea, Thailand, Singapore and China (all once regarded as being on a par with India), Indian perceptions of where they are in a hierarchy of nation-states have undergone significant changes. While in the 1950s and 1960s, Indian "under-development" was expiated through deferred expectations ("we are gearing up for take-off and we'll get there"), today there is a palpable sense of impatience and frustration at India's position in the competitive order of nation-states. Such feelings manifest themselves in intriguing ways in the public culture of a society, and they are useful in understanding postcolonial identity.

A casual reader of Indian newspapers and news magazines would be struck by two, related, obsessions in the country usually phrased in the form of questions: Is India a "soft state"? and, Do we lack a "killer instinct"?[10] The "soft state" – a term coined by Gunnar Myrdal (1968) to describe inefficient South and Southeast Asian states – is a crucial conceptual node in the discourse on developmentalism. It is defined as a state that is easily corruptible, lacks "unity" or sense of purpose, is unable to make "hard" decisions and stick by them, and is contrasted unfavorably with the efficient and committed states of the First and "Second" worlds. To put it simply, a soft state is congenitally incapable of making the trains run on time. The questions are rhetorical – they are expected to be answered only in the affirmative. For instance, type the words "India Soft State" or "India Killer Instinct" into the cognoscenti's search engine for the world wide web, Google, and you will be surprised at the number of hits. Most are from recent newspapers and magazines, but these obsessions have been staples of Indian journalism and middle class life from at least the early 1970s. The usual occasion for self-flagellation over the lack of a killer instinct is a sporting defeat at the hands of Pakistan (in cricket or field hockey), or, more generally, of coming close to but failing to win in international competitions (the Olympics, various World Cups, tennis matches, and the like). Yet, such articles quickly move from sports to a general lament on the malaise of Indian civilization and its inability to deliver the goods. The "soft state" debate is generally triggered off by acts of "terrorism," by hijackings of the national airline and the tendency to capitulate to the demands of the hijackers, in discussions of the "population problem," and in the inability of the Indian state to police its borders effectively and prevent Pakistani incursions into Kashmir. At a more general level, India's "soft state" was supposedly exemplified by the inability to decide whether we were a capitalist society or one committed to socialism. Instead, we muddled on with unconvincing claims to a hybrid strategy for a "socialistic pattern" of society.

Notwithstanding the triggers, the narratives produced thereupon repeat certain standard themes. These include: as a country we lack the "discipline" or the "toughness" to achieve our goals; our history has been one of pusillanimity in face of external conquerors; we lack the "unity" and single-mindedness of the Pakistanis/the Israelis/the western countries/the Chinese (depending on the context, these are interchangeable spaces); our disunity and lack of "team spirit" have resulted in/from centuries of external conquest and subjugation; we have never shaken off the mindset of a colonized people; Hinduism and vegetarianism have sapped us; it's the caste system's fault. Together, these add up to a postcolonial obsession about perceived effeminacy and a corresponding desire for hyper-masculine "hardness" in all domains of life.[11]

Hitherto, such a fear of "softness" and a despair over the lack of a "killer instinct" was confined to a fairly narrow middle-class: the urban, educated, media-aware segments of society. Today, with the increasing ambit of newspapers and news magazines, the ubiquity of television, the reach of advertising, rapid urbanization, growing literacy, and the increased mobility of people across the national and international landscape, these ideas have a reach that is unprecedented. One could argue that the recent surge in popularity of the Hindu fundamentalist Bharatiya Janata Party is one indicator of the spread of ideas about the soft Indian state and the lack of a killer instinct. The political–cultural ideology of the BJP and its fellow travelers foregrounds a narrative that relies heavily on the need to reverse the softness of the Indian state and to acquire a killer instinct. On the nuclear issue, in its attack on the "pseudo-secular" Constitution of India, in its exhortations for Hindu pride and masculinity, in its attitude toward minorities in India, on relations with Pakistan, on Kashmir, in the critique of the decades of planned economic development under Nehru, and on a variety of issues, the ruling party in today's India expounds a rhetoric closely structured around these themes. The BJP has also been hard-line on the need to defend the frontiers of India and in its border negotiations with Pakistan, China, Bangladesh, and other neighbors. In a classic instance of the victim ingesting the values of an oppressive system, postcolonial India has come to believe that acquiring a killer instinct and becoming a hard state are crucial steps toward decolonization, whereas in reality they move us decisively in the opposite direction.

If globalization has contributed to a hardening of aggressive nationalism in India, what about the much-vaunted claim of many recent works that have prematurely written the obituary of the nation-state?[12] Such works suggest that the idea of the nation is *passé*, that we are moving into postnational realms, that the acceleration of time and the annihilation of space *qua* globalization has rendered the nation porous, and that the spread of global capitalism under a transnational bourgeoisie has made national economies less relevant. Yet, in India, one can argue that the nation as an imagined community is very much on the ascendant, and further, that globalization has, if anything, accelerated this process. Wider swathes of India are now within the ambit of television and other technologies that mediate communities. An expanding middle class has served as the site on and through which the global is mediated within India, and this class's consumption of certain iconic commodities has become a primary index of "development." These include computers, soft drinks, televisions, music systems, washing machines, cell phones and a range of other goods that render Indian middle-class space increasingly indistinguishable from that in Bangkok or Singapore or Hong Kong. As a model of development, this one simply moves the vast majority of Indians (still struggling for "essentials" such as clean drinking water, sanitation, employment, and adequate nourishment) out of the picture. The implicit premise is that with a freer economy, increased domestic and foreign investment, a larger market, and higher growth rates, the urban middle class will enlarge to the point where it suffuses the entire country. This new India is premised on the desire to overcome inferiority and backwardness: politically, by recourse to a hyper-masculine and univocal discourse of citizenship, and economically through acts of consumption that constitute us as iconic consumers in a globalized marketplace in world-time.[13]

The nation-building efforts of the era of decolonization, especially in societies such as India, emphasized pluralism, multiethnicity, and, despite limitations, an incipient critique of capitalism. Contemporary postcolonial nationalism in the era of globalization is marked by ethnic or religious majoritarianism, an aggressively competitive attitude toward "others" in an anarchic international milieu, and a reconstitution of the modal citizen primarily as a middle-class consumer in a national/ global capitalist space.

Conclusion

This essay has suggested that the geopolitical imaginaire that governs our thinking about space has different entailments depending on your location. A postcolonial locus of enunciation, in contrast to postmodernism, is sensitive to certain aspects of a Third-World reality that often escapes the metropolitan eye. It has tried to map out some of the anxieties and obsessions of those in postcolonial space who perceive themselves to be disadvantageously positioned in the "national order" of things. These anxieties have, if anything, been accentuated by the globalization of the capitalist mode of production: the accelerated flow of commodities, capital, and ideas, alongside the selective mobility of labor – all occurring within an increasingly homogenizing and singular metric of world-time. Contrary to widespread expectations about the impending demise of the nation-state, we may be witnessing precisely the opposite in postcolonial space: a resurgent and chauvinistic celebration of national identity. Which would, of course, render the post of postcoloniality the most dubious post of them all.

ENDNOTES

1. One of the hallmarks of our time is that the threshold for caricature has been ratcheted impossibly high thanks to the absurdity of the "real." Wallerstein's satirical "learned scholars" are fleshed out in reality by Akbar Ahmad who posits a genealogy for Pakistan that harkens back to the twelfth century Muslim ruler Saladin, and by Ahmad's compatriot Aitzaz Ahsan who discerns a subcontinental divide between what he terms "Indic" culture (India) and the culture of the Indus (Pakistan) as far back as 6000 years. To Ahsan and Ahmad then, India and Pakistan have been entities destined for a bloody partition literally from "time immemorial." See Ahmed (1997) and Ahsan (1996).
2. Arjun Appadurai (1996, p. 31) succinctly defines an "imaginaire" as "...a constructed landscape of collective aspirations...mediated through the complex prism of modern media."
3. Even a cursory familiarity with the evolution of most nation-states in the Western world would tell us that their boundaries are no less derivative, accidental, and either a fallout of disintegrating empires or premised on the genocide of the native peoples of "new" continents. As Nevzat Soguk observes in the course of a fine analysis of how the figure of the "refugee" allowed for the emergence and consolidation of a system of sovereign nation-states in Europe in the early twentieth century,

   > Nation-statehood, envisioned in the destruction of polyglot imperial-dynastic systems...became central to the imagination of the future organization of the polities across Europe. In nearly all cases, the newly found alignment between nations and states in organizing the polity was less experienced and more imagined, through complex mediations and representations....The logical implication of trying to create a continent neatly divided into coherent national territorial states, each inhabited by a separate ethnically and linguistically homogeneous population, was the mass expulsion or extermination of minorities and other unwanted populations (Soguk, 1999, p. 114).

   Soguk reminds us here that nation-building and ethnic cleansing have been intertwined processes from the very beginning.
4. This is an intentionally broad-brush summarization of the story of the emergence of the modern state system. Much of the literature in mainstream international relations is amnesiac about the relative novelty of its central category of analysis, the sovereign nation-state, and intent on reifying a conjunctural disposition of spatial understandings into an enduring, abstract and ahistorical system. Recently, certain works have critically re-politicized the narratives on the emergence of this modern geopolitical imaginaire. A brief list of such works that underpin this essay would include Agnew and Corbridge (1995), Anderson (1991), Balibar and Wallerstein (1991), Campbell (1998), Said (1993) and Walker (1993).
5. I have called this desire one based on the "fiction of homogeneity" in Krishna (1999). For an extensive discussion of the same theme or desire to align territory with identity, drawing heavily from Derrida, see the exposition of ontopology in Campbell (1998).
6. For an exposition of the numerous comportments toward space visible in these other genres that are largely illegible to the world of international relations, see Michael Shapiro's essay (chapter 18) in this volume.

7. Trying to specify differences between nebulous and often ill-defined intellectual currents is a thankless task at the best of times. To attempt to do so in an abbreviated essay is to invite disaster as one is forced to exaggerate differences of emphasis. I reiterate that this is a compressed version of an argument with more nuance that is elaborated in my *Postcolonial Insecurities*. For more on the same issue, also see Said (1988), Appiah (1992), Chakrabarty (1992), and Chakravorty-Spivak (1988).
8. As Ghosh (1996, p.49) observes, "There are thousands of putative nationalities in the world today; at least sixteen of them are situated on Burma's borders. It is hard to imagine that the inhabitants of these areas would be well served by becoming separate states. A hypothetical Karenni state, for example, would be landlocked, with the population of a medium-sized town: it would not be less dependent on its larger neighbors simply because it had a flag, and a seat at the UN."
9. I am aware of the dangers of reifying First and Third worlds in such an argument. It is, moreover, a fact that the tensions associated with deconstruction in these instances is by no means unique to Third-World spaces even if they have greater salience there. As bell hooks (1990, p. 28) reminded us in the context of the deconstruction of identity politics in the United States, "It never surprises me when black folks respond to the critique of essentialism, especially when it denies the validity of identity politics by saying 'Yeah, its easy to give up identity when you got one'. Should we not be suspicious of postmodern critiques of the 'subject' when they surface at a historical moment when many subjugated people feel themselves coming to voice for the first time?" In an almost identical passage, Nancy Hartsock attacks the anti-essentialism of postmodern feminism when she asks: "why is it that just at the moment when so many of us who have been silenced begin to demand the right to name ourselves, to act as subjects rather than as objects of history, that just then the concept of subjecthood becomes problematic?" (Hartsock, 1990, p. 163). Nancy Hartsock and bell hooks amplify the point I have tried to make in the context of Myanmar and Sri Lanka: deconstruction and anti-essentialism have different imports depending on context, and these differences in import are by no means confined to postcolonial spaces. Of course, both bell hooks and Nancy Hartsock, in their critiques of postmodernism, valorize a stable, subject-centered notion of politics that is itself limiting and problematic. For an insightful critique along these lines, see Judith Butler (1990). My thanks to the incomparable Geo-free Whitehall for bringing this latter debate to my attention.
10. In Indian public discourse, the "killer instinct" is the ability to see a task ruthlessly through to its successful completion, to never let up once in command, and to show no mercy for one's opponent on the playing field or the killing fields of international relations. It is supposedly reflective of a single-minded and disciplined commitment to clearly articulated goals. India's genial tennis player of yesteryear, Vijay Amritraj, bore the brunt of the Indian press' excoriation for his supposed tendency to snatch defeat from the jaws of victory and for generally being an underachiever. These were deemed to arise from his lack of a killer instinct. Ironically, Amritraj's record when playing for his country in the Davis Cup far exceeded anything he ever accomplished on the professional circuit when playing for himself – a fact rarely noticed by the Indian media. His sportsmanship on court and his appreciation of a good shot by his opponent often drew their wrath as it was interpreted as a sign of his "softness." Amritraj's demeanor situated tennis as a game, and not a place where the self-worth of an entire nation was to be secured, and this no doubt contributed to the vehemence with which he was often attacked for lacking a killer instinct.
11. These reasons are repeated *ad nauseam* in the various writings on these issues in the Indian media. Recently, "India's number one website," www.samachar.com, had an opinion poll which asked a straightforward question: "Do Indians Lack Killer Instinct?"

Of the 6,669 responses received, 5,110 (76.6%) replied "yes" while 1,256 (18.8%) said "no." The website has transcripts of the reasons afforded by many respondents for their answers, and there is much valuable material there which confirms the narrative that I have pieced together here.

12. For example, see Appadurai (1993), Guehenno (1995), Ohmae (1995), and Habermas (1998).
13. My arguments here are greatly abbreviated owing to lack of space. On the affinities between the process of economic liberalization in India, the changing character of public space, and the rise of the BJP variety of Hindu fundamentalism see Rajagopal (2001). For two sensitive inquiries into the ways in which globalization, economic liberalization, an aggressive and hyper-masculine Hindu cultural nationalism, and the citizen-as-consumer, are imbricated in India, see Fernandes (2000a, b).

BIBLIOGRAPHY

Agnew, J. and Corbridge, S. 1995. *Mastering Space: Hegemony, Territory and International Political Economy*. London: Routledge.

Ahmad, A. 1992. *In Theory: Classes, Nations, Literatures*. London: Verso.

Ahmed, A. S. 1997. *Jinnah, Pakistan and Islamic Identity: The Search for Saladin*. London: Routledge.

Ahsan, A. 1996. *The Indus Saga and the Making of Pakistan*. Karachi: Oxford University Press.

Anderson, B. 1991. *Imagined Communities: Reflections on the Origin and Spread of Nationalism*. London: Verso.

Appadurai, A. 1996. *Moderntiy at Large: Cultural Dimensions of Globalization*. Minneapolis, MN: University of Minnesota Press.

Appiah, A. K. 1992. *In My Father's House: Africa in the Philosophy of Culture*. New York: Oxford University Press.

Balibar, E and Wallerstein, I. 1991. *Race, Nation, Class: Ambiguous Identities*. London: Verso.

Butler, J. 1992. Gender trouble, feminist theory and psychoanalytic discourse. In L. J. Nicholson (ed.) *Feminism/Postmodernism*. New York: Routledge, 324–40.

Campbell, D. 1998. *National Deconstruction: Violence, Identity, and Justice in Bosnia*. Minneapolis: University of Minnesota Press.

Chakrabarty, D. 1992. The death of history? Historical consciousness and the culture of late capitalism. *Public Culture*, 4(2), 57–68.

Chakravorty-Spivak, G. 1988. Subaltern Studies: deconstructing historiography. In R. Guha and G. Chakravorty-Spivak (eds.) *Selected Subaltern Studies*. New York: Oxford University Press, 3–34.

Chatterjee, P. 1993. *The Nation and its Fragments: Colonial and Post-Colonial Histories*. Princeton, NJ: Princeton University Press.

Dirlik, A. 1992. The Postcolonial Aura: Third World criticism in the age of global capitalism. In A. McClintock et al. (eds.) *Dangerous Liaisons: Gender, Nation and Postcolonial Perspectives*. Minneapolis, MN: University of Minnesota Press, 501–28.

Fernandes, L. 2000a. Nationalizing the global: media images, cultural politics and the middle class in India. *Media, Culture, and Society*, 22(5), September 611–28.

Fernandes, L. 2000b. Rethinking Globalization: gender and nation in India. In M. de Koven (ed.) *Feminist Locations: Global/Local/Theory/Practice in the Twenty First Century*. New Brunswick: Rutgers University Press.

Ghosh, A. 1996. A reporter at large: Burma. *New Yorker*, 12 August 1996.
Guehenno, J. M. 1995. *The End of the Nation-State*. Transl. V. Elliott. Minneapolis, MN: University of Minnesota Press.
hooks, b. 1990. *Yearning: Race, Gender and Cultural Politics*. Boston: South End Press.
Habermas, J. 1998. The European Nation-State: on the past and future of sovereignty and citizenship. *Public Culture*, 10(2), 297–416.
Hartsock, N. 1990. Foucault on Power: a theory for women? In L. J. Nicholson (ed.) *Feminism/Postmodernism*. New York: Routledge, 157–75.
Krishna, S. 1999. *Postcolonial Insecurities: India, Sri Lanka and the Question of Nationhood*. Minneapolis, MN: University of Minnesota Press.
Menon, A. 1987. Time for a political solution. *Frontline* (Chennai), May 16–29.
Myrdal, G. 1968. *Asian Drama: an Inquiry into the Poverty of Nations*. New York: Pantheon Books.
Ohmae, K. 1995. *The End of the Nation-State: The Rise of Regional Economies*. New York: Free Press.
Rajagopol, A. 2001. *Politics After Television: Hindu Nationalism and the Reshaping of the Public in India*. New York: Cambridge University Press.
Rushdie, S. 1991. *Imaginary Homelands: Essays and Criticism 1981–1991*. London: Granta.
Said, E. 1988. Identity, negation, violence. *New Left Review*, 171, 46–59.
Said, E. 1993. *Culture and Imperialism*. New York: Knopf.
Soguk, N. 1999. *States and Strangers: Refugees and Displacements*. Minneapolis, MN: University of Minnesota Press.
Walker, R. B. J. 1993. *Inside/Outside: International Relations and Political Theory*. Cambridge: Cambridge University Press.
Wallerstein, I. 1991. *Unthinking Social Science*. Cambridge: Polity.

Chapter 21

Entrepreneurial Geographies of Global–Local Governance

Matthew Sparke and Victoria Lawson

In the context of globalization and in the wake of the Cold War there has been much talk, much hype, and much political argument based on the view that the nation-state is in terminal decline. While more cautious scholarly accounts of globalization tend to suggest that these claims are overblown (see Dicken, 1998, for a useful survey), and while nation-states still clearly remain the primary actors on the global stage, this has not prevented the ongoing reactivation of the "decline" hype in political debate, nor has it stopped the changes to policy-making and political discourse that have developed in reaction to the "end of the nation-state" arguments. One particularly notable strand of these changes in policy-making and political discourse has been the rise of highly entrepreneurial approaches to re-presenting the positions of political communities amidst the global–local nexus. Our academic understanding of the global–local nexus has been advanced considerably over the last decade by studies of how global economic forces lead to the production of new localization patterns or the so-called "glocalization" of regional development (e.g. Swyngedouw, 1992, 1997) and, by all kinds of research into how local landscapes and communities mediate and/or resist the changes brought about by global interdependency (e.g. Hines, 2000; Pred and Watts, 1991). However, in popular and political discourse the "global" and "local" are by contrast usually represented in far more vague and unexamined ways, and it is generally the force of their juxtaposition (rather than the actual character of the mediating links) that is taken-up and deployed discursively in attempts to shape the direction of political debate. It is just such instrumental approaches to the global–local nexus that lie at the heart of the entrepreneurial political geographies that concern us in what follows. We are calling these geographical discourses about the global and local "entrepreneurial" because they primarily serve the basic capitalist interests of businesses and business elites, and because, at the core, they are essentially based on a series of neoliberal commitments to the free market, deregulation, privatization, and competition. This is not to suggest that the resulting representations of a political community's positioning are always exactly the same. Sometimes the entrepreneurial approach can be *promotional* – advertising the virtues of a particular place as a crucial node or gateway between the local and global – and at other times it can be *managerial* – controlling the direction of political debate by drawing dividing lines between the local and global. In either case, though, the entrepreneurial approach to remapping the global–local nexus is becoming a dominant and highly effective way of extending and entrenching the interests of global businesses and their political representatives around today's world.

In this chapter, we advance the argument that a useful bracket term for describing these entrepreneurial approaches to representing and remapping political communities is "geoeconomics." We argue that in contrast to orthodox geopolitics and its concerns for soldiers and citizens, geoeconomics elevates the entrepreneurial interests of investors and customers; in contrast to the geopolitical focus on national borders and place, geoeconomic discourse privileges networks and pace; and instead of concentrating international politics on building alliances for "security" against supposed "evil empires," geoeconomics is primarily concerned with building international partnerships that advance "harmonization," "efficiency," "economic leverage" and "growth" against the supposed threats of political "radicalism," "anachronism" and "anarchy." In drawing these distinctions, our critical approach to the political discourse of geoeconomics follows the analytical example of the now extensive literature on "critical geopolitics" developed by Gerard Ó Tuathail (1996), Simon Dalby (1990) and others over the last decade (see also chapters 5, 6, and 12–16, this volume). Like these theorists of critical geopolitics, we do not take geoeconomic claims at face value, but rather see them as representational power moves which, notwithstanding their discursive inventiveness, can still have powerful real world effects. As a potent form of representing and managing politics and space, then, we are arguing that geoeconomics maps places, political communities and even the protocols of political accountability into and out of particular positions on the global–local interface. In so doing, geoeconomic representations effectively produce political geographies that are political both in the sense of favoring the particular political interests of entrepreneurial elites as well as in the sense of forcing through new political arrangements, regulations, and deregulations on the ground.

In order to substantiate these arguments about the power of geoeconomic discourses to produce new political geographies we introduce two main examples. The first example, which focuses on the development of a cross-border region between Canada and the USA, illustrates the promotional side of geoeconomics in action. The region has been dubbed "Cascadia" by its local boosters. By promoting Cascadia as a "gateway" region on the Pacific Rim, the boosters strive to represent the region's transnationalism as a sign of its special capacities for capitalizing on the global–local capitalist nexus. The second example, which also focuses on North American developments, illustrates the managerial aspect of geoeconomics put to work as political crisis management. This example concerns the ways in which an increasingly successful broad-based activist campaign against sweatshop labor has been contained and controlled by the development of a government and industry-backed organization called the Fair Labor Association (FLA). This association, we argue, has effectively remapped the dividing line between the global and local so as to mute and spatially contain local protests in the USA while simultaneously providing US elites with new disciplinary controls over global commodity chains of textile production and consumption. In both of our examples, therefore, the new political geographies of the global and local produced by geoeconomic discourse clearly serve entrepreneurial interests. In order, though, to make these claims more telling, we need to specify at the outset how our definition of geoeconomics departs from that of others.

Geoeconomics and entrepreneurial governance

The most well-known and dominant usage of the term geoeconomics in recent years has been by the US security consultant Edward Luttwak (1990, 1993, 1999). His approach to the concept is different, though not totally disconnected from the use of the term we are advancing here. Luttwak, described by Ó Tuathail (1996, p. 231) as the "quintessential defense intellectual," still very much believes in the logic of interstate rivalry that dominated the Cold War. His state-centric argument is simply that the languages and logics of this rivalry are now most routinely predicated on "the grammar of commerce" (Luttwak, 1990). For Luttwak, this new grammar demands abandoning the old vocabulary of geopolitics. In its place, he argues, comes the new vocabulary of "geo-economics" where, he says, "the authority of state bureaucrats can be asserted anew, not in the names of strategy and security this time, but rather to protect 'vital economic interests' by geo-economic defenses, geo-economic offensives, geo-economic diplomacy, and geo-economic intelligence" (Luttwak, 1993, p. 19).

Though largely uncritical, Luttwak's observations about the rising significance of commercial struggles are in themselves indisputable. The moves towards global free trade and the end of the Cold War have indeed led to the increasingly obvious eclipse of many traditional geopolitical disputes over borders and regions of dominance by new concerns with economic positionality in the global economy. Quite legitimately in our view, he also argues that this new regime, while generally taken for granted, is also equally bound-up with quasi-imperial US interests. But Luttwak's assumptions and assertions about the unchanging elemental feature of territorial states as "inherently adversarial" (1999, p. 128) prevent him from noticing the complex ways in which geoeconomics is also coeval with the increasing eclipse of governments by new free trade regimes and, within them, public–private networks of quasi-governmental authority. In the language of the international relations theorist James Rosenau, it stops him from coming to terms with new, often transnational, forms of "governance without governments" (Rosenau, 1992). Following others as well as Rosenau (see Jessop, 1995; and Brenner, 1999a), we understand governance in this sense to mean any enduring regime of systemic state-like control and authority (more on this definition in a moment when we come to the specific topic of entrepreneurial governance). It is because Luttwak only envisions governance in a strategic, sovereigntist, nationally bounded form that he is unable to theorize the transnational stretching of governance wrought by geoeconomics. In other words, his state-centrism prevents him from addressing the rising dominance of both state-like regulative arenas that transcend traditional state lines (such as NAFTA) and the devolution of central state authority to local governments and diverse public–private consortia (such as the sorts of public–private partnerships used increasingly to build and run everything from railways to sports stadia). It is this combined transcendence and hollowing out of the central authority once held by national governments in the

West during the mid-twentieth century that defines the so-called "leveled playing-field" in which governance is now negotiated (Jessop, 1995, 1997a). Existing national governments may well be pursuing many of the geoeconomic strategies of rivalry Luttwak describes, but it is the more generalized emergence of geoeconomic common sense as a means for negotiating governmental-type authority in a variety of other venues that is the more influential and remarkable aspect of the contemporary order. From the World Bank and the World Trade Organization to private think-tanks, city halls, and local business clubs, geoeconomic assumptions about the need to position communities most competitively in the global–local nexus are now basic. Moreover, this geoeconomic stress on competitive positionality also comes with a number of attendant assumptions about how the world works as a global marketplace, assumptions that in turn dictate a set of neoliberal principles for navigating the leveled landscapes of free trade. Thus if the geoeconomic emphasis on positionality comprises the new grammar, the neoliberal common denominators of *competition*, *liberalizing the market*, *increasing efficiency*, *regulatory harmonization*, *networking*, and *nodality* provide the vocabulary for day-to-day geoeconomic speech.[1]

In order to provide a clearer sense of geoeconomics as a newly dominant mode of articulating and mapping the global–local interface in political discourse, we have provided below the following table of contrasts between geoeconomics and geopolitics (see table 21.1). The table is only designed for heuristic purposes. It should not be viewed as an historical statement designating a strict temporal passage from one era to another; clearly, as 9.11 reminded us, geopolitics persists today. The table is rather a survey of certain *dominant tendencies in the political–geographic representation of political–economic space*, and it attempts to name these tendencies only as a prelude to forging a better understanding of how they play-out in the case studies that are considered later.

Both Sum (1999) and Jessop (1997a) use geoeconomics in a similarly critical, postnational way, applying it largely to the macro, continental-scale dynamics associated with the relations between the so-called "triad" regions of the EU, NAFTA, and the Yen-bloc. Our suggestion here is that this emerging tendency towards imagining territory in terms of the struggle over positionality in the global economy also applies equally well to sub-national and locally-transnational regional developments. This is not a terribly novel argument in itself. Popular globalists like Kenichi Ohmae (1995) who have been boldly writing epitaphs to the nation-state over the last decade also point to these phenomena repeatedly. They may not use the language of geoeconomics and they may well disagree with Luttwak's assertions about the ongoing centrality of national territorial communities in global affairs, but in their promotional tributes to region states and in their advocacy of neoliberal managerial strategies, Ohmae and others (see also Orstrom-Moller, 1995) provide ample illustration of the geoeconomic discourses centered on struggles over the imagining and scripting of economic positionality in global–local networks. The question, however, that these various commentators nevertheless dodge in all their normative populism surrounds the political structuring of these new struggles over position in the global economy. In particular, they fail to theorize or, as in Ohmae's case, simply celebrate the entrenchment of entrepreneurialism that the new geoeconomic meta-narratives support. The recent literature on entrepreneurial governance moves us beyond this failing by theorizing the transformation of governance associated with the rise of geoeconomics.

Table 21.1 Contrasting geopolitics with geoeconomics

Geopolitics	Geoeconomics
Both political geo-graphical dynamics involve processes of:- Managing complex changes through territorial representation, thereby spatializing political-economic processes and struggles.	
But they differ in the following ways…	
Develops at the end of C19th.	Develops at the end of C20th.
Origins in the end of empire.	Origins in the end of the Cold War.
Develops at a time of declining free trade and increasing national autarchy.	Develops at a time of increasing free trade and decreasing national autonomy.
Reflects struggle between territorial states over hegemony *over* the world system.	Reflects struggles for nodality *within* a global hegemonic system within which the USA has dominance but not imperial-type control.
Comes (by mid C20th) to take shape in territorialized Fordist political-economies where belonging is underwritten by: (a) the centralization of governance; (b) commitments to welfare equalization; (c) the management of competition.	Takes paradigmatic shape in deterritorialized post-Fordist political-economies where belonging is underwritten by: (a) the decentralization of governance; (b) socioeconomic polarization; (c) the deregulation of competition.
Forged in a context where state and market are relatively distinguished from one another.	Forged in a context where state and market are "networked" together in a complex array of public–private partnerships.
Ideologically preoccupied with alternatives to global capitalism.	Ideologically preoccupied with the idea that there are no alternatives to global capitalism.
Theoretically propounded by writers linked to the military: e.g. Halford Mackinder, Friedrich Ratzel.	Theoretically propounded by writers linked to business: e.g. Edward Luttwak, Kenichi Ohmae.
Demarcates the domestic and foreign with a language of sovereignty, allies and enemies.	Blurs the foreign/domestic distinction with a language of "intermestic" politics, perforated sovereignty, and joint ventures.
High politics focused on dominance and alliance-building for "security."	High politics focused on competition and partnering for "economic leverage."
Idealized subjects: citizens and soldiers.	Idealized subjects: customers and investors.
Territorial imaginary organized around blocs, nation-states and boundaries like the "iron curtain."	Territorial imaginary organized around nodality, region-states and linkages like "the web."
Fixates on borders and frontiers.	Fixates on borderless-ness and networks.

In particular the work of the political scientist Bob Jessop provides a series of valuable reflections on the transformation of governance in the context of a world where actual governmental authority is increasingly superceded or hollowed-out by the emerging trends towards market-driven and market-oriented policy-making (but see also Harvey, 1989, and the useful geographical overview of MacLeod and Goodwin, 1999). According to Jessop, the resulting "tendential de-statization of politics" does not create a power vacuum. Instead, he argues, it provides for the possibility of new modes of governance which in turn actually construct new objects to govern. Pertinent examples, he notes, "include: Porterian industrial clusters; flexible industrial districts; cross-border regions; and 'negotiated economies'" (Jessop, 1997b, p. 105). Such a list indicates that the new modes of governance are themselves *heterogeneous*, *poly-centric*, and *emergent* as opposed to deliberately and singularly-planned. This means, as Jessop further argues, that major stakes in these new modes of economic governance concern how "the boundaries of the economy are discursively constructed and materially instituted and the extent to which this 'spatial imaginary' corresponds in some significant sense to real economic, juridico-political and social processes" (Jessop, 1997b, p. 115). These are very large stakes indeed, of course, and, as we shall see in both the Cascadia case and the Fair Labor struggles, it is not at all pre-ordained that this kind of correspondence between the new spatial imaginaries (or political geographies as we are calling them here) and actual places and processes is possible. However, this is precisely where we see the meta-narratives of geoeconomics making a discursive difference. They keep the new entrepreneurial governance strategies afloat even while the problem of correspondence renders the plausibility of the new political geographies suspect. As a result they allow the strategists and promoters of the new modes of governance to go on producing their visions, and, as we shall now show, it is these visions that are shot through with entrepreneurial ideas about *how* precisely to go about positioning political communities in global networks.

Cascadia: Promotional Political Geographies of Cross-Border Regionalism

Cascadia is an idealized transnational space on the Pacific Coast of North America, transcending the 49th parallel and linking the Canadian province of British Columbia and the US states of Washington and Oregon (see figure 21.1). Such a clear cut geographical description of the cross-border region, though, should be understood as a cartographic still-shot of a much more dynamic process of regional invention – a process which, at its most grandiose, has been extended to include the whole of the so-called Pacific Northwest Economic Region, including Alaska, Alberta, Montana and Idaho. Cascadia is thus, as its boosters often claim, the product of "active evolution." Indeed, having taken the template for the cross-border region from an anti-growth ecotopian vision of bio-regionalists, the promoters "shared" and "evolved" the environmentalist vision into a pro-growth rationale for sustainable business growth. The genius of geoeconomics is that it helps make even this cooptation of environmentalism seem natural. Paul Schell, the ex-Mayor of Seattle, and John Hamer, a fellow of the politically conservative Discovery Institute in Seattle, put it like this:

> [Cascadia] is a shared notion, and one in active evolution. We're still inventing ourselves as a regional culture. Cascadia is a recognition of emerging realities, a way to celebrate commonality with diversity, a way to make the whole more than the sum of its parts. Cascadia is not a State, but a state of mind. But a state of mind can have important practical consequences (Schell and Hamer, 1993, p. 12).

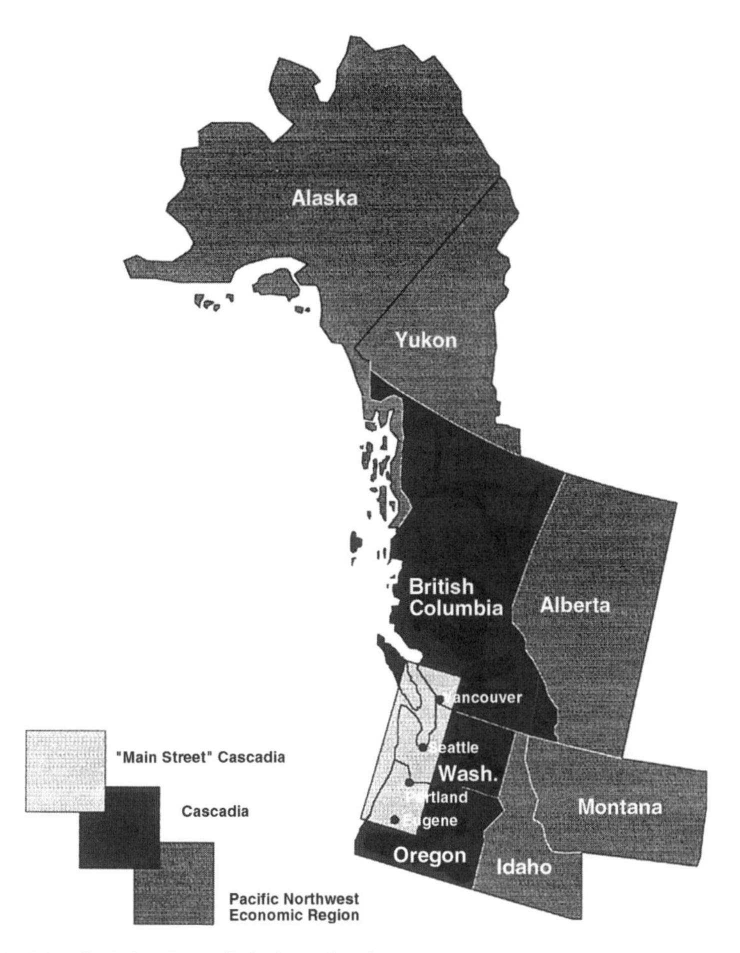

Figure 21.1 Map depicting Cascadia in its regional context.

Almost like dream images that are constantly condensing and displacing diverse determinations, the political geographic mappings of the cross-border region are thus far from being territorially fixed, static, and state-like. Instead, Cascadia appears to its promoters first and foremost as a moving "state of mind." However, as a vision, an idea, a discourse, a dream image, a space-myth and a state of mind, to list only a few of the words used to describe Cascadia, this cross-border regional development concept has continued to be sustained by both Canadian and US think-tanks, visionaries, and policy-makers. As well as being noted in numerous academic articles (e.g. Alper, 1996; Blatter, 1996; Courchene, 1995; Edgington, 1995; Relyea, 1998; Swanson, 1994), and as well as attracting the attention of mainstream policy-oriented think-tanks like the Carnegie Endowment for International Peace (Papademetriou and Meyers, 2000), it has also continued to surface in popular media outlets ranging from *B.C. Business* (Hathorn, 1993) to *The Economist* (1994) to the *Atlantic Monthly* (Kaplan, 1998) to *The Seattle Times* (Agnew, 1998) to *The Christian Science Monitor* (Poterfield, 1999).

It is the expanding discourse around Cascadia that we present here as an example of geoeconomics in action. While the main goals of the visionaries – a bullet train between Vancouver, Seattle and Portland, a bi-national Olympics, a "bull-dozing" of the border check-points on the 49th parallel, the development of an integrated high-tech industrial region, and so on – have so far failed, geoeconomics helps us explain why the visions nevertheless continue to be produced. Likewise, it is as an entrepreneurial political geographic remapping of the global–local nexus that we need to understand the claims about Cascadia constituting an example of an emerging trans-border region state.

A good illustration of the way in which the promoters conceive of Cascadia as a region blessed with the necessary capacities to serve as a "gateway" in global–local networks is provided by another promotional essay by Schell and Hamer. Their geoeconomic argument goes as follows:

> The lines imposed over 100 years ago have simply been transcended by contemporary cultural and economic realities. . . . Cascadia is organizing itself around what will be the new realities of the next century – open borders, free trade, regional cooperation, and the instant transfer of information, money and technology. The nineteenth- and twentieth-century realities of the nation-state, with guarded borders and nationalistic traditions are giving way (Schell and Hamer, 1995, p. 141).

Publishing this celebration of Cascadia a year after the implementation of NAFTA, and after six years of Canada–USA free trade, Schell and Hamer here make the geoeconomic case that the cross-border region is organized around and embodies the new borderless realities of globalization. Transcending the border, benefiting from free trade, exemplifying regional cooperation among key nodes of the new knowledge economy, the region appears to them as regional realization of the new world order.

Other versions of Cascadian geoeconomics directly cite Kenichi Ohmae as academic authority for the claims that in transcending the anachronistic obstacle of the border, the transnational region would somehow release hitherto untapped economic potentialities. John Miller, a former Republican member of Congress, observed in an opinion piece for the Canadian *Vancouver Sun* entitled "Riding the Cascadia Express":

> As the 21st century approaches we are entering the era of the region. This is not to say that nations with all their political, security, monetary and cultural concerns will not remain prominent – they will. But when it comes to economic and environmental concerns, global currents are already lifting the region into prominence. As Japanese economist Kenichi Ohmae has pointed out, sometimes the region involves part of a country – northern Italy; sometimes parts of several countries – the Asian city triangle on the Malacca straits of Medan, Pennang and Phuket; or sometimes parts of two countries – Hong Kong and southern Guandong province in China. But always the same phenomena are present: a geographically coherent market where millions of people have common economic and environmental interests, as well as large ports which provide links with the global economy. That's Cascadia or at least the main street of Cascadia from Vancouver to Eugene, Oregon. We are an internal market of seven million people all living between the Cascade and Pacific Coast mountains, all sharing an interest in trade and the environment far exceeding our eastern and southern neighbors (Miller, 1994).

Quite how the full formation of the region would happen did not need to be specified in this article, nor in others like it. Geoeconomics *à la* Ohmae's arguments already answered the question tautologically in terms of embodying the global–local nexus: "[region-states] make such effective ports of entry into the global economy because the very characteristics that define them are shaped by the demands of that economy" (Ohmae, 1995, p. 7). Created as a borderless market in the very image of free trade, region-states like Cascadia could be thus represented geoeconomically as having a privileged future precisely because of their capacity to internalize the liberalized logic of the global marketplace. Doing so, the argument ran, would enable them to become key nodes and portals in the new global networks. Such, at any rate, was the geoeconomic script of embodying globalization, and it was one that Cascadia's promoters read from repeatedly.

Despite all their appeals to the impact of borderless free trade, though, Cascadia's promoters have faced a major predicament: they are unable to point to any widespread regionalizing impact of the new trade flows. Certainly, truck crossings of the border on the Cascadia corridor have increased dramatically in both directions since the start of the impacts of free trade in the early nineties. But these increases in north–south flows do not indicate the rise of *regionalizing* tendencies in supply networks that cross the border and actually integrate Cascadia economically. Instead, the trucks cross the border and then drive on to many other distant places beyond the supposed region. The main British Columbia (B.C.) exporters export to the whole of the USA not just Washington and Oregon, and US companies like Microsoft and Boeing based in Seattle deal more with Ontario and places like Winnipeg than with B.C. There are none of the densely intermeshed input–output networks that have comprised the much-studied agglomerative affects in regions like Silicon Valley, Baden-Wurtenburg, and the Third Italy, the other regions highlighted by Ohmae and studied by researchers of the new economic and regional geography (see Storper, 1995 for an overview). In other words, while the promoters add-up the GDP figures for all the component parts of Cascadia and wax lyrical about its economic size and clout as a global–local gateway, they cannot point to an integrated economy or even a set of economic complimentarities across the border (see also Helliwell, 1998; Wall, 1999).

Lacking strong empirical evidence of their claims, the promoters nevertheless press on with further projects geared to marketing Cascadia in the global competition for investment and consumption spending. This it seems is where the force of geoeconomics as political discourse makes its mark. While the economy may not be regionalizing in quite the way it is meant to, the work of producing further entrepreneurial remappings can persist under the cover of geoeconomic common sense. In this way, they argue that the region also somehow embodies the political spirit of globalization: chiefly, the spirit of the neoliberal dogma of smaller, less interventionist government. As a corollary to the spatial supposition that suggests Cascadia's eclipse of the 49th parallel enables it to capitalize on the benefits of free trade, this argument asserts that because B.C., the most western province in Canada, and Washington and Oregon states have all shared a similar experience of historical alienation from faraway federal capitals they are all also inclined towards a distrust of big government. Bruce Agnew, for example, the director of the Cascadia Project at the Discovery Institute, put it like this: "We are finding borders and national government policies increasingly irrelevant and even crippling" (quoted in Schodolski, 1994, p. 1). No wonder then that perhaps the other most significant promotional use of the Cascadia name and concept has not been to launch a movement for more meaningful regional democracy but rather to brand a regional stock fund, the *Cascadia Equity Fund*, managed by the Aquila investment firm (Halverson, 1996).

In addition to the arguments about embodying global–local networking and neoliberal alienation from government, yet another line of geoeconomic promotional discourse has sought to position and prepare Cascadia as a global player by focusing on the need for local, cross-border cooperation. The slogan of cooperating regionally in order to compete globally would appear to be a commonplace of entrepreneurial planning and, as such, a sign of a rising geoeconomic commonsense behind strategizing regional governance in many parts of today's world. However, in the case of Cascadia it has been repeated with peculiar force and with very particular entrepreneurial visions of cooperation in mind (e.g. Chapman, 1996). Part of the forcefulness stems from a certain sort of naturalism. Here, for example, is the Canadian planner Alan Artibise's version of the argument:

> As nations have responded to the restructuring of the global economy, natural regional alliances have been stimulated. In a North American context, for example, the Pacific Northwest/Alaska is a small player. If that regional market is expanded to include British Columbia and Alberta, however, it then ranks as one of the largest in North America. On an international scale the same principle applies. The two nations and the two regions can bring complementary strengths to the international marketplace (Artibise, 1994, p. 4).

Part of the naturalism of this appeal for cooperation would also appear to spring from the logic of what might be described as a geoeconomic form of social Darwinism. The global economy is a harsh wilderness, this script seems to read, but by hanging together as some kind of bi-national regional wolfpack, Cascadia can develop enough so-called "critical mass" to defeat competitors and win a larger slice of planetary resources.

Again, with unfortunate implications for the promoters, it has been much harder to practice meaningful cross-border cooperation on the ground than it has been to preach it in geoeconomic discourse. Across the economic board in sector after sector from fishing to tourism, in the areas where the economies of B.C., Washington and Oregon share most, the competition is felt most fiercely. The ports of Vancouver and Seattle-Tacoma vie for each other's container and luxury liner trade. Likewise, the airports of Vancouver and Seattle, the very "gateways" to the supposedly integrated global–local regional nexus, are highly competitive with one another, a competition that would only be heightened if a high-speed rail link were to be put in place between the two cities (giving travelers the option of landing in either Seattle or Vancouver even when it might not be their final destination). The visionaries of Cascadia are aware of this. Indeed, for some promoters of cross-border integration such as Roger Bull, the former director of the Pacific Northwest Economic Region in Seattle, increasing intra-regional competition is precisely the point of building an integrated Cascadia. In an interview, he illustrated this point by noting that pitting the two airports into a still fiercer competition for long distance custom by introducing a high-speed rail link, would make it easier to push through new developments like the third runway planned for Seattle airport (interview April 1998). This runway construction is still being fought by the local communities afraid about increased noise, but with the threat of losing business to Vancouver hanging over their heads – argued Bull – this resistance could easily be overcome. This, it seems, is where the potency of geoeconomic promotional discourse shows its transformative power. While the envisioned political geographies may not correspond with actual economic dynamics, they can still nevertheless be counted on to coerce business-friendly entrepreneurial reforms on the ground. As we shall show in the next section, the same is also true for more managerial attempts to remap political communities in the global–local nexus.

Locating Fair Labor: Managerial Political Geographies of a Global Commodity Chain[2]

The Fair Labor Association (FLA) emerged from a White House initiative, the Apparel Industry Partnership in 1996. This was a political response to progressive anti-sweatshop campaigns launched since the early nineties by a broad coalition of human rights workers, labor activists, students, and non-governmental organizations (NGOs) (e.g. Global Exchange). Such campaigns have in turn been part of a much wider union of feminist, labor, and environmentalist struggles against the excesses of neoliberal free trade. In these early campaigns, "fair trade" was used as a slogan by progressives in the USA and elsewhere against "free trade" as a way of arguing for a whole series of more worker- and environmentally-friendly initiatives ranging from improved pay and working conditions to the support for more worker-controlled cooperative production. Fair trade has therefore taken on a range of meanings: from the radical anti-globalization arguments put forward by popular coalitions of activists, researchers, and groups such as the International Forum on Globalization (Roberts, 2000), to the more accommodationist union meanings through which international federations of unions work to establish labor clauses within WTO and associated trade treaties (Herod, 1998), to the "conscientious capitalist" meanings such as the pro-environment marketing stance of the Body Shop. Fair trade activism has not been a revolutionary vision, but more a social welfarist, or ameliorist vision – one that has represented a rallying point for critics of neoliberal hegemony. When articulated with broader global struggles for democracy, fair trade campaigns provide a model for alternative, emancipatory systems of global governance predicated on the ideals of reducing suffering and increasing justice and democracy (see Falk, 1999; Hines, 2000). Always at risk of being co-opted and compromised, Fair Trade has stood for something quite different from what it has come to mean under the auspices of the FLA. In this section of the chapter we show how a managerial form of geoeconomics has played a role in this domestication and cooptation of fair trade campaigning.

In applying the "Fair" moniker to its approach to sweatshop labor practices in the apparel industry (a moniker clearly "borrowed" from earlier campaigns), the FLA fashions a geographically encompassing image of a united and concerned global community adopting universal codes of conduct and monitoring procedures. Despite the image-oriented aspects of this upbeat universalistic rhetoric, and despite its obvious inadequacy in the face of complex and differentiated local labor problems, the FLA seeks to counter or at least sugarcoat the critiques of free trade and sweatshop labor, such as the anti-WTO protests in Seattle and Geneva, and various rallies at Niketowns from New York to San Francisco (Shaw, 1999). While critics at these rallies seek to re-map the global–local nexus in the interests of revealing the links between exploitation "there" and profits "here," the managerial geoeconomics of the FLA appropriates the language of fairness to downplay these very links. Thus, just as the promoters of Cascadia have used the bioregional concept as a trojan horse for entrepreneurial boosterism, the corporation-dominated FLA has appropriated the progressive language of "fair trade not free trade" so as to better control consumer concerns about sweat shop labor. Likewise, just as the boosters of Cascadia turn the notion of a cross-border ecology into a marketing device, the FLA similarly turns the notion of Fair Trade into a branding mechanism by setting up a protocol for certifying the brands of member companies. And just as Cascadia has been promoted by a series of public–private consortia, so too does the FLA represent another blurring of the state/business divide, operating as a public–private network with quasi-governmental authority. As such, and as a political response to the increasing student pressure exerted through the Students Against Sweatshops movement and their Worker Rights Consortium (WRC), the FLA produces political cover for corporations by providing a company-focused, rather than locally-based, labor-focused mechanism for addressing sweatshop labor conditions. As Appelbaum and Bonacich (2000, p. B5) argue, there is

> a fundamental flaw in the organization: It was never designed to change the industrial system dynamics that produce sweatshops. Nothing in the FLA's rules requires manufacturers to pay more money to their contractors, money which could then be passed along to workers. Nothing provides even a hint that manufacturers could be held legally accountable for abuses that occur in the factory in which they contract.

Indeed, the stated goals of the FLA all focus broadly on *companies themselves* rather than on the concerns and actions of workers *in specific places* in the context of globalized production. There appear to be three ways in which this company-focus of the FLA results in a geoeconomic remapping of the global–local nexus, and in the reworking of progressive agendas around fair trade and fair labor practices.

First, the membership structure and the priorities of the FLA operate to de-link the global from the local. Whereas the promotional geoeconomics of Cascadia promotes a particular place through an appeal to borderless networks and the opportunities therein for investors and consumers, the crisis-managing geoeconomics of the FLA focuses on borderless corporations and their investors so as better to hide from consumers particular places of exploitation. The fourteen-member governing board comprises six industry representatives, six non-governmental members (such as National Consumers League and Lawyers Committee for Human Rights), one university representative, and the executive director. US labor unions have refused to participate and there is no board representation of labor unions from any country, international worker federations or non-union workers. Both this organizational structure, and the emphasis upon the actions of globalized companies, means that place-specific struggles by workers are not directly represented inside the FLA. As in geoeconomic discourse more generally, the primary concern of the association is instead with investors and consumers. If a company has met the FLA standards

> ... then, the Participating Company shall be entitled to communicate to the public that such brands have been produced in [c]ompliance with the Fair Labor Association Standards and *shall be entitled to use the service mark of the Association in product labeling, advertising and other communications to consumers and shareholders* (Charter Document: Fair Labor Association, June 1999, our emphasis).

Workers barely register in this sort of legalese. The legally narrowed approach to certification within the FLA further serves to pre-empt the possibility of even noticing the particularity and diversity of workplace exploitation. It is an approach that is designed only to certify the globally-produced brands of member companies, rather than particular production sites. This framework has been adopted even though scholarship on production chains clearly demonstrates that much production in Asia and the Americas occurs in informal production sites *outside* of formally-owned corporate plants. Research in settings as diverse as Mexico (Beneria and Roldan, 1987), Ecuador (Lawson, 1995, 1999), Taiwan (Tai-Li, 1983), and the USA (Becker, 1997) demonstrates, in fact, that global corporations subcontract to domestic firms, which then frequently fan out production into unregistered sweatshops and homeworking sites. The conditions of apparel work in these diverse sites are often harsh and, given the formal company focus of the FLA, they fail even to qualify for the association's monitoring. Indeed, the complex local geography of apparel manufacturing in unregistered sites effectively ensures the incomplete and thus inaccurate nature of the certification claims made by the FLA. The association's approach to certification is thus fundamentally flawed. It only "succeeds" insofar as it enables the FLA to delink the locally specific character of apparel work and exploitation from the global claims of engaging in fair labor practices.

If the first geoeconomic aspect of the FLA's crisis-management has been to delink the global from the local, the second has more to do with the way in which the association has further narrowed the meaning of the "local" by establishing a discrete US focus on a particular niche market for apparel – universities. This narrow focus does not represent a geoeconomic scripting of territory so much as a containment of the space of political dissent. All the same, this strategy is a form of political geographic reterritorialization that is fully concordant with the geoeconomic mindset that places the interests and desires of investors and consumers above the principles of citizenship and equality. In order to limit damage to sales in the world's largest market (the USA), the certification system proposed by the FLA is tightly targeted at student-led, US-based anti-sweat shop protests. The FLA has thus worked hard to incorporate one of the most vocal and organized consumer constituencies – universities – by actively promoting itself at campuses across the USA. The organization has also targeted universities because, even though they represent a tiny fraction of apparel production and consumption, through their licensing agreements with apparel companies they are a visible market for apparel as well as constituting a crucial site of activism against sweatshop conditions.

There is, of course, a broader political struggle that also accounts for the FLA focus on universities. The student group University Students Against Sweatshops

(USAS) demonstrated the potential of consumer power not only through their campus organizing, but also through the formation of an alternative, labor-focused Workers' Rights Consortium (WRC). This alternative oversight body has represented a much more worker-accountable monitoring mechanism, and as such, a real threat to corporate business as usual. The FLA's own brand-certification process, by contrast, appears to address consumer concerns all the while pre-empting the further development and adoption of the WRC's more aggressive, pro-worker oversight system. Only twelve companies are actually participating in the FLA, but the organization's discourses of fair labor and of a collaborative approach that unites key actors, combined with substantial funding from the US State Department, have allowed the organization to dominate the consumer protection agenda by bringing universities into its fold (university members number 147 in FLA and 62 in WRC).[3] This internal institutional structure also allows local and global issues to be further delinked. Thus despite the fact that approximately 60% of the FLA budget has been coming from the US State Department ($750,000 in 2000, $1.1 million in 2001), representatives from the department have a very low profile within the FLA. As with other geoeconomic challenges to democratic governance, we also see here a shift of crucial public issues – such as safe and just working conditions in a globalized economy – out of the formal government sphere and into a private–public organization in which obviously relevant governmental agencies such as the Department of Labor have no vote on the governing board.

The third and perhaps most clearly geoeconomic aspect of the FLA's activities has been the way it has provided powerful US apparel manufacturers with new forms of control over global commodity chains of production and consumption. Progressive voices such as USAS argue that public disclosure of production sites is an essential first step in public accountability and meaningful monitoring of working conditions. However, FLA member companies are not required to make their production sites public.[4] Effectively, the FLA allows companies to develop a global competitive strategy based on the prestige and networking possibilities attached to being an FLA-certified company. This FLA-sponsored control over global markets is provided through managing and shaping consumer demand for their certified brands, despite the empirical impossibility of creating meaningful certification through FLA procedures. Notwithstanding these failings, the procedures lend prestige to member companies, bolstering their capacity to shape globalized networks of apparel production and consumer demand. Moreover, the FLA also privileges member companies in another way through its overarching global agenda of promoting US-style neoliberal discourses of economic development and democratization. In this respect it should be noted that it is through organizations such as FLA, that the US State Department, and its office of Democracy, Human Rights and Labor continue to entrench and expand the reach of what has been critically dubbed the "Washington Consensus" (Chomsky, 1999). This refers to a set of neoliberal principles of entrepreneurial governance predicated on trade and finance liberalization and privatization of state activities. Speaking at the World Economic Forum, Clinton clearly enunciated this position, "[W]e have to reaffirm unambiguously that open markets are the best engine we know of to lift living standards and build shared prosperity" (quoted in Dollar and Kraay, 2000). Embedded within these logics, the US State Department promulgates the notion that US investment in poorer countries has progressive impacts on standards of living, democratization and individual freedom. State Department investments in organizations engaged in entrepreneurial governance such as the FLA, and the increased global competitiveness of member companies is thus consistent with a neo-colonial history that has already been widely critiqued (e.g. Briggs and Kernaghan, 1994, p. 39).

Insofar as it represents an extension of these new modes of entrepreneurial governance from afar, we are arguing that the FLA serves to legitimate and sanitize the brutal impact of neoliberal policies. It thereby bolsters the moral hegemony of US business practices abroad and further helps to socialize foreign capitalists into US practices and codes of conduct. The association's efforts can in this sense be seen as a strategic geoeconomic two-step that at once reigns in foreign corporations by deploying consumer power in the dominant US market, while simultaneously, functioning to advance US business interests and influence in locating manufacturing in Southern states. In fact, the organization legitimates overseas investment by member companies and gives them a justification, if not legally binding rights, to dictate production practices in other countries whilst not engaging at all with formal political channels, local or national governments to change the larger social/political dynamics surrounding their own initial investments. Such is the way that the FLA helps to manage the global–local nexus in the interests of US-style entrepreneurial governance.

Overall, then, the FLA's managerial form of geoeconomics reworks the political geographies of fair trade protests by dividing workers, politicized communities and business practices across global–local lines, by therefore containing and limiting the scope of campus-based activism, and, most spectacularly, by proceeding to turn that very activism into another marketing ploy, a ploy that at the same time further serves to protect entrepreneurial activities from democratically accountable oversight both at home and abroad. Indeed it seems structurally impossible for the universities to demand more stringent, labor-friendly accountability measures from within the existing FLA mechanisms that allow companies to self-monitor.[5] As Randy Rankin, partner with PricewaterhouseCoopers, candidly puts it: "[A]lthough universities have the ability and obligation to *challenge* their partners to improve working conditions worldwide, they must have *realistic expectations for change*" (Rankin, 2000, our emphasis). Just as the architects of traditional geopolitics once spoke in the language of realism (see Ó Tuathail, 1996), so too then it seems do today's practitioners of geoeconomics claim the high ground of being realistic. In the process, however, they clearly gloss over all kinds of global–local connections and power relations in the interests of neoliberal crisis management. As we have shown, the FLA in this way effectively obfuscates the very links between exploitation and consumption practices that the original student-led campaigns for Fair Trade sought to highlight. In the end, the FLA's managerial geoeconomics amounts thus to an attempt to co-opt consumer concerns by certifying member company brands as Fair Labor Certified so as to reduce more effectively the ability of politicized consumers to diminish corporate exploitation. The entrepreneurial interest of this divide-and-conquer political geography is clear. It drastically curtails the possibilities for worker justice that a proactive worker-accountable system of monitoring might start to provide. Against hopes that consumer power can thus serve as a political tool for advancing workers' rights, the managerial response of the FLA puts consumer concerns at the top of its agenda while simultaneously ignoring the complex local geographies of apparel manufacturing in unregistered sites. Despite all this, and despite the very small number of companies actually participating, the FLA is nevertheless positioned by business elites and the US State Department as *the* answer to sweatshop labor abuses. This is how, helped along by a number of geoeconomic assumptions, representations and practices, the FLA manages the discourses of protest against sweatshop production practices in the interests of protecting and expanding entrepreneurial governance.

Conclusion: Resisting Geoeconomics

Defined as broadly as we have defined and used it here, geoeconomics begins to seem ubiquitous. From the speeches of local politicians arguing that their community is the best place in the world to locate new businesses, to borderless-world themes in adverts for cellphones, airlines, and new recordings, to the more aversarial geoeconomics of mining and logging TNCs telling local communities that globalizing environmentalists threaten local life, to the attempts of so-called "business geographers" to market themselves as consultants in geoeconomics, the trend appears all-encompassing. To see it in such a totalizing and overweening way, however, would be a mistake. To be sure, this market-oriented and competitive approach to imagining and representing the positioning of political communities is part of a much larger series of worldwide changes associated with the globalization of contemporary capitalism. As such it cannot be easily contested or independently reversed. But just as scholars of globalization have taught us not to treat the patterns of accelerated global interdependency as anonymous, unstoppable forces, so too is it important to see geoeconomic tendencies as profoundly political and thus inherently resistable and transformable. In both the cases of promotional and managerial geoeconomics highlighted in this chapter examples of these opportunities for resistance can be easily found. In the Cascadian case, for example, while the boosters may have co-opted the outline and name of Cascadia from bioregionalists, this has not stopped local environmentalists from continuing to articulate radically different policy proposals for transboundary ecosystem protection in the region. Many groups, including the People for Puget Sound, the Georgia Strait Alliance, and the Northwest Ecosystem Alliance, continue thus to imagine other alliances and futures for the area which, while borderless, establish clear boundaries for business. In the FLA case, the examples and opportunities for resistance are still more significant, and it is with these that we will conclude.

From the launching of the FLA in 1998, the student organization USAS resisted the association's entrepreneurial inclinations by injecting a pro-worker agenda into the organization. In the fall of 1999, after having no serious impact on the FLA and after appealing directly to the White House and receiving no response, the students abandoned this inside approach and instead launched a nationwide activist campaign and their own organization, the WRC. As a result, there are now activities on over 150 university campuses nationwide and USAS has sparked the biggest wave of student activism in years. In 1999 and 2000 there have been a flurry of demonstrations and sit-ins on college campuses by students demanding Codes of Conduct that guarantee the protection of the basic human rights of workers producing college gear worldwide. Successful campaigns have been waged at large public universities such as the University of Michigan, University of North Carolina, University of Washington, University of Arizona and University of Wisconsin at Madison. In all of these cases, the students requested university administrators to leave the FLA and join the WRC. Students advocate the WRC because it embraces four important provisions that focus directly on workers and the protection of their basic rights: full public disclosure, enforcement of the rights of women, independent monitoring, and a living wage. This student activism has won swift and significant progress. Indeed, the popular movement it has spawned has even disseminated to the high-school level (see http://www.nlcnet.org/student/usas for more details on the ever-changing status of student activism). These sincere attempts to link politically the local and global would clearly need to be democratically transnationalized to be effective as long-term resistance. But they nevertheless indicate a way through which entrepreneurial governance *can* be contested, and as such provide some hope that the dominance of today's geoeconomics might be challenged before it inflicts on our new century the violent legacy traditional geopolitics inscribed on the last.

ENDNOTES

1. We should note in this regard that we are not using geoeconomics in the same way that Peter Dicken uses the phrase "the new geo-economy" to specify the uneven spatial development effects of global economic interdependency (1998, pp. 1–15). For him, the phrase serves as a broad analytical category; for us "geoeconomics" is by contrast a descriptive bracket term that usefully captures the transformation of certain key geopolitical tropes and discursive preoccupations in the context of global economic interdependency.
2. In summer 2000, Vicky Lawson participated in extended committee deliberations regarding University of Washington membership in the Fair Labor Association and the Workers Rights Consortium. This committee, brought together in response to student against sweatshops activism, included faculty, students, administrators, and alumni. In addition to extensive reading on the issues surrounding globalization, sweatshop labor practices and these organizations, we also engaged in detailed conversations with Ron Blackwell of the WRC and Bob Durkee of the FLA. Our analysis draws in part on our interpretations of these proceedings. We also note that the FLA is evolving at the time of this writing, and we base our analysis on the status of the organization in fall 2000.
3. According to Sam Brown, Executive Director of the FLA, approximately 60 percent of the FLA budget came from the Department of Labor and the Agency for International Development in 2000. In 2001, these funds are expected to come from the office of Democracy, Human Rights and Labor at the Department of State.
4. We note, however, that approximately 40 of the 147 member universities have chosen to require public disclosure of production sites for their licensed products (Durkee, 2000). Nonetheless, production for these universities represents a tiny proportion of total production by these companies. As such, compliance with this university request is a small price to pay for membership by this highly visible constituency of producers.
5. FLA procedures allow that in any given year, 70–90 percent of factory monitoring is carried out by the companies themselves. Critics have argued that the notion that companies will meaningfully monitor themselves is not compelling (Appelbaum and Bonacich, 2000; Shaw, 1999; Weiss, 2000).

BIBLIOGRAPHY

Alper, D. 1996. The idea of Cascadia: emergent transborder regionalisms in the Pacific Northwest-Western Canada. *Journal of Borderland Studies*, XI(2), 1–22.

Appelbaum, R. and Bonacich, E. 2000. Choosing sides on the campaign against sweatshops. *The Chronicle of Higher Education*, April, B4–5.

Artibise, A. 1994. *Opportunities of Achieving Sustainability in Cascadia*. Vancouver: International Center for Sustainable Cities.

Becker, L. 1997. *Invisible Threads: Skill and the Discursive Marginalization of the Garment Industry's Workforce*. Doctoral Dissertation, Department of Geography, University of Washington.

Beneria, L. and Roldan, M. 1987. *The Crossroads of Class and Gender: Industrial Homework, Subcontracting and Household Dynamics in Mexico City* Chicago: Chicago University Press.

Blatter, J. 1996. *Cross-border Cooperation and Sustainable Development in Europe and North America*. Unpublished paper, Fakiltaet fuer Verwaltungswissenschaft, Universitaet Konstanz, Germany.

Brenner, N. 1999a. Beyond state-centrism? Space, territoriality, and geographical scale in globalization studies. *Theory and Society*, 28, 39–78.

Briggs, B. and Kernaghan, C. 1994. The US Economic Agenda: a sweatshop model of development. *NACLA Report on the Americas*, 27(4), 37–40.

Chapman, B., 1996. Cooperation not competition, key to Cascadia Region Success. *The Seattle Post Intelligencer*, 6.14.96, A16.

Chomsky, N. 1999. *Profit Over People: Neoliberalism and Global Order*. New York: Seven Stories.

Courchene, T. 1995. Globalization: the regional/international interface. *Canadian Journal of Regional Science*, XVIII.1, 1–20.

Dalby, S. 1990. American security discourse: the persistence of geopolitics. *Political Geography Quarterly*, 9(2), 171–88.

Dicken, P. 1998. *Global Shift: Transforming the World Economy*, 3rd edn. New York: Guildford.

Dollar, D. and Kraay, A. 2000. *Growth is Good for the Poor*. Development Research Group, The World Bank, Washington, DC. At http://www.worldbank.org/research, accessed 10/100.

Durkee, R. 2000. A new movement breaks a sweat. *Trusteeship*, July/August, 17–23.

The Economist. 1994. Welcome to Cascadia, 5.21.94, 52.

Edgington, D. 1995. Trade, investment and the new regionalism: Cascadia and its economic links with Japan. *Canadian Review of Regional Science*, XVIII.3, 333–56.

Fair Labor Association, 1999. *Fair Labor Association Charter Document, Amended Agreement*. At http://www.fairlabor.org, accessed 10/100.

Falk, R. 1999. *Predatory Globalization: A Critique*. New York: Polity.

Harvey D., 1989. From managerialism to entrepreneurialism: the transformation of urban governance in late capitalism. *Geografiska Annale*, 71B, 3–17.

Halverson, G. 1996. Regional Road Maps Guide Some Mutual Funds. *The Christian Science Monitor*, 10.8, A10.

Helliwell, J. 1998. *How Much Do National Borders Matter?* Washington, DC: Brookings Institution Press.

Herod, A. 1998. Of Blocs, Flows and Networks: the end of the Cold War, cyberspace, and the geoeconomics of organized labor at the *Fin de Millenaire*. In A. Herod et al. (eds.) *An Unruly World? Globalization, Governance and Geography*. London: Routledge, 44–59.

Hines, C. 2000. *Localization: A Global Manifesto*. London: Earthscan.
Jessop, B. 1995. The regulation approach, governance and post-Fordism: alternative perspectives on political-economic change? *Economy and Society*, 24, 307–33.
Jessop, B. 1997a. The Entrepreneurial City: re-imaging localities, redesigning economic governance, or restructuring capital. In N. Jewson and S. Macgregor (eds.) *Transforming Cities: Contested Governance and New Spatial Divisions*. New York: Routledge, 28–41.
Jessop, B. 1997b. The governance of complexity and the complexity of governance: preliminary remarks on some problems and limits of economic guidance. In A. Amin and J. Hauser (eds.) *Beyond Market and Hierarchy: Third Way Approaches to Transformation*. Aldershot: Edward Elgar, 111–47.
Kelly, C. 1994. Midwifing the new regional order. *The New Pacific*, Spring, 6.
Lawson, V. 1995. Beyond the firm: restructuring gender divisions of labor in Quito's garment industry under austerity. *Environment and Planning D: Society and Space*, 13, 415–44.
Lawson, V. 1999. Tailoring is a profession, seamstressing is work! Resiting work and reworking gender identities among artisnal garment workers in Quito. *Environment and Planning A*, 31, 209–27.
Luttwak, E. 1990. From Geopolitics to Geo-Economics: Logic of Conflict, Grammar of Commerce. *The National Interest*, 20, 17–23.
Luttwak, E. 1993. The Coming Global War for Economic Power: There Are No Nice Guys on the Battlefield of Geo-Economics. *The International Economy*, 7(5), 18–67.
Luttwak, E. 1999. *Turbo Capitalism: Winners and Loosers in the Global Economy*. Perennial: New York.
MacLeod, G. and Goodwin, M. 1999. Reconstructing an urban and regional political economy: on the state, politics, scale, and explanation. *Political Geography*, 18, 697–730.
Miller, J. 1994. Riding the Cascadia Express. *Vancouver Sun*, 8.18.94.
Ohmae, K., 1995. *The End of the Nation-State: The Rise of Regional Economies*. New York: Free Press.
Orstrom-Moller, J. 1995. *The Future European Model: Economic Internationalization and Cultural Decentralization*. Westport, CT: Praeger.
Ó Tuathail, G. 1996. *Critical Geopolitics: The Politics of Writing Global Space*. Minneapolis, MN: University of Minnesota Press.
Papademetriou, D. and Meyers, D. 2000. Of Poetry and Plumbing: the North American Integration Project. Paper presented to the workshop *Managing Common Borders: North American Border Communities in the 21st Century*, Carnegie Endowment for International Peace, Washington, D.C.
Poterfield, E. 1999. Emerging Cascadia: Geography, economy bring Northwest cities ever-closer. *The Christian Science Monitor*, 7.26.99, 3.
Pred, A. and Watts, M. 1992. *Reworking Modernity: Capitalisms and Symbolic Discontent*. New Brunswick, NJ: Rutgers University Press.
Rankin, R. 2000. Manufacturers can be part of the solution to the sweatshop issue. *Trusteeship*, July/August, 20–1.
Relyea, S. 1998. Trans-state entities: postmodern cracks in the Great Wesphalian Dam. *Geopolitics*, 3(2), 30–61.
Roberts, S. 1998. Geo-governance in trade and finance and political geographies of dissent. In A. Herod et al. (eds.) *An Unruly World? Globalization, Governance and Geography*. London: Routledge.
Rosenau, J. 1992. Governance, order and change in world politics. In J. N. Rosenau and E-O. Czempiel (eds.) *Governance without Government: Order and Change in World Politics*. Cambridge: Cambridge University Press, 1–29.
Schell, P. and Hamer, J. 1993. What is the future of Cascadia? *Discovery Institute Inquiry*.

Schell, P. and Hamer, S. 1995. Cascadia: the new binationalism of Western Canada and the U.S. Pacific Northwest. In R. Earle and J. Wirth (eds.) *Identities in North America: The Search for Community.* Palo Alto, CA: Stanford University Press, 140–56.

Schodolski, V. 1994. Northwest's economy defies national borders. *Chicago Tribune*, 8.1.94, 1.

Schoonmaker, P., von Hagen, B., and Wolf, E. 1997. *The Forests of Home: Profile of a North American Bioregion.* Washington, DC: Island Press.

Shaw, R. 1999. *Reclaiming America. Nike, Clean Air, and the New National Activism.* Berkeley, CA: University of California Press.

Storper, M. 1995. The resurgence of regional economies, ten years later: The region as a nexus of untraded interdependencies. *European Urban and Regional Studies*, 2(3), 191–221

Sum, N. 1999. Rethinking Globalization: rearticulating the spatial scale and temporal horizons of trans-border spaces. In K. Olds et al. (eds.) *Globalisation and the Asia-Pacific: Contested Territories.* New York: Routledge, 129–45.

Swanson, L. 1994. Emerging transnational economic regions in North America under NAFTA. In M. Hodges (ed.) *The Impact of NAFTA: Economies in Transition.* London: LSE, 64–95.

Swyngedouw, E. 1992. The Mammon quest. Glocalization, interspatial competition and the monetary order: the construction of new scales. In M. Dunford and K. Kafkalas (eds.) *Cities and Regions in the New Europe: the global–local interplay and spatial development strategies.* Wiley: New York, 39–67.

Swyngedouw, E. 1997. Neither global nor local: "glocalization" and the politics of scale. In K. Cox (ed.) *Spaces of Globalization.* New York: Guildford, 137–66.

Tai-Li, H. 1983. The emergence of small-scale industry in a Taiwanese rural community. In J. Nash and M.P. Fernandez-Kelly (eds.) *Women, Men and the International Division of Labor* Albany, NY: SUNY Press, 387–406.

The New Pacific, 1992. The Power of One. Summer, 7.

Wall, H. 1999. How Important is the US-Canada Border? *International Economic Trends*, August, 1.

Weiss, L. 2000. Fair Labor group doing little to stop sweatshops. *Minneapolis Star Tribune*, 7.8.2000, 17A.

Part V Geographies of Political and Social Movements

Chapter 22

Representative Democracy and Electoral Geography

Ron Johnston and Charles Pattie

With the majority of the world's population living in countries with governments elected through some form of representative democratic process, Fukuyama (1992) identified the late twentieth century as the "end of history", the triumph of a form of government best suited to economic liberalism. But the "end of history" is not associated with an "end of geography", however much the latter may be claimed for other concurrent changes in economic, social and cultural organization (Johnston, 1994). Places remain central to the structuration of the political elements of contemporary life, as illustrated in this discussion of political parties and the geography of their operation.

Parties and the Democratic Process

Political parties are modern creations; few are as long-established as the British Conservative party, which traces its origins to 1832 (Blake, 1985). The majority emerged alongside franchise extension and then dominated the politics of the twentieth century in most states (even in single-party dictatorships). Some argue that parties are obsolescent, outflanked by cyber-democracy, new social movements, and so on (Mulgan, 1994). But the evidence is decidedly mixed: parties may seem tired in the West, but they are multiplying rapidly in new democracies of the former Eastern bloc. Even in the established democracies, they remain key actors (Seyd, 1998), changing but far from redundant.

Parties occupy the interface between the civil society and the state, providing most of the successful candidates for elected office. Parties present policies to voters in election manifestos. Parties form governments and governing parties are held accountable by the electorate. Voters form allegiances (sometimes long-lasting) with parties. And parties campaign to mobilize votes, encouraging public democratic participation.

Why parties?

Why should parties provide these mechanisms? The answer is partly linked to the development of representative democracy as the pre-eminent model for democratic governance (Held, 1987). Although often taken for granted in contemporary discourse, democracy is in fact a slippery and contested concept, which has meant different things to different societies. During the twentieth century, for instance, communist regimes claimed to be democracies – but democracies in which rights were conferred not to individuals but to social classes: more often than not, this "collective democracy" was associated with human rights abuses. Nor would contemporary understandings of democracy be comprehensible to the first democrats. For the ancient Athenians, the originators of the concept, democracy meant the direct involvement of all citizens in government decision making: anything which restricted a citizen's ability to take part in deliberation was seen as a denial of democracy. "Direct democracy" along these lines was possible in small city-states. Citizenship was restricted to a relatively few free-born adult males (women, slaves, and non-Athenians were all disbarred from citizenship) and all citizens could, in principle, meet together to deliberate. For much of history, Athenian direct democracy was seen as synonymous with democracy as a whole. For much of history, too, democracy was widely distrusted as a route by which demagogues and "the mob" might rule. Only with the rise of liberal ideals in the seventeenth and eighteenth centuries was democracy "rehabilitated".

In mass societies, however, direct democracy becomes impossible. Not only are there too many citizens to deliberate together in any meaningful way, but most citizens are too busy with their daily lives to take a close interest in policy. So, citizens elect politicians to take decisions on their behalf: accountability is assured by the power to remove the elected representatives at a subsequent election if their performance is deemed inadequate. (This is, of course, quite at odds with classic Athenian direct democracy.) Arguably, representative democracy, by its very nature, makes the emergence of powerful political parties more likely, as illustrated by Downs' (1957) classic "economic model of democracy".[1]

Downs postulates a political world with two sets of actors – parties and voters. Both are rational, maximizing their benefits relative to costs. Voters maximize the chance that their preferred bundle of policies is enacted; parties maximize the chance of winning office. Voters should support the candidate whose policy position is closest to their own. But information is not cheap. It takes time for voters to think about the major issues; to work out personal costs and benefits arising from each potential policy position; to weigh up candidates' positions in order to choose the one closest to their own view; to calculate where other voters stand on the issues; and to assess the likelihood that an individual candidate will win election. For most voters, the costs of becoming a "perfectly informed" citizen are so extensive as to outweigh any potential benefits that might accrue from election of their favored candidate. With high information costs, therefore, rational voters should assume that all other voters will bear the cost of keeping abreast, and therefore abstain, knowing that their decisions not to vote will have only the most marginal impact on the election outcome. But if all voters are so rational, none will vote, and the political system will cease to function (or only a few will vote and "capture" the system). Ideologically-based parties offering distinct and well-known policy packages offer a way out of this dilemma. By developing their own distinctive ideological ground, parties offer voters a "cheap" means of establishing where each party stands on any given issue. But to work, parties must stick to their side of the bargain: their ideological images must be reflected in both party statements and in policies advanced by the party when in office. The party's ideological image must reflect a real substance.

Under some circumstances, parties may converge onto similar ideological ground. In Downs' hypothetical example, most voters are near the center of the ideological spectrum, and there are only two parties. Under these conditions, vote-maximizing parties converge on the policy ideals of the median voter. However, they do not "cross": a previously left-wing party will not suddenly become more right-wing than its right-wing opponent because to do so would not only move it further from the median voter, but it would also risk alienating existing supporters. Even when the ideological gap between parties narrows, a sufficient gap should be maintained to sustain the parties' respective ideological "images" in the voters' minds.

The importance of a stable support base

In representative democracies, parties succeed or fail through their ability to mobilize votes. A party's job is eased considerably if it is able to develop a large and loyal support base within the electorate, on which it can rely. Indeed, for some, this is almost an absolute prerequisite for a successful representative democracy (Lipset, 2000; Osei-Kwame and Taylor, 1984). The party must create and then mobilize core support – electors who normally vote for it through thick and thin. The core vote (sometimes called the "normal vote": Converse, 1966) provides a long-standing support foundation on which the party can rely. It cannot be taken for granted, however. In the short term, the party must encourage as many of its "guaranteed" supporters to vote as possible, especially in close contests. In the longer term, it must deliver sufficient incentives (in the form of policy success) to persuade its core voters that continued support is worthwhile. Failure to develop or nurture a core vote makes electoral life difficult for a party, since it must build support from scratch at each new election. This is not uncommon in developing world democracies and may stimulate a "politics of failure" (Osei-Kwame and Taylor, 1984), often related not only to electoral instability but also to wider economic and political instability too. Alternatively, as the New Zealand Labour party discovered in the late 1980s, repeatedly ignoring the demands of the core vote can generate a sudden and catastrophic loss of support. The party abandoned its social democratic program in favour of neo-liberal policies after electoral victory in 1984. It was re-elected in 1987, on the same neo-liberal platform. At both contests, the party retained its core voters' support, partly because they had no alternative, partly because they hoped Labour would return to its original left-of-centre policies (though turnout in 1987 was low in Labour heartlands: Johnston and Honey, 1988). But when the party failed to do so, its core support fell away at the 1990 election, and it lost.

There are two main approaches to the analysis of parties' core support. *Partisan identification* emphasizes the long-term development of lasting loyalties to particular parties (Campbell et al., 1960; Miller and Shanks, 1996). Party loyalties are learned from parents in childhood, and reinforced through adult experience (including evaluations of parties' effectiveness in government: Fiorina, 1981). Partisan identification helps voters make sense of the plethora of information available in the political market place. It stands near the start of the "funnel of causality" which explains the voting decision. Voters are influenced first by their long-term loyalties, then by their political attitudes, and only then by their evaluations of how particular parties have performed recently in government – with both political attitudes and evaluations influenced by long-term loyalties (Campbell et al., 1960).

Social cleavage models emphasize the sociology of party support. Lipset and Rokkan's (1967) seminal study traced European electoral cleavages in the 1950s and 1960s to two major nineteenth century waves of social change – the national and industrial revolutions (see also Johnston, 1990; Lipset, 2000). Each engendered two sets of conflict within society, and each of the four conflicts provided fertile ground for parties to mobilize support.

The *national revolution*, encapsulating conflicts surrounding the development of modern nation-states, created center–periphery and church–state cleavages. The former was a product of the tensions between the centralizing, homogenizing processes of state-formation and the claims of regional autonomy. Modern "regional" parties, such as the Scottish Nationalists and the Italian Lega Nord, are products of this cleavage – although the Lega claims to represent a core rather than a peripheral region (Agnew, 1987, 1995; Giordano, 2000). And in post-unification Germany, the PDS (the successor to the former East German Communist party) has re-established itself as a "regional" party representing the eastern periphery of the new German state (Hough, 2000). The church–state cleavage, meanwhile, contrasts the secular claims of the emerging nation-state with the universalist claims of the Roman Catholic Church. As such, it is more common in southern Europe, where the Reformation had little impact, than in the north. The Christian Democrat parties that characterize the European center-right draw much support from the "church" side of this cleavage, especially among active Roman Catholics.

The *industrial revolution* resulted in cleavages between agricultural and industrial interests, and between employers and workers (the class cleavage). The agriculture–industry cleavage arose out of the two sectors' different needs. In mid-nineteenth century Britain, for instance, there was conflict over the Corn Laws, with agricultural interests defending, and industrial interests opposing, subsidies for British grain production. In continental Europe, peasants' parties were common manifestations of the cleavage. Finally, the class cleavage between employers and workers proved the major electoral divide in northern Europe (although less so in southern Europe, and less so again in the USA). With the extension of the franchise to large sections of the (male) working class in the late nineteenth and early twentieth centuries, new parties formed to represent proletarian interests; the German Social Democratic party and the British Labour party were early examples.

The key cleavages in each polity were those active when major extensions of the franchise took place. In Britain, for instance, the last major franchise extension was largely class-based, so the class cleavage came to dominate politics there (Butler and Stokes, 1969; Pulzer, 1967). Through campaign activities, parties strengthened their support bases in the key social groups and helped institutionalize the cleavages, often long after the original conflict had faded. Europe's party system in the mid-twentieth century was "frozen" in the conflicts of the late-nineteenth. The USA had few national cleavages, however, and parties needed to form alliances of different groups if they were to win the Presidency and Congressional majorities: the Democratic party combined the support of southern, white-supremacists with that of north-eastern working-class voters (many of them from European stock), for example, to create their New Deal hegemony; the "coalition" collapsed in the 1970s, as did previous "normal voting patterns" (Archer and Taylor, 1981)

No sooner had Lipset and Rokkan argued that electoral cleavages were the "frozen" outcomes of the national and industrial revolutions, however, than evidence of new cleavages emerged. World War II was followed by a quarter century of unprecedented prosperity in the "developed world". Poverty and unemployment, widespread during the pre-war Depression, seemed vanquished and Inglehart (1971, 1977) announced the arrival of a "silent revolution". Political attitudes, he argued, were shifting from materialism to postmaterialism as the new prosperity removed worries over economic well-being. Whereas the "old" politics was typified by debates over the management of scarcity (who gets what, and who pays), the "new" politics was concerned with the quality of life (enhanced democratic participation, environmentalism, gender politics, etc.). Furthermore, support for "postmaterialist" values was greatest among the highly-educated young and much lower among older voters, who had direct experience of pre-war scarcity. To some extent, the rise of Green parties occurred on the back of the "postmaterialism" wave and by the end of the century 20–30 percent of voters shared postmaterialist views (Dalton, 2000).

Another "new" cleavage reflected the growth of the welfare state in most Western countries. This stimulated the development of two "camps": state employees and those reliant upon the state for major goods and services (public sector housing being an important example); and those reliant on the private sector (Dunleavy, 1979). While the interests of the former would be served by expanding state provision, the interests of the latter would not. As some parties were associated with raising, and others with cutting, state spending this created perfect conditions for the development of both sectoral and consumption cleavages (documented in Britain by Dunleavy and Husbands, 1985; and Edgell and Duke, 1991). The sectoral cleavage arose from the growth of state employment: among the middle class, state employees in the UK were more likely to vote Labour and less likely to vote Conservative than those working in the private sector. Consumption cleavages, meanwhile, arose from the consumption of goods and services, particularly housing: among the working class, homeowners were more likely to vote Conservative and less likely to vote Labour than were those who rented homes in the public sector. (The latter cleavage became a political battleground in the 1980s, as the Conservative government encouraged public sector tenants to buy their homes: Heath et al., 1991.) However, several analysts argue that consumption and sectoral cleavages are, in practice, insufficiently differentiated from the class cleavage (Devine, 1996; Franklin and Page, 1984).

Both partisan identification and social cleavage approaches suggest a stable, strongly aligned electorate, with parties able to rely on the support of key sections of the electorate. But whereas this seems to have been a good description of Western electorates until the 1960s, over the last thirty years most Western democracies have experienced weakening voter–party attachments, a process often referred to as *dealignment* (Dalton, 1996a, b; Särlvik and Crewe, 1983). Partisan alignment is declining: fewer voters report identifying strongly with a particular party. Class and other social cleavages have apparently become weaker too. In Britain, for instance, Mrs Thatcher's 1980s successes were built in part on winning over a significant proportion of previously Labour-supporting skilled-manual workers, and by the 1997 election (when Labour extended its support far into the middle classes), class voting was, for some, dead (Franklin, 1985; Sanders, 1997; but see Evans, 1999; Heath et al., 1985, 1991).

Various explanations have been offered for dealignment. Converse (1969) argued that the strengths of voter alignments would wax and wane through a process of generational replacement: younger, weaker partisans replace older, stronger ones. However, research on the UK electorate suggests that all generations dealigned at roughly the same time, in the late 1960s and early 1970s (Abramson, 1992; but see Cassell, 1999). The explanation, rather, would seem to be political. The period of most rapid dealignment in most Western democracies coincided with the end of the long postwar boom and the onset of a protracted period of recession. Voters lost faith in "their" party. Furthermore, they became less likely to be expressive in their party choice, and more likely to be instrumental; rather than saying "which party represents people like me?", they were asking "which is most likely to improve my standard of living?" (Devine, 1992; Goldthorpe et al., 1968).

While the general consensus is that voter alignment has weakened in the West, that conclusion is not universally accepted (Evans, 1999). Furthermore, the aftermath of the 1989 fall of Communism in Eastern Europe seems to have created conditions for new electoral cleavages there. As new (and some older) parties have struggled to make their mark in the new democracies, they have nurtured support from particular sectors of society (Evans and Whitefield, 1998; Whitefield and Evans, 1999). But stable cleavages have been slower to emerge in Russia, with its still chaotic party system (McAllister and White, 1995; White *et al*, 1997; but see Evans and Whitefield, 1999).

Winning new voters

Under normal circumstances, parties cannot rely solely on their core vote being sufficient to guarantee electoral victory, however. In part this is because few cleavages are absolute; even at the height of British class voting, for instance, a significant minority supported the "wrong" party with the Conservatives, the most successful electoral force in twentieth century Britain, relying on an ability to win some working-class votes (McKenzie and Silver, 1968). And extensive dealignment has occurred since then. The "core vote" is probably smaller now, and certainly less reliable, than in the past. Electors are more volatile.

To win elections, therefore, parties must also mobilize support among floating voters, those with no clear commitment to a particular party. To some extent, the convergence between parties of the left and right in Britain and the USA during the 1990s is a product of this. Furthermore, parties must demonstrate their competence in government. Particularly where the electoral system tends to produce clear government by one party, there is strong cross-national evidence that voters reward governments which are judged to deliver economic prosperity, and punish those associated with economic failure (Chappell and Viega, 2000; Fiorina, 1981; Key, 1966; Lewis-Beck, 1990). The loss of a reputation for competence in economic management can be catastrophic, as the British Conservatives found in 1997 (Sanders, 1999). But equally, to the extent that governments can manipulate the "feelgood factor" to coincide with the electoral cycle, they can engineer conditions which increase re-election chances. As both the Falklands and Gulf Wars showed, military victory might not be enough for a government to be re-elected, but economic "victory" often is (Nicklesburg and Norpoth, 2000; Sanders et al., 1987).

Campaigning, increasingly sophisticated, is also important in ensuring the party's message reaches the electorate (Farrell, 1996). Parties identify and target floating voters in their campaign efforts. At the 1997 UK general election, for instance, Labour focused much of its campaign on appeals to uncommitted voters in 90 key marginal constituencies (out of 659), just 925,000 individuals in an electorate of nearly 43 million (Denver et al., 1998, p. 178).

Changing parties

Alongside altered relationships between parties and voters, parties themselves have also changed (Katz and Mair, 1995). In the first half of the twentieth century, most democracies were dominated by "mass parties" with large memberships, drawing support from particular, organized, sectors of society: the parties that mobilized around the class cleavage were archetypal. The mass party model was successful under conditions of strong voter alignment, in an environment where parties relied more on local activists than the national media to reach their supporters.

But with dealignment, mass parties were less likely to succeed. Rather, some argued, parties had to develop a "catch-all" profile, appealing to voters beyond particular, circumscribed social groups (Evans et al., 1999; Kirchheimer, 1966). Furthermore, the rapid development of electronic media meant that party leaders could appeal quickly and easily to the entire electorate, in ways not possible before (Jamieson, 1992; Kavanagh, 1995). The "third way" politics espoused in the 1990s by Bill Clinton in the USA, Tony Blair in Britain, and Gerhard Schroeder in Germany exemplify such a "catch-all" agenda.

Katz and Mair (1995) argued that a further organizational form has emerged: the "cartel party". Like catch-all parties, these draw support widely across the electorate. But they are also more likely to be typified by an approach to politics as a profession, and to enjoy a closer relationship with the state (perhaps receiving direct financial aid from it).

At the same time, and in some ways related to organizational changes, party membership is in long-term decline in many democracies (Katz, 1996; for Britain, see Seyd and Whiteley, 1992; Whiteley et al., 1994). The membership of the British Conservative party, for instance, is rapidly ageing: in 1994, the average member was 62 (Whiteley et al., 1994, p. 40). Parties still play an important part in political life, but fewer citizens are willing to donate their time to them. For some, declining membership is not, in itself, a problem. The rise of electronic campaigning via TV (and, increasingly, the Internet) provides a direct means of communication between party elites and voters. But local campaigning remains important, especially in close-run elections, and where targeting of floating voters is to be essayed. And this requires members. Parties are as necessary now as they were a century ago, but they face challenges of legitimacy and human resources.

Parties, Elections, and Geography

Electoral mobilization

In order to mobilize electoral support, political parties must contact individual electors, to inform them of the benefits (personal and other) that support and votes might yield. At the minimum they need votes – a willingness to turn out. But much more is needed to win that support, especially from floating voters. Workers are needed to mobilize election-day support – individuals prepared to give time to canvassing and campaigning. Parties also need money – plus people prepared to spend time and energy canvassing potential donors (corporate as well as individuals; Fisher, 1999) and conducting "profit-making" functions. (On political finance generally, see Alexander, 1989; Gunlicks, 1993.) Such commitment to long- and short-term goals involves finding people who not only identify with a party but also are prepared to work for it – if not continually, then especially during immediate pre-election periods. Political parties are permanent organizations of individuals committed to the cause, prepared to give time and money (most have a membership fee) – from which they should gain politically delivered benefits alongside individual satisfaction.

Most party organizations are spatially structured in mass political systems, with substantial grass-roots components. At larger spatial scales, most have paid staff but more locally, unless the party is well-funded (perhaps in some places only; Pattie and Johnston, 1996), such staff are voluntary and very much part-time. Local organization depends on local willingness to work for the party.

Political parties in a liberal democracy are unlikely to attract members and activists in equal proportion everywhere. Most draw much of their support from particular sections of the population defined according to the salient electoral cleavages. Given the maps of uneven economic and social development that characterize all countries, plus associated cultural variations – some pre-dating institution of the electoral cleavage – there are geographies of party strength reflecting the cleavage geography. Each party is better able to mount substantial electoral campaigns in some areas than others, especially important if long as opposed to short campaigns are crucial in support mobilization. A substantial minority of people make electoral decisions in the intensive campaigning period immediately preceding an election, calling for resources to be built-up for those short bursts of activity, when campaign intensity can significantly influence turnout rates (Johnston and Pattie, 2001). But with a significant number of people making decisions well before the election campaign, local parties must maintain regular contact with voters, informing them of (and perhaps involving them in determining) their policy agenda. Substantial resources are needed to sustain such "permanent campaigns" – which will favor the areas of parties' greatest strength; many local parties, especially small ones, are moribund between elections because their activists are unwilling to devote time and energy to the cause almost permanently. (In the UK people are increasingly more willing to donate money than to work for parties.)

Party organizations are the main vehicles for mobilizing electoral support, but many are significantly assisted by other locally based organizations, either affiliated to or strongly supportive of their ethos and goals. Many trades unions support left-of-center parties, for example, and provide financial and other resources; most are spatially structured, with local branches – many at individual workplaces – mobilizing voter interest. Other institutions have less formal links to parties but nevertheless provide much implicit if not explicit support: many right-of-center Christian Democrat parties are supported by churches, for example, whose local representatives may not canvass for votes but offer general "guidance".

To win elections, therefore, most parties have local organizations, comprising committed members who work to mobilize and deliver support. New parties, or established parties seeking votes in areas where they formerly had no presence, establish such organizations – as in Burgenland, transferred to Austria from Hungary after World War I. The first elections in 1922 were contested by three parties, with the Christian (Catholic) Party expected to perform best in what was largely a rural area. But the Socialist Party was more successful in establishing organizational networks in some districts (based on factory workers who commuted to Vienna) and it gained 39 per cent of the votes compared to the Christian Party's 31 percent (Burghardt, 1964). The Christian Party subsequently established networks and won increasing support as the "natural" party for such rural areas, but the Socialist Party retained a strong second place thanks to its initial efforts at establishing a grass-roots organizational structure (see also Giordano, 2000, on the Lega Nord in Italy).

An established foothold in a district can provide the foundation for further electoral mobilization. As a party's strength builds there, its workers may "colonize" adjacent areas to recruit members. Support spreads out from its initial cores in a diffusion-like process (as Dorling et al., 1998, show for the British Liberal Democrats). If its candidates are successful in local and regional government elections, competent performances by its representatives (plus the publicity gained in local media) may mobilize further support. And if the party wins control of the local authority its credibility as an "alternative government" at higher levels may be aided, with more people consequently joining and working, and others voting, for it: local success can breed national achievements.

Parties thus recruit members and win voters in areas offering them the greatest potential: the British Labour Party gained much of its early support and Parliamentary representation in the country's coalfields, for example, aided by both the trades unions and, in some areas such as the South Welsh valleys, dissenting religions (Tanner, 1990): in rural Wales, the Liberal Party won support among the landless population opposed to the (Conservative-supporting, English) landed gentry (Cox, 1970). But parties may be unable to mobilize in some areas, despite apparent potential, as in the Dukeries coalfield in the English north Midlands, opened in the 1920s, where a paternalistic ownership structure, closed village communities, and a working system that discouraged large-scale mobilization (reflected in a relatively weak, "'bosses' union") blocked Labour Party efforts and produced a relatively conservative local political culture which included compulsory church attendance. This culture remained influential sixty years later, during the 1984–5 National Union of Mineworkers' strike (Griffiths and Johnston, 1991), and was reflected in Labour's poor performance there at the 1987 general election (Jones et al., 1992). Dukeries' miners learned what it was to be a member of the working class in a different context from most of their contemporaries in other fields: class may be a general concept, but its meaning is structured in local milieux (Agnew, 1987).

The geography of voting: spatial polarization

Given these arguments regarding electoral cleavages and mobilization, it should be relatively straightforward to appreciate the geography of a party's electoral support from the geography of the relevant cleavage(s). In a class-dominated structure, for example, knowledge of the geography of class membership should allow one to "read-off" the geography of party support.

This is frequently not so, however. Instead, the geography of a party's support is commonly spatially more polarized than the geography of its underpinning cleavage base. Where a party has strong potential support it tends to win an even larger percentage of the votes cast than predicted for it; where its potential support is weak (relatively few of its "natural supporters" live in the area), its vote is even lower than expected. Attempts to "predict" the geography of voting at Great Britain's 1979–87 general elections from knowledge of parties' support nationally across the various classes (information derived from sample surveys) and census data on the class structure of each constituency, for example, were unsuccessful. Each of the two main parties (Conservative and Labour) won many more votes than predicted in constituencies where its support base was strong, but many less where it was weak; members of the working class were less likely to vote Labour if they lived in middle-class than working-class constituencies (Johnston et al., 1988).

A variety of accounts have been offered for this polarization. The *neighborhood effect* is firmly grounded geographically. It has several variants; an original formulation presented it as a friends-and-neighbors effect, with candidates winning stronger support in their home areas than elsewhere (as in Key's, 1949, seminal detailed analyses of US intra-party primary contests; see also Johnston, 1972, on multi-member contests in New Zealand). More generally, the neighborhood effect has been associated with a process of "conversion by conversation" resulting from the biased flows of politically-relevant information through locally-focused social networks, as formalized in Cox's (1969) classic essay. Most social networks are dominated by people of a particular political persuasion, whose net influence on their fellow-members exceeds that of those having a minority persuasion. Thus, over time the dominant view wins more converts than the minority, producing voting polarization; "people who talk together, vote together" (Miller, 1977, p. 65). As a consequence, working-class people living among members of the middle class are more likely to identify themselves as middle class than those who live in predominantly working-class areas and vice versa (Eagles, 1990; Johnston et al., 2000; MacAllister et al., 2001; on the efficacy of conversations, see Pattie and Johnston, 2000).

There are exceptions to this "dominance by the majority". In Great Britain, for example, a "deferential worker" culture has been identified in rural areas, whereby the majority – the rural working class – were strongly influenced by the small, but politically-powerful, minority – the landed gentry (Jessop, 1974; Newby, 1977). Despite their class status, many of the former were socialized to believe that rule by the gentry was "the natural order", and so voted for the Conservative party. It was difficult for parties to mobilize them to other political positions – they were dispersed across many thousands of small workplaces which were not easily unionized, for example, and they lived in small settlements, many as their employers' tenants, with occupancy of their homes tied to their jobs. Thus they were socialized into a culture in which they assessed their position in the "national" class structure differently from those in similar employment situations but who lived in different locales – urban housing and labor markets, for example.

Other processes contribute to geographical polarization in voting patterns without involving informal social interaction. If the strength of local political party organizations varies and is reflected in available resources for mounting election campaigns (Denver and Hands, 1997), they should be better able to mobilize "floating voters" where they are strong than where they are relatively weak. But whatever the combination of processes, there is strong evidence of geographically-polarized support for political parties consistent with the neighborhood effect concept, along with a great deal of sustaining circumstantial evidence. (Recent British studies using small-scale neighborhoods separately defined for each respondent in a large sample survey of voting behavior have shown that similar people vote differently according to their local context: MacAllister et al., 2001; Tunstall et al., 2000.) Parties, and the political projects and attitudes that they represent, are differentially embedded in places (Agnew, 1987); they contribute to their cultures and the socialization of new voters (including incomers). Polarization becomes self-perpetuating.

"It's the economy, stupid"

With dealignment and greater recent fluidity of political attitudes, many individuals are only weakly committed to any party and their voting decisions more likely to be influenced by other factors. Cleavage structures are fading, as are the local networks (both formal and informal – as Putnam's, 2000, work on social capital argues) which both sustain them and help to create the polarized voting patterns.

Some of the most successful models of contemporary voting behavior relate party choice to electoral issues, with economic performance and the government's record salient in many contests. If people think a government has delivered economic prosperity, both nationally and to them as individuals, they are more likely to vote for it than are those who think its economic policies have failed (Johnston and Pattie, 2000). But for many the national economy is an abstraction; the important economic events occur in their local economies and boom in one area may be accompanied by slump elsewhere. Changes in local labor and housing markets impinge on individuals: they experience them either personally or vicariously, through the experiences of family, friends and neighbors; they observe them in local milieux; and they hear of them in local media. In the UK in the 1990s, perceptions of the local economy were unrelated to those of national or individual situations, with all three strongly linked to voting choices (Johnston and Pattie, 2001; Pattie and Johnston, 1998): some who believed they were prospering because of government policies voted (altruistically) against it because they perceived that their neighbors were not (Johnston et al., 2000).

Micro- and macro-scale spatially polarized voting patterns can be produced in a variety of ways, therefore, reflecting different aspects of the voters' milieux (Books and Prysby, 1991). Voters are not influenced equally by these processes. Those strongly committed to a party are less likely to be weaned from it than those with either a relatively weak or no such identity. And it is not just a question of being influenced by either general or local issues; the great majority are attached to a number of networks providing politically-relevant information, some larger-scale in their structuring than others (Savage, 1996). Interactions among these networks are crucial; the local may reinforce the more general, or it may crosscut it, with voters having to evaluate the various sources of information and determine which cues to follow (Huckfeldt and Sprague, 1990, 1995).

Territories, Campaigns, and Representation

This discussion of local milieux, their constitution by political parties and other agents, and their influence on individuals' voting, has said little about the places involved. Local labor and housing markets are imprecise spatial conceptions; they may be approximated using administrative divisions of the national territory, but most have fuzzy, and shifting, boundaries (and may comprise multiple sets of overlapping areas; Coombes, 2000). Almost all local social networks lack clear territorial divisions – as do many formal social and related organizations. The milieux for much political mobilization are fuzzy and shifting, part of an ever-changing complex mosaic of structuration locales.

There is one major exception to this generalization. In many countries election of at least some members of the relevant legislature is to represent a defined territory – such as a UK Parliamentary constituency and a US Congressional District. These territorial containers are salient elements in the structuration of local political activity. The parties evaluate their political complexion, focusing their campaigning on areas they are best able to win; the more intensive the local campaign the better the candidate's/party's chances of victory, which contributes to the polarization of voting patterns (Denver and Hands, 1997; Jacobson, 1978; Johnston and Pattie, 2001; Pattie et al., 1995).

Almost all electoral systems use territorially-defined constituencies in some ways, but their spatial definition is of greatest significance in those employing single-member constituencies and plurality rules (i.e. the candidate with most votes wins, irrespective of whether they form a majority of those cast). Constituency definition is thus crucial to the parties; they want boundaries drawn so that they are likely to win as many seats as possible – and their candidates want constituencies where they are known and can develop "friends and neighbors" support. Redistricting is a major exercise in political cartography especially where, as in the USA, it is largely undertaken by the political parties; fairness in their construction is rarely attainable (Johnston, 1999)

Two forms of electoral abuse characterize many political cartography exercises; both give a better return on its votes to a party than to its opponents. *Malapportionment* involves creating small constituencies in areas where a party is strong, and larger ones elsewhere; *gerrymandering* involves drawing a constituency's boundaries so that it contains a majority of the party's supporters who may be rewarded by geographically-focused political benefits (the "pork barrel": Johnston, 1980). Many countries have outlawed malapportionment, and seek to preclude gerrymandering: the latter is difficult because of the absence of normative criteria to assess maps against – and it remains common in the redistricting after every US census (see Clark, 1991).

Both malapportionment and gerrymandering were common electoral strategies in the USA until the mid-twentieth century. Malapportionment occurred because electoral maps for Congress and State Legislatures were not redrawn to take account of population changes – notably the rural–urban shift. Thus in many States the urban areas were substantially under-represented, whereas the rural areas had very low ratios of population to representative: one of the extremes was Connecticut, where the urban areas housed 23 percent of the population in 1950 but returned only eight of the 279 members of the Legislative Assembly (on which, see Jewell, 1962). Such practices were outlawed by a series of rulings from the US Supreme Court in the 1960s, which found that legislative districts of unequal population violated the 14th Amendment of the Constitution which guarantees equal treatment for individuals. As a consequence, each State now has a legal requirement to redistrict for both the US House of Representatives and its own State Legislature within two years of every decennial census.

Whereas malapportionment was relatively easily dealt with once the judicial will was there, this was not the case with gerrymandering. Even with equal-sized districts it is possible to produce a configuration favoring one party over another – as Morrill (1973) showed in his analyses of plans for the State of Washington after the 1972 Census: for the State House, for example, the Republican plan had 48 safe Republican seats (out of 98) and 38 safe Democrat seats, whereas the respective numbers under the Democrat plan were 36 and 50. Judges were reluctant to rule that a plan was gerrymandered, however, because they lacked clear criteria against which to evaluate it, so that where parties controlled a State's redistricting process (through their power in its legislature and the governorship) they were usually able to produce a plan which favored their candidates over their opponent's. Partisan cartography was still feasible (Morrill, 1981).

One area in which the Courts did find it possible to rule against apparent gerrymanders concerned the distribution of racial groups across districts. Under the *Voting Rights Act 1965* (and subsequent amendments) those controlling redistricting processes were constrained from creating districts such that black voters formed a minority in all (or most) and so had their voting strength diluted: the presumption was, for example, that if blacks formed 15 percent of a state's population they should be in a majority in 15 percent of its Congressional/Legislative districts. This led to the creation of some very oddly shaped districts in several southern States, which were contested in the courts – as in Georgia where two very elongated Congressional Districts were created in 1992 (one of which linked the main black ghettos of Atlanta, Augusta and Savannah: Leib, 1998; Lennertz, 2000). The Court ruled that such explicit gerrymandering was inconsistent with normal practice for drawing district boundaries, for which compact shapes were considered the norm, and race could not be employed as the overriding consideration when defining one or more districts: a more "shapely" configuration of districts was thus imposed on the State. There are still problems in defining criteria for redistricting on grounds other than malapportionment (Rush, 2000), however, despite academic work on both shape considerations (though see Altman, 1998) and race as an over-riding consideration (Cirincione et al., 2000).

Even where explicit malapportionment and gerrymandering are not available strategies for political parties – there are independent redistricting bodies that conduct redistricting exercises on a regular, timetabled basis, as in the UK (Rossiter et al., 1999) – the outcome can be consistent with one or both. Single-member constituency systems almost invariably produce disproportional election results, with each party's percentage of the votes cast nationally differing from its percentage of the legislative seats (see King and Browning, 1987). The reasons for this are almost entirely geographical, as Gudgin and Taylor (1979) demonstrated in a seminal work: the consequence of placing a grid of constituencies over an uneven geographical distribution of party supporters is that the outcome will almost certainly advantage some parties and disadvantage others – what they term a non-partisan gerrymander. Furthermore, there may be biases favoring one of the advantaged parties, not because the system is partisan but simply because the interacting geographies advantage one more than others – as demonstrated for the UK, where the non-partisan system nevertheless allows some political maneuver for electoral gain (Johnston, 2001; Johnston et al., 2001).

Conclusions

Parties are integral to major aspects of political life. They win support through socialization/mobilization strategies, which involve them becoming parts of – and competing for support in – the cultural geography of places. Major changes in the nature of modern politics notwithstanding, they are key players in representative electoral democracy. Despite claims to the contrary, there is little sign that either political parties or representative electoral democracy are in decline.

Yet, as we have argued here, they are currently both evolving rapidly. The geography of electoral competition between parties is a crucial part of that evolution. As parties shift from mass to catch-all and cartel forms, and as voter alignments continue to weaken, for instance, permanent and sophisticated campaigning becomes ever more important. That requires a spatial as well as a political strategy, to ensure that the "right" voters are targeted in the "right" places. Furthermore, the growth of instrumental, economic voting also carries implications for party competition. Competence and an ability to deliver prosperity are now possibly more important for electoral success than maintaining a core of lifelong loyal voters. But voters' perceptions of government competence can be influenced by their experiences in their local milieux. Electoral geography remains a key concern, both for political actors and for the academic analysts of politics.

ENDNOTE

1. Downs' analysis is rooted in rational choice theory, a set of concepts drawn from economics. The rational choice approach is to build theoretical models based on the assumption that each actor in a social system behaves to maximize his or her benefits and

to minimize his or her costs. Critics argue that this is an overly restrictive assumption that precludes the possibility of altruistic behavior. Proponents of rational choice, however, would counter that the assumption is not a description of how society actually is, but is, rather, a simplification to ease formal modeling

BIBLIOGRAPHY

Abramson, P. R. 1992. Of time and partisan instability in Britain. *British Journal of Political Science*, 22, 381–95.

Agnew, J. 1987. *Place and Politics: The Geographical Mediation of State and Society.* London: Allen and Unwin.

Agnew, J. 1995. The rhetoric of regionalism: the Northern League in Italian politics, 1983–1994. *Transactions, Institute of British Geographers, NS*, 20, 156–72.

Alexander, H. E. (ed.). 1989. *Comparative Political Finance in the 1980s.* Cambridge: Cambridge University Press.

Altman, M. 1998. Modeling the effect of mandatory district compactness on partisan gerrymanders. *Political Geography*, 17, 989–1012.

Archer, J. C. and Taylor, P. J. 1981. *Section and Party.* Chichester: John Wiley.

Blake, R. 1985. *The Conservative Party from Peel to Thatcher.* London: Fontana.

Books, J. W. and Prysby, C. L. 1991. *Political Behavior and the Local Context.* New York: Praeger.

Burghardt, A. F. 1964. The bases of support for political parties in Burgenland. *Annals of the Association of American Geographers*, 54, 372–90.

Butler, D. and Stokes, D. E. 1969. *Political Change in Britain.* London: Macmillan.

Campbell, A., Converse, P., Miller, W.E., and Stokes, D.E. 1960. *The American Voter.* New York: Wiley.

Cassel, C. A. 1999. Testing the converse party support model in Britain. *Comparative Political Studies*, 32, 626–44.

Chappell, H. W. and Viega, L. G. 2000. Economics and elections in western Europe: 1960–1997. *Electoral Studies*, 19, 183–9.

Cirincione, C., Darling, T. A., and O'Rourke, T. G. 2000. Assessing South Carolina's congressional districting. *Political Geography*, 19, 189–212.

Clark, W. A. V. 1991. Geography in Court: expertise in adversarial settings. *Transactions, Institute of British Geographers, NS*, 16, 5–20.

Converse, P. E. 1966. The concept of a normal vote. In A. Campbell et al. (eds.) *Elections and the Political Order.* New York: Wiley.

Converse, P. E. 1969. Of time and partisan stability. *Comparative Political Studies*, 2, 139–71.

Coombes, M. G. 2000. Defining locality boundaries with synthetic data. *Environment and Planning A*, 32, 1499–1518.

Cox, K. R. 1969. The voting decision in a spatial context. *Progress in Human Geography*, 1, 81–117.

Cox, K. R. 1970. Geography, social contexts and voting behaviour in Wales, 1861–1951. In E. Allardt and S. Rokkan (eds.) *Mass Politics.* New York: Free Press.

Dalton, R. J. 1996a, *Citizen Politics: Public Opinion and Political Parties in Advanced Western Democracies*, 2nd edn. Chatham, NJ: Chatham House.

Dalton, R. J. 1996b, Political cleavages, issues and electoral change. In L. LeDuc et al. (eds.) *Comparing Democracies: Elections and Voting in Global Perspective.* Thousand Oaks, CA: Sage.

Dalton, R. J. 2000. Value change and democracy. In S. J. Pharr and R. D. Putnam (eds.) *Disaffected Democracies: What's Troubling the Trilateral Countries?* Princeton, NJ: Princeton University Press, 252–69.

Denver, D. and Hands, G. 1997. *Modern Constituency Electioneering: Local Campaigning in the 1992 General Election*. London: Frank Cass.

Denver, D., Hands, G., and Henig, S. 1998. Triumph of targetting? Constituency campaigning in the 1997 election. In D. Denver et al. (eds.) *British Elections and Parties Review 8: the 1997 General Election*. London: Frank Cass.

Devine, F. 1992. *Affluent Workers Revisited*, Edinburgh: Edinburgh University Press.

Devine, F. 1996. The "new structuralism": class politics and class analysis. In N. Kirk (ed.) *Social Class and Marxism: Defences and Challenges*. London: Scolar.

Dorling, D. F. L., Rallings, C., and Thrasher, M. 1998. The epidemiology of the Liberal Democrat vote. *Political Geography*, 17, 45–70.

Downs, A. 1957. *An Economic Theory of Democracy*. New York: Harper Collins.

Dunleavy, P. 1979. The urban basis of political alignment. *British Journal of Political Science*, 9, 409–43.

Dunleavy, P. and Husbands, C. 1985. *British Democracy at the Crossroads: Voting and Party Competition in the 1980s*. London: Allen and Unwin.

Eagles, M. 1990. An ecological perspective on working class political behaviour: neighbourhood and class formation in Sheffield. In R. Johnston et al. (eds.) *Developments In Electoral Geography*. London: Routledge, 100–20.

Edgell, S. and Duke, V. 1991. *A Measure of Thatcherism: A Sociology of Britain*. London: Harper Collins.

Evans, G. 1999. Class voting: from premature obituary to reasoned appraisal. In G. Evans (ed.) *The End of Class Politics? Class Voting in Comparative Context*. Oxford: Oxford University Press.

Evans, G. and Whitefield, S. 1998. The structuring of political cleavages in post-communist societies: the case of the Czech Republic and Slovakia. *Political Studies*, 46, 115–39.

Evans, G. and Whitefield, S. 1999. The emergence of class politics and class voting in post-communist Russia. In G. Evans (ed.) *The End of Class Politics? Class Voting in Comparative Context*. Oxford: Oxford University Press.

Evans, G., Heath, A., and Payne, C. 1999. Class: Labour as a catch-all party? In G. Evans and P. Norris (eds.) *Critical Elections: British Parties and Voters in Long-term Perspective*. London: Sage.

Farrell, D. 1996. Campaign strategies and tactics. In L. LeDuc et al. (eds.) *Comparing Democracies: Elections and Voting in Global Perspective*. Thousand Oaks: Sage.

Fiorina, M. 1981. *Retrospective Voting in American National Elections*. New Haven, CT: Yale University Press.

Fisher, J. 1999. Modelling the decision to donate by individual party members: the case of British parties. *Party Politics*, 5, 19–38.

Franklin, M. 1985. *The Decline of Class Voting in Britain*. Oxford: Oxford University Press.

Franklin, M. and Page, E. 1984. A critique of the consumption cleavage approach in British voting studies. *Political Studies*, 32, 521–36.

Fukuyama, F. 1992. *The End of History and the Last Man*. Harmondsworth: Penguin.

Giordano, B. 2000. Italian regionalism or "Padanian" nationalism – the political project of the Lega Nord in Italian politics. *Political Geography*, 19, 445–71.

Goldthorpe, J., Lockwood, D., Bechhofer, F., and Platt, J. 1968. *The Affluent Worker: Industrial Attitudes and Behaviour*. Cambridge: Cambridge University Press.

Griffiths, M. J. and Johnston, R. J. 1991. What's in a place? An approach to the concept of place as illustrated by the British National Union of Mineworkers strike, 1984–85. *Antipode*, 23, 185–213.

Gudgin, G. and Taylor, P. J. 1979. *Seats, Votes and the Spatial Organisation of Elections*. London: Pion.

Gunlicks, A. B. (ed.). 1993. *Campaign and Party Finance in North America and Western Europe*. Boulder, CO: Westview.

Heath, A., Curtice, J., Jowell, R., Evans, G., Field, J., and Witherspoon, S. 1991. *Understanding Political Change: the British Voter 1964–1987*. Oxford: Pergamon.

Heath, A., Howell, R., and Curtice, J. 1985. *How Britain Votes*. Oxford: Pergamon.

Held, D. 1987. *Models of Democracy*. Cambridge: Polity.

Hough, D. 2000. Societal transformation and the creation of a regional party: the PDS as a regional actor in Eastern Germany. *Space and Polity*, 4, 57–75.

Huckfeldt, R. and Sprague, J. 1990. Social order and political chaos: the structural setting of political information. In J. A. Ferejohn, and J. H. Kuklinski (eds.) *Information and Democratic Processes*. Urbana, IL: University of Illinois Press.

Huckfeldt, R. and Sprague, J. 1995. *Citizens, Politics and Social Communication: Information and Influence in an Election Campaign*. Cambridge: Cambridge University Press.

Inglehart, R. 1971. The silent revolution in Europe: inter-generational change in post-industrial societies. *American Political Science Review*, 65, 991–1017.

Inglehart, R. 1977. *The Silent Revolution: Changing Values and Political Styles Among Western Publics*. Princeton, NJ: Princeton University Press.

Jacobson, G. C. 1978. The effects of campaign spending in Congressional elections. *American Political Science Review*, 72, 469–91.

Jamieson, K. H. 1992. *Packaging the Presidency: A History and Criticism of Presidential Campaign Advertising*, 2nd edn. Oxford: Oxford University Press.

Jessop, B. 1974. *Traditionalism, Conservatism and British Political Culture*. London: George, Allen and Unwin.

Jewell, M. E. (ed.). 1962. *The Politics of Reapportionment*. New York: Atherton.

Johnston, R. J. 1972. Spatial elements in voting patterns at the 1968 Christchurch City Council election. *Political Science*, 24, 49–61.

Johnston, R. J. 1980. *The Geography of Federal Spending in the United States of America*, Chichester: Research Studies Press.

Johnston, R. J. 1990. Lipset and Rokkan revisited. In R. J. Johnston et al. (eds.) *Developments in Electoral Geography*. London: Routledge, 121–42.

Johnston, R. J. 1994. One world, millions of places: the end of history and the ascendancy of geography. *Political Geography*, 13, 111–21.

Johnston, R. J. 1999. Geography, fairness and liberal democracy. In J. D. Proctor and D. M. Smith (eds.) *Geography and Ethics: Journeys in a Moral Terrain*. London: Routledge, 44–58.

Johnston, R. J. 2001. Manipulating maps and winning elections: measuring the impact of malapportionment and gerrymandering. *Political Geography*, 20.

Johnston, R. J. and Honey, R. 1988. Political geography of contemporary events X: the 1987 general election in New Zealand: the demise of the electoral cleavage? *Political Geography Quarterly*, 7, 363–68.

Johnston, R. J. and Pattie, C. J. 2001. Dimensions of retrospective voting: economic performance, public service standards and Conservative party support at the 1997 British General Election. *Party Politics*, 7, 469–90.

Johnston, R. J. and Pattie, C. J. 2001. Is there a crisis of democracy in Great Britain? Turnout at general elections reconsidered. In K. M. Dowding et al. (eds.) *The Challenge to Democracy*. London: Macmillan.

Johnston, R. J., Pattie, C. J., and Allsopp, J. G. 1988. *A Nation Dividing? The Electoral Map of Great Britain 1979–1987*. London: Longman.

Johnston, R. J., Pattie, C. J., Dorling, D. F. L., MacAllister, I., Tunstall, H., and Rossiter, D. J. 2000. The neighbourhood effect and voting in England and Wales: real or imagined?' In P. J. Cowley et al. (eds.) *British Elections and Parties Yearbook 10*. London: Frank Cass, 47–63.

Johnston, R. J., Pattie, C. J., Dorling, D. F. L., and Rossiter, D. J. 2001. *From Votes to Seats: The Operation of the UK Electoral System since 1945*. Manchester: Manchester University Press.

Jones, K., Johnston, R. J., and Pattie, C. J. 1992. People, places and regions: exploring the use of multi-level modelling in the analysis of electoral data. *British Journal of Political Science*, 22, 343–80.

Katz, R. S. 1996. Party organization and finance. In L. LeDuc et al. (eds.) *Comparing Democracies: Elections and Voting in Global Perspective*. Thousand Oaks: Sage.

Katz, R. S. and Mair, P. 1995. Changing models of party organization and party democracy: the emergence of the cartel party. *Party Politics*, 1, 5–28.

Kavanagh, D. 1995. *Election Campaigning: the New Marketing of Politics*. Oxford: Blackwell.

Key, V. O. 1949. *Southern Politics in State and Nation*. New York: Alfred A. Knopf.

Key, V. O. 1966. *The Responsible Electorate: Rationality in Presidential Voting, 1936–1960*. Cambridge, MA: Harvard University Press.

King, G. and Browning, R. X. 1987. Democratic representation and partisan bias in Congressional elections. *American Political Science Review*, 81, 1251–73.

Kirchheimer, O. 1966. The transformation of west European party systems. In J. LaPalombara, and M. Weiner (eds.) *Political Parties and Political Development*. Princeton, NJ; Princeton University Press.

Leib, J. 1998. Communities of interest and minority districting after *Miller v Johnson*. *Political Geography*, 17, 683–700.

Lennertz, J. E. 2000. Back in their proper place: racial gerrymandering in Georgia. *Political Geography*, 19, 163–88.

Lewis-Beck, M. S. 1990. *Economics and Elections: The Major Western Democracies*. Ann Arbor, MI: University of Michigan Press.

Lipset, S. 2000. The indispensability of political parties. *Journal of Democracy*, 11, 48–55.

Lipset, S. and Rokkan, S. 1967. Cleavage structures, party systems and voter alignments. In S. Lipset and S. Rokkan (eds.) *Party Systems and Voter Alignments*. New York: Free Press.

McAllister, I. and White, S. 1995. Democracy, parties and party formation in post-communist Russia. *Party Politics*, 1, 49–72.

MacAllister, I., Johnston, R. J., Pattie, C. J., Tunstall, H., Dorling, D. F. L., and Rossiter, D. J. 2001. Class dealignment and the neighbourhood effect: Miller revisited. *British Journal of Political Science*, 31, 41–60.

McKenzie, R.T. and Silver, A. 1968. *Angels in Marble: Working Class Conservatives in Urban England*. London: Heinemann.

Miller, W. E. and Shanks, J. M. 1996. *The New American Voter*. Cambridge, MA: Harvard University Press.

Miller, W. L. 1977. *Electoral Dynamics in Britain Since 1918*. London: Macmillan.

Morrill, R. L. 1973. Ideal and reality in reapportionment. *Annals of the Association of American Geographers*, 63, 463–77.

Morrill, R. L. 1981. *Political Redistricting and Geographic Theory*. Washington, DC: Association of American Geographers.

Mulgan, G. 1994. *Politics in an Antipolitical Age*. Cambridge: Polity.

Newby, H. 1977. *The Deferential Worker: A Study of Farm Workers in East Anglia*. London: Allen Lane.

Nickelsburg, M. and Norpoth, H. 2000. Commander-in-chief or chief economist? The president in the eye of the public. *Electoral Studies*, 19, 313–32.

Osei-Kwame, P. and Taylor, P. 1984. A politics of failure. *Annals of the Association of American Geographers*, 74, 574–89.
Pattie, C. J. and Johnston, R. J. 1996. Paying their way: local associations, the constituency quota scheme and Conservative party finance. *Political Studies*, 44, 921–35.
Pattie, C. J. and Johnston, R. J. 1998. The role of regional context in voting: evidence from the 1992 British General Election. *Regional Studies*, 32, 249–63.
Pattie, C. J. and Johnston, R. J. 2000. People who talk together vote together: an exploration of the neighborhood effect in Great Britain. *Annals of the Association of American Geographers*, 90, 41–6.
Pattie, C. J., Johnston, R. J., and Fieldhouse, E. A. 1995. Winning the local vote: the effectiveness of constituency campaign spending in Great Britain. *American Political Science Review*, 89, 969–83.
Pulzer, P. 1967. *Political Representation and Elections in Britain*. London: George, Allen and Unwin.
Putnam, R. D. 2000. *Bowling Alone: The Collapse and Revival of American Community*. New York: Simon and Schuster
Rossiter, D. J., Johnston, R. J., and Pattie, C. J. 1999. *The Boundary Commissions: Redrawing the United Kingdom's Map of Parliamentary Constituencies*. Manchester: Manchester University Press.
Rush, M. E. 2000. Redistricting and partisan fluidity: do we really know a gerrymander when we see one? *Political Geography*, 19, 249–60.
Sanders, D. 1997. The new electoral battleground. In A. King (ed.) *New Labour Triumphs: Britain at the Polls*. Chatham, NJ; Chatham House.
Sanders, D. 1999. Conservative incompetence, Labour responsibility, and the feelgood factor: why the economy failed to save the Conservatives in 1997. *Electoral Studies*, 18, 251–70.
Sanders, D., Ward, H., Marsh, D., and Fletcher, T. 1987. Government popularity and the Falklands War: a reassessment. *British Journal of Political Science*, 17, 281–313.
Särlvik, B. and Crewe, I. 1983. *Decade of Dealignment*. Cambridge: Cambridge University Press.
Savage, M. 1996. Space, networks and class formation. In N. Kirk (ed.) *Social Class and Marxism: Defences and Challenges*. London: Scolar.
Seyd, P. 1998. In praise of party. *Parliamentary Affairs*, 51, 198–208.
Seyd, P. and Whiteley, P. 1992. *Labour's Grass Roots: the Politics of Party Membership*. Oxford: Clarendon.
Tanner, D. 1990. *Political Change and the Labour Party, 1900–1918*. Cambridge: Cambridge University Press.
Tunstall, H., Rossiter, D. J., Pattie, C. J., MacAllister, I., Johnston, R. J., and Dorling, D. F. L. 2000. Geographical scale, the "feel-good factor" and voting at the 1997 general election in England and Wales, *Transactions, Institute of British Geographers, N.S.*, 25, 51–64.
White, S., Rose, R., and McAllister, I. 1997. *How Russia Votes*. Chatham, NJ: Chatham House.
Whitefield, S. and Evans, G. 1999. Class, markets and partisanship in post-Soviet Russia: 1993–96. *Electoral Studies*, 18, 155–78.
Whiteley, P., Seyd, P., and Richardson, J. 1994. *True Blues: the Politics of Conservative Party Membership*. Oxford: Clarendon.

Chapter 23

Nationalism in a Democratic Context

Colin H. Williams

Nationalist movements are often enigmatic in their aims and legitimacy, because they can oscillate between *individualist* conceptions of democratic nationalism and those that are more *communitarian* in orientation.[1] This dualism has implications for conceptions of citizenship, group membership, participation, and social inclusion or exclusion. In many nationalist movements there are philosophies of inclusion or group membership which, if they were to be couched in terms of race rather than national culture, would be deemed abhorrent to liberal democrats. And yet, questions such as, "Who may be represented by nationalist parties?" "In whose name is the claim of autonomy made?" and, "For whom do nationalists speak?" are fundamental elements of the contemporary battle of ideas and political affiliation in many plural polities. Clearly, despite its often exclusionary and reactive framings, the language of nationalism is a widely acceptable discourse among modern democratic states such as those of Europe.

For some, this is a matter of the underlying meaning rather than the specific language which is used in reproducing nationalist ideology. For others it goes even deeper, for it problematizes the very legitimacy of nationalist mobilization in what are increasingly multicultural democracies. Elements of the tension between individual and group consciousness are also at the heart of the current epistemological and methodological debates within the social sciences (McIntyre, 1984), and underlie the difficulties of finding the appropriate institutional expression of nationalism in practice.[2]

The literature on nationalism is voluminous.[3] Although geographers have made significant contributions, few have analysed the democratic context of nationalism, or the internal dynamics of nationalist movements, whether in terms of conflicting aims, ideological disputes or support base. Most geographical analysis has focused on comparing the electoral fortunes of nationalist movements within various voting systems (Johnston and Taylor, 1989).[4] A second emphasis has been on nationalism and uneven development. Other work has explored the notion of imagined communities, the iconographic representation of nationalist symbols in the landscape, conflict analysis, and the use of violence by minority nationalist movements (see, e.g., Blaut, 1987; Johnson, 1995).[5] Some of the outstanding examples of geographical analyses of majoritarian nationalisms include Zelinsky (1988), who offers a penetrating account of the American experience. Additionally, Kaiser (1994) and Smith (1996a, b; Smith et al., 1998) disentangle the Russian and Soviet experience, and Jisi (1997) and Zhao (1997, 2000) offer cogent perspectives on nationalism in China, focusing on nativism, anti-traditionalism, and pragmatism.

In this chapter, my focus will be an analysis of the scope for nationalist movements within the context of the contemporary European political system. I believe this system has yet to come to terms with the nation-state nexus, based as it is on differing principles of membership and inclusion.[6] Here, I argue that for liberal nationalism to succeed in any measure, current European conceptions of the nation and of nationalism must grapple more effectively with the implications of three simultaneous processes: globalization, regionalization and state adaptation (see also Jönsson et al., 2000).

The European Context

The main struggle in contemporary European politics is how to best represent the interests of local, regional, state, and supra-state authorities within an effective system which commands mutual respect. Nationalist and separatist movements have traditionally sought increased autonomy, and ultimately of course, complete independence as free and equal members of the international system.[7] Given the interdependent and multi-level character of governance we may ask whether pristine ideals of autonomy, autarchy, and self-determination are less realizable now than in the past. If so, what can liberal nationalism still hope to achieve within an integrated quasi-federal Europe? Experts doubt that there will soon be a federal Europe resembling present-day federal states, or a Europe of the Regions where nation-states will have disappeared. But they are certain that there will be a regionalized Europe where decentralized and regionalized states will be at an advantage (Loughlin, 1996b). If so, what scope is there for various realizations of nationalism within a democratic system?[8]

Throughout the nineteenth century the coupling of "nation" with "state" led to the development of *nationalism* as an ideology, and to the rise of nationalist political movements. History shows us that nationalism remained a powerful, if highly negative, political force when linked to imperialism, fascism and Nazism, rather than to the creation of liberal democracy. Nationalism owes much to the emergence of the nation-state ideal, even if the latter has been contested throughout its existence – most recently by globalization, and more specifically in this case, by European (EU) integration. What has not been examined systematically are the consequences of this transformation from the point of view of democratic practice, especially at the *sub*-state level, the locus of nationalist angst and mobilization.

A contemporary approach argues that new forms of governance have direct consequences for the regional and local levels, which are becoming increasingly important alongside and implicated within the national and supranational levels.[9] In fact, they may be even more important as the locus of the exercise of power shifts away from national governments representing the territorial state, both upwards to European and other transnational types of organization, but also downwards to nations, regions and local authorities.[10] The privileging of the territorial state, especially in the literature on international political economy, has been criticized by geographers, most effectively in Agnew and Corbridge's *Mastering Space* (1995). In a similar vein Jönsson et al. (2000) argue that theories which privilege the state at the expense of other organizational forms are of limited value in understanding the dynamism of change. Their magisterial study of the role of evolving networks in the organization of European space points towards the possibility of governance without government as the operational norm for the superstructural level. And as this supra-national dimension has largely escaped control of any kind, this makes it all the more imperative to strengthen regional and local levels and design new institutional forms of expression.[11] The old nation-state system, however, will not simply disappear. It will adapt itself as the "negotiating state," capable of dealing with both "glocalization" and "fragmentation."[12]

Nationalist movements have adopted a variety of forms of resistance to the incursions of central state agencies. The most violent expressions, as in Northern Ireland, Corsica and the Basque Country, dominate both media headlines and academic analysis alike (Cox et al., 2000). But to the more conventional constitutional electioneering, as happens in Catalonia or Scotland, we may add yet other expressions, as in Christian pacifism or widespread conscientious objection, which were used by nationalists as instruments of opposition to warfare in the name of imperial defense and the subjugation (genocide) of minority peoples. There has also been an acute sense of divine destiny in the mobilization of minority cultures, whether in Catholic Catalonia and Euskadi (Conversi, 1997), Ireland (Goldring, 1993), or Nonconformist Wales (Llywelyn, 1999). Equally revealing has been the concern with the localism, with self-reliance and with nonalignment, in order to distance supporters from the hegemony of the superpowers and strong states. Environmental and minority autonomist rights movements seek to place the individual within a wider communal framework, stressing interdependence and a shared destiny. Feelings of belonging, of shared responsibility and of rootedness all figure prominently, in contrast to the possessive individualism stressed by the post-Fordist culture with its mobility, individual advancement and regime of flexible accumulation.

Below, I identify ten structural characteristics that re-appear in most nationalist struggles. These consistently involve a fear of loss of identity, exploitation of the cultural and physical resources of the nation, resistance to external intervention and control, reinterpretation of the nation's plight in terms of the leadership's mission-destiny view of their own transcendent existence, and the ultimate securing of freedom in a world of free and equal nations (Williams, 1994). I believe that transgenerational concepts of the nation defined as a community of communities are essential to an understanding of European minority nationalism.

1. Nationalists insist on the defense of their unique homeland and of the protection of their valued environment. Nationalism almost always has a specific piece of territory to which it must relate as a response to a particular set of circumstances. This reflects the classic element of place-centered politics described in Agnew (1987), MacLaughlin (1986), Mar-Molinero (1994), and Williams (1982). It is not an autonomous force. It is evident that resistance, struggle, and the politics of collective defense over land and territory dominate the

relationship between a minority movement and its incursive, hegemonic state power. More recently, ecologists have taken up this concern alongside environments, and in some cases caused common alliance with nationalists to protect threatened spaces.[13] This seems set to become a major issue of the century where conflict over the control of space and resources will quicken our awareness of the commonality of defensive movements, even if their ideological rationale remains distinctly separate.

2. Nationalists also insist on the defense and promotion of their cultural identity, whose diacritical markers are usually a distinct language, religious affiliation, separate social existence, and historical experience. Nationalist movements often attach the utmost symbolic significance to the preservation of a separate identity. It does not follow that the majority of its target constituents share such markers as a common language, or religious persuasion. Indeed one of the inherent paradoxes of nationalism is that by searching for a distinct cultural infrastructure to differentiate itself from state-wide political parties or powerful neighbors within the polity, nationalist rhetoric often serves to alienate the very people in whose name the claim to liberty and equality is being made.[14]
3. Resistance to trends which integrate the national territory into the core state apparatus is a significant feature of nationalism. Dismissed as antiprogressive or recidivist, nationalists have faced difficulties in constructing an economic argument to counter the economies of scale justification for state integration. The increasing scale and complexity of political units renders anathema any conscious return to small nations as the basic building block of the international political system. It is argued that the "liberation" of the Baltic States, and of Slovenia and Croatia cannot offer an exemplar to Scotland, Euskadi, and Québec, for they were born out of civil war and the threat of physical force in the wake of imperial dissolution. But this reasoning undervalues the historical integrity of territorial identities, as found in Slovenia and Croatia, which have an authenticity and integrity pre-dating modern conceptions of the nation-state.[15]
4. Nationalists deny the claim that many putative nations are not economically viable in an increasingly globalized system. They would stress that Euskadi, Québec and the like have been systematically exploited and underdeveloped as "internal colonies" (Hechter, 1975). Persistent structural discrimination can only be halted and reversed once the goals of political sovereignty and economic autarchy have been achieved. On realizing their plight, the masses will do what all other colonial victims do – rise up and evict their oppressors in the name of national liberation. It matters less that nationalism *per se* offers little that is prescriptive for the postindependence economic recovery, for they believe that the advent of independence itself creates a new social reality, where truth, prosperity, and development can be constructed largely from within the nation, rather than being denied from without. Scotland and Wales manifest this, for since May 1999, they have a Parliament and National Assembly, occasioning a concomitant increase in national confidence and ambitions to play a more active role within the international system.
5. A more persistent tendency is for local communities to resist population transfer and demographic changes which adversely affect the majoritarian position of co-nationals. Thus, resistance to outsiders, immigrants, settlers, and colonizers as agents of the hegemonic state and its associated culture is as widespread in rural Macdeonia and Wales as it is in urban Euskadi. It is manifested in agonizing fears over the survival of a threatened language where each child born to the

foreigner and the death of each native speaker is logged in a mythical but pervasive national register and the balance of probability against the survival of Welsh or Euskerra weighed anxiously anew every morning.

6. Fear of loss of dominance and influence is expressed through cultural attrition and campaigns to save local communities, sacred sites, and valued land from rapacious developers, whether they be private companies or government departments conducting legally binding state business. Several campaigns in Wales, such as at Rhandirmwyn, Tân yn Llyn, "Nid yw Cymru ar Werth" (Wales is not for sale), or in Brittany at Plogouf and second home/tourist development, and in Scotland in relation to nuclear installations and defense establishments, demonstrate how the socioeconomic infrastructure is transformed to suit external interests. Often dangerous and obnoxious industries such as nuclear power plants, oil terminals, or defense establishments are situated within the minority's territory, who are relatively powerless to stop such exploitation of their land and social fabric. Equally the minority's resources are expropriated, be they natural such as water, or human, a well-educated but relatively poorly paid labor force exploited in the service sector with its seasonal and tourist-dependent characteristics. All of this induces a dependency situation, which is reactive rather than purposive, defensive rather than self-confident, and outer- rather than inner-directed.
7. In selected circumstances this perception of being exploited and subdued leads to open conflict and sustained violence as in Euskadi, Ulster, and Corsica (Loughlin and Letamendia, 2000). Attempts to defuse such violence through various stages of regional autonomy, power sharing and state restructuring do not appear to be wholly successful. Once a culture of violence is established, it tends to have an internal dynamic, self-energizing and sustaining, undeterred by piecemeal reform.
8. With the rise of print capitalism and mass communication came the possibility of constructing alternative versions of historical reality (Williams, 1988). Minority nationalist intelligentsias sought to influence group learning by stressing those aspects of history which explain inequalities in the light of significant acts of oppression. Reconstructed historical discourses contend with orthodox interpretations of popular history and state development. When centralizing elites seek to suppress alternative discourses, as Franco sought to ban the political use of Basque because it told the story of the vanquished, then a fresh round of conflict is unleashed. Centralists oppose formal education in the minority's mother tongue precisely because they fear their inability to control the messages circulated through the minority languages.[16]
9. An insistence on the immorality and political illegitimacy of the *status quo* provides a political rationale for the emergence of nationalist movements. The desire of most nationalist movements is to be a fully accepted nation among other nations. In political terms this necessitates the attainment either of sovereignty or a great deal of devolved autonomy within the international system.
10. Finally, minority nationalists seek evidence which serves to redefine their situation in the light of reforms, concessions, political accommodation, and gains for the beleaguered minority within the dominant system. In Western Europe, the nationalist appeal of the 1960s was couched on the basis of a universal trend towards decolonization. What was good enough for Nigeria or Niger was good enough for Scotland and Brittany. In the 1970s it was as a

reaction to the overloading of governmental structures in Westminster or Paris, so that developed government became good government, conferring the twin blessings of participatory democracy and enhanced efficiency in government services. In the 1980s it was an appeal to localism, to valued environments, to the appeal of place over placelessness and anomie. In the 1990s, it was an appeal to reconstruct a Europe of the nations, to overthrow the state-centric and hegemonic nineteenth-century state system and to return to Europe's organic, constituent nations. Unperturbed by the difficulties in squaring the reality of a multicultural Europe within a nationalistic order, nationalists seek to bypass their state cores, London, Madrid or Paris and allow Edinburgh, Barcelona or Rennes to entreat directly with the Brussels–Strasbourg axis in an increasingly federal Europe.[17]

Nationalist mobilization within the various state forms

Orridge and Williams (1982) have demonstrated how the triggering factor of nationalist mobilization is related to the structural pre-condition of the state system. We now need to relate such features to the contemporary context of West European democracy, i.e. the variety of ways in which it is *understood*, *expressed*, and *practiced* in different countries as summarized by Loughlin and Peters (1997) in table 23.1.[18]

Each of these state traditions has given rise to distinct political and administrative cultures, forms of state organization and kinds of state-society relationship. Yet within each of these "families", there exist distinct *national* traditions.[19] Different state traditions also express subnational political systems in distinct ways. The two extremes are the Napoleonic tradition, as in France, which allows little variation across its national territory and the Anglo-Saxon tradition which tolerates wide variations, as in the United Kingdom.

Nationalism in the Context of Regional and Local Democracy

Loughlin et al.'s (1999) study of "Regional and Local Democracy in the EU" illustrates the functioning of democracy at sub-state levels through a variety of central–local relations and differing electoral systems (table 23.2).[20]

In the UK, three different electoral systems are in operation: the "first-past-the-post" or plurality system at the state level; the single transferable vote (STV) system of proportional representation in Northern Ireland; and the "additional-member" system for the Scottish Parliament and the Welsh National Assembly alongside the "first-past-the-post" system. In federal and regionalized states, such as Germany, Belgium, Italy, and Spain there may be different party systems at the subnational level with the existence of regionally-based parties, such as the Bavarian *Christlich Sozial Union* (CSU, Christian Social Union), the Catalan *Convergencia i Unío* (CiU, Convergence and Union), and the Basque *Partido Nacionalista Vasco* (PNV, Basque Nationalist Party). Such regionally-based parties may also play a role at the national level. In the UK, there has always been a different party system in Northern Ireland, where, until recently, the British parties did not organize (in recent years, the Conservative Party has permitted a Northern Irish branch). With devolution, the tendency to develop regionalized political cultures and party systems in Scotland and

Table 23.1 State traditions

	Tradition			
Feature	Anglo-Saxon	Germanic	French	Scandinavian
Is there a legal basis for the state?	No	Yes	Yes	Yes
State–society relations	Pluralistic	Organicist	Antagonistic	Organicist
Form of political organization	Union state/ limited federalist	Integral/ organic federalist	Jacobin, "one and indivisible"	Decentralized unitary
Basis of policy style	Incrementalist, "muddling through"	Legal corporatist	Legal technocratic	Consensual
Form of decentralization	"State power" (USA), devolution/local government (UK)	Cooperative federalism	Regionalized unitary state	Strong local autonomy
Dominant approach to discipline of public administration	Political science/ sociology	Public law	Public law	Public law (Sweden), organization theory (Norway)
Countries	UK, USA, Canada (not Québec), Ireland	Germany, Austria, Netherlands, Spain (post-1978), Belgium (post-1988)	France, Italy, Spain (pre-1978), Portugal, Quebec, Greece, Belgium (pre-1988)	Sweden, Norway, Denmark, Finland

Source: Loughlin and Peters (1997).

Table 23.2 Central–local relations in EU member states

Type of state	State	Political region[a]	Administrative/ planning regions[b]	Right of regions to participate in national policy-making	Right of regions to conclude foreign treaties[c]	Political/ legislative control over subregional authorities
Federal	Austria	*Länder* (10)		Yes	Yes (but limited)	Yes (not absolute)
	Belgium	Communities[d] (3)		Yes	Yes (but limited)	No
	Germany	Regions (3)		Yes	Yes (but limited)	Yes (not absolute)
		Länder (16)			Yes (but limited)	Yes (not absolute)
Regionalized	Italy[e]	Regioni[g] (20)		Consultative	No	Yes
	France	Régions[h] (21)		Consultative	No	No
Unitary	Spain	*Comunidades autonomas* (17)		No	No	Yes
	United Kingdom[f]	Scottish Parliament; Welsh National Assembly; Northern Ireland Assembly	English standard regions	No with regard to English regions, still unclear with regard to Scotland, Wales, and NI	No at present, but may evolve	Yes in Scotland and NI, no in Wales (so far)
Decentralized unitary	Denmark	Faroe Islands	Groups of Amter	No	No	No
	Finland	Aaland Islands	Counties have a regional planning function	No	No (but has a seat in the Nordic Council)	Yes
	Netherlands	Rijnmond region[i]	Landsdelen	Consultative	No	No

(*continued*)

Table 23.2 *(continued)*

Type of state	State	Political region[a]	Administrative/ planning regions[b]	Right of regions to participate in national policy-making	Right of regions to conclude foreign treaties[c]	Political/ legislative control over subregional authorities
	Sweden	Region[i]	Regional administrative bodies	No	No	No
Centralized unitary	Greece		Development regions (13)	No	No	No
	Ireland		Regional authorities (8)	No	No	No
	Luxembourg		Potential planning regions	No	No	No
	Portugal		Island regions[j]	—	—	—

Source: reproduced with permission from Loughlin et al. (1999), p. 16.
[a]This refers to regions and nations (as in Scotland, Wales, Catalonia, the Basque Country, and Galicia) with a directly elected assembly to which a regional executive is accountable.
[b]This refers to regions without a directly elected assembly, which exist primarily for administrative/planning purposes.
[c]There is a sharp distinction between the federal and nonfederal states in this regard; however, the majority of nonfederal states may engage in international activities with the approval of, and under the control of, the national governments.
[d]The Flemish linguistic community and the Flanders economic region have decided to form one body; the French-speaking community and the Walloon region remain separate.
[e]Italy is currently undergoing a process of political reform that involves the transformation of the old state into a new kind of state with some federal features. However, although the position of the regions will be strengthened, this will not be a federal state such as Germany or Belgium.

[f]The United Kingdom was, until the referendums in Scotland and Wales in September 1997, a highly centralized "Union" state. However, the positive outcome of the referendums means that there was a Scottish Parliament and a Welsh National Assembly by 1999. A referendum in 1998 on a Greater London Authority with an elected mayor was also successful, and this is seen as a precursor to possible regional assemblies in England. The successful outcome of the Northern Ireland peace process means there will be a Northern Ireland Assembly as well as other new institutions linking together the different nations and peoples of the islands.

[g]In Italy there are seventeen "ordinary" regions and five regions with a special statute because of their linguistic or geographical peculiarities: Sicily, Sardinia, Trentino-Alto Adige (South Tyrol, large German-speaking population), Val d'Aosta, and Friuli-Venezia Giulia.

[h]There are twenty-one regions on mainland France. However, to this one must add Corsica and the overseas departments and territories (the DOM and TOM). Since 1991 Corsica has had a special statute and is officially a *Collectivité territoriale* rather than a region. The TOM too have special statutes, and one of them, New Caledonia, has recently (May 1998) been permitted to accede to independence within a period of twenty years.

[i]In 1991 it was decided to set up a new metropolitan region with an elected government in the Rotterdam area to replace the *Gemeente* of Rotterdam and the Province of South Holland. However, this was rejected by a referendum held in Rotterdam.

[j]Portugal, while making provision in its Constitution for regionalization, has so far only granted autonomy to the island groups of the Azores and Madeira. The mainland remains highly centralized.

Wales will intensify making nationalists' issues more significant. In Scotland, future political cleavages will concern the political expression and relationship of the Scottish nation to various options, concerning the *status quo*, independence within Europe, or a federal United Kingdom. In Wales there is no sense of "settled nationhood" and even less of a civil society, and the cleavage is between those who support and oppose devolution. In Northern Ireland, even if the Good Friday settlement seems to have bade "A Farewell to Arms?" for the time being, there remains the fundamental dispute between unionists and nationalists as to the interpretation of the Agreement (Cox et al., 2000).

Similar regionalization and differentiation of political systems elsewhere is reinforced by the development of regional or subnational branches of national parties, as for example, the Catalan and Basque sections of the Spanish Socialist Party or the Scottish Conservatives who favor devolution more than their English counterparts do, or the Welsh Labour Party which has adopted much of the symbolism of a national (at times, nationalist) party. In some cases the regional system may develop in an autonomous manner, as in the Iberian experience (Conversi, 1997; Mar-Molinero and Smith, 1996; Moreno, 1995). In others, the regional and local are completely dominated by the centralist national parties as in Ireland and France and, even in the decentralized UK, the central organs of the Labour Party have difficulty in letting go of control over the parties in the periphery.[21]

Europeanization, nationalism, and the "new" regionalism

A second set of relationships concerns the emergence of the EU at the superstructural level and the refashioning of nationalist movements as inclusive regionalist parties. The relaunch of Europe in the 1980s, with the 1992 Single European Market project, is often interpreted as a response by national governments and business elites to the perceived threat of a global economy dominated by the USA and Japan (O'Loughlin, 1993; Ó Tuathail, 1993; Williams, 1993a). However, it was also the result of intense lobbying by governmental actors and other elites who wished to strengthen federal elements of an integrated Europe (Loughlin, 1996b; Pinder, 1995) and promote the notion of a "Europe of the Regions."[22] The deepening of integration through the 1987 Single European Act and the revision of the EC treaties at Maastricht and Amsterdam has created a new administrative and legal environment for local and regional authorities. This European system of governance that has both state-like and federal-type characteristics creates challenges and opportunities for regional and local authorities.

In this changed context, new forms of nationalism and regionalism have emerged as part of a "modernizing" project strengthening regional and structural action policy and the reinforcement of the principles of subsidiarity and partnership (Keating, 1998). However, this does not amount to the establishment of a "Europe of the Regions" because it is unclear as to *which* regions would form the constituent parts, and state governments still remain the dominant actors. Regional representation through the Committee of the Regions or via DG XVI of the European Commission or the European Parliament still remains weak. What is new is the sense of a "Europe *with* the Regions," in that subnational authorities have more opportunities to interact at a European scale through the Assembly of European Regions and the

Conference of Peripheral Maritime Regions (Hooghe and Marks, 1996). The influence of the Committee of the Regions is set to grow (Loughlin and Seiler, 1999) and with it a greater transparency and accountability. As Loughlin comments, it is ironic that, despite the success of the EU in incorporating former dictatorial regimes,[23] the Union has itself failed to develop into a fully democratic system. A major criticism leveled by nationalists and others is its serious "democratic deficit." The challenge is to strengthen democratic institutions within the member states and to encourage trends such as the European Parliament's participation in the decision-making processes of the Council of Ministers, thereby increasing democratic control at the European level.[24] The resultant overlapping and nested systems of governance involving European, national, regional and local actors, in new networks may either be viewed as providing a special focus for nationalist programs, for they bring the national region back into the picture of governance, or they may be seen as undermining nationalist rhetoric, for they threaten to present its message of group exclusivity as fundamentally undemocratic.

Globalization and European Integration

An additional development compounding the logic of nationalist ideology is the process known as globalization. Together with European integration, globalization changes the context within which civil society is mediated. This poses a challenge to conventional territorial relationships, and simultaneously opens up new forms of inter-regional interaction. Ethno-linguistic minorities, the most prominent advocates of nationalist ideas, have reacted to these twin impulses by searching for Europe-wide economies of scale in broadcasting, information networking, education, and public administration. They are also establishing their own EU institutions and forming new alliances to influence EU decision-making bodies. They believe that by appealing to the superstructural organizations for legitimacy and equality of group rights, they will force the state to recognize their claims for varying degrees of political/social autonomy within clearly identifiable territorial/social domains.

These trends must be set against the EU's policies, which strengthen majority language regimes within a refashioned European network of nodal sites (Jönsson et al., 2000). But the wider question of the relative standing of official languages makes political representatives wary of further complicating administrative politics by addressing the needs of roughly fifty million citizens who have a mother tongue which is not the main official language of the state which they inhabit. Recent expansion of the EU has increased the difficulties in translating multicultural communication and guaranteeing access to information, and hence power, for all groups. The real geolinguistic challenge is to safeguard the interests of all the nonstate language groups, especially those most threatened with imminent extinction. A critical aspect of constructing these safeguards is access to knowledge, thus we need to ask and act upon the answers to questions such as who controls access to information *within the mother tongue and the working languages of* European minorities? Are such languages destined to occupy a more dependant role because of superstructural changes favoring dominant groups or will they achieve relative socio-cultural autonomy by adopting aspects of mass technology to suit their particular needs?

Additional issues concern the adaptation of lesser used language speakers to the opportunities afforded by changes in global–local networks, the growth of specialized economic segments or services and of information networks which are accessed by language-related skills. Accessibility to or denial of these opportunities is the virtual expression of real power in society that must be taken on board in any discussion of the changing geography of regional economic development and cultural representation. Advanced regions, such as the "Four Motors" group of Baden-Würtemberg, Catalonia, Lombardy, and Rhône-Alpes, have become even stronger while weaker regions, such as those in southern Italy and Spain or in Greece or in parts of Great Britain, have become weaker. Some previously peripheral countries such as Ireland and Portugal and parts of Spain have managed to use the new developments to achieve economic performances that are quite spectacular.[25] What is clear is that the old conceptualization of centers and peripheries no longer holds, and that spatial and territorial concepts need to be radically reformulated in the search for new representational spaces (Williams, 1997a).

Regional-level challenges to globalization have to cope with the fact that current approaches to public action are more sympathetic to competition and markets than to solidarity and equalization among territories. A downgrading of regional development priorities results in the issues being seen in purely economic or financial terms rather than in a wider sense as social, political and cultural development. Economic actors thus become crucial in conceptions of new forms of governance, which has less real purchase in addressing issues of fragmentation and social exclusion.

Technological developments and the deterritorialization of society and space

European-level institutions are also reacting to trends such as telematic networking and global communication systems which have reinforced the dominance of English and promoted a Pan-European, Trans-Atlantic melange of culture, values, and entertainment. Other international languages, such as French, German, and Spanish, are re-negotiating their positions within the educational, legal and commercial domains of an enlarged Europe. English has been strengthened by the admission of Nordic members to the EU, but there is no agreement as to whether other major languages are necessarily weakened by enlargement. Neither do we know how existing lesser-used organizations will fare, given the imminent enlargement of the EU with its effect on the management of ethno-linguistic and regional issues.[26]

An abiding concern is that autochthonous language groups, such as the Basques, Bretons, Irish, and Welsh will be further marginalized in an increasingly complex and competitive social order (Williams, 2000). Their main hope lies in establishing regional bilingualism as the dominant pattern. Limited success in introducing bilingual practices in domains such as education, public administration and the law offers encouraging signs of a more equitable future. Yet even in the midst of success, some groups are experiencing the ambiguous effects of mass technology, for they suffer the erosion of their traditional strength in heartland areas and key cities whilst simultaneously harnessing the potential of mass communication and electronic networking in education, broadcasting, and leisure service provision. In Central and Eastern Europe, comparable but poorer ethno-linguistic groups face a more difficult future in seeking to reproduce their culture and identity, and this is likely to reinvigorate nationalist tendencies (Kuzio, 2000, 2001; Landsbergis, 2000). If territorial loss and relative economic decline continue for these more vulnerable groups straddling major cultural fault lines, then it is likely that border tensions will spill over into EU and neighboring states. Currently we have little detailed foreknowledge of how such trends will impact on the "opening up the frontiers of Europe." Neither do we know what role intractable ethnic conflicts will play in triggering major regional clashes, or how the security architecture of Europe will react to such conflagrations.

A related source of conflict will be the differential access groups enjoy to innovative regional impulses, information space and power networks. Equally intriguing is to ask what effect globalization will have on the regional–local infrastructure upon which European ethnic minority groups depend? The established transfer of manufacturing from peripheral locations in Western Europe to Eastern European, Asian or Central American states mirrors today similar changes in for example the textile industry of North-west Europe in the mid-nineteenth century. First, core–periphery differentials are maintained because surplus regional capital is re-invested elsewhere. Galicia, Wales, and Brittany find it difficult to sustain vibrant ethno-linguistic communities in the face of emigration, relative deprivation, and regional infrastructure decline. Secondly, there is the overcoming of temporal and spatial discontinuities in "real-time" communication and economic transactions, which are increasingly independent of the limitations of specific locations (Brunn and Leinbach, 1990; Castells, 1997). Thirdly, globalization processes have a simultaneity of *both* increased uniformity and increased diversity, giving birth to alternative identities, practices, and preferences which have little if anything to do with conceptions of the nation. Nowhere is this more evident than in the cultural infrastructure of world cities, suffused as they are with multicultural choices and exotic consumption, quite distinct from most of the state's remaining territory. In such a milieu, emerging or re-born linguistic identities are nurtured and expressed. So, also, are their opponents, who wish to impose a pristine cultural order on dissenting ethnic activists, resulting in tension, hostility, and racial/ethnic violence. Fourthly, superficial global homogenization deserves special scrutiny because the spread of English in particular contributes to the link between globalization and postmodernity. Many European minorities, despite being bi- or trilingual, face extreme pressures as a result of superstructural changes. Fifthly, there is the countercurrent of increased religious and/or ethnic identification and confrontation, within and across national frontiers and often in violent and emotional forms (Mlinar, 1992; Williams, 1993b).

Globalization also influences cultural patterns and modes of thought because as a constant interactive process it is always seeking to break down the particular, the unique, and the traditional so as to reconstruct them as a local response to a general set of systematic stimuli. This is the threat of the *deterritorialization of society and space*. "The collapse of both space and time" demand a fresh appreciation of global interdependence for we have been quick to characterize the advantages which accrue to well-placed groups and regions. We have been less careful to scrutinize the impact such transitions might have on minorities and the disadvantaged. Some of the changes wrought by globalization include the process of deterritorialization and its obligation to redefine spatial relationships; a shift in the old certainties of global strategic relationships; a change in the nature of the state and its legitimizing philosophy of national self-interest enshrined in the sovereignty of the citizen; the direction of world development and our common hopes and fears as we face a succession of global environmental crises.

As with modernization in past times, globalization is an ideological program of thought and action, for it is not merely an account of how the world is changing, but also a *prescription* of how it *should* change. As yet we do not have global economic change; rather, we have macro-regional functional integration in Western Europe, North America, and, to a lesser extent, in parts of South and East Asia. But the cumulative impact of these trading blocs is to establish a new regime whereby barriers to capital, trade, and influence are reduced and the race for resources, uniform product standards, manufacturing, and technology transfer is increased. Social and cultural change are deeply implicated in this world vision, and we have enough evidence to recognize that some groups and regions will be advantaged, and others marginalized as globalization is entrenched (Cerny, 1999). However, globalization is neither an inevitable nor an uncontested process. We need more information before practical measures can be designed to specify how minorities may cope with these new challenges. We also need data on how the recent strengthening of the EU influences the impact of these diverse processes in selected regions.

One final element stressed by nationalists and regionalists is the democratic involvement of those who live in a region or locality. This is particularly the case with regard to contested issues such as the environment, urban policy, spatial planning, or tourism (Judge et al., 1995). The answer to this problem may lie in devising new forms of institutional design more appropriate to a system of governance than a system of government. It may be that the latter, based on hierarchy, routine and slow responses, needs to be complemented by a system that is more flexible, horizontal, and open, and that can respond to the ever increasing challenges of a turbulent environment.

Conclusion

The key questions posed in this chapter concern how the processes summarized as "globalization" and "deterritorialization" affect the maintenance of regional, nationalist identities and the transformation of corresponding regional spaces. We are not yet at the dawn of a "Europe of the Regions" despite a campaign which has lasted for well over a generation. But we daily witness the emergence of new regional actors whose cumulative impact on the territorial state-nation will be profound (Keating and Loughlin, 1997). Tension and conflict between all the processes at different levels will be inevitable; such is the nature of our competitive system. But I also detect a strong yearning for mutual understanding, rapprochement, and partnership which may yet prevail if the appropriate supportive political and economic infrastructures can be constructed. Chauvinistic nationalism will not entertain constructive partnerships, and democrats are right to be alarmed by its growth in Europe. But liberal, communitarian nationalism may yet prove an effective anchor for some who appear to have lost control and purchase in a system dominated by ever-increasing currents of globalization and systemic integration.

Acknowledgment

I am grateful to my colleague at Cardiff University, Professor J. Loughlin, for his gracious assistance in adapting the insights gained from the "Regional and Local Democracy in the European Union" project and for his permission to reproduce tables 23.1 and 23.2.

ENDNOTES

1. That is, whether a democracy is in essence the free expression of the rights of individual citizens, regardless of putative socio-cultural origins or whether it is an expression of the existence of "communities," which may, in certain respects, override individual interests as "individual" autonomy is sublimated to "communitarian" autonomy (Agnew, 1987; Lapidoth, 1997).
2. Epistemologically, as Loughlin (1999) argues, the debate concerns the manner in which our minds grasp reality, either analytically or synthetically, or at least which of these two aspects is dominant. Methodologically, the debate is between "methodological individualism," which usually takes the form of rational choice approaches, and approaches that are more structuralist, culturalist or institutionalist. Normatively, in terms of public policy, the question is whether democratic practice can be based on the notion that the individual citizen is a member of a collectivity or whether h/she should be seen as a consumerist, rational individual making choices with regard to the use of public services. If the latter model is adopted, "public" services are increasingly redefined in a more "privatized" fashion.
3. Fascinating accounts of nationalism may be found in Anderson (2000), Guibernau and Hutchinson (2001), Hutchinson and Smith (1994), McCrone (1998) and Smith (2000). For an excellent overview of geographic perspectives see Kaiser (2000).
4. The two notions of nation as *demos* and nation as *ethnos* are closely related. Demos can be defined as a legitimizing principle based on equal citizenship regardless of ancestry or geographical origin. While ethnos gives primacy to collective cultural identity, the ethnic community, and is heavily imbued with arguments drawing on perennialist rather than modernist conceptions of the state. In France, membership of the French nation obliged individuals to assimilate to French culture and language. In Germany, cultural nationhood found its expression in a democratic system or nation as demos – consolidated only in the second half of the twentieth century. However, in all countries and nationalist movements there is a tension between these two aspects of nationhood and this is one of the reasons for the difficulties many contemporary states and nationalist movements have in defining the meanings of nationality and citizenship (Alter, 1994; Loughlin, 2001).
5. In multiethnic polities, ethnic separatism often derives from an acute concern over the erosion of a group's identity and resource base. Separatists, such as ETA (Basque Homeland and Liberty), assert that ethnic discrimination can only be halted through the transformation of their territory to form a sovereign state co-equal with all other states (Clark, 1994). Regionalists assert that one need not go so far as to break up the state, but insist that its internal affairs should reflect its plural character. In tandem, both regionalists and separatists pose fresh challenges to the territorially fixed nature of monopolistic sovereign space.
6. In analysing the nature of those trends which conduce to a regionalized Europe, we need to differentiate between *regionalization*, identified as the regional application of state policy and *regionalism*, defined as the attempt to optimize the interests of a region's

population through the manipulation of the political process. In this essay the interrelationship of globalization and localism, holistic and ecological ideas, and reformed conceptions of space occasioned by telematic revolutions will figure strongly as we seek to advance our understanding of the role of nationalism in European politics.

7. On typologies of nationalism see Breuilly (1982), Orridge and Williams (1982), Smith (1981, 1983); and for a political geographic overview see Williams (1994), pp. 1–53.
8. For a series of insightful geographic essays on the rebirth of the nationalities question in Europe, see van Amersfoort and Knippenberg (1991).
9. This new approach is represented in the work of scholars such as Jönsson et al. (2000) and Tägil (1999), and the research of the ECPR "Territorial Politics and Regionalism" group (see Keating and Loughlin, 1997; Loughlin, 2001).
10. Democracy is also about citizens holding decision-makers accountable for their activities and this is becoming increasingly difficult given this shift in power.
11. Thus, it is almost certain that expressions of nationalism, rather than wither away under the pressures emanating from a stronger EU will in fact re-vivify themselves, most often perhaps in conjunction with other regional or environmentally based political movements.
12. Jönsson et al. (2000) have used the network metaphor to capture the essence of the three simultaneous processes of globalization, regionalization, and state adaptation. They warn against repeating the three most common mistakes when speculating about the future: "the first is to assume that the future will be entirely different from the past; the second is to believe that it will be just the same; and the third, and most serious, mistake is not to think about it at all" (p. 189).
13. Nationalism and environmentalism represent themselves as also holistic in thought and deed. The universal desideratum is the ideal nation-state in an international community of free and equal states. Its local manifestation may be at any one of the varying stages in a continuum from full statehood through to a nonstate nation in the making, but whatever its exact position it is always possible to relate the local to the global blueprint and back again. So it is with environmentalism. The unique character of the local environment is only given purchase by the general context of the global milieu, but it is given its urgency by the near-cataclysmic refraction of fundamental global issues, namely the greenhouse effect, the destruction of the tropical rain forest, the degradation of soil and sustenance, and the disappearance of species and habitat. Intrinsic to this coupling of the immediate and local to the evolutionary and global is a sense of shared involvement and responsibility, summed up as "Our common future." For an excellent account of such movements, see Galtung (1986).
14. However, most minority nationalisms justify their existence by a claim to separateness, usually on the basis of a unique cultural heritage.
15. Klemencic and Klemencic (1997) offer a systematic, cartographic analysis of the historical identification of the northeastern Adriatic peoples with key locations and illustrate just how rooted are conceptions of land, language, and nationhood in the European imagination.
16. How often in Europe have we heard the charge that minority schools are the breeding ground for a new generation of dissenting nationalists? And throughout Western Europe and North America we have new demands for a centrally controlled educational curricula rather than tolerate regional variations which may, in part, perpetuate group divisions and sectarian animosities.
17. It is a moot point, however, and one which demands considerable restructuring, to appreciate how such small nations will fare better when faced with a plethora of competing claims in Europe, than in seeking redress of their grievances within the existing state structure. Similarly in Québec, as the movement towards greater auton-

omy gathers apace, critics argue that increased cultural erosion will accompany independence because the Québécois state will not have the federal system to protect it in its dealings with foreign powers and agencies.

18. The UK and the Scandinavian countries passed directly to liberal democracy, without major revolutionary upheavals or reversals. Most other states – France, Germany, Italy, Spain to name just a few of the large states – alternated between democratic and nondemocratic forms of regime before finally settling down in the democratic family. Finland has also progressed smoothly to liberal democracy, although it did experience a civil war in the 1920s.
19. There is also a limited number of distinct state forms such as federal and unitary states capable of further differentiation (Loughlin, 1996a). Thus, there exist different types of federations such as "dual federalism," where the different levels of the federation operate independently of each other, as in Belgium or the USA, and "co-operative federalism," where the levels operate in close conjunction with each other, as in Germany (Hix, 1998). However, "unitary" states are also differentiated. First, the UK has been described as a "union rather than a unitary" state given that its formation occurred through a series of Acts of Union, with Wales, Scotland and Ireland respectively (Urwin, 1982). Among the rest, there are a variety of central–local relationships giving rise to distinct models of unitary state: centralized unitary, decentralized unitary, regionalized unitary (Loughlin, 1996b, 1998). There thus emerges a complex picture of the variety of the expression of democracy at the national and subnational levels within the member states of the EU as illustrated in table 23.1. All of these state forms, operating within state traditions, would claim to be "nation-states" but, clearly, this general concept has been interpreted in a variety of ways. The importance of state traditions for the understanding and expression of nationalism lies in the ways in which the institutions of liberal democracy and the practices of policy-making have developed.
20. The different ways of electing political parties affects the probability as to whether or not nationalist movements will be represented within the electoral system. In some cases, the regional and local political systems replicate the national, in others they are different. The key question, which few have addressed, is whether nationalist movements are more or less successful if they adopt a form of electoral strategy that mimics that of state-centric parties. A second key question is whether or not the electoral system allows a form of proportional representation, a reform which certainly improved the representation of Scottish and Welsh nationalist members within the devolved bodies of Scotland and Wales.
21. A further point made in Loughlin et al.'s study with regard to changing local patterns of governance is that there is in Western and Central Europe a tradition of local self-government and autonomy that predates the emergence of the nation-state. Cases in point are the cities of the Hanseatic League or the Italian city states such as Florence and Venice. Although these communes were not fully democratic, being ruled by local oligarchies, the tradition of communalism may nevertheless be regarded as a forerunner of local democracy. Without accepting the thesis of a neo-mediaevalism (because of the very "modern" or "postmodern" nature of these developments), it is clear that one of the consequences of the changing nature of the nation-state and the loosening of central–local bonds is that this tradition is reasserting itself today.
22. This notion originally referred to a federal Europe in which the constituent units would be not the nation-states, but the regions (at least those that possessed a strong identity such as Corsica, Brittany, Flanders, Wales, Scotland, etc.). Clearly, such a federal Europe is highly unlikely today, as nation-states will remain the key levels of government within the Union. Nevertheless, the term is valuable as an indicator of the new importance of regions in the new Europe.

23. Countries such as Spain, Portugal, and Greece which became members not long after having experienced nondemocratic systems, found that membership helped them to consolidate their democratic systems.
24. The British government has suggested that a new body representing the national parliaments might be created to oversee the activities of the Commission.
25. Ireland is the most spectacular of all and is sometimes referred to as the "Celtic Tiger" by analogy with the Asian tigers.
26. The most significant development was the establishment of the European Bureau for Lesser Used Languages in 1984. Located initially in Dublin but now centralized in Brussels, this small but effective organization has sought to coordinate and nurture inter-linguistic experience and transfer good practice from one group to another (O'Riagain, 1989; Williams, 1993b). Other initiatives involve the Conference on Local and Regional Authorities of Europe with its Charter on European Regional and Minority Languages (Council of Europe, 1992). Politically the most important reinvigorated actor is the Council for Security and Cooperation in Europe (CSCE) which has increased its involvement in minority group rights since 1989. Although still evolving as the reconstituted Organisation for Security and Cooperation in Europe (OSCE) it has detailed the rights and obligations of both minorities and host governments throughout Europe and has put into practice several of the key ideas advanced by specialist and minority rights agencies alike (Kymlicka, 2000; Minority Rights Group, 1991; Williams, 1993a).

BIBLIOGRAPHY

Agnew, J. A. 1987. *Place and Politics: The Geographical Mediation of State and Society.* London: Allen and Unwin.

Agnew, J. A. and Corbridge, S. 1995. *Mastering Space.* London: Routledge.

Alter, P. 1994. *Nationalism*, 2nd edn. London: Edward Arnold.

van Amersfoort, H. and Knippenberg, H. (eds.). 1991. *States and Nations: The rebirth of the "nationalities question" in Europe.* Amsterdam: Netherlands Geographical Studies.

Anderson, M. 2000. *States and Nationalism in Europe since 1945.* London: Routledge.

Blaut, J. 1987. *The National Question, Decolonising the Theory of Nationalism.* London: Zed.

Breuilly, J. 1982. *Nationalism and the State.* Manchester: Manchester University Press.

Brunn, S. and Leinbach, T. R. (eds.). 1991. *Collapsing Space and Time: Geographic Aspects of Communication and Information.* London: HarperCollins Academic.

Castells, M. 1997. *The Rise of the Network Society.* Oxford: Blackwell.

Cerny, P. 1999. Globalization and the erosion of democracy. *European Journal of Political Research*, 36(1), 1–26.

Clark, R. P. 1984. *The Basque Insurgents, ETA, 1952–1980.* Madison, WI: University of Wisconsin Press.

Conversi, D. 1997. *The Basques, the Catalans and Spain.* Reno, NV: University of Nevada Press.

Council of Europe, 1992. *European Charter for Regional or Minority Languages.* Strasbourg: Council of Europe.

Cox, M., Guelke, A., and Stephen, F. (eds.). 2000. *A Farewell to Arms?: From "long war" to long war in Northern Ireland.* Manchester: Manchester University Press.

Galtung, J. 1986. The Green Movement: a socio-historical exploration. *International Sociology*, 1(1), 75–90.

Goldring, M. 1993. *Pleasant the Scholar's Life: Irish Intellectuals and the Construction of the Nation State.* London: Serif.

Guibernau, M. and Hutchinson, J. 2001. *Understanding Nationalism.* Cambridge: Polity.

Hechter, M. 1975. *Internal Colonialism: The Celtic Fringe in British National Development, 1536–1966*. Berkeley, CA: University of California Press.
Hix, S. 1998. Elections, parties and institutional design: a comparative perspective on European Union Democracy. *West European Politics*, 21(3), 19–52.
Hobsbawm, E. 1992. *Nations and Nationalism since 1780. Programme, Myth, Reality*, 2nd edn. Cambridge: Cambridge University Press.
Hooghe, L. and Marks, G. 1996. "Europe with the Regions". Channels of interest representation in the European Union. *Publius*, 26(1), 73–91.
Hutchinson, J. and Smith, A. D. (eds.). 1994. *Nationalism*. Oxford: Oxford University Press.
Jisi Wang, 1994. Pragmatic nationalism: China seeks a new role in world affairs. *Oxford International Review*, Winter 94, 27–38.
Johnson, N. 1995. Cast in stone: monuments, geography and nationalism. *Environment and Planning D; Society and Space*, 13, 51–65.
Johnston, R. J. and Taylor, P. J (eds.). 1989. *A World in Crisis?* Oxford: Blackwell.
Jönsson, C., Tägil, S., and Törnqvist, G. 2000. *Organizing European Space*. London: Sage.
Judge, D., Stoker, G., and Wolman, H. (eds.). 1995. *Theories of Urban Politics*. London: Sage.
Kaiser, R. J. 1994. *The Geography of Nationalism in Russia and the USSR*. Princeton, NJ: Princeton University Press.
Kaiser, R. J. 2000. Geography and nationalism. In A. Motyl (ed.) *Encyclopaedia of Nationalism*. San Diego, CA: Academic Press.
Keating, M. 1995. Size, Efficiency and Democracy: consolidation, fragmentation and public choice. In D. Judge et al. (eds.) *Theories of Urban Politics*. London: Sage, 117–34.
Keating, M. 1998. *The New Regionalism in Western Europe: Territorial Restructuring and Political Change*. Cheltenham: Edward Elgar.
Keating, M. and Loughlin, J. (eds.). 1997. *The Political Economy of Regionalism*. London: Frank Cass.
Klemincic, M. and Klemincic, V. 1997. The role of the border region of the northern Adriatic in Italy, Croatia and Slovenia in the past and the process of European integration. *Annals for Istrian and Mediterranean Studies*, 10, 285–94.
Kuzio, T. 2000. Nationalism in Ukraine: towards a new framework. *Politics*, 20(2), 77–86.
Kuzio, T. 2001. "Nationalizing states" or nation-building? A critical review of the theoretical literature and empirical evidence. *Nations and Nationalism*, 7, 135–54.
Kymlicka, W. 2000. Nation-building and minority rights: comparing West and East. *Journal of Ethnic and Migration Studies*, 26(1), 183–212.
Landsbergis, V. 2000. *Lithuania Independent Again*. Cardiff: University of Wales Press.
Lapidoth, R. 1997. *Autonomy; Flexible Solutions to Ethnic Conflicts*. Washington, DC: University of Washington Press.
Llywelyn, D. 1999. *Sacred Place, Chosen People*. Cardiff: University of Wales Press.
Loughlin, J. 1996a. Representing regions in Europe: the Committee of the Regions. *Regional and Federal Studies*, 6(2), 147–65.
Loughlin, J. 1996b. "Europe of the Regions" and the federalization of Europe. *Publius*, 26(4), 141–62.
Loughlin, J. 2001. *Subnational Democracy in the European Union: Challenges and Opportunities*. Oxford: Oxford University Press.
Loughlin, J. and Letamendia, F. 2000. Lessons for Northern Ireland: peace in the Basque Country and Corsica? *Irish Studies in International Affairs*, 11, 147–58.
Loughlin, J. and Peters, B. G. 1997. State traditions, administrative reform and regionalization. In M. Keating and J. Loughlin (eds.) *The Political Economy of Regionalism*. London: Frank Cass.
Loughlin, J. and Seiler, D. 1999. Le Comité des Régions et la supranationalité en Europe. *Etudes Internationales*, Décembre 99.

Loughlin, J. et al. 1999. *Regional and Local Democracy in the European Union*. Luxembourg: Committee of the Regions/Office of Official Publication of the European Union.
MacIntyre, A. 1984. *After Virtue; a Study in Moral Theory*, 2nd edn. Indiana: University of Notre Dame.
MacLaughlin, J. 1986. The political geography of nation-building and nationalism in the social sciences: structural versus dialectical accounts. *Political Geography Quarterly*, 3(4), 299–329.
MacLaughlin, J. 1993. Defending the Frontiers : the political geography of race and racism in the European Community. In C. H. Williams (ed.) *The Political Geography of the New World Order*. London: Belhaven/Wiley, 20–46.
Mar-Molinero, C. 1994. Linguistic nationalism and minority language groups in the "NEW" Europe. *Journal of Multilingual and Multicultural Development*, 15(4), 319–29.
Mar-Molinero, C. and Smith, A. (eds.). 1996. *Nationalism and the Nation in the Iberian Peninsula*. Oxford: Berg.
McCrone, D. 1998. *The Sociology of Nationalism*. London: Routledge.
Mendras, H. 1997. *L'Europe des Européens*. Paris: Gallimard.
Minority Rights Group. 1991. *Minorities and Autonomy in Western Europe*. London: Minority Rights Group.
Mlinar, Z. (ed.). 1992. *Globalization and Territorial Identity*. Aldershot: Avebury.
Moreno, L. 1995. Multiple ethnoterritorial concurrence in Spain. *Nationalism and Ethnic Politics*, 1(1), 11–32.
O'Loughlin, J. 1993. Fact or fiction? The evidence for the thesis of US relative decline, 1966–1991. In C. H. Williams (ed.) *The Political Geography of the New World Order*. London: Belhaven/Wiley, 148–80.
Ó Riagain, D. 1989. The EBLUL: its role in creating a Europe united in diversity. In T. Veiter (ed.) *Federalisme, regionalisme et droit des groupes ethnique en Europe*. Vienna: Braümuller.
Ó Tuathail, G. 1993. Japan as threat: geo-economic discourses on the USA-Japan relationship in US civil society, 1987–1991. In C. H. Williams (ed.) *The Political Geography of the New World Order*. London: Belhaven/Wiley, 181–209.
Orridge, A. W. and Williams, C. H. 1982. Autonomist nationalism: A theoretical framework for spatial variations in its genesis and development. *Political Geography Quarterly*, 1, 19–39.
Smith, A. D. 1981. *The Ethnic Revivial in the Modern World*. Cambridge: Cambridge University Press.
Smith, A. D. 1983. *Theories of Nationalism*. London: Duckworth.
Smith, A. D. 2000. *The Nation in History*. Cambridge: Polity.
Smith, G. (ed.). 1996a. *The Baltic States*. Basingstoke: Macmillan.
Smith, G. (ed.). 1996b. *The Nationalities Question in the Post-Soviet States*. London: Longman.
Smith, G., Law, V., Wilson, A., Bohr, A., and Allworth, E. (eds.). 1998. *Nation-building in the Post-Soviet Borderlands*. Cambridge: Cambridge University Press.
Tägil, S. (ed.). 1999. *Regions in Central Europe*. London: Hurst.
Urwin, D. 1982. Territorial structures and political developments in the United Kingdom. In S. Rokkan and D. Urwin (eds.) *The Politics of Territorial Identity*. London: Sage.
Williams, C. H. 1988. Minority nationalist historiography. In R. J. Johnston et al. (eds.) *Nationalism, Self-determination and Political Geography*. London: Croom Helm, 203–21.
Williams, C. H. 1993a. Towards a New World Order: European and American perspectives. In C. H. Williams (ed.) *The Political Geography of the New World Order*. London: Belhaven/Wiley, 1–19.

Williams, C. H. 1993b. The European Community's lesser used languages. *Rivista Geografica Italiana*, 100, 531–64.
Williams, C. H. 1994. *Called unto Liberty: On Language and Nationalism*. Clevedon, Avon: Multilingual Matters.
Williams, C. H. 1997. European regionalism and the search for new representational spaces. *Annales: Anali za istrske in mediteranske študje*, 14, 265–74.
Williams, C. H. (ed.). 2000. *Language Revitalization: Policy and Planning in Wales*. Cardiff: The University of Wales Press.
Zelinsky, W. 1988. *Nation into State: The Shifting Symbolic Foundations of American Nationalism*. Chapel Hill, NC: University of North Carolina Press.
Zhao, S. 1997. Chinese Intellectuals' Quest for National Greatness and Nationalistic Writing in the 1990s. *The China Quarterly*, 152, 725–45.
Zhao, S. 2000. Chinese nationalism and its international orientations. *Political Science Quarterly*, 115(1), 1–33.

Chapter 24

Fundamentalist and Nationalist Religious Movements

R. Scott Appleby

Despite religion's capacity for inspiring nonviolent social change, as well as forgiveness and reconciliation among divided peoples, extremist modes of religion have wielded unprecedented political influence around the world since the 1970s. Extremist religious actors view the defeat and, where possible, the elimination of their religious and political enemies as a sacred duty in the quest for justice (Liebmann, 1983). The power of religious movements willing to employ violence to purify the religiopolitical community and to defeat political competitors was demonstrated by the 1978–9 Islamic revolution in Iran; the dual emergence of mirror-image radical Jewish and Islamic movements in Israel and Palestine; the birth and growth of Islamist political parties and radical cells in northern Africa, the Middle East and South Asia; and the often violent campaigns of Hindu nationalists, Sikh extremists, Christian and Hindu Tamils, and Sinhalese Buddhist nationalists in India and Sri Lanka. The rise of the New Christian Right in the USA exemplified political religion's aspirations to cultural hegemony where political dominance was impossible. Indeed, it is noteworthy that many extremist religious movements have developed a pragmatic mode, often expressed through the establishment of a political party or lobbying group, that ensures both their longer-term presence on the national political landscape and the tempering of their exclusivist religious zeal (see discussion in Appleby, 2000).

Extremist Movements: Sacralizing Religion, Land, and Nation

In November 1993 the *Ichud Rabbanim L'Ma'an Am Yisrael V'Eretz Yisrael* (Union of Rabbis for the People and Land of Israel), a new organization of religious Zionist rabbis founded to coalesce opposition to the Oslo accords, issued a rabbinic *psak*, a binding judgment based on halakha, which declared the accords null and void, without legal or moral force. According to the laws of the Torah, they judged, "it is forbidden to relinquish the political rights of sovereignty and national ownership over any part of historic Eretz Yisrael to another authority or people." The rabbis warned of a loss of Jewish identity, as a result of the government's policies: "We are extremely concerned over the present trend that aims to create a secular culture here which is to blend into 'a new Middle East'—a trend which will lead to assimilation. We have a sacred obligation to strengthen and deepen our people's connection to the Torah and to Jewish tradition as passed down through the generations.... We support the continuation of protests, demonstrations, and strikes within the framework of the law. In addition, we encourage educating and informing the masses in order that they may realize the falseness of this 'peace'"(quoted in Heilman, 1987, p. 381). Fighting for the disputed territories occupied by Israel after the Six Day War of 1967, they ruled, is required by "*pikuach nefesh*," a situation threatening Jewish existence.

According to halakha, virtually all laws may be suspended, all actions taken when *pikuach nefesh* exists. This "emergency" rationale for extra-legal activism is not unique to Jewish extremism, however; in the last years of his life, the Ayatollah Khomeini, supreme ruler of the Islamic republic of Iran, famously suspended Shari'a (Islamic law) and concentrated power in his own hands – for the sake of preserving Islamic society, he explained, in a time of economic and social crisis. The cloaking of political power-plays in religious rationales, while hardly a new phenomenon, has become a standard feature of political cultures where religion remains a vibrant cultural force. In their opposition to the Oslo accords, the Zionist rabbis typified this trend, Samuel Heilman notes, when they "wrapped their political stand in the mantle of religion and a rabbinic *da'at Torah* (Torah wisdom)" (ibid., p. 347).

Alarmed by the rabbinic declaration, the Rabin Government worried that it might be interpreted by politically mobilized circles of religious Jews as a justification for insurrection. The situation worsened when Israel withdrew from Jericho and Gaza. The Ichud Rabbanim warned in May 1994 that the "false peace process" was "creating deep spiritual and social divisions... [and] an atmosphere of civil war." In July 1995 another ruling declared that the Torah should be interpreted as prohibiting the evacuation of Israeli army camps in the territories. Many of the soldiers in the camps, including officers in charge of evacuating them, had studied in yeshivas under these rabbis. They now faced a crisis of authority, with their spiritual guides on one side, and the secular government and the army on the other. Israel's President Ezer Weitzman, echoing the outrage of the general public, warned that the ruling "undermines the basic principles on which the Israeli Defense Force is based and could invalidate the democratic foundation of the state." Rabin called on the attorney general to determine whether the rabbis had breached the law and were fomenting sedition. Even the National Religious Party, the rabbis' natural political ally, was divided by a ruling that seemed designed to foment civil war; its parliament members rejected the ruling, as did the head of a prominent yeshiva that educated and formed Jewish soldiers (*Jerusalem Post* News Service, July 13, 1995).

Meanwhile, some rejectionist rabbis and their students were arguing that the withdrawals from the territories were endangering Jewish life, and that any Jew who authorized or implemented them might be defined as a *rodef* (one who threatens the life of a Jew, or puts the life of a Jew in danger) or a *moyser* (one who hands Jews over to their enemies). The law asserts that one may kill the rodef and punish the moyser. Should Rabin be classified halakhically as a *rodef*? Members of B'nai Akiva, the religious Zionist youth movement, discussed the prospect of shooting at Israeli soldiers attempting to move Jews out of settlements. In the fall of 1995, at Ramat Gan's Orthodox Bar Ilan University, students debated the question of whether Rabin deserved "to die for his deeds" (Avineir, 1995).

On November 4, 1995 Yigal Amir, a Bar Ilan student, army veteran and former "yeshiva boy," assassinated Rabin. "My whole life, I learned Jewish law," Amir commented later. His studies convinced him, he said, that killing the enemy – in this case, the prime minister – was permitted under the law. "We had to stop the [peace] process," explained Dror Adani, a B'nai Akiva member and one of the suspects in the conspiracy to murder Rabin. "Inside of me, I feared that our actions may bring about a civil war... [but] both Rabin and Peres were classified as *rodefim* and therefore had to be killed" (Adani, 1995).

This episode of Jewish extremism is difficult to categorize. One could argue that it was primarily an expression of Jewish *fundamentalism*, that is, a fight to preserve religious orthodoxy and "purity" against compromises introduced by misguided or malevolent religious or secular actors. Certainly the Union rabbis' legitimation of violence against fellow Israelis was predicated upon their "fundamentalist" intepretation of what they called "our sacred obligation to strengthen and deepen our people's connection to the Torah and to Jewish tradition as passed down through the generations."

One may also contend that this was a case of *religious nationalism*, in light of the rabbis' claim to represent "true Zionism" and thus to be the arbiters of the legal and moral warrants for Government policy. From this religious Zionist perspective, *the Jewish nation* is the focus of messianist fervor and expectation; secular conceptions as well as religious interpretations of Israel, be they Orthodox or haredi (ultra-Orthodox), are evaluated according to their plausibility within, and support, a religious nationalist ideology.

A vibrant civil society bolstered by a democratic state, Israel provides public space for observant Jews to debate the appropriate Jewish response to developments in state and society. The rulings of the Ichud Rabbanim were repudiated by many other rabbis. Yet a strong democratic society, the case of Israel also reminds us, may diminish the prospect of religious violence but certainly does not eliminate it. The rabbis of the Ichud Rabbanim made shrewd use of the freedom and public space guaranteed them within a democratic and largely secular society, in order to advance the cause of an intolerant Jewish nation ruled according to halakhic norms.

Closely related to religious nationalism is irredentism, the equation of a people's sovereign identity with a particular plot of land – in this case, the Biblical "Land of Israel, which includes "Judea and Samaria" (the occupied territories). Jewish irredentists believe the nation is sacred because it is rooted in land promised to the chosen people by Yahweh, soil sanctified by the blood of martyrs. The land *in its entirety* is to be secured and defended as the homeland of God's chosen people and the site of the messianic kingdom.

Fundamentalism, religious nationalism, religiously motivated irredentism – each of these interpretive lenses is useful in analysing religious movements that seek political power. Such terms must be employed carefully and presented as constructs that facilitate comparison among broadly similar movements, however, rather than as precise labels for any particular movement.

Sacralizing Religion: Fundamentalism

"Fundamentalism," as I use the term, refers to a pattern of religious militance by which self-styled true believers attempt to arrest the erosion of religious identity, fortify the borders of the religious community, and create viable alternatives to secular structures and processes. The majority of practitioners of the religion in question, be it Christianity, Islam, or Judaism, are not fundamentalists, however orthodox their observance or dogmatic their beliefs. Unlike most orthodox or conservative believers, fundamentalists seek political change and sometimes use coercion and violence in pursuit of their goals; typically, they compete with secular institutions and philosophies for resources and allegiances.

The penetration of the religious community by secular or religious outsiders stimulates the rise of fundamentalist movements. Western businessmen bring their luxury hotels, casinos, wine, and women to Cairo, heedless of Islamic codes and culture. Despite its seeming subversion of the traditional Christian notion of creation and divine providence, the theory of evolution receives prominent attention in school textbooks and gains intellectual respectability in North America. The Shah of Iran inaugurates and accelerates a cultural revolution apparently designed to glorify a Persian and pagan past at the expense of Shi'ism. The secular Zionist government of Israel abandons and then bulldozes the Jewish settlements of Yamit on the Sinai penninsula in order to preserve the peace agreement with Egypt. In an increasingly secularized Israel support for Jewish religious education and strict Sabbath observance erodes.

Such developments betoken the marginalization of religion, transplanted to the developing world from the secularized West. Among the religious a corresponding sense of being in exile in one's own land has given rise to groups that seek to bolster and reshape the religious community and its social environment in order to retain membership and reverse decline. Islamists from West Africa to South Asia, segments of the Christian Right, and the religious Zionists and haredim of Israel have emerged from and seek to preserve orthodox and conservative religious environments. Prepared to give their lives to the struggle for justice, some turn to fundamentalist movements when the religious establishment repeatedly fails to stem an increasingly aggressive, secular, religiously plural, materialist, amoral, feminist tide.

Devotion to the will of God and to the demands of divine law is thereby turned to extremist ends – naming the infidel, demonizing the other, expelling the lukewarm. Fundamentalists select one aspect of the law, elevate it above others, and equate its observance to the achievement of concrete political objectives. Thus, the religious Zionists of the Ichud Rabbanim and their counterparts in the settler movement Gush Emunim emphasize one of the 613 Torah *mitzvot* and subordinate the remaining 612 religious duties to its observance – "the commandment to settle the land is tantamount to all other commandments" (Aran, 1991, p. 309). This is classic "fundamentalism:" the selection and reinterpretation of politically charged doctrines or precepts, around which a political movement is built.

Similarly, the traditional rituals and devotions that valorize self-sacrifice become in extremist hands a means of preparing the devout cadres for physical warfare. The self-flagellation of Iranian or Lebanese Shi'ites during the Ashura ritual commemorating the martyrdom of Imam Husayn is a form of militant religiosity, as are the fasts, prayers, rosaries, and candlelight vigils of Christians who mount militant campaigns to protect the unborn fetus threatened by abortion. Neither of these rituals is inherently extremist; but they become sources of violence for the Ashura penitent who exacts vengeance on Husayn's contemporary persecutors, or the "pro-lifer" who guns down an abortionist. Such prescribed prayers and rituals, interpreted by an extremist preacher, locate the believer in a sacred cosmos that rewards martyrdom or imprisonment endured in a divine cause. The ability of religion to inspire ecstasy – literally, to lift the believer psychologically out of a mundane environment – stands behind the distinctive logic of such political violence. As unpredictable and illogical as this violence may seem to outsiders, it falls within a pattern of asceticism leading to the ecstasy of self-sacrifice which runs as a continuous thread through most religions.

Initially, the seeker of religious purity is tempted to build an enclave, a social and cultural system dedicated to the fortification of communal boundaries (Sivan, 1995, pp. 16–18). The state is identified as the enemy to the extent that it is perceived as the enabler or agent of the infiltration of the religious community and the family units at its core. While other military, legal, or economic policies of the state may aggravate the situation, fundamentalist movements target political figures, economic policies and social practices that weaken the religious community. The religious outsider is their *bête noire*. But fundamentalists inevitably define "outsiders" to include lukewarm, compromising, or liberal co-religionists, as well as people or institutions of another or no religious faith. Foreign troops stationed on sacred ground, missionaries, Western businessmen, their own government officials, sectarian preachers, educational and social service volunteers, relief workers, and professional peacekeepers – any or all of these might qualify at one time or another as intentional or inadvertent agents of secularization. And secularization, in turn, is seen as a ruthless, but by no means inevitable process by which traditional religions and religious concerns are gradually relegated to the remote margins of society where they can die a harmless death—eliminated by what the Iranian intellectual Jalal Al-e Ahmad called the "sweet, lethal poison" of "Westoxication."

Organizationally, fundamentalist movements form around male charismatic or authoritarian leaders.[1] The movements begin as cells and local or regional cadres, but they are increasingly capable of rapid functional and structural differentiation and of international networking with like-minded groups from the same religious tradition. They recruit rank-and-file members from professional and working classes and both genders, but draw new members disproportionately from among young, educated, unemployed or underemployed males (and, in some settings, from the universities and the military); and they impose strict codes of personal discipline, dress, diet, and other markers that set group members apart from others.

Ideologically, fundamentalists are both reactive against and interactive with secular modernity; and they tend to be absolutist, inerrantist, dualist, and apocalyptic in cognitive orientation. That is, fundamentalists see sacred truths as the foundation of all genuine knowledge, and religious values as the base and summit of morality. This in itself does not differentiate fundamentalists from traditional believers. Formed by secular modernity, or in reaction against it, however, fundamentalists present their sacred texts and traditions – their intellectual resources, so to speak – as inherently free from error and invulnerable to the searching critical methods of secular science, history, cultural studies, and literary theory. Having subordinated secular to sacred epistemology, fundamentalists feel free to engage and even develop new forms of computer and communications technology, scientific research, political organizations, and the like.

The tendency of fundamentalists to imagine the world divided into unambiguous realms of light and darkness peopled by the elect and reprobate, the pure and impure, the orthodox and the infidel, helps explain how a religious tradition that normally preaches compassion, forgiveness, and hospitality turns intolerant and violent. Set within an apocalyptic framework, the world is seen as mired in spiritual crisis, perhaps near its end, when God will bring terrible judgment upon the children of darkness. The children of light are depicted in such millenarian imaginings as the agents of divine wrath. They believe themselves to be living in a special dispensation – an unusual, extraordinary time of crisis, danger, apocalyptic doom, the advent of the Messiah, the Second Coming of Christ, the return of the Hidden Imam, etc. This "special time" is exceptional not only in the sense of being unusual; its urgency requires true believers to make exceptions, to depart from the general (nonviolent and tolerant) practices of the host religious tradition.

Thus, the religious Zionist elders of the Ichud Rabbanim invoked the halakhic norm of *pikuach nefesh* in ruling that the Oslo accords threatened the very existence of Israel – and Judaism itself. Thus, the Sikh extremist Jarnail Singh Bhindranwale – likewise claiming that the Sikh faith was under mortal threat in the early 1980s – added the motorcycle and revolver to the traditional Sikh symbols, and recruited a ferocious minority of his fellow Sikhs into an updated version of the seventeenth-century order of baptized Sikhs, the Khalsa Singh (the "purified" or "chosen" lion race).[2]

Fundamentalist movements choose to battle secular modernity on its own turf. Their weapons thus include radio, television, audio cassettes, faxes, the Internet, Stinger missiles, black markets, think tanks, paleontological "evidence" for the young earth theory, identity politics, modern marketing techniques, terrorist tactics. With success (or with failure for lack of clout) comes an expansion of the agenda to include the attainment of greater political power and the transformation of the surrounding political culture, the moral purification of society and, in some cases, secession from the secular state and/or the creation of a "pure" religious homeland.

By any reckoning, Islam produces more contemporary fundamentalist movements than any other religious tradition. The phenomenon is over-determined. Mass media has increased popular awareness of the social, economic, and political inequalities and injustices which abound in many Muslim societies, and the corruption and mismanagement which bedevils their governments and state-run institutions. The growing popular sense of "relative deprivation" compared with other societies coincided with exhaustion and disgust at the string of failed secular or liberal "solutions," from Arab nationalism to Islamic socialism. Islamists blame the failures (including vulnerability to the Western colonial powers and, especially, military defeat at the hands of the Israelis) on the abandonment of Islam as the basis for the ordering of society. The glorious Islamic empires and civilizations of the medieval past serve as precedents. Formed well before the rise of the modern secular nation-state, Islam's political imagination conceives of a transnational religious community of believing Muslims as the fundamental political entity.

In addition, Islam has been remarkably resistant to the differentiation and privatization processes accompanying secularization. Islam has not undergone a Reformation like the one experienced by Christianity, which led to a pronounced differentiation of sacred and secular, religious and political spheres. In general, religion remains prior and privileged in Islamic societies. Islamism, or Islamic fundamentalism, is in part a nervous reaction to the de-Islamization of the political sphere (see discussion in Hibbard and Little, 1997, pp. 3–28 and 107–13).

Not all of the Islamist movements are extremist, however; the Muslim Brotherhood organizations in Egypt and Jordan, for example, have renounced violence and pursue power through strictly political means. Equally ill-founded is the notion that Islam somehow constitutes a monolithic "civilization" that spans and effaces the particularities of local cultures. Muslims living in diverse cultural and social contexts produce significantly different political expressions; both the Muslims of South Asia and the Muslims of Turkey, for example, have different political cultures than Arab Muslims. In Turkey, perhaps the most secularized Muslim nation, the pro-Islamist Virtue Party runs a distant third in Parliament behind the Nationalist Action Party and the Democratic Left Party. During its brief period in power, its Islamist predecessor, the Welfare (Refah) Party, was faced repeatedly with charges that it was supported by "radical Islamic circles" bent on "fundamentalizing" the school system and thereby Islamizing a reluctant nation. The all-powerful military suspended the party in 1996.[3]

In Pakistan, the Jamaat-i-Islami today is a far cry from the organization envisioned by its founder, Maulana Sayyid Abul Ala Maududi (1903–79). Maududi, a native of Hyderabad, India, was a brilliant systematic thinker, a prolific writer, a charismatic orator, and a shrewd politician. He single-handedly created the modern Islamist discourse; his works elaborate the social and legal implications of concepts such as "Islamic politics," "Islamic economics," and "the Islamic constitution." Maududi's core concept, based on the traditional affirmation of Islam as a complete and comprehensive way of life, was *iqamat-i-deen* (literally, "the establishment of religion") – the total subordination of the institutions of civil society and the state to the authority of divine law as revealed in the Qur'an and practiced by the Prophet. By the 1980s, however, the Jamaat-i-Islami had become less a movement and more an entrenched institution, a lobbying group for Islamist causes (Ahmad, 1991, pp. 464–6).

Other Islamist movements have maintained their militias even while developing a political party to contest elections. In Lebanon, Hizbullah, the infamous "Party of God" formed in 1982 after a series of traumatic external events shook the Shi'ite community (i.e. Israel occupied the Shi'ite south, Maronite militiamen in league with Israel massacred Palestinian refugees, and American and French troops were deployed near the Shi'ite slums of Beirut).[4] Young Lebanese religious scholars, who had been expelled from theological academies in Iraq and spurned by the Shi'ite clerical establishment, became Hizbullah's leaders. Prosperous Shi'ite clans of the Bekaa Valley, whose wealth came from illicit drug trafficking, supported the movement; and young Shi'ite militiamen who had worked for Palestinian organizations found jobs, weapons, and a sense of divine purpose within its ranks. Hizbullah also won followers among the Lebanese populace, especially the Shi'ite refugees in the southern suburbs of Beirut. "Iran's emissaries moved quickly to offer food, jobs, loans, medicine, and other services to the teeming masses of impoverished Shi'ites in Beirut's slums," Martin Kramer writes. "In return, they gave Hizbullah their loyalty" (1993, p. 554).

Shaykh Muhammad Hussayn Fadlallah, spiritual leader of Hizbullah, was not a religious nationalist: Islam transcended the nation-state. The aim was not the establishment of Islam in one country, but the creation of an "all-encompassing Islamic state" embodying the *umma* (worldwide Islamic community). An apocalyptic messianism animated Fadlallah's vision of a sweeping global triumph of Islam. "The divine state of justice realized on part of this earth will not remain confined within its geographic borders," one of his disciples predicted. That achievement "will lead to the appearance of the Mahdi, who will create the state of Islam on earth." To Hizbullah Fadlallah assigned the heroic role of purifying a province of Islam to create "the divine state of justice" (1986, pp. 4–13).

Thus began Hizbullah's campaign of political violence, embraced by thousands of Lebanese Shi'ites as a noble endeavor that transcended the boundaries of family, clan, sect, and state. By joining Hizbullah, Kramer explains, the poor village boy became a true Muslim, a member of a religiopolitical community spanning three continents, and a soldier in a global movement led by the Imam Khomeini for redressing the imbalance between Islam and infidelity (Kramer, 1993, p.546).

Although Muslims did not have the same weapons of war as their enemies, Fadlallah admitted, power did not reside only in quantitative advantage or physical force, but in strikes, demonstrations, civil disobedience, preaching – and in the disciplined use of violence. Hizbullah violence in the 1980s included campaigns to rid the Shi'ite regions of Lebanon of all foreign presence, operations to free Hizbullah members imprisoned by enemy governments in the Middle East and Europe, and battles against rival movements for control of neighborhoods in Beirut and villages in the South. The campaigns against foreigners included assassinations and massive bombings committed by "self-martyrs" who destroyed the command facilities of Israeli forces in the occupied South in 1982 and 1983, exploded bombs in the barracks of American and French peacekeeping troops in 1983, and demolished the American embassy and its annex in two separate attacks in 1983 and 1984. Hundreds of foreigners died in these suicide missions, including 241 US Marines. Eventually, American and French forces retreated from Lebanon in the face of "Muslims who loved martyrdom," as one Hizbullah leader put it (ibid., p. 547).

Hizbullah employed violence for a distinct political purpose – to bring the movement greater power to implement and enforce religious law. But political power also came by way of Hizbullah's social service work among the suffering poor of southern Lebanon. In 1989 the Taif accords ended Lebanon's 16-year civil war; the settlement attempted to maintain a balance among Lebanon's four main confessional groups. In response to Taif, Hizbullah decided to make itself more Lebanese and shed many of its Iranian trappings in order to enter mainstream politics. The movement won eight Parliamentary seats in the 1992 elections and held its own in the 1996 elections. By the mid-nineties, the social service wing of the movement had evolved into a bona fide political party, notwithstanding the fact that by 1996 its former coalition partners, Amal and the Progressive Socialist Party, were calling

Hizbullah "extremist" and accusing it of "damaging the nation's welfare" through its "exclusivist resistance to Israel in the south" (*The Economist*, 1996).

Islamists in Lebanon and elsewhere push for the full implementation of Islam's normative legal system; they chafe at Shari'a's lack of influence over governmental behavior in many purportedly Islamic nations. Achievement of the measure of uniformity desired by the Islamists is unlikely, however. In Islamic Iran, after almost two decades of governmental efforts to project a comprehensive and stable Islamic identity, most Iranians would acknowledge that the contents of such an identity remain ambiguous. Even deeply committed Islamist groups can adopt political positions poles apart; for compelling evidence one need only review the broad spectrum of Islamist reaction to the Gulf Crisis of 1990–1 (see essays in Piscatori, 1991).

Although consistent coordination of purpose and program has eluded them, however, Islamist movements increasingly cooperate with one another, transcending national boundaries and the traditional Sunni–Shi'ite divide (e.g. Iranian Revolutionary Guards have trained both Sudanese militia and Hizbullah extremists). Individual movements exercise disproportionate influence in their respective societies through the manipulation of modern technology and mass communications. In the absence of alternative popular channels for effectively battling social injustice and expressing discontent, extremist movements use religious outlets to mobilize segments of the population.

Indeed, extremist Islamist movements have multiplied in recent years. Prominent among these are the Taliban, the Sunni and Pashtun movement that ruled 97 percent of Afghanistan by late 2000 (and sought recognition of its Government by the United Nations); the Harkat Mujahedeen of Pakistan; the Armed Islamic Group [GIA] of Algeria, which has waged a terrorist campaign against the Government of Algeria since 1992; and the terrorist networks sponsored by Osama bin Laden, the wealthy Saudi exile accused of masterminding the 1998 bombing of the US embassies in Kenya and Tanzania and the 9.11.01 attacks in New York and Washington.

Sacralizing the Nation: Religious Nationalism

Religious actors may identify their tradition so closely with the fate of a people or a nation that they perceive a threat to either as a threat to the sacred. In this indirect sense, such religious actors are concerned, like fundamentalists, with the marginalization of religion. Ethnoreligious nationalists may demonize their enemies, consider their own religious sources inerrant and their religious knowledge infallible, and interpret the crisis at hand as a decisive moment in the history of the faith – a time when exceptional acts are not only allowed but required of the true believer. But enemies, collaborators, motivations, objectives, and timing and use of violence are different when religion is subordinated to and placed at the service of ethnic or nationalist forces.

While they are intensely concerned with the ills of the larger society, its corrupt morality and ungodly political culture, fundamentalists are convinced that its transformation will come only at the hands of the purified religion. By contrast religious nationalists feel that the most direct route to purifying or strengthening the host religion is the establishment of a political collective within which the religion is privileged and its enemies disadvantaged.

This dynamic shaped Serbian Orthodox extremists' alliance with Slobodan Milosevic on the eve of the Bosnian war of 1991–5. Neither the Serbian president nor his ultranationalist political allies were motivated by religious concerns, but they shrewdly tapped fanatical religious tendencies within the Serbian Orthodox community. Yet this was a case not of fundamentalism, but of the manipulation of "folk religion" to construct ethno-national legitimations for violence. The seeds of Serbian Orthodox (as well as Croatian Catholic) religiosity were not stamped out during the period of communist rule of Yugoslavia, even among the so-called secularized masses; but neither were they nurtured. Scattered and left untended, they were eventually planted in the crude soil of ethno-nationalism, ultimately coming to terrible fruition in the Bosnian genocide (Sells, 1996, p. 25).

South Asia demonstrates a quite different pattern of religion turned to ultranationalist causes. The nation-building project of the postcolonial era provided opportunities for some communal groups in the region to monopolize the state apparatus and to dominate, incorporate, or diminish other groups. Communalism attracts both majorities and minorities, elites and masses, who complain that the post-Independence secular order has left them "victimized" and grasping for their share of educational opportunities, capital assets, occupational training, and jobs. The proliferation of communal politics, in turn, has arisen out of the political arithmetic of majority rule: the competition for resources and benefits requires the formation of coalitions of "ethnic concerns and interests acting as a monolithic principle, vertically integrating a people differentiated by class" (Tambiah, 1996, p. 12). In such settings religion can become a powerful means of binding together racial, linguistic, class and territorial markers of identity – if the religion in question cultivates its boundary-setting capacities and sharpens its discriminatory edge.

Majoritarian dominance, experienced as exploitation and oppression by ethno-religious minorities, lies at the root of the conflicts in India and Sri Lanka, for example.[5] Allied with secular politicians, the Hindutva movement in India seeks to establish a representative structure resembling the secular nation-state but pursuing a policy of civic intolerance toward "outsiders" (i.e. non-Hindus). As the host religion for the nationalist movement flying the saffron flag of Hindustan, the imaginary Hindu nation, Hinduism is a weak vessel for religious fundamentalism. It lacks a strong historical sense of itself as an organized religion, with a body of revealed religious law and a concept of God acting dramatically within history to bring it to a definitive conclusion. Perhaps for the same reasons Hinduism lends itself to the cause of nationalist movements constructed around the fluid categories of "religion" and "ethnicity" and drawing on a mix of secular and religious symbols and concepts, religious and nonreligious actors.[6]

The banner of Hindu nationalism is carried by three organizations: the Rashtriya Svayamsevak Sangh (RSS) (National Union of Volunteers), the Vishwa Hindu Parishad (VHP) (World Hindu Society), and the Bharatiya Janata Party (BJP) (Indian People's Party). These groups are descended from the Hindu Mahasabha (Hindu Great Council), founded in 1915 in reaction to the formation of the Muslim League. V. D. Savarkar, the leader of the Hindu Mahasabha and author of the book *Hindutva*, formulated the doctrinal basis and ideological tenets of Hindu nationalism around the notion of Hindu racial, cultural, and religious superiority. In 1925 Keshav Baliram Hedgewar, a Maharashtrian Brahman, founded the RSS in response, he said, to the ineffectiveness of Gandhi's tactics of nonviolence in the face of what Hedgewar described as India's long history of domination and exploitation by the Muslims and the British. Hedgewar and his successor, M. S. Golwarkar, led the organization from its founding until 1973; under their tutelage the RSS became a highly organized brotherhood established through a network of local paramilitary cadres called *shakhas*. The young RSS recruits – more than two million strong by the early 1990s – submit to demanding schedules of indoctrination and physical training. They wear saffron-colored uniforms, conduct military drills at sunrise, and undergo intensive training in forest encampments. Called *svayamsevaks*, by the early nineties the recruits were organized in twenty-five thousand shakhas, in some eighteen thousand urban and rural centers across the country. The activities of these groups were supervised by three thousand professional organizers – primarily celibate young men.[7]

Founded in 1964 at the initiative of the RSS, the VHP is a cultural organization led by seasoned officials of the RSS. Through the staging of huge religious processions designed to arouse popular fervor for "Hindu causes" and to intimidate Muslims and other "outsiders," the VHP promotes Hindu revival in the remote corners of India and among the Hindu diaspora overseas. It boasts three hundred district units and some three thousand branches throughout India. Outside of India it claims to have several thousand branches in twenty-three countries. In 1994 the VHP reported more than 100,000 members, with 300 full-time workers, each dedicated to re-affirming "Hindu values." By sketching a broad and somewhat vague definition of "Hindu values," the VHP seeks to transcend internal differences among Hindus, to bring secularized Indians back to the fold, and to reclaim the Untouchables to Hinduism. The VHP strategy is to propagate a coherent modern version of Hinduism as the national religion of India. Thus, it downplays local differences in Hindu religious doctrine and represents Hinduism, ahistorically, as a single all-embracing ethnonational religious community including Jains, Buddhists, and Sikhs.

The Bharatiya Janata Party emerged in 1980 out of the Janata coalition that displaced the Indira Gandhi regime in 1977. Most of the BJP leaders were formed in the RSS, but as a political party contesting nationwide elections, the BJP has attempted to appeal broadly to all Indians, including Sikhs and Muslims. In the 1996 national elections the BJP won the largest bloc of seats (160) in the Lok Sabha (the lower house of Parliament), thereby helping to topple the Congress Party from power for only the second time in the 49 years of Indian statehood. Atal Bihari Vajpayee, the head of the BJP, served as Prime Minister of India during the two weeks in May 1996 in which the party attempted unsuccessfully to form a minority government to rule the nation's 930 million people (Burns, 1996, p. A4). In March 1998 he became Prime Minister a second time after the BJP successfully contested the February 1998 elections, winning 178 of the 543 elected seats in the Lok Sabha. Again, Vajpayee was forced to form a coalition government that included regional parties opposed to the religious nationalists' doctrine of Hindu supremacy.[8]

The Hindu nationalist movements employ an ethno-nationalist ideology that employs the rhetoric and imagery of blood, soil, and birth (see Embree, 1990; Frykenberg, 1994). Like "Hinduism" itself, Hindu nationalism is clearly a construct.

It has borrowed from the Abrahamic traditions both an eschatology of ultimate destiny (with the Hindu nation depicted as the realization of the mythical Kingdom of the Lord Ram) and the notion of the elect, righteous ones (applied generally to the Aryan race and specifically to the celibate and highly disciplined staff of the RSS and VHP).

The inflammatory and diffuse appeals to "Hindu national pride" in the face of perceived Muslim encroachments have produced a great deal of uncontrolled mob violence. Hindu nationalists calculate the potential advantages of such violence, including the opportunity a crisis situation presents for recruiting and mobilizing young men (Gold, 1991). They seek platforms for disseminating their ideology and create "events" that publicize their cause. Characteristically they redefine sacred land and sacred space in a controversial way, using the mass media coverage of their activism as a means of grabbing attention and mobilizing followers. In this they resemble many of the movements profiled in this essay.

Conclusion: Sacralizing the Land for Religion and Nation

Whether fundamentalist or religious nationalist in character, extremist religious movements base their political and territorial claims upon a prior claim to holy ground, sacred space. More often than not, holy land is contested land, sanctified by the blood of martyrs or warriors who sacrificed their lives in its defense. Citing historical or scriptural precedent, ethnic and religious groups compete for hegemony over the same square footage.

In the former Yugoslavia, it was the province of Kosovo, site of a decisive battle between Serbs and Turks in 1389 – and of Milosevic's appearance at a massive rally in 1989, a ritualized performance calculated to stoke the flames of religious hatred against the "Turk" (Bosnian Muslims). In India it was a plot of land in Ayodhya, India, the site of the legendary birthplace of the Hindu god Ram, which was also the site of the oldest Islamic mosque in India – until young Hindu militants destroyed it in December 1992, following a series of fiery speeches given by RSS leaders.[9] For a hundred years, the site had produced no more than a local contestation over facts and proprietorship. But it took on a national focus in 1985 when the VHP announced its intention to have Hindus conduct regular *puja* (worship) in the mosque. After this goal was achieved in February 1986, the VHP demanded that the mosque be replaced with a temple.[10] Such rhetoric led not only to the destruction of the Babri Masjid, but also to some of the bloodiest pogroms against India's Muslims since independence.

In the case of Jerusalem, the Temple Mount remains in principle the most sacred site in Judaism, almost two millennia after the Romans razed the Jewish Temple itself in the year AD70. In practice, Jews for centuries have come to worship at the base of one of King Herod's retaining structures, the Western Wall, also known as the Wailing Wall, in reference to Jews mourning for ruined sanctuary. Yet the same plot of land, known to Muslims as Al-Haram Al-Sharif, is the third holiest site in Islam – home to the Al-Aksa mosque and the shrine known as the Dome of the Rock, whence the Prophet Muhammad made his "night journey" to heaven. The Wailing Wall, the supporting wall of Temple Mount/Haram Al-Sharif, is positioned directly beneath the Islamic shrines (Gorenberg, 2000, pp. 11–12). In September 2000, Ariel

Sharon, leader of Israel's opposition Likud bloc, accompanied by hundreds of armed police, visited Temple Mount in what appeared to be a defiant assertion of Israel's sovereignty over the sacred sites. If provocation was Sharon's goal, he was distressingly successful. Months of renewed violence halted negotiations on the status of Jerusalem and all but shattered the Oslo accords.

The prominence of extremist religious movements in world affairs has led some commentators to declaim "the new tribalism," to predict "a clash of civilizations," or to declare Islam the new enemy of the West, replacing the "evil empire" of the Soviet Union (Huntington, 1993). Such sweeping judgments trade in hyperbole and conveniently overlook the fact that radical or extremist religious movements, even those rooted in transnational "world" religions such as Christianity or Islam, are fundamentally local in character. Radical Islam, it is true, has come to power in Iran, Sudan, and Afghanistan, and has threatened the regimes in Egypt, Algeria, and Pakistan. Hindu nationalism, as mentioned, exercises considerable influence on the Indian state and its current policies. But religious extremism must be analysed and countered not as a civilizational or transnational force, but on a case-by-case basis. The framework for such analysis is an understanding of the distinction between the two major contemporary forms of religio-political extremism – "fundamentalism" and "ethno-religious nationalism" – and the various ways they exploit the profound human attachment to place.

ENDNOTES

1. The description of fundamentalism's ideological and organizational properties is taken from Almond et al. (1995), pp. 402–9.
2. The traditional Sikh teaching on outsiders is summarized in the legend attributed to Guru Nanak: "Take up arms that will harm no one; let your coat of mail be understanding; convert your enemies into friends; fight with valor, but with no weapon but the word of God." Yet Bhindranwale argued, in effect, that exceptional times require extreme measures. "The Hindu imperialist rulers of New Delhi" seek to annihilate the Sikh people, he warned; thus he retrieved Guru Hargobind's doctrine of temporal power wedded to spiritual authority (*miri-piri*) and Guru Gobind's concept of righteous war (*dharma yuddha*). He defended this move by citing Gobind's maxim that "when all else fails, it is righteous to lift the sword in one's hand and fight." In effect, Bhindranwale announced a new dispensation: "For every village you should keep one motorcycle, three young baptized Sikhs and three revolvers. These are not meant for killing innocent people. For a Sikh to have arms and kill an innocent person is a serious sin. But Kahlsaji [O, baptized Sikh], to have arms and not to get your legitimate rights is an even bigger sin." Quoted in Tully and Jacob (1985); see also Juergensmeyer (1988).
3. In early 1996, the two non-Arab parliamentary democracies in Muslim lands led by women prime ministers – Pakistan and Bangladesh – alone had a combined population of approximately two hundred and twenty million, significantly larger than that of the Arab world. Turkey, a parliamentary democracy until recently also led by a woman prime minister, has a population of sixty million.
4. On the emergence of Hizbullah and the role of Iran, see Ramazani (1986, pp. 175–95) and Norton (1990, pp. 116–37).
5. On Sri Lanka, see the essays on Buddhist justifications of ethnic chauvinism in Smith (1978).

6. Larson (1995) enumerates common features of communal violence in India. Also see Juergensmeyer (1993)
7. "The parallels with Hitler Youth and the Young Communist League are obvious," writes Gabriel Almond. He summarizes the history of the Hindutva organizations in Almond et al. (1995).
8. Vajpayee, a high-caste Brahmin, former Marxist and member of the RSS, was perceived as a moderate who had led his party's turn to more inclusive policies and language. His public discourse usually employed the ambiguous rhetoric of a veteran politician but occasionally projected an aggressive religious nationalism that appealed to his militant Hindu followers. On the one hand, he condemned Gandhi's assassination by a Hindu nationalist as a "terrible crime," and criticized the December 1992 destruction of the Babri mosque, the oldest Muslim shrine in India, by a Hindu mob as a "blunder of Himalayan proportions." At political rallies he made it clear that discrimination on the basis of religion "is not our way in India . . . not in our blood, or in our soil." On the other hand, he often described India as essentially a Hindu nation that should enshrine "Hindu culture" at its core (Burns, 1998).
9. The VHP–BJP–RSS claimed that the site of the mosque is the exact spot where the Lord Ram was born; in commemoration of his birth, a temple called Ramjanambhoomi once stood at the site. For a history of the shrine see Srivastra (1991).
10. In order to rally public support for the cause, Advani, then president of the BJP, launched his infamous *rath yatra* (pilgrimage by chariot) on September 25, 1990. He announced that he would journey from Somnath in the province of Gujarat to Ayodhya in the state of Uttar Pradesh, covering a distance of some 10,000 kilometers. His arrival in Ayodhya on October 30 was to coincide with the construction of the proposed Ramjanambhoomi temple. Before Advani could finish his pilgrimage, the Bihar state government arrested him.

BIBLIOGRAPHY

Ahmad, M. 1991. Islamic Fundamentalism in South Asia: the Jamaat-i-Islami and the Tablighi Jamaat. In M. E. Marty and R. S. Appleby (eds.) *Fundamentalisms Observed*. Chicago: University of Chicago Press, 464–6.

Allah, Ayatollah M. H. F. 1986. Islam and Violence in Political Reality. *Middle East Insight*, 4(4–5), 4–13.

Almond, G. A., Sivan, E., and Appleby, R. S. 1995. Fundamentalism: genus and species and examining the cases. In M. E. Marty and R. S. Appleby (eds.) *Fundamentalisms Comprehended*. Chicago: University of Chicago Press, 402–9 and 464–9.

Appleby, R. S. 2000. *The Ambivalence of the Sacred: Religion, Violence and Reconciliation*. Lanham, MD: Rowman & Littlefield, chapter 3.

Aran, G. 1991. Jewish Zionist fundamentalism: the Bloc of the Faithful in Israel (Gush Emunim). In M. E. Marty and R. S. Appleby (eds.) *Fundamentalisms Observed*. Chicago: University of Chicago Press, 309.

Augustus, R. N. 1990. Lebanon: the internal conflict and the Iranian connection. In J. L. Esposito (ed.) *The Iranian Revolution: Its Global Impact*. Miami: Florida International University Press.

Avineir, S. 1995. Letter to Edmond I. Esq., 20 December 1995. Quoted in Heilman (1997), 354.

Burns, J. F. 1996. Debate Spells Likely Defeat for India's Hindu Militants. *The New York Times*, 5.28.1996, A4.

Burns, J. F. 1998. Sworn in as India's leader, ambiguity in his wake: Atal Bihari Vajpayee. *The New York Times*, 3.20.1998, A3.

Embree, A. 1990a. The Function of the Rashtriya Swayamsevak Sangh: to define the Hindu Nation. In M. E. Marty and R. S. Appleby (eds.) *Accounting for Fundamentalisms*. Chicago: University of Chicago Press, 641.

Embree, A. 1990b. *Utopias in Conflict: Religion and Nationalism in Modern India*. Berkeley, CA: University of California Press, 130.

Frykenberg, R. E. 1994. Accounting for Fundamentalisms in South Asia: ideologies and institutions in historical perspective. In M. E. Marty and R. S. Appleby (eds.) *Accounting for Fundamentalisms*. Chicago: University of Chicago Press, 601–2.

Gold, D. 1991. Rational action and uncontrolled violence: explaining Hindu communalism. *Journal of Religion*, 21, 357–70.

Gorenberg, G. 2000. *The End of Days: Fundamentalism and the Struggle for the Temple Mount*. New York: The Free Press, 11–12.

Heilman, S. 1987. Guides of the Faithful: contemporary religious Zionist rabbis. In R. S. Appleby (ed.) *Spokesmen for the Despised: Fundamentalist Leaders of the Middle East*. Chicago: University of Chicago Press.

Hibbard, S. W. and Little, D. 1997. *Islamic Activism and U.S. Foreign Policy*. Washington, DC: United States Institute of Peace Press, 3–28, 107–13.

Huntington, S. P. 1993. The clash of civilizations? *Foreign Affairs*, 72(3), 22–49.

Jerusalem Post. 1995. News Service, 7.13.1995.

Juergensmeyer, M. 1988. The logic of religious violence: the case of the Punjab. *Contributions to Indian Sociology*, 22(1), 70.

Juergensmeyer, M. 1993. *The New Cold War? Religious Nationalism Confronts the Secular State*. Berkeley, CA: University of California Press.

Kramer, M. 1993. Hizbullah: the calculus of Jihad. In M. E. Marty and R. S. Appleby (eds.) *Fundamentalisms and the State: Remaking Polities, Economies and Militance*. Chicago: University of Chicago Press, 544.

Larson, G. J. 1995. *India's Agony Over Religion*. Albany, NY: State University of New York Press, 274.

Liebman, C. 1983. Extremism as a religious norm. *Journal for the Scientific Study of Religion*, 22(1), 75–86.

Piscatori, J. (ed.). 1991. *Islamic Fundamentalisms and the Gulf Crisis*. Chicago: American Academy of Arts and Sciences.

Ramazani, R. K. 1986. *Revolutionary Iran: Challenge and Response in the Middle East*. Baltimore, MD: Johns Hopkins University Press.

Sells, M. A. 1996. *The Bridge Betrayed: Religions and Genocide in Bosnia*. Berkeley, CA: University of California Press, 25.

Sivan, E. 1995. The enclave culture. In M. E. Marty and R. S. Appleby (eds.) *Fundamentalisms Comprehended*. Chicago: University of Chicago Press, 16–18.

Smith, B. L. (ed.). 1978. *Religion and Legitimation of Power in Sri Lanka*. Chambersburg, PA: Anima.

Srivastava, S. 1991. *The Disputed Mosque*. New Delhi: Vistaar.

Tambiah, S. J. 1996. *Leveling Crowds: Ethnonationalist Conflicts and Collective Violence in South Asia*. Berkeley, CA: University of California Press, 12.

The Economist. 1996. Hizbullah in politics, 340(7982), 7.9.96, 38.

Time Magazine. 1995. Killing for God, 146(23), 12.4.95, 27.

Tully, M. and Jacob, S. 1985. *Amritsar: Mrs. Gandhi's Last Battle*. New Delhi: Rupa, 114.

Chapter 25

Rights and Citizenship

Eleonore Kofman

In the past two decades, there has been a revival of interest in citizenship in Western societies. In the 1990s citizens of Eastern European states and South Africa began to exercise their rights. The transition from military regimes in Latin America has also opened up the practice of democracy and citizenship. And in societies, such as those in Asia (Davidson and Weekley, 1999), which had previously rejected Western liberal democracy, debates were conducted about the relevance of the idea of citizenship. Furthermore, in a world of greater economic, social and cultural interaction, and migratory movements, the deepening and extension of human rights in conjunction with citizenship seems more than ever desirable. It has raised issues concerning the protection of migrants by international conventions as opposed to national governments and the availability of dual citizenship in a world where individuals and families increasingly move between two or more states. The vulnerability of undocumented immigrants and the attempt to withdraw social rights from them and their children, as in Proposition 187 passed in 1994 in California, has caused concern.

The study of citizenship had mainly been pursued by sociologists and political theorists, but geographers began to direct their attention to the issue of rights and citizenship in the late 1980s as the potential basis of a human geography for the new times (Smith, 1989). Citizenship would steer a middle course between the New Right and Old Left and incorporate other forms of social divisions than class, such as "race" and gender. The "new post-Fordist" times meant that other scales of spatial organization than the nation-state had to be taken into account in understanding the uneven entitlement to rights. Smith emphasized the locality as an important site for political mobilization and negotiation. Several special issues, such as *Environment and Planning A* (1994) and *Political Geography* (1995) examined new areas of citizenship and the ways in which spaces at different scales were used to make claims and extend rights. An emphasis on the locality in terms of struggles against restructuring and expansion of rights has been notable in geographical contributions to citizenship (Brown, 1994; Miller, 1995; Pincetl, 1994; Staeheli, 1994).

A number of economic, social, and political reasons can be put forward for the resurgence in the interest in citizenship. First, in states such as the USA and the UK, the conservative governments of the 1980s implemented neo-liberal programmes that undermined the role of the state in favor of private provision (Hall and Held, 1988) and sought to redraw the boundaries between the state and civil society in favor of the active citizen who would take more responsibility for their life (Kearns, 1992). Secondly, it stemmed from critiques of older models of political participation which were no longer seen to be effective in including large sections of the population, such as the working class and migrants. The call for a new and more inclusive citizenship was particularly marked in societies anchored in a civic republican tradition (see below), such as France (Bouamama et al., 1992; Jennings, 2000).

Thirdly, the social movements of the 1960s had been emancipatory, calling for recognition of their differences and needs, but by the 1980s many of them, such as feminists and sexual dissidents, were demanding fuller and more rapid inclusion within society (Weeks, 1998). Fourthly, feminist political theorists (Pateman, 1988) had drawn attention to the historical construction of the public/private dichotomy and consequently the difficulties women encountered in making their presence felt in the public sphere. Citizenship, it was argued, was gendered and differential in its application (Lister, 1990; Walby, 1994). Fifthly – and this emerged more strongly in the 1990s – there was the impact of globalization and its consequences for rights anchored in nation-states. The age of global migration (Castles and Miller, 1998) had drawn attention to the position of migrants who found themselves between states. Lastly, a whole host of new rights were being added to the political agenda, often also involving some form of international conventions or mobilization, such as cultural rights, rights of indigenous peoples, environmental rights (Kymlicka, 1995; Turner, 1990), and, more recently, new technologies and cyber-citizenship (Vandenberg, 1999).

In this chapter I first outline the origins of thinking about citizenship, in particular examining the work of T. H. Marshall, a British sociologist whose conceptualization of citizenship has been highly influential. In the second part, I highlight recent geographic contributions to issues of citizenship. In the third section, I examine two key areas in which citizenship has been re-evaluated: gender and citizenship, and migrants and the emergence of postnational citizenship.

Origins of Citizenship and the Incorporation of New Claims and Rights

Historically citizenship has been closely tied in with the city-state as in Greece, Rome, and the Italian Renaissance. The modern revival of citizenship goes back to the period of revolutionary activity at the end of the eighteenth century in France and the USA, a crucial period in the emergence of the modern state. The civic republicanism model drew its key ideas from the formation of a political community in which the citizen is conceived as a political actor who exercises civic duty in the public sphere (Heater, 1990). Aptitude for citizenship was premised on a mode of civility, or on how to behave within the public spaces of the polis, and called for acceptance of agreed values underpinning the organization of the nation-state (Oldfield, 1990). Civic republicanism is not a rights-based conception but sees citizenship as a practice. The citizen shares an identity with other individuals and is expected to defend the political community. Education of youth is vital to develop this sense of participation and engagement (Oldfield, 1990). In the USA for example, "civics" is a ubiquitous element of the school curriculum where schools are considered to be responsible for teaching about democracy and preparing students to be effective democratic citizens (Frazer, 2000).[1] Not all elements of this model are always present. Switzerland, with its decentralized governance, referendums, and compulsory male military service, is probably one of the best contemporary examples. It is also a significant aspect of French and American conceptions of citizenship. Liberalism provided the philosophical underpinning of the state in the USA with the republican model providing the political mechanism based on the idea of a set of individuals who shared a common interest in public affairs (Marston, 1990, p. 451).

The liberal model, derived from seventeenth-century political theorists such as Hobbes and Locke, focuses on rights accorded to individuals and the obligations they owe to society and the state. The notion of the liberal individual stresses autonomy, rationality, and freedom from interference; in theory the individual is equal and without qualities, identities, and geographic situatedness. Liberal individualism does not postulate what constitutes the good life but offers the procedures and rules and an institutional framework within which individuals pursue their goals (Oldfield, 1990). As with civic republicanism, many states incorporate elements of another model. Hence, Britain has combined the liberal with the social democratic and its more collectivist provision of services. Many of the settler societies in the British Commonwealth also adopted a blend of liberalism and social democracy.

Radical models (Mouffe, 1992; Young, 1990) critique the individual as the bearer of rights (liberal) and the notion of a pre-constituted political community (civic republicanism) as well as the rigidity of the public/private divide. They emphasize the plurality of identities of individuals, the relationships and responsibilities between individuals, and the complex interaction between public and private spheres. In liberal and radical models, the gendered basis of citizenship (Pateman, 1989), and cultural and group rights have been ignored or marginalized (Isin and Wood, 1999; Kymlicka, 1995; Young, 1990). Another key critique is the linking of citizenship with nationality and the state, which some theorists have argued has become increasingly untenable (Davidson, 2000).

Probably the best known and influential theorist of citizenship is T. H. Marshall (Marshall and Bottomore, 1996), who after the advent of the welfare state, outlined the historical development of rights in the UK. Citizenship represented a widening incorporation and participation of different segments of society. Each century was associated with the evolution and deepening of a different set of rights with specific institutions and practices allied to particular rights. Civil rights emerged in the eighteenth century at a time of a highly individualistic phase of capitalist development and the delineation of a bourgeois public sphere. These consisted of the rights necessary for individual freedom, such as liberty of the press, freedom of speech, thought and faith, the right to own property, to conclude valid contracts, and the right to justice. The latter meant the right to assert oneself on terms of equality with others and by due process of law. By the nineteenth century, individual economic freedom was accepted. Today we might want to add to the practice of citizenship the right not to be discriminated on grounds of gender, sexuality, race, ethnicity, age or disability. One group for a long time denied civil rights were women, and especially married women. Indeed, the French declaration of the Universal Rights of Man at the end of the eighteenth century was challenged by Olympe de Gouges who penned a Declaration for the Rights of Women.

Political rights pertain to the right to participate in the exercise of political power at all levels, either as a member of a body invested with political authority or as an elector, as well as the right to association and access to information which enables the individual to exercise them. These rights began to be extended to wider economic classes in the early nineteenth century. Nevertheless, it was not until the twentieth century that political rights were separated from the economic status of the person, at least for men. Social rights were the last to be won in the twentieth century and have been the most contested in recent years. These range from the right to economic security and welfare to the right to live as a civilized being according to the prevailing standards of the society. Originally, social rights were bestowed through membership of a local community and functional associations but, from the late nineteenth and especially the twentieth centuries and with the centralization and the increasing territorial reach of the state, these were increasingly transferred to the national level and its institutions. Marshall argued that citizenship was a status bestowed on those who were full members of a community, which served to incorporate different sectors of society, and reduce social inequality, thus ensuring a degree of class abatement and social consensus. Citizenship should be seen therefore as a form of governance of diverse populations (Procacci, 2000). Although social rights have been the most hotly debated, especially with the restructuring of the welfare state – which has been most pronounced in states with a liberal regime of welfare, such as the USA and the UK (Esping-Andersen, 1990) – in many countries civil and political rights still form the basis of claim making.

Given that the context in which he was writing has changed in the last half century, it is not surprising that Marshall's analysis has been subjected to many criticisms. British society of the postwar years was previously far more homogeneous, having become more differentiated and multicultural in recent decades. Women are now more visible and active in the public sphere, thus modifying the relationship between the public and private spheres. International conventions and global processes have altered the role of the nation-state in relation to citizenship claims and guarantees, and new issues have arisen around the body and the environment. Fundamental issues, such as the relationship between citizenship exercised at different scales from the local, to the national and the international, the gap between formal and substantive citizenship, and the relationship between individual and groups rights, have also been raised.

Probably the most common critique of Marshall has been his conception of evolutionary stages in the unfolding of rights. Writers have argued that the chronological stages vary between countries and groups within society (Turner, 1990). For example, in many Asian countries, social rights have been granted before political or civil rights. For women, full civil and political rights were withheld until the twentieth century, often only being won slowly in the past century. Marshall took for granted the existence of formal rights, yet some groups were excluded from citizenship until the civil rights movements of the 1960s. African-Americans, especially in Southern States were segregated and denied civil and political rights. Indigenous peoples, as in Australia, were not admitted as citizens until 1967 and have invoked international conventions in order to acquire more extensive rights, especially in their struggle to gain control of land and resources from which they had previously been dispossessed (Peterson and Saunders, 1998). The implementation of the International Convention on the Elimination of Racial Discrimination into Australian law in 1975 facilitated some landmark decisions, in particular the Mabo decision in 1992 (Mercer, 1997). *Terra nullius* denied aboriginal people rights because they did not have a Western conception of property and were thus not considered civilized.

The United Nations Draft Declaration on the Rights of Indigenous Peoples (1984–93) is another instrument in undermining national discriminatory legislation, but national governments have set it aside. In 1996, the Wik decision versus the state of Queensland (Australia) declared that pastoral leases did not extinguish Native claims to territory but this was reversed by the Australian (Conservative) government in 1998, who championed the interests of large-scale corporations and pastoralists (Isin and Wood, 1999, p. 65).[2] Seeking control over land is particularly threatening in a capitalist society, involving as it does the redistribution of resources. Native peoples also challenge the liberal state through their demands for the recognition of group rights.

Geographic Contribution to Issues of Citizenship

The geographic contribution to citizenship issues has focused on the different scales at which citizenship is practiced (Painter and Philo, 1995; Smith, 1989), and the use of different kinds of spaces, such as the private and public, especially in relationship to gendered and sexual citizenship (Binnie, 1996; Kofman, 1995). Geographers (Cox, 1996; Marston, 2000; Smith, 1993, 1996; Staeheli, 1994) have considered the significance of scale for social and political action and organization. Much of the interest in scales stems from discussions about the re-scaling of the nation-state upwards (globalization and world regions such as the EU) and downwards (regions within states and cities). Geographers have tended to emphasize the playing out of citizenship, empowerment, and resistance at the local scale and in cities. They have also examined the ways in which social movements can move across scales (Miller, 1994) and the significance of making claims in particular kinds of places (Blomley, 1994).

Global cities, in particular, with their highly diverse and often polarized populations are thought to hold some promise for moving away from a citizenship anchored in political identification and membership of a nation-state (Garcia, 1996; Holston and Appadurai, 1996; Isin, 2000). We should, however, be careful in ascribing new forms of citizenship primarily to the recent re-scaling and reorientation of the state. Although the nation displaced parishes and cities as the locus of citizenship, many of the social rights associated with modern citizenship in the twentieth century, e.g. social housing, health services, and public facilities, were first promoted in large cities, such as Amsterdam, Berlin, London, Paris, and Vienna. Here, rural and international migrants flocked in search of employment and opportunities. After World War II and the emergence of the national provision of welfare, the state took over many of these services, although provision remained uneven depending on the political identity, social structure, and history of different localities.

From the 1960s, the redevelopment of cities, the expansion of social movements, the right to inhabit the city and define the use of public spaces became more significant as marginal populations were expelled to the periphery (Kofman, 1998; Lefebvre, 1996). In the 1980s, struggles over the built environment intensified under neo-liberalism and the effects of economic restructuring, while fortress spaces patrolled by private agents were constructed to protect the privileged. The inclusion of the marginalized may, however, be most effective through actions taken at neighborhood and community levels (Marston and Staeheli, 1994). Those who are denied legal status are nonetheless able to engage in political activism and influence local policies, as in the case of undocumented Latino workers in Los Angeles (Pincetl, 1994). In using a wide range of protest strategies in different spaces (demonstrations, occupation of theaters and churches, hunger strikes), the undocumented have succeeded in forging alliances, leading to a reconsideration of their status. An example was the *sans papiers* movement in France in 1996–7 (Davidson, 2000; Fassin et al., 1997) which bought together undocumented migrants, who found themselves without social rights and threatened with deportation, and who made demands for recognition of their rights to residence, work and welfare.

Since the 1960s, a number of groups have made themselves more visible and asserted their rights within a framework that recognizes equity through difference. In some cases, as with gays, they had to struggle against criminalization before demanding equal treatment in areas of economic and social life and their exclusion from state institutions, especially those ensuring social order such as the police and armed forces. The public life of gay men and women is severely constrained by laws and social norms in many states. In the USA, only a handful of states and municipalities protect their civil rights – employment, housing, education, etc. At the other end of the spectrum, same-sex marriage has been permitted since April 2001 in the Netherlands, August 2001 in Germany and inheritance, pension and other rights aligned with heterosexual couples. Domestic Partner registration now exists in Nordic states as well as in France. The attempt at redrawing the spaces of sexual citizenship and the distinction between the public and private has been pursued through forms of sexual dissident activities, such as consumerism, AIDS activism, and queer politics (Bell, 1995). New spaces of citizenship (Brown, 1994) have also been found in the shadow state spaces which are often less regulated by the state, operating with the help of volunteers. Civil rights have also been extended through the use of international human rights. Since the Labour Government came to power in 1997 in the UK, gays have seen the age of consenting sex reduced to 16 years and gays and lesbians have won the right in October 2000 to remain in the armed forces at a time when the Human Rights Bill was being incorporated into British law.

New Agendas

Gender and citizenship

Historically, women were exiles from citizenship, which was constituted on the basis of independence and of male norms and attributes. In the Greek city state, women and slaves could not participate in the *polis* or the public sphere; they belonged to the household. Critiques of the male bias and premises of citizenship have ranged from the model of the male worker acting as the breadwinner for his dependants, the independent male in possession of his body, to the soldier, who is prepared to sacrifice himself for his country (Yuval-Davis, 1991). In the liberal model, autonomy and independence are valued and dependency is viewed negatively. Some feminists have propounded an ethic of care (Sevenhuijsen, 1998; Tronto, 1993), that is grounded less in a vocabulary of rights and duties, as in the liberal conception of the atomistic and disembedded individual, than in one of responsibility and responsiveness which values the needs of others situated in specific contexts.

The stages of citizenship have often been quite different for women to that indicated by Marshall, as Walby (1994) points out. Civil rights were essentially enacted as part of the fraternal contract (Pateman, 1988), whereby the sovereign or the absolute father was replaced by a band of brothers who collectively gained access to women's bodies. In fact, in France and in many other countries women's civil rights actually regressed during the first part of the nineteenth century with the reinforcement of private patriarchy (Walby, 1990). The lives of married women in particular were severely constricted by their husbands. For First World women, political citizenship was typically achieved before civil citizenship, the reverse of the order for men. In the UK the struggle for the vote and political citizenship was part of a wider struggle over issues that fit into Marshall's category of civil citizenship, such as education, ownership of property, divorce, bodily integrity, and access to professional employment. Some were won before suffrage, others in the decades after it was obtained. Walby (1994) argues that political citizenship was more important than has often been considered in analyses of changes in gender relations and was the basis of the transformation from private to public patriarchy.

Feminist theorists have outlined very different solutions to regendering citizenship. According to Pateman (1989), women have been caught in Wollstonecraft's dilemma in their attempt to gain full citizenship – that is, the tension between difference and equality and the private versus the public sphere. They have chosen either a differential path from that of men, locating female citizenship in the home and bound up with the qualities of maternal care, or else they have modeled themselves on male qualities striving to enter work and the political arena. The first position espousing distinctive female qualities was derived from motherhood and caring, and has been advocated by some radical feminists. Elshtain (1981) has castigated the immorality of the public sphere and extolled the virtues of the home and the family. It is easy to critique her stance. Not all women are or wish to be mothers nor is the family necessarily the haven from conflict and violence that she cherishes. Others have suggested that maternal thinking will infuse the public sphere in the interest of children and peace (Ruddick, 1989) and bring attitudes of realism and attentiveness. Dietz (1985) contends, however, that the mother–child relationship and qualities of love and compassion are not sufficient for acting politically as engaged citizens. Putting the private sphere on such a lofty pedestal may discourage women from seeking to participate, reform, and reinvigorate public life, thus impoverishing it for all.

The second strategy of equality with men may be decidedly difficult for many women to achieve. They have different patterns of work, often part-time and with interrupted careers, which reduce their income and social entitlements. This is most clearly in evidence in liberal welfare regimes (Esping-Andersen, 1990). The alternatives put forward have concentrated on rethinking the worlds of work and welfare regimes and decoupling social rights and the family, such that minimum incomes and entitlements are linked to the individual, independent of family status. We need also to examine the relationship between the public and private spheres and the resources that enable an individual to participate in different economic, social, and political activities. The obligations to care for the young, the disabled, and the elderly are not distributed evenly throughout society; they are disproportionately located in the private sphere where the burden continues to fall heavily on women. Fulfilling these obligations may reduce resources and access to social rights and the ability to participate in collective life

The dichotomy described above is not inevitable; proponents of a pluralist regendering of citizenship have advocated the bridging of the spheres and of care, work and political participation (Lister, 1997, 2000). Pateman (1989) has suggested that a way through the Wollstonecraft dilemma is to break down the opposition between women's dependence and men's independence. This would involve looking at men's and women's responsibilities in society, shifting the sexual division of labor in relation to care and paid work, and disrupting the divide between the public and private. The degree to which some of these changes have been effected varies between societies. Nordic states have generally tried to provide parents with "time to care" and have high levels of publicly financed provision of childcare. The UK, Ireland, and Netherlands have low levels as does the USA where public support for childcare, other than through tax credits, is usually confined to anti-poverty and welfare-to-work programs (Lister, 1997, p. 183). Of course, in states where state provision for care is thin or inadequate, migrant domestic labor has been brought in to fill the gap. Yet feminist writing on citizenship has been slow to take into account the implications of this development (Kofman et al., 2000; Lister, 1997). Migrant women's rights may be further constrained by their immigration status, which restricts their conditions of work and access to welfare provision. More equitable citizenship rights are also likely to require a social security system that is not based on male employment patterns, although the chequered trajectory is also experienced increasingly by men. Part-time employment is still largely filled by women and women's earnings tend in many countries to be lower than men's. In the UK, 25 years after equal pay legislation, women in general earn 82% of men's earnings. Thus, female patterns of employment result in inferior entitlements to pensions. In the face of these inequalities can we achieve a woman-friendly citizenship?

Growing class polarization, which is particularly marked in liberal welfare states, intersects with other forms of division. Linda Gordon (1990), for example, has pointed out the racialized nature of the welfare state in the USA, where welfare debates have taken place within a uniquely white set of political, economic, and familial assumptions. So, too, is women's relationship to the welfare state mediated by life course, age, sexuality, disability (Morris, 1993), and geographical location (Lister, 1997, p. 175).

In recent years, attention has also turned to equality of political representation. Although the domain of the political had been enlarged by feminism in the 1970s, notably through the slogan "the personal is political," most effort was devoted to acquiring social rights in the early years of organized second-wave feminism. Women in many states have been largely absent from the higher echelons of formal political and corporate institutions. They may have gained the right to political participation, but in reality their actual participation rates are in most states fairly low, especially in the higher echelons of formal politics. In the mid-nineties only Scandinavian countries and the Netherlands had over 30% women in their parliaments; while France had just 6% although this had reached 10.2% following the Socialist victory in the 1997 legislative elections (Mossuz-Lavau, 1998). In France in the 1990s, a concerted effort was made to break down sexual apartheid in the political sphere which feminists argued required the enforcement of parity or equal representation of women in electoral lists (Mossuz-Lavau, 1998). Since the 2001 elections, electoral lists based on proportional representation (local, regional, and European but not legislative) have had to be constructed on principles of parity. Women's involvement in transitions to democracy in the former Communist states and in South Africa have not, however, been translated into their political presence. They are generally more involved in community politics, especially in relation to health, education and care of children. Northern Ireland exemplifies the almost total absence from formal politics in contrast to the involvement of women's groups in community politics. African-American women, too, are more involved in community activism. Women have sometimes adopted less traditional ways of capturing political attention. Argentinian and Chilean women's occupation of public spaces drew attention to the disappearance of their children

Migration, cultural diversity, and citizenship

The growing cultural and ethnic diversity of many societies has led to discussions of multicultural citizenship. Multiculturalism has different interpretations and practices in a variety of increasingly culturally diverse societies. For example, in the USA it is about the contribution of traditionally excluded groups to the dominant canon of US history and culture; in Australia, it has been about public policy designed to ensure the socioeconomic and political participation of all members of an increasingly diverse population (Castles, 1997). In European states, it concerns the right to cultural recognition, although in some states this has been reinforced with social justice policies. Castles (1997) raises four problematic aspects in the relationship of immigration, citizenship, and democracy: formal inclusion; substantial citizenship that is the effective exercise of rights as a member of a community; recognition of collective cultural rights; and the appropriateness of political institutions.

Others have suggested that the growing presence of migrants in many societies means we should reconceptualize rights around principles of human rights attached to personhood, which are seen as more appropriate (Bottomore, 1996; Soysal, 1994; Turner, 1993). These writers question the suitability of the nation-state to accommodate citizenship rights as well as pointing out the ways in which citizenship for migrants is no longer anchored exclusively in the nation-state (Sassen, 1996), thereby resulting in a multi-leveled or postnational citizenship.

Soysal's thesis of postnational citizenship (1994) contends that long-term residents or denizens in the EU now enjoy almost all basic economic and social rights irrespective of their formal membership of the nation-state, i.e. rights have been decoupled from membership of the nation-state and embedded in territorial residence and personhood. It is argued that these rights have been obtained through an international discourse and regime of rights that are binding on states, thereby reducing their control of migration policy. The only major form of rights denied to denizens are political, especially voting rights. Soysal's subsequent formulation of the postnational thesis (2000) has emphasized the multilevel dimensions of migrant claims and organizations. It may appear paradoxical that particular claims for the recognition of cultural difference are being formulated in terms of universal discourse of basic human rights, i.e. the right to cultural difference. She illustrates this paradox with Muslim demands and organizations in the EU.

In relation to the core element of the postnational thesis, international conventions do not bind states into protecting migrants. While endorsing the principle of postnational citizenship, many writers point out that the reality does not yet match the normative statement (Stasilius, 1997) and that the influence of international instruments has often been very slow and highly variable between states (Guiraudon, 1998). Postnational theorists have in effect derived their analysis from the de-territorialization of social rights in the past two decades. However this varies considerably between states. A study of migrant legal rights (Waudrach and Hofinger, 1997), which the authors believed were a precondition for social and cultural integration, revealed that Switzerland and Austria had granted far fewer rights than Nordic countries or the Netherlands. The latter group of countries have granted local voting rights which remain highly contested in other states such as France. There remain significant areas from which denizens are excluded, especially in the economic field. In France, a third of employment in a very generously defined public sector is barred to non-Europeans (Bataille, 1997; Kofman, 2003). In Germany, Switzerland and Austria, with their low rates of naturalization for migrants, large sectors of employment cannot be entertained by migrant noncitizens, who may even constitute a third generation. The right to set up businesses is also severely constrained in some countries.

European citizens do partake of postnational citizenship under the Maastricht Treaty enacted in 1993 but denizens do not. Rights to mobility within the European Union are probably one of the most significant that denizens still do not enjoy. These include the right to reside and work in another European country and to bring in close family members without fulfilling conditions not required of citizens.

The existence of an international regime of migrant rights is hard to sustain. None of the major receiving countries has agreed to ratify the *UN Convention for the Rights of All Migrant Workers and their Families* passed in 1990, largely due to their unwillingness to include guarantees for undocumented workers. Yet this has been one of the fastest growing group of migrants in Europe and the USA, comprising labour migrants, those seeking family reunion, and asylum seekers, whose request for refugee status has been rejected.

Furthermore, the gap between citizens, denizens, and aliens has increased in the 1990s both in the EU and in settler societies, generating a complex system of civic stratification (Kofman, 2003). Temporary statuses and entry without accompanying social rights have been increasingly used. In settler societies there has been an attack on the social rights of denizens in their initial years of settlement, as in the 1996 Immigration Act in the USA. Subsequent congressional bills have curtailed services to legal, permanent resident immigrants and rendered family reunification more difficult. Indeed, most settler societies and the UK have in the 1990s barred family-sponsored migrants from access to welfare benefits for periods of up to 10 years. Attacks on social spending on migrants have followed the failure of attempts to reduce immigration through strengthened, and sometimes militaristic border controls, as in the case of the US–Mexican border. Yet migration between two states has become a survival strategy for many transborder families. These migrants too raise issues of transnational rights and political loyalties (Jonas, 1996). However, the increasing numbers of undocumented Latino migrants have led to a series of attempts to curtail their basic rights. Although rescinded by the federal court, many of the provisions of Proposition 187 passed by the Californian electorate in 1994 have stimulated campaigns in other states of high immigration, such as Florida and Texas, to deprive undocumented migrants of their constitutional rights and access to basic social services (Stasilius, 1997, p. 205).

A disturbing development has been the undermining of the Geneva Convention, one of the earliest postwar international instruments to protect those fleeing from persecution. Sovereignty continues to be enshrined in international conventions, especially with regard to entry into a state. States cannot be forced to accept asylum seekers but they are obliged under the Convention to give them a hearing, a principle many states have increasingly flouted either on grounds that they are bogus "economic migrants," that they have passed previously through another safe country, or could have access to safe zones, as with the Iraqi Kurds. For asylum seekers who do manage to enter another state, the recognition rate has slumped dramatically in developed countries, generating what has been called an informal regime of asylum (Brachet, 1997), whereby asylum seekers are tolerated but deprived of any social rights. There has also been an extension in the use of temporary statuses and recognized asylum seekers have been forced to return to their homelands, for example in the case of Albanian Kosovan refugees in Australia or Bosnians in Germany. Many asylum seekers are also prevented from gaining entry to territory in order to be able to request protection and this in turn has led to the growth of an industry in human trafficking (Salt, 2000).

While there has been some advance in what is accepted as persecution, especially in relation to gender guidelines which have begun to influence asylum determination in Canada, the USA, Australia and the UK (Crawley, 2001), other social groups, such as the Roma, have still to contend with the racism of the countries which they have fled as well as those in which they seek a haven. Victims of generalized racism are not seen to conform to the individual persecution demanded by a strict interpretation of the Geneva Convention. It is also likely that the Roma encounter such high levels of hostility as a result of their nonconforming territoriality (Kofman, 1995). Their nomadism and territorial practices connote shiftiness and untrustworthiness, and a metaphor of their social existence as outsiders in settled societies. They have become the outcasts of European societies (Sivanandan, 2001).

Despite the growing gap between denizens, those on temporary statuses, and the undocumented, I would not wish to deny the significance of an international discourse of rights. Nonetheless the thesis of postnational citizenship remains highly utopian and normative and has not adequately considered substantive citizenship whose practices may undermine the effective exercise of rights. In order to work towards more progressive policies, we need to understand the mechanisms and institutions by which citizenship is extended and maintained, on the one hand, and rights are revoked, on the other. The state or a quasi state, such as the EU which is extending the rights of legal long-term migrants, is likely to remain the main guarantor of migrant rights.

Conclusion

Citizenship has attracted so much attention in recent years owing to the widespread economic, social, cultural, and political changes that have occurred at different scales and the emergence of new claims for individual and group rights. States, however, remain the main focus of claim making. Although international conventions are helping to deepen and extend claims for recognition and justice, states are able to ignore and circumvent them unless the conventions are enshrined in domestic law. In complex and multicultural societies, citizenship has to encompass the diversity of practices and needs. The challenge today is to promote equality through a framework that recognizes different ways of belonging. In particular, this requires a rethinking of the spaces in which individuals and groups can engage as citizens without locking them into national spaces and the rigid public/private divide. Geographers have drawn attention to the complexities and ambiguities of citizenship practices at different scales and places and the ways in which social and political movements can move across scales. There is much to be done in examining the making of claims across scales and in different types of spaces and places at a time when there has been a multiplication of scales of governance and with growing resistance to the inevitability of global processes.

ENDNOTES

1. Frazer (2000) argues that Britain has resisted citizenship education because of widespread doubts about political education in schools, the nature of British culture and politics which identifies citizenship as a foreign concept, and the absence of any consensus of a body of key events and texts that form the narrative of US history.
2. Pastoralists are those who use land for the grazing of animals.

BIBLIOGRAPHY

Bataille, P. 1997. *La discrimination au travail*. Paris: La Découverte.

Bell, D. 1995. Pleasure and danger: the paradoxical spaces of sexual citizenship. *Political Geography*, 14(2), 139–53.

Binnie, J. 1996. Invisible European: sexual citizenship in the New Europe. *Environment and Planning A*, 28, 3–13.

Blomley, N. 1994. Mobility, empowerment and the rights revolution. *Political Geography*, 14(5), 407–22.

Bottomore, T. 1996. *Citizenship and Social Class, Forty Years On*. London: Pluto, 55–93.

Bouamama, S., Cordeiro, A., and Roux, M. (eds.). 1992. *La citoyenneté dans tous ses états*. Paris: CIEMI/ l'Harmattan.

Brachet O. 1997. L'impossible organisgramme de l'asile en France. Le développement de l'asile au noir. *Revue Européenne des Migrations Internationales*, 13(1), 7–36.

Brown, M. 1994. The work of city politics: citizenship through employment in the local response to AIDS. *Environment and Planning A*, 26, 873–94.
Castles, S. 1997. Multicultural citizenship: the Australian experience. In V. Bader (ed.) *Citizenship and Exclusion*. London: Macmillan, 113–38.
Castles, S. 1998. *The Age of Migration*. London: Macmillan.
Cox, K. 1996. Editorial: the difference that scale makes. *Political Geography*, 15, 667–70.
Crawley, H. 2001. *Refugees and Gender. Law and Process*. Bristol: Jordans.
Davidson, A. 2000. Fractured identities: citizenship in a global world. In E. Vasta (ed.) *Citizenship, Community and Democracy*. London: Macmillan, 3–21.
Davidson, A. and Weekley, K. (eds.). 1999. *Globalization and Citizenship in the Asia-Pacific*. London: Macmillan.
Dietz, M. 1985. Citizenship with a feminist face: the problem with maternal thinking. *Political Theory*, 13(1), 19–37.
Elshtain, J. 1981. *Public Man, Private Woman*. Princeton, NJ: Princeton University Press.
Esping-Andersen, G. 1990. *Three Worlds of Capitalism*. Cambridge: Polity.
Fassin, D., Morice, A., and Quiminal, C. (eds.). 1997. *Les lois de l'hospitalité. La politique de l'immigration a l'épreuve des sans papiers*. Paris: La Découverte.
Frazer, E. 2000. Citizenship education: anti political culture and political education in Britain. *Political Studies*, 48(1), 88–103.
Garcia, S. 1996. Cities and citizenship. *International Journal of Urban and Regional Research*, 20, 7–21.
Gordon, L. 1990. *Women, the State and Welfare*. Madison, WI: University of Wisconsin Press.
Guiraudon, V. 1998. Third country nationals and European law: obstacles to rights' expansion. *Journal of Ethnic and Migration Studies*, 24(4), 657–74.
Hall, S. and Held, D. 1989. Left and rights. *Marxism Today*, June, 16–23.
Heater, D. 1991. *Citizenship: the Civic Ideal in World History, Politics and Education*. Harlow: Longmans.
Holston, J. and Appadurai, A. 1996. Cities and citizenship. *Public Culture*, 8, 187–204.
Isin, E. and Wood, P. 1999. *Citizenship and Identity*. London: Sage.
Jennings, J. 2000. Citizenship, republicanism and multiculturalism in contemporary France. *British Journal of Political Science*, 30, 575–98.
Jonas, S. 1996. Rethinking immigration policy and citizenship in the Americas: a regional framework. *Social Justice*, 23(3), 68–85.
Kearns, A. 1992. Active citizenship and urban governance. *Transactions of the Institute of British Geographers, NS*, 17, 20–34.
Kofman, E. 1995. Citizenship for some but not for others: spaces of citizenship in contemporary Europe. *Political Geography*, 14(2), 121–37.
Kofman, E. 1998. Whose City?: gender, class and immigrants in globalizing European cities. In R. Fincher and J. Jacobs (eds.) *Cities of Difference*. New York: Guilford, 279–300.
Kofman, E. 2003. European migrations, civic stratification and citizenship. *Political Geography*, in press.
Kofman, E., Phizacklea, A., Raghuram, P., and Sales, R. 2000. *Gender and International Migration: Employment, Welfare and Politics*. London: Routledge.
Kymlicka, W. 1995. *Multicultural Citizenship*. Oxford: Oxford University Press.
Lefebvre, H. 1996. *Writings on Cities*. Transl. and Ed. E. Kofman and E. Lebas, Oxford: Blackwell.
Lister, R. 1997. *Citizenship: Feminist Perspectives*. London: Macmillan.
Lister, R. 2000. Citizenship and gender. In K. Nash and A. Scott (eds.) *The Blackwell Companion to Political Sociology*. Oxford: Blackwell, 6.
Marshall, T. and Bottomore, T. 1996. *Citizenship and Social Class*. London: Pluto.

Marston, S. 1990. Who are the people?: gender, citizenship and the making of the American nation. *Environment and Planning D: Society and Space*, 8(4), 449–58.
Marston, S. 2000. The social construction of scale. *Progress in Human Geography*, 24(2), 219–42.
Marston, S. and Staeheli, L. 1994. Citizenship, struggle, and political and economic restructuring. *Environment and Planning A*, 26, 840–8.
Mercer, D. 1997. Aboriginal self-determination and indigenous land titles in post-Mabo Australia. *Political Geography*, 16, 189–212.
Miller, B. 1994. Political empowerment, local-central state relations and geographically shifting political opportunity structures: strategies of the Cambridge, Massachusetts peace movement. *Political Geography*, 13(5), 393–406.
Morris, J. 1993. Feminism and disability. *Feminist Review*, 43, 57–70.
Mossuz-Lavau, J. 1998. *Femmes/Hommes pour la parité*. Paris: Presses des Sciences Politiques.
Mouffe, C. (ed.). 1992. *Dimensions of Radical Democracy*. London: Verso.
Oldfield, A. 1990. *Citizenship and Community. Civic Republicanism in the Modern World*. London: Routledge.
Painter, J. and Philo, C. 1995. Spaces of citizenship. *Political Geography* 14(2), special issue.
Pateman, C. 1988. *The Fraternal Contract*. Cambridge: Polity.
Pateman, C. 1989. *The Disorder of Women*. Cambridge: Polity.
Peterson, N. and Saunders, W. (eds.). 1998. *Citizenship and Indigenous Australians: Changing Conceptions and Possibilites*. Melbourne: Cambridge University Press.
Pincetl, S. 1994. Challenges to citizenship: Latino immigrants and political organizing in the Los Angeles area. *Environment and Planning A*, 26, 895–914.
Procacci, A. 2000. Governmentality and citizenship. In K. Nash and A. Scott (eds.) *The Blackwell Companion to Political Sociology*. Oxford: Blackwell.
Ruddick, S. 1989. *Maternal Thinking. Towards a Politics of Peace*. London: Women's Press.
Salt, J. 2000 Human trafficking. *International Migration*, 38(1), 30–7.
Sassen, S. 1996. Beyond sovereignty: immigration policy making today. *Social Justice*, 23(3), 9–20.
Sevenhuijsen, S. 1998. *Citizenship and the Ethics of Care. Feminist Considerations on Justice, Morality and Politics*. London: Routledge.
Sivanandan, A. 2001. UK refugees from globalism. *Race and Class*, 42(30), 87–91.
Smith, N. 1993. Homeless/global: scaling places. In J. Bird et al. (eds.) *Mapping the Futures*. London: Routledge, 87–119.
Smith, N. 1996. Spaces of vulnerability: the space of flows and the politics of scale. *Critique of Anthropology*, 16, 63–77.
Smith S. 1989. Society, space and citizenship: a human geography for the new times. *Transactions Institute of British Geographers*, 14(1), 144–56.
Soysal, Y. N. 1994. *Limits of Citizenship: Migrants and Postnational Membership in Europe*. Chicago: University of Chicago Press.
Soysal, Y. N. 2000. Citizenship and identity: living in diasporas in post-war Europe. *Ethnic and Racial Studies*, 23(1), 1–15.
Staeheli, L. 1994. Empowering political struggle: spaces and scales of resistance. *Political Geography*, 13(5), 387–91.
Stasilius, D. 1997. Citizenship and immigration: pathologies of a progressive philosophy. *New Community*, 23(2), 173–96.
Tronto, J. 1993. *Moral Boundaries. A Political Argument for an Ethics of Care*. London: Routledge.
Turner, B. 1990. Outline of a theory of citizenship. *Sociology*, 24(2), 189–217.
Turner, B. (ed.). 1993. *Social Theory and Citizenship*. London: Sage.

Vandenberg, A. (ed.). 1999. *Citizenship and Democracy in a Global Era*. London: Macmillan.
Walby, S. 1990. *Theorising Patriarchy*. Oxford: Basil Blackwell.
Walby, S. 1994. Is citizenship gendered? *Sociology*, 28(2), 379–95.
Waudrauch, H. and Hofinger, C. 1997. An index to measure legal obstacles to the integration of migrants. *New Community*, 23(2), 271–85.
Weeks, J. 1998. The sexual citizen. *Theory, Culture and Society*, 15(3), 35–52.
Young, I. 1990. *Justice and the Politics of Difference*. Princeton, NJ: Princeton University Press.
Yuval-Davis, N. 1991. The citizenship debate: women, the State and ethnic processes. *Feminist Review*, 39, 58–68.

Chapter 26

Sexual Politics

Gill Valentine

Introduction

"The idea of citizenship refers to relationships between individuals and the community (or State) which impinges on their lives because of who they are and where they live" (Smith, 1989, p. 147). While members of a state have certain duties and obligations towards it, in return they can expect certain rights and benefits. In particular, understandings of citizenship have been developed from the work of Marshall (1950). He defined citizenship in terms of *civil or legal rights* (freedom of speech, assembly, movement, equality in law, etc.), *political rights* (right to vote, hold office, engage in political activity, etc.), and *social rights* (rights to social security, welfare, basic standard of living) (Muir, 1997).

Despite the fact that the language of citizenship implies inclusion and universality, it is also an exclusionary practice (Lister, 1997). Historically, only select groups – notably white property-owning men – have been entitled to full citizenship (see critiques of the gendered and racialized notions of citizenship: Anthias and Yuval-Davis, 1992; Walby, 1997). Other groups (such as women and ethnic minorities) have had to fight to have this extended to them. Indeed, some social groups remain effectively only partial citizens because they are excluded from particular civil, political or social rights (Smith, 1989). Furthermore, as Muir (1997) points out, even though members of minority groups may enjoy citizenship they may still not "feel" as if they are citizens if they experience discrimination and harassment and regard themselves as less able to exercise their rights before the law.

This is particularly true of lesbians and gay men whose claims to citizenship rights and political legitimacy are not fully established in most modern states (Corviono, 1997). Through the law and criminal justice system the state defines, represses, and penalizes particular sexual practices and identities and in the process confers, limits or withholds citizenship (Isin and Wood, 1999). In some states homosexuality *per se* is illegal. For example, under Italian law same sex acts which are defined as against the common sense of decency in the Criminal Code may be punished with a prison sentence of between three months to three years. Likewise, in some states of the US sodomy, oral sex, and "unnatural sex acts" are criminalized (Isin and Wood, 1999), while in the UK the prosecution of men on assault charges for engaging in consensual same-sex sado-masochistic activities in the privacy of their homes demonstrate the limits of British sexual citizenship (Bell, 1995). Foucault termed such interactions between the individual and nation whereby certain bodies are defined as "normal" and others as "deviant" as "biopolitics."

Writing about the British parliamentary process through which the legal age of consent for gay male sex – originally set at 21, five years higher than the heterosexual age of consent – was reduced to 18 (it has subsequently been brought in line with the heterosexual age of consent), Epstein et al. (2000, p. 14) argue that: "parliamentary territorialization signals a double movement whereby sexualities are nationalized and nation is sexualized. In this context, parliamentary process is a powerful engine for recognitions and misrecognitions, for strengthening, marginalising or disorganising sexual/national identities."

Here, Epstein et al. (2000, p. 14) identify citizenship as both *formal* and *symbolic*. "Formal citizenship includes, for example, the rights to vote, to marry and to bring a non-national partner into the country as a permanent resident."

"Insofar as marriage between persons of the same sex is not allowed by most legislation, acquisition of citizenship by way of marriage is impossible for lesbian and gay couples of different nationalities" (Tanca, 1993, p. 280). In such ways the movement of nonheterosexual bodies across national borders is regulated (Kofman, 1995). As a result campaigns have been held across Europe to draw attention to the predicament faced by lesbian and gay couples of different nationalities because of discriminatory partnership legislation (Valentine, 1996a).

Australia, New Zealand, the Netherlands, Denmark, Norway, Sweden, and Iceland are among the few states that do recognize same-sex couples in the area of immigration. Yet, even in some of these states lesbians and gay men are not treated on a par with heterosexual couples. For example, to obtain citizenship in New Zealand a heterosexual man or woman must have lived with a partner who holds New Zealand nationality for two years. In contrast, a lesbian or gay man needs to have lived with a New Zealander for four years to gain the same rights (Binnie, 1997).

By refusing lesbians and gay men basic civil rights, states not only exclude them from full citizenship, but also from a range of wider social rights and material benefits. For example, married couples receive tax benefits, inheritance rights, custody and adoption rights, the right to succeed to their partner's tenancy, and so on. All of these rights are denied to same-sex couples because they cannot legally marry.

Moreover, because legal, health, and welfare systems are all founded on the basis of the hegemonic heterosexual family unit (Wilton, 1995), lesbians and gay men also experience a range of inequalities and forms of discrimination as a result of this institutionalization of heterosexuality (Cooper, 1993; Herman, 1993, 1994; Richardson, 1998). Yet, in most modern states individuals or groups have no rights to antidiscrimination protection on the grounds of sexual orientation in relation to employment, housing, education, and so on (Andermahr, 1992; Betten, 1993; Valentine, 1996a; Waaldijk, 1993). It is not surprising, therefore, that the term "dissident," which refers to those who disagree with state ideology, is sometimes used to describe sexual minorities.

Symbolic citizenship is defined by Epstein et al. (2000, p. 14) in terms of "whether one is considered to be a full member of the nation state." Just as lesbians and gay men are excluded from formal citizenship, so too they are also excluded from symbolic membership of the nation in a number of ways.

Anderson (1983) has described nations as "imagined communities." Other writers have highlighted the importance of cultural symbols, rituals, locations, and icons in fostering such imaginings of sameness or communion amongst members of a nation, and thus in producing "the nation" (see Donald, 1993; Sharp, 1996). Yet in these imaginings the emphasis is often upon the traditional heterosexual nuclear family. In this way, Sibley (1995, p. 108) points out that "key sites of nationalistic sentiment, including . . . the suburbs and the countryside, implicitly exclude" lesbians and gay men as well as other minorities from the nation.

In debates about the introduction of section 28 of the British *Local Government Act, 1988* (which bans local authorities from intentionally promoting homosexuality, or teaching the acceptability of homosexuality as a pretended family relationship in state schools) the point was frequently made that materials on nonheterosexual relations would be a waste of tax-payers' money. Such "discursive distinctions" between sexual dissidents and taxpayers "symbolically dislocate lesbian and gay people from full civil status, even though empirically, lesbian and gay people pay taxes" (Epstein et al., 2000, p. 14). In turn such "[s]ymbolic disenfranchisements are," Epstein et al. (2000, p. 14) argue, "invariably reinvested in material exclusions."

An example of this relationship between the symbolic and the material is evident in the census. Brown (2000) suggests that the census is a way of thinking about the nation. He writes "[T]hrough the complex web of state data on its nation, reliable categorisation provides a way of representing and knowing that channels state power with state knowledge of its citizens (and non citizens!). Citizens, and their problems, come to be known, defined and recognised along certain descriptive lines . . . " (Brown, 2000, p. 94). Yet, neither the US, nor the UK national population surveys asks a specific question about same-sex relationships or sexuality. As a consequence, lesbians and gay men are not directly counted in these surveys and so are not only symbolically missing from these snapshots of the nation, but are also invisible when census data is used to understand population changes, to identify areas of need, and to ensure representational equity in the legislature (Brown, 2000).

Indeed, states often acknowledge/deny lesbians' and gay men's membership of the nation in complicated and contradictory ways (Brown, 2000). A case in point is provided by the debate in the USA about whether lesbians and gay men should be allowed to serve in the military. Under pressure from right-wing opponents to maintain a ban on lesbians and gay men serving in the armed forces on the grounds that homosexual military personnel undermine the moral and morality of serving units and therefore threaten the defense of the nation; yet reluctant to deny their presence in the military and sanction discrimination on the grounds of sexual orientation, former President Bill Clinton's famous solution was: "Don't ask, don't tell."

The question of formal and symbolic citizenship is even more complex for those who define themselves outside the heterosexual/homosexual and male/female binaries. For transsexuals and transgendered people, citizenship depends on the question of their right to self determination given state definitions of their identity as male or female which are determined at birth (Evans, 1994; Isin and Wood, 1999).

Not surprisingly, perhaps, lesbians and gay men have fought many battles to achieve the limited citizenship rights which they currently hold, and continue to struggle to achieve full membership of the nation-states in which they live. The following sections of this chapter trace some of the different political trajectories that these campaigns have taken in the UK and highlight the spatialities (particularly in relation to "public" and "private" space) which are implicit within them.

The Sexual Politics of (in)Tolerance

In Britain, sex in private between men over 21 was decriminalized by the *Sexual Offences Act* of 1967 (sex between women was never criminalized because Queen Victoria did not believe that it was possible). However, "[h]eterosexual supporters of the Act quickly warned that it was not a mandate for equality but an 'act of toleration'" (Wilson, 1993, p. 173). The Act, it was pointed out, was designed to protect homosexuals from blackmail and violence (which had been highlighted in a government inquiry known as the Wolfenden Report) and was therefore a necessary evil rather than a stamp of approval for gay sexuality. The phrase repeatedly used in relation to the Act was not one of equality, but of tolerance. This difference is significant. As Wilson points out (1993, pp. 174–5), whereas equality belies approval and acceptance, "[t]oleration ... necessarily rests on a bed of disapproval.... Toleration has limits. When those limits are reached the tolerator has the power to criminalize and punish the tolerated."

In the Britain of the late 1960s these limits were defined in spatialized terms. Coupled with the extension of tolerance of homosexual practices was an expectation of "privacy:" that gay men would continue to be discreet in public about their identities, sexual practices, and relationships. One of those who supported the introduction of *The Sexual Offences Act*, Lord Arran, told gay men to: "comfort themselves quietly and with dignity and to eschew any forms of ostentatious behaviour or public flouting" (Wilson, 1993, p. 175).

Implicit in such expectations lies two key assumptions. First, that the State itself is located in the "public" realm and embodies so-called masculine qualities such as rationality, intellect, culture, and impartiality in contrast to sexuality which is linked with characteristics that are associated with the feminine such as the body, untamed emotion, irrationality, and nature, and therefore with domestic or private space (Cooper, 1993). Through such discursive constructions, sexuality is located as "other" in relation to the state. As Cooper (1993, p. 207) explains: "Ostensibly unsexed, it [the state] governs the boundaries of the sexual and asexual worlds" and regulates the division between public and private space.

Secondly, the expectations of "the tolerant" also implicitly assume that public space is the space of the "normal" (although at the same time it is rarely acknowledged that this space is sexualized or rather heterosexualized) (Valentine, 1993, 1996b). As such, when homosexuality is articulated in "private" and is therefore hidden or invisible it is not regarded as threatening. However, when it spills over into "public" space lesbians and gay men are accused of invading and polluting the space of the "normal" and of undermining the social order, by corrupting innocent young people, destroying the family, eroding the nation's morality, and even threatening its competitiveness in a global economy (Cooper, 1993; Epstein et al., 2000). As Cooper (1993, p. 209) explains, "lesbians and gays are . . . discursively constructed as so sexual that even their mere presence is erotically saturated. The threat of contamination means no space where others are present can be safe."

This spatialized framework of the normal/public and the perverse/private in turn splits the category of lesbians and gay men into the "good," "responsible," "safe" homosexual subject who stays quietly within the confines of the private space of the closet and on the margins of society, with the "bad," "irresponsible," "dangerous," homosexual subject who is publicly "out" and proud and disrupts the center (Epstein et al., 2000; Sedgwick, 1994).

Ironically perhaps, it was the State's tolerance of "private" homosexual practices in the UK that aroused a desire amongst some lesbians and gay men to be more publicly visible and to fight for equal rights with heterosexuals. So began a continuing series of campaigns aimed at the right to be incorporated into the hegemonic project.

The Sexual Politics of (in)Equality

While sexual politics in the 1960s were focused on challenging legislation which penalized sexual acts between men, in the 1970s campaigns for sexual rights shifted away from a concern with *practices per se* towards a focus on sexual *identities* (Richardson, 2000). "These campaigns were not asking for the right to engage in same-sex activity couched in terms of respect for privacy, but were expressing opposition to social exclusion on the basis of sexual status" (Richardson, 2000, p. 106). They have been articulated first in the language of sexual freedom and liberation and more recently in the language of citizenship.

In the early 1970s, various gay liberation manifestos were published in the UK, USA, and elsewhere. The London Gay Liberation Front, for example, defined itself as aiming: "to defend the immediate interests of gay people against discrimination and social oppression" (leaflet cited in Evans, 1993, p. 116). Its demands included: "that all discrimination against gay people, male and female, by the law, by employers, and by society at large should end," "that all people who feel attracted to a member of their own sex should know that such feelings are good and natural," "that psychiatrists stop treating homosexuality as though it were a problem or sickness, and thereby giving gay people senseless guilt complexes," "that gay people should be legally free to contact other homosexuals, through newspaper ads, on the streets, and by any other means they wish, as are heterosexuals and that police harassment should cease right now," and "that gay people be free to hold hands and kiss in public as are heterosexuals" (op. cit.).

As such, these identity based sexual rights campaigns included not only a demand for basic civil rights but also for the right to self definition, and the right of self expression (Richardson, 2000). In other words, the campaigns were about individuals reclaiming a sexual identity which had previously been denied to them and overcoming the shame or stigma attached to the label homosexual by defining or naming themselves (Weeks, 1977). They were also about lesbians and gay men collectively contesting medical and psychiatric practices which labeled them as "diseased" or "sick," and fighting for social, economic, and political, as well as legal change. At both an individual and collective level, this form of sexual politics therefore represented what Richardson (2000, p. 35) has termed a "politics of self discovery." Above all, gay liberation movements were about the right for sexual dissidents to be gay not only in private, but also in public.

Gay liberationists in this period therefore placed great importance on encouraging lesbians and gay men to "come out" and publicly proclaim their sexual identity. "Consciousness raising" became a political buzzword of the time. The first gay pride march – a collective coming out – took place in London in July 1972.

While many of these initial campaigns were couched in terms of "gay" rather than lesbian and gay, this soon changed. The 1970s was also the period when a second wave of feminism emerged. The women's liberation movement began to challenge women's sexual passivity and to recognize women's autonomous sexuality, their rights to sexual pleasure, and their rights to freedom from male sexual violence and harassment (Jefferys, 1994). "More specifically, lesbian feminists [also] insisted on the right to *be* lesbian, as a specific identity and practice and the freedom (for all women) to be able to choose to have relationships with other women" (Richardson, 2000, p.258). The consequent emergence of lesbian consciousness and identity politics resulted in the notion of "gay" liberation being contested.

From the mid-1980s onwards, lesbian and gay politics have changed tack. The radical, often spontaneous and somewhat crude "come out" and take to the streets campaigns of the gay liberation movement have been substituted by slick, professionally organized campaigns employing mainstream political tactics (fund raising, lobbying, commercial advertisements, and so on) and being fronted by popular celebrities. The emphasis is no longer on liberation through confrontation but rather on winning full citizenship rights for lesbians and gay men through persuasion, the aim being to claim the space of citizenship and public participation. To use a spatial metaphor, lesbian and gay politics have moved from the margins to the center (although, in part it is worth remembering that this style of politics is only possible because of the more public profile, and with it acceptance, that the liberation movements achieved for lesbians and gay men).

For example, Stonewall, a British lesbian and gay rights group established in 1989, defines its own agenda as: "To work for equality under the law and full social acceptance for lesbians and gay men. Our approach is an innovatory one for lesbian and gay rights – professional, strategic, tightly managed, able and willing to communicate with decision-makers in a constructive and informed way (Interim Report, Stonewall Group, 1990, cited in Smyth, 1992, p. 19).

The citizenship issues which have, and are being, pursued by groups like Stonewall include: legal rights in terms of the repeal of anti-gay statutes and the extension of anti-discrimination protection to lesbians and gay men, protection from discriminatory law enforcement and policing practices (particularly in relation to public sex), and challenges to the infringement of lesbian and gay social rights (in relation for example, to HIV testing, drug experimentation, insurance rights, travel and migration restrictions, etc.) (see Evans, 1993).

According to Richardson (2000, p. 110), "although the language of rights largely speaks to the freedoms and obligations of the citizen, many citizenship rights are grounded in sexual coupledom rather than rights granted to us [lesbians and gay men] as individuals." She identifies three key areas of relationship-based rights: the right of consent to sexual practice in personal relationships, the right to freely choose sexual partners, and the right to publicly recognized sexual relationships. Such demands are evident in recent high-profile campaigns in the USA, UK, and other European countries for the right of lesbians and gay men to "marry" or form what has been termed state-registered partnerships (Bech, 1992; Rankine, 1997), a practice which is now legal in Denmark and the Netherlands.

A recognition of the uneven contours of citizenship rights between countries (Evans, 1993) has led to the formalization of transnational networks such as the International Lesbian and Gay Association into a coordinated Federation of Lesbian and Gay Organisations. For gay activists such as Peter Tatchell (1992, p. 75), "it is through collective solidarity, overriding national boundaries and sectional interests, that we [lesbians and gay men in the European Union] have our surest hope of eventually winning equality."

Such international campaigns based on imaginings of sameness (in these cases on the basis of same-sex desire) have led some to claim that such forms of social identification might threaten the emotional attachment and sense of loyalty people feel to the state and its authority, potentially "disuniting the nation" (Morley and Robins, 1995; Schlesinger, 1992). Hobsbawn (1990, p. 11) observes "we cannot assume that for most people national identification – when it exists – excludes, or is always or ever superior to, the remainder of the set of identifications which constitute the social being. In fact it is always combined with identification of another kind, even when it is felt superior to them." However, Littleton (1996, p. 1, cited in Isin and Wood, 1999) is more forthright, claiming that "[i]nstead of regarding themselves as citizens of sovereign nation-states, much less citizens of the world, many people have come to see themselves primarily as members of a racial, ethnic, linguistic, religious or gender groups."

Indeed, Watney (1995) has even gone so far as to suggest that there is a queer diaspora. Comparing the position of lesbians and gay men to that of Jewish people who are often regarded as "foreigners" inside the nation, he argues that lesbians and gay men are also in a form of internal exile. He writes "Wherever the nation is popularly envisaged as if it were a closed family unit, homosexuality may also be perceived as similarly threatening, a refusal of homogeneity and sameness understood as indispensable aspects of properly 'loyal' national identity" (op. cit., pp. 60–1). While recognizing the dangers of promoting a homogenizing transnational sexual identity which fails to recognize the way that sexuality "might intersect in unpredictable ways with...contingent national identities," Watney (1995) argues that nonetheless the notion of a queer diaspora holds a certain appeal for lesbians and gay men. He writes: "if only because it is likely to accord to our direct experience of overseas travel, as well as of queer culture and its constitutive role in our personal life. Few heterosexuals can imagine the sense of relief and safety which a gay man or lesbian finds in a gay bar or dyke bar in a strange city in a foreign country. Even if one cannot speak the local language, we feel a sense of identification" (op. cit., p. 61).

Despite the fact that both national and transnational campaigns to extend full citizenship rights to lesbians and gay men have achieved some degree of success, they have also been widely criticized. Epstein et al. (2000) observe that many of these struggles take place in parliamentary space which is itself heterosexual and masculinist. As such legal debates can create a space or platform for the moral right to promote anti-lesbian and gay views, which, in turn, can help to produce a climate of opinion in which homophobia and discrimination are legitimated. Likewise, Cooper (1993) points out that the very structures which are required to implement any legal reforms are also (hetero)sexualized in ways which are not addressed by equal rights campaigners. Police and judiciary who are less sympathetic to progressive laws may therefore undermine this legislation. Indeed, by imposing equality the state can precipitate a conservative backlash (Cooper, 1993). Commenting on the state's lack of omnipotence, Cooper (op. cit., p. 203) observes that "the ability of the state to reshape sexual ideologies and practices is limited. Just as conservative forces cannot eradicate sadomasochism through criminal prosecutions and convictions, lesbian and gay forces cannot achieve social parity with heterosexuality simply by law reform and municipal policies."

Indeed, by locating the source of lesbian and gay oppression in the public spheres of employment, policing, and service provision law reform campaigns can also unwittingly foster a false public–private division (Herman, 1993). As Herman (1993, p. 253) explains, "a focus on 'public sphere' discrimination may leave unsaid and therefore unaddressed one of the primary sites of the construction and enforcement of heterosexuality – home and family relations."

Indeed, he goes on to argue that the whole rights framework (for example of registered partnerships and antidiscrimination legislation) merely serves to pull in "new" identities, assimilating or integrating them into the norms and practices of heterosexual institutions "thereby regulating them, and containing their challenge to dominant social relations" (Herman, 1993, p. 251).

In the process, lesbians and gay men are both homogenized and divided. They are homogenized in the sense that sexual dissidents are implicitly constructed as a minority group with a common purpose in ways which can come dangerously close to essentializing this group identity and which obscures the fact that sexual identities are fluid and shifting rather than stable and fixed. And they are homogenized in the sense that the category "lesbian and gay" conceals a very complex and diverse range of people with multiple social identities and different politics, and who also experience different and sometimes conflicting forms of oppression (not only in terms of traditional axes of difference such gender, class, "race," ableism, and age, but also in terms of parenthood, sexual attitudes/practices, lifestyle, and so on). For example, lesbians have critiqued gender-neutral notions of citizenship which fail to differentiate between the experiences of lesbians and gay men, and have pointed out the tendency of rights campaigns to focus on men's issues such as equalizing the age of consent for gay male sex. As Phelan (1994, p. xi) observes, the "paradigm of 'lesbian and gay' too often heralds a return to male-dominated politics."

At the same time as homogenizing lesbians and gay men, the equal rights framework can also accentuate divisions between those lesbians and gay men who aspire to "the norm" and therefore want to be assimilated into heterosexual institutions, and those who stress the radical differentness of lesbians and gay men from heterosexuals.

The rights model of sexual citizenship usually carries with it certain obligations or duties. Notably, in return for particular civil, political, and social rights the sexual citizen is expected to conform to the (hetero)norm with all the expectations of responsibility and privacy, etc. that this entails rather than to publicly challenge and disrupt heterosexual institutions and ways of life. While some groups of lesbians and gay men are happy to settle for this, others argue that sexual dissidents should not be trying to gain rights in a framework that privileges heterosexuality and that defines lesbians and gay men as the deviant "other" (thereby upholding it). Rather, they argue sexual dissidents should be challenging the very normalcy of heterosexuality and communicating a new progressive sexual order to other groups. In particular, they question the desirability and necessity of monogamous couple-based sexual relationships that are modeled on heterosexual gender norms, and highlight instead other ways of having sexual relationships (Delphy, 1996). For example, "public (homo)sex runs against many societal constructs of intimacy, with the casual anonymous encounter being thought of as the very antipathy to the romantically charged (and heteronormative) model of sexual love" (Bell, 1995, p. 306). Giddens (1992) suggests that without the asymmetrical power relationships which frame heteronormative constructions of love and sex, such (homo)sexual relationships offer a possibility for more autonomous, democratic and liberatory forms of sexual intimacy. Indeed, other writers argue that rather than claiming rights in the law, sexual dissidents should be (out)laws, developing their own approaches to relationships, ways of living, the law, and citizenship. The following section explores one such attempt to do so.

The Sexual Politics of Subversion

In April 1990, Queer Nation was established by a group of sexual dissidents determined to fight lesbian and gay bashing in the East Village, New York, USA. Borrowing confrontational tactics from AIDS activists, feminism, and black liberation, these activists aimed to challenge heterosexual hegemony and to make the nation a safe space for "queers" (Munt, 1998). A few weeks later a similar group known as Outrage was set up in London, UK.

Smyth (1992, p. 20) claims that: "queer marks a growing lack of faith in the institutions of the state, in political procedures, in the press, the education system, policing and the law. Both in culture and politics, queer articulates a radical questioning of social and cultural norms, notions of gender, reproductive sexuality and the family." She continues: "We are beginning to realise how much of our history and ideologies operate on a homo-hetero opposition, constantly privileging the hetero perspective as normative positing the homo perspective as bad and annihilating the spectrum of sexualities that exists." As such, the term "queer" refers not only to those who define themselves as gay or lesbian, but also to anyone whose practices, identities or sympathies fall outside the hegemonic heterosexual and gender regimes of the mainstream.

In contrast to the sexual politics of equality practiced by those who want to achieve full citizenship rights for lesbians and gay men within the existing social and political framework, queer represents a more radical form of sexual politics. Whereas equal-rights activists stress the sameness of lesbians and gay men to heterosexuals, queer activists highlight the differences. Whereas equal-rights activists seek assimilation or incorporation into the center, queer activists aim to disrupt, destabilize, and subvert the mainstream.

Under slogans such as "We're here! We're Queer! Get used to it!" activists have publicly performed queer sexualities in the streets as a means of highlighting heterosexual hegemony while also parodying it. For example, in the 1990s Outrage held KISS-INs at the statue of Eros in Piccadilly Circus, and a Queer Wedding in Trafalgar Square. Activists have also abseiled into the House of Lords chamber during a debate, chained themselves to the railings of Buckingham Palace and invaded the studio of the BBC television six o'clock evening news program during a live broadcast (Smyth, 1992).

As these examples suggest colonizing and occupying space has proved an important queer tactic (Munt, 1998). One queer slogan runs: "Let's make every space a Lesbian and Gay space. Every street a part of our sexual geography. A city of yearning and then total satisfaction. A city and a country where we can be safe and free and more' (cited in Richardson, 2000, p. 48). To this end, the sort of spaces that have been targeted for occupation are national institutions that are not only seats of traditional power but are also spaces of national sentiment that have important symbolic meanings in the production of the nation (Munt, 1998).

Whereas equal rights activists have marched through key public spaces to raise the visibility of the lesbian and gay "community," queer aims to actually "reterritorialise public space" (Richardson, 2000, p. 49). Through performative strategies queer activists – like feminists – not only challenge heterosexual/homosexual binaries but "disrupt the ways in which certain issues (marriage, domestic violence, incest) as well as subjects (women, gay men, heterosexuals) are constituted as belonging in public or private spheres" (Richardson, 2000, p. 49).

However, despite being founded on a principle of inclusiveness, embracing transvestites, bisexual people, sadomasochists, and transsexuals and celebrating subversion, ambiguity, and sexual freedom, queer has still been unable to rid itself of accusations that it is a predominantly white, middle-class masculinist movement (Bell, 1995). Fissures have particularly opened up along the usual lines of gender, "race," and ableism (see Smyth, 1992). As Munt explains (1998, p.15), "[i]n its rush to affirm a new political moment Queer Nation, like many nationalisms, refused and 'forgot' the complex lessons of history, and broke apart over the same social divisions evident in . . . other single-issue projects."

Conclusion

This chapter has traced the contours of sexual politics, focusing on three different campaigns through which lesbians and gay men have sought tolerance from mainstream society, equality with straight citizens, and to subvert heterosexual hegemony. Each of these forms of sexual politics has been articulated through different spatialities. While campaigns for tolerance were based on respect for gay men's privacy, those of equality have focused on the rights of sexual minorities to be lesbian and gay in both public and private space, while queer politics has attempted to challenge and disrupt public/private binaries.

The extent to which each of these political strategies might be understood as successful, or potentially successful, is hotly disputed by sexual dissidents. In particular, contemporary tensions are evident between those who favor pursuing equality for lesbians and gay men within contemporary society and those who, in the words of Outrage, want to "fuck up the mainstream." These clashes in approach have raised questions, about who are sexual dissidents, how they can be represented and how any form of sexual politics can be truly inclusionary. Watney (1995, p. 62), for example, comments: "it remains far from clear on what basis current lesbian and gay political 'leaders' claim to represent their constituency, or how that constituency itself might best be conceptualised, beyond exhausted and inadequate notions of lesbian and gay 'community'." Likewise, there is little agreement over what tactics are most successful in advancing the cause of sexual dissidents. Watney (1995) goes so far as to suggest that it is consumption rather than any form of confrontational politics – be they campaigns for tolerance, equality or radical change – which has proved the most effective means through which spaces of sexual citizenship have been constituted. He writes: "it is only too apparent in many countries, including the UK, that some twenty years of puritanical leftist and feminist lesbian and gay politics have achieved less for most self-defining lesbians and gay men, and queers than has resulted from changes and expansion within the commercial scene, including the dramatic expansion of social facilities, cable TV, the gay and lesbian press and so on" (Watney, 1995, p. 63).

Indeed, while conventional political parties across Western Europe are concerned that young voters are more interested in the politics of lifestyle and consumption than they are with the traditional party political democratic process, similar fears are apparent within sexual politics too. The commercialization of lesbian and gay space and the commodification of lesbian and gay lifestyles are being blamed for undermining both rights based, and more radical forms of politics, as lesbians and gay men have "never had it so good." Yet, at the same time, sexuality continues to be a means of social control and regulation and sexual issues, from AIDS to homophobic violence, remain as pressing as ever.

Acknowledgments

I am very grateful for the support of the ESRC (award no: L134251032) and The Philip Leverhulme Prize which enabled me to write this chapter.

BIBLIOGRAPHY

Andermahr, S. 1992. Subjects or citizens? Lesbians in new Europe. In A. Ward et al. (eds.) *Women and Citizenship in Europe: Borders, Rights and Duties: Women's Differing Identities in a Europe of Contested Boundaries*. London: Trentham.

Anderson, B. 1983. *Imagined Communities: Reflections on the Origin and Spread of Nationalism*. London: Verso.

Anthias, F. and Yuval-Davis, N. 1992. *Racialized Boundaries: Race, Nation, Gender, Colour and Class and the Anti-Racist Struggle*. London: Routledge.

Bech, H. 1992. Report from a rotten state: "marriage" and "homosexuality" in "Denmark". In K. Plummer (ed.) *Modern Homosexualities*. London: Routledge.

Bell, D. 1995. Perverse dynamics, sexual citizenship and the transformation of intimacy. In D. Bell and G. Valentine (eds.) *Mapping Desire: Geographies of Sexualities*. London: Routledge.

Betten, L. 1993. Rights in the workplaces. In K. Waaldijk and A. Clapham (eds.) *Homosexuality: A European Community Issue: Essays on Lesbian and Gay Rights in European Law and Policy*. Dordrecht: Martinus Nijhoff.

Binnie, J. 1997. Invisible Europeans: sexual citizenship in the New Europe. *Environment and Planning A*, 29, 237–48.

Brown, M. 2000. *Closet Geographies*. London: Routledge.

Cooper, D. 1993. An engaged state: sexuality, governance, and the potential for change. In J. Bristow and A. Wilson (eds.) *Activating Theory: Lesbian, Gay and Bisexual Politics*. London: Lawrence & Wishart Ltd, 190–218.

Corviono, J. 1997. *Same Sex: Debating the Ethics, Science and Culture of Homosexuality*. Lanham, MD: Rowman and Littlefield.

Delphy, C. 1996. The private as a deprivation of rights for women and children. Paper presented at the International Conference on Violence, Abuse and Women's Citizenship, Brighton, November 10–15.

Donald, J. 1993. How English is it? Popular literature and national culture. In E. Carter et al. (eds.) *Space and Place: Theories of Identity and Location*. London: Lawrence & Wishart.

Epstein, D., Johnson, R., and Steinberg, D. L. 2000. Twice told tales: transformation, recuperation and emergence in the Age of Consent Debates 1998. *Sexualities*, 3, 5–30.

Evans, D. 1993. *Sexual Citizenship: The Material Construction of Sexualities*. London: Routledge.

Giddens, A. 1992. *The Transformation of Intimacy: Sexuality, Love and Eroticism in Modern Societies*. London: Polity.

Herman, D. 1993. The politics of law reform: lesbian and gay rights struggles into the 1990s. In J. Bristow and A. Wilson (eds.) *Activating Theory: Lesbian, Gay and Bisexual Politics*. London: Lawrence & Wishart, 245–62.

Herman, D. 1994. *Rights of Passage: Struggles for Lesbian and Gay Equality*. Toronto: University of Toronto Press.

Hobsbawm, E. J. 1990. *Nations and Nationalism since 1780*. Cambridge: Cambridge University Press.

Isin, E. F. and Wood, P. K. 1999. *Citizenship and Identity*. London: Sage.

Jefferys, S. 1994. *The Lesbian Heresy: A Feminist Perspective on the Lesbian Sexual Revolution*. London: The Women's Press.

Kofman, E. 1995. Citizenship for some but not for others: spaces of citizenship in contemporary Europe. *Political Geography*, 14, 121–37.

Lister, R. 1997. *Citizenship: Feminist Perspectives*. New York: New York University Press.

Marshall, T. H. 1950. *Citizenship and Social Class*. Cambridge: Cambridge University Press.

Morley, D. and Robins, K. 1995. *Spaces of Identity: Global Media, Electronic Landscapes and Cultural Boundaries*. London: Routledge.

Muir, R. 1997. *Political Geography*. Basingstoke: Macmillan.

Munt, S. 1998. Sisters in exile: the lesbian nation. In R. Ainley (ed.) *New Frontiers of Space, Bodies and Gender*. London: Routledge.

Phelan, S. 1994. *Getting Specific: Postmodern Lesbian Politics*. Minneapolis, MN: University of Minnesota Press.

Rankine, J. 1997. For better or for worse? *Trouble and Strife*, 34, 5–11.

Richardson, D. 1998. Sexuality and citizenship. *Sociology*, 32, 83–100

Richardson, D. 2000. Claiming citizenship? Sexuality, citizenship and lesbian/feminist theory *Sexualities*, 3, 255–72.

Schlesinger, A. 1992. *The Disuniting of America*. New York: Norton.

Sedgwick, E. K. 1994. *Tendencies*. London: Routledge.

Sharp, J. 1996. Gendering nationhood: a feminist engagement with national identity. In N. Duncan (ed.) *Bodyspace: Destablising Geographies of Gender and Sexuality*. London: Routledge.
Sibley, D. 1995. *Geographies of Exclusion: Society and Difference in the West*. London: Routledge.
Smith, S. J. 1989. Society, space and citizenship: a human geography for "new times". *Transactions of the Institute of British Geographers*, 14, 144–56.
Smyth, C. 1992. *Lesbians Talk Queer Notions*. London: Scarlet.
Tanca, A. 1993. European citizenship and the rights of lesbians and gay men. In K. Waaldijk and A. Clapham (eds.) *Homosexuality: A European Community Issue: Essays on Lesbian and Gay Rights in European Law and Policy*. Dordrecht: Martinus Nijhoff.
Tatchell, P. 1992. *Europe in the Pink*. London: Gay Men's Press.
Valentine, G. 1993. (Hetero)sexing space: lesbian perceptions and experiences of everyday spaces. *Environment and Planning D: Society and Space*, 11, 395–413.
Valentine, G. 1996a. An equal place to work? Anti-lesbian discrimination and sexual citizenship in the European Union. In M. D. Garcia-Ramon and J. Monk (eds.) *Women of the European Union: The Politics of Work and Daily Life*. London: Routledge.
Valentine, G. 1996b. (Re)negotiating the heterosexual street. In N. Duncan (ed.) *Bodyspace: Destablizing Geographies of Gender and Sexuality*. London: Routledge.
Waaldijk, K. 1993. The legal situation in member states. In K. Waaldijk and A. Clapham (eds.) *Homosexuality: A European Community Issue: Essays on Lesbian and Gay Rights in European Law and Policy*. Dordrecht: Martinus Nijhoff.
Walby, S. 1997. *Gender Transformations*. London: Routledge.
Watney, S. 1995. AIDS and the politics of Queer Diaspora. In M. Dorenkamp and R. Henke, (eds.) *Negotiating Lesbian and Gay Subjects*. London: Routledge.
Weeks, J. 1977. *Coming Out. Homosexual Politics in Britain from the 19th Century to Present*. London: Quartet.
Wilson, A. 1993. Which equality? Toleration, difference or respect. In J. Bristow and A. Wilson (eds.) *Activating Theory: Lesbian, Gay and Bisexual Politics*. London: Lawrence & Wishart, 171–89.
Wilton, T. 1995. *Lesbian Studies: Setting An Agenda*. London: Routledge.

Part VI Geographies of Environmental Politics

Chapter 27

The Geopolitics of Nature

Noel Castree

Introduction

Did the three short years between the fall of the Berlin Wall and the convening of the United Nations "Earth Summit" symbolize the demise of one geopolitical order and the tentative emergence of another? Several politicians, diplomats, and strategic analysts think so. With the eclipse of a Cold War geopolitics in which two opposed "worlds" – the West and the communist bloc – vied for control of another – the so-called "Third World" (witness the infamous wars in Vietnam and Afghanistan, for example) – commentators have "been searching ever since for a new global drama to replace it..." (Ó Tuathail and Dalby, 1998, p. 1). Opinion is divided over how to make sense of this new era of intergovernmental relations. For many geopolitical "experts," be they academic analysts or policy practitioners, the "new world dis/order" is one where the decline of US hegemony, the withering of state communism, rampant economic globalization, the "hollowing out" of the nation-state and the telecommunications revolution have conspired to render "classical geopolitics" obsolete. In a hyper-integrated "borderless" world, it is often argued that the tenets of twentieth-century statecraft – wherein dominant states sought to occupy and encircle strategically important areas of land, sea, and resources – make little sense at the dawn of the new millennium.[1] Other commentators, while agreeing that the significance of the "geographic factor" in world politics has changed, place less emphasis on the globalization of production, trade, and infomatics. For them, the emerging geopolitical order is (or should be) founded in significant part on the unprecedented challenges posed by an array of new environmental problems. In a world where the ecological interdependence of states is more obvious and acute than at any time in modern history, these analysts see these problems as a new "ordering principle" for what might be called a post-Cold War "environmental geopolitics."

This explains my reference above to the so-called Earth Summit. This event, held in Rio de Janeiro, was the largest gathering of government officials, diplomats, civil servants, and non-governmental organizations (NGOs) the world had ever seen (over

25,000 delegates attended). Yet it was not about the venerable intergovernmental questions of trade, economics or armed conflict. Nor was it about how to deal with the fall-out of the dramatic "revolutions" in Eastern Europe and the USSR during the late 1980s/early 1990s. Instead, it was devoted to considering a range of local, regional, and global environmental problems that had become more numerous and apparent since the early 1970s. This is not to say that environmental issues were wholly irrelevant to geopolitics earlier in the twentieth century.[2] But it is to say that these issues were never deemed sufficiently serious and widespread to alter the whole pattern of intergovernmental relations on a continental or global scale. In Rio, by contrast, it was clear that the environmental problematic was going to be a key element in any post-Cold War geopolitical order. For by the early 1990s environmental problems had become "globalized:" that is, they were marked by a new *extensiveness* (geographical reach), *intensity* (depth and seriousness) and *velocity* (speed of development and spread) (Held et al., 1999, p. 376). In other words, local environmental problems were becoming global environmental problems (and vice versa) in ways that readily transcended international political boundaries. What the Rio delegates were forced to confront were two broad classes of environmental problem that together implicated states – however geographically distant from one another – in each other's environmental practices (Benedick, 1999). The first type was transnational environmental problems, such as greenhouse warming. These are problems that are international or global in scale, simultaneously affecting several nation-states. The second type was "local;" environmental problems that were "leaky" – that is, problems that arose in one country but which have serious implications (moral, financial, aesthetic, health, etc.) for a whole set of other countries.[3] An example, discussed at Rio, is the loss of biodiversity. Biodiversity is concentrated in a few poor, tropical countries but is valued by nation-states worldwide because many commercial crops and medicines derive from biological materials harvested in these biodiverse countries.[4]

A decade on from Rio can we identify a new geopolitical order organized largely as a response to these two broad classes of environmental problems? Probably not. We are still very much in a period of post-Cold War transition characterized by geopolitical experimentation, institutional learning and new interstate power plays. Nonetheless, what we can say is that environmental questions have never mattered so much in world politics as they do now. For instance, since Rio the number and scale of intergovernmental meetings and agreements about environmental matters has increased dramatically. At the same time, other nonstate actors in world politics (such as the World Bank and international NGOs such as Greenpeace) are confronting environmental issues with unprecedented vigor. In short, although we do not live in a new "environmental geopolitical order," we arguably do live in a world where "environmental geopolitics" is very much part of the larger interstate order that is slowly solidifying.

In this chapter I propose to describe, explain and evaluate this emergent environmental geopolitics. As we'll see, this is a fascinating but far from easy task. There is no single God's eye view of interstate relations that will offer us a neutral understanding of environmental geopolitics today. I say this for two reasons. First, all analyses of geopolitics themselves have a geography and a politics. Until recently, it was presumed that geopolitical analysts described – or at least sought to describe –

world politics as it was and is. In other words, it was long believed that geopolitics was a relatively dispassionate "science" which could offer an objective, Olympian, "big-picture" view of trends in international politics. However, a new generation of political geographers and international relations scholars has shown that all geopolitical knowledge is both situated and inherently value-laden (Ó Tuathail and Dalby, 1998). It is situated because it is crafted by individuals, organizations, and institutions that are embedded in particular geographic contexts. And it is value-laden because, implicitly and explicitly, the contexts in which geopolitical analysts think and write decisively affect how they view interstate relations. Moreover, once produced and circulated, geopolitical knowledge can affect the geopolitical "realities" it purportedly describes: it is constitutive not merely reflective. Accordingly, geopolitics is best seen as a field of power-knowledge. It does not innocently describe world politics. It is knowledge of world politics *by* certain situated analysts *towards* very particular geopolitical ends (whether or not those analysts are aware of the fact). As Ó Tuathail (1998a, p. 3) puts it, we cannot "... assume that 'geopolitical discourse' is the language of truth; rather, ... [we] understand it as a discourse seeking to establish and assert its own truths." When analysing any form of geopolitical reasoning we therefore have to ask: what kind of geopolitical knowledge for what kind of geopolitics?

As we will see, asking this key question allows us to offer a complex and sophisticated account of environmental geopolitics today.[5] It gets us away from taking the claims of the new environmental geopoliticians – such as former US Vice-President Al Gore[6] – at face value. It allows us to recognize that there is more than one valid perspective on environmental geopolitics today. And it enables us to identify which of these perspectives is hegemonic and which, for better or worse, is marginalized by those in positions of political power. There is, however, a rub (and this leads me to my second point about the non-neutrality of geopolitics). For if all geopolitical knowledge is value-laden it follows that the knowledge produced here is too. It is therefore incumbent upon me to declare my particular biases at the outset of this chapter.

My tack is to combine ideas drawn from "geopolitical economy" and "critical geopolitics"[7] in order to argue the following: that currently dominant visions of the pattern of environmental geopolitics are a form of power-knowledge which help perpetuate global inequality and environmental degradation. As such, these visions need to be resisted and overturned (see Bernstein, 2000). Accordingly, this chapter should be read as one small attempt to challenge these visions. Environmental geopolitics – as both a set of practices and a set of discourses about those practices – is not simply the "high politics" practiced by agents of statecraft. It also involves the words and deeds of myriad actors outside the formal apparatuses of the state – including "critical" academics and members of the citizen-public. Governments are only able to accomplish their goals in the environmental arena with the full or partial support of an array of non-governmental constituencies. If, by the end of this chapter, I've encouraged my readers to question the taken-for-granted "realities" of environmental geopolitics today then something worthwhile will have been accomplished. Whether or not readers actually agree with the arguments I make, I hope it will at least be clear just how much is at stake in the ongoing attempts by states to grapple with contemporary environmental change.

Environmental Geopolitics: a Framework for Analysis

As noted, geopolitical discourses and practices are multiple and contingent: there is no "essence" to geopolitics. Since the term was coined by Swedish political scientist Rudolph Kjellen in 1899, geopolitics has been reworked in its various historical–geographic contexts of knowledge and practice.[8] A century ago the term was associated with the imperialist power politics of countries such as Germany and Great Britain. Framed by key intellectuals such as Karl Haushofer, Halford Mackinder and Alfred Mahan,[9] the imperialist geopolitical imagination focussed on controlling, containing or limiting access to what were seen as strategically important spaces (land, sea or resources). Then, once the Second World gave way to a Cold War "three worlds" geopolitics, analysts like American George Kennan envisioned a global political map in which the ideological conflict between "blocs" (East and West) was to be fought out with nuclear deterrents and on the killing fields of developing countries like Vietnam.[10] However, for all the specific differences between imperialist and Cold War geopolitics both shared an overarching – and what John Agnew (1998) calls specifically "modern" – worldview. This worldview, which stretched back to the European Renaissance, imagined the world as a set of distinct parcels of political space that could be surveyed, and strategized about, from on-high. In this worldview, according to Agnew, states were seen as being the most important political actors. They were also seen as sovereign actors, such that "domestic" and "foreign" affairs could be separated and national identities readily identified. Finally, the modern geopolitical worldview saw interstate relations as a contest between sovereign entities whose power rested on control of resources, land or sea.

I mention all this because the globalization of environmental problems – along with the globalization of production, trade, and telecommunications – has brought the modern geopolitical imagination into a state of crisis (Ó Tuathail, 1998b). As will be seen, transnational and "leaky local" environmental problems have together posed a major challenge to modern geopolitical discourses and practices. It is a challenge of both substance and mode. Substantively, the globalization of environmental problems has had a number of effects. Among other things it has: (i) eroded the seeming inviolability of national boundaries; (ii) meant that seemingly domestic matters (such as using coal-fired power stations) are in fact transnational issues (since burning coal locally contributes to global warming); (iii) compelled states to cede some of their sovereignty "upwards" to transnational quasi-governmental actors such as the Global Environmental Facility; (iv) led states to participate in a plethora of international environmental agreements and actions; and (v) brought non-state actors into world politics, such as the major environmental NGOs like Greenpeace. In terms of mode, the fact that many environmental problems have a supranational dimension has meant that an older geopolitics of militaristic conflict and diplomatic standoff over "key spaces" has necessarily (but not exclusively) given way to more consensual, communal, and negotiation-based forms of interstate relations. However, it should be noted that this "consensus" politics is often based on wealthy, powerful states – like the USA, Germany, and the UK – applying diplomatic, financial, and other pressures on weaker states in order to further their own international agendas.

This said, it would be wrong to think that the environmental problematic ushers in a new "postmodern" form of geopolitics. It is sometimes said the two types of environmental problems identified above hail a novel era of "threats without enemies" (Prins, 1993) since they do not always originate with one or another nation-state which can be held to account. But these problems are, in fact, only *partly* "deterritorialized." After all, phenomena like greenhouse warming may indeed be "global" in scale, but the fact remains that they must be dealt with by over 170 differently empowered national governments possessed of legal sovereignty over particular parcels of space. Likewise, an ostensibly international "resource" like biodiversity is, in practice, only international in name; *de facto*, it must be managed with the consent (and within the politically constructed borders) of those few tropical countries that are so-called "biodiversity hotspots" (like Cameroon and Costa Rica).

In geopolitical terms, then, the environmental challenges of the twenty-first century represent a dialectic of territorialization/deterritorialization, a mixture of spatial fixity and unfixity. It's here, though, that things start to get really interesting and complicated. For one of the arguments of this chapter will be that *there is a geopolitics to how environmental problems are represented*. We cannot, in other words, "read-off" how interstate relations will be changed by the environmental problems of the new millennium as if we can all agree on what the "facts" of the case are. While these problems are undoubtedly real, there is no objective perspective on their nature, causes, and solutions. Instead, we have an array of actors – such as states, NGOs, quasi-governmental bodies, and environmental scientists – all claiming to know the "truth" about these problems (to the extent that what is defined by some actors as an environmental "problem" is not seen as one by others). As Dalby (1998a, p. 180) puts it, "how these [problems] . . . are described and who is designated as either the source of the problem, or provider of the potential solution . . . , is an important matter in how environmental themes are argued about and who gets to make decisions about what should be done by whom. . . ."

Adapting the ideas of Agnew and Corbridge (1995), we can think of different actors in the geopolitical arena offering different *geopolitical environmental discourses* which may – or may not – engender specific *geopolitical environmental practices*. The former are sets of statements about the contemporary "realities" of environment–state relations in specific continents or worldwide. These statements "frame" environment–state relations and identify (i) what the supposed "problems" are, and (ii) which states are responsible for causing and solving specific environmental problems. The framing process works through a process of discursive inclusion/exclusion, as some things are deemed "legitimate" subjects for debate while others are simply off the agenda.[11] Geopolitical environmental discourses are related to actual or potential geopolitical environmental practices. The latter are sets of established intergovernmental practices (diplomatic, administrative, financial, legal, etc.) which organize the way states deal with environmental problems. They may be consensual, coercive or, more rarely, even violent.

In theory, one discourse–practice cluster or regime could be hegemonic globally at any one time if one or more key state supports it (that is, the regime becomes a set of "ruling ideas and practices."[12] Following Cox (1981), this hegemony can be secured if this/these key state/s possess/es the right material capability (e.g. economic power) and institutional access (e.g. to suprastate organs like the UN or World Bank). However, given the geopolitical flux of the early twenty-first century, it is more likely that an array of contested global environmental discourse–practice regimes will exist within and between different world regions.[13] The geopolitical environmental discourse–practice relation is recursive or two-way. Particular actors with particular political interests frame their discourses in ways that are interesting as much for what they do not say as what they say. In turn, if enough other actors can be persuaded that a specific discourse is "true" (or at least credible), a set of new *geopolitical environmental practices* can be established in this basis. But in turn, these practices may function in ways that cause states and others implicated in world politics (directly and indirectly) to alter their discourses about the global environment.

Who, then, is involved in this contest of geopolitical environmental discourses and practices? At the broadest level, we can think of geopolitics as operating in three interrelated domains or sites (Ó Tuathail and Dalby, 1998, pp. 4–5). Geopolitics is not just about the thoughts and deeds of state officials and others (like the UN or Greenpeace) active in world politics (what we might call *practical geopolitics*). It also involves the *formal geopolitics* of non-governmental theorists and stategists (e.g. academics) and the *popular geopolitics* to be found in the mass media (see figure 27.1). The relations between these three domains – within and between

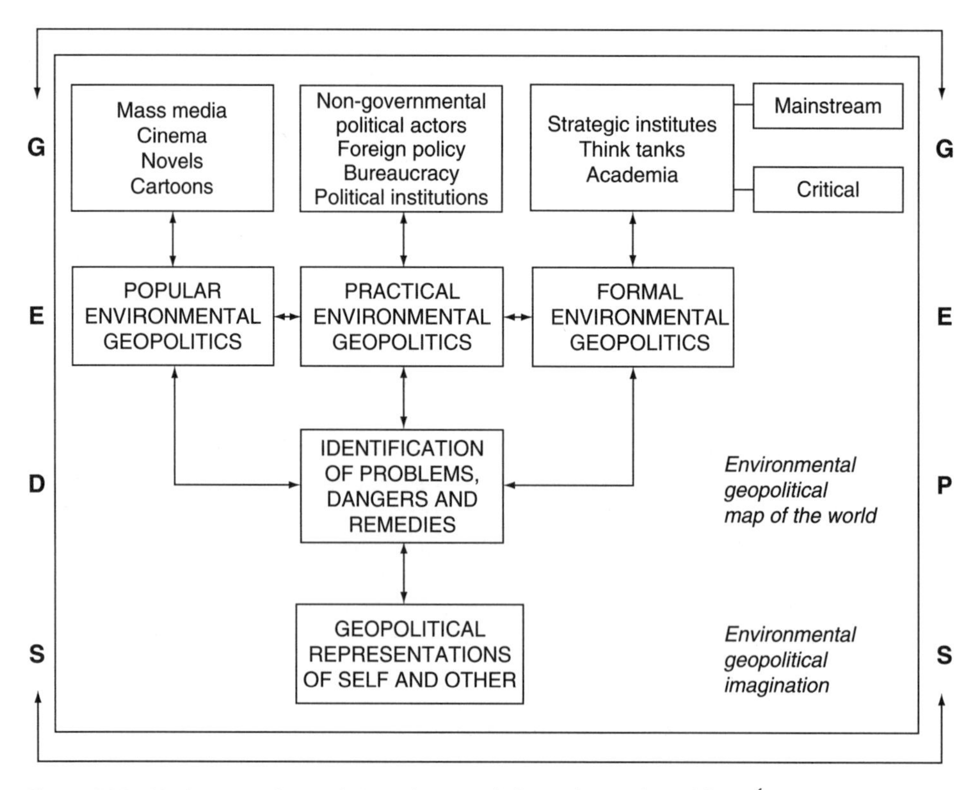

Figure 27.1 Environmental geopolitics: a framework for analysis (adapted from Ó Tuathail and Dalby, 1998, p. 5).

countries – are exceedingly complex. Practical and formal geopoliticians may use the media to affect public perceptions so that specific foreign policy goals are deemed acceptable. By the same token, however, "critical," non-governmental geopoliticians may question the geopolitical environmental discourses and practices advocated by states and their formal allies, using books such as this one to shape how the public thinks about intergovernmental/environment relations.

In the rest of the chapter I shall focus on the geopolitical environmental discourses and practices that are both cause and effect of practical and formal geopolitics today.[14] In other words, I will examine the major actors, practices and discourses involved in giving shape to environmental geopolitics in the post-Cold War world. My argument will be that while these arrangements do have some virtues when judged by *their own* criteria, they can be found wanting when judged by criteria that are, alas, presently marginalized by those in political power worldwide (see also Dalby, chapter 28, this volume).

Representations, Practices and Interpretations: "Global Environment Management" and "Environmental In/security"

The importance of environmental issues to intergovernmental relations is unprecedented. Even before the Cold War was over, states worldwide had begun to recognize the seriousness of regional, continental and global scale environmental problems. The UN Stockholm Conference on Environment and Development (1972, which created the UN Environmental Programme), the Convention on International Trade in Endangered Species (1973), the UN Convention on the Law of the Sea (1982), the Bruntland Report (which popularized the concept of "sustainable development" in 1987), the Montreal Protocol on chlorofluorocarbons (CFCs, 1987), and the international protocols on sulfur dioxide and nitrous oxide emissions (1988) were just a few of the intergovernmental activities indicating the environment's new-found importance in world political affairs. In retrospect, the Earth Summit was something of a turning point for environmental geopolitics for, with the disappearance of Cold War preoccupations (ideological and militaristic), there has been greater space for environmental matters to intrude on world political agendas.[15]

Simplifying, we can see that over the last decade two related sets of geopolitical environmental discourses and practices have risen to prominence in the G7 countries – that is, the major economic and political players in world affairs. The first is the discourse and practice of "global environmental management" (and its sub-global variants). Its advocates, such as Al Gore, see it as arising from the supposed imperatives of transnational environmental problems, which fall into two categories: those relating to environmental sinks (i.e. pollution problems, such as acid rain, greenhouse warming, ozone layer destruction, and the disposal of hazardous wastes in the high seas) and those relating to environmental resources (i.e. depletion problems, such as overfishing, whale hunting, the trade in endangered species, and the diversion of water from international rivers). In his book *The Earth in Balance* (1993), Gore outlines a multipoint plan that will, he argues, allow states worldwide to deal with these and other pressing transnational environmental problems. It hinges on a major commitment of the G7 countries to redistribute wealth to the developing world – through institutions like the Global Environmental Facility (the GEF, run on behalf of the international community by the UN and World Bank) – so that they can deal effectively with problems they are currently too poor to address.[16] This kind of global managerial vision has been realized – albeit in more restrictive forms – in many of the new institutions and agreements stemming from the Rio conference. For instance, guided by the Intergovernmental Panel on Climate Change (a creation of the UN and the World Meteorological Organisation), most world governments have now signed up to a legally binding Convention on Climate Change (CCC) committing them to a significant reduction in present and future carbon dioxide emissions. The wealth redistribution needed for former communist and "Third World" countries to pay for this reduction will, in theory, come from the GEF and a new "carbon tax" regime which allows Western countries to purchase carbon credits from less wealthy nations.[17]

Related to the global managerial GED/GEP is that of "environmental in/security." From the early 1990s, a number of commentators began to see environmental problems as new "security threats" to nation-states, replacing Cold War threats such as nuclear arsenals. Emblematic of this form of geopolitical praxis are the influential essays written by the American geopolitical strategists Robert Kaplan and Thomas Homer-Dixon. Kaplan's apocalyptically titled "The coming anarchy" (published in *Atlantic Monthly* in 1994 and widely read in Washington, DC) sees a world of "rogue states" (such as Iraq) and "failed states" (such as Afghanistan) using the dependence of countries like the USA on their resources (such as oil) against them. In a related polemic, Homer-Dixon (1996) argues that resource scarcity and natural hazards in many non-Western states (think of Ethiopia's successive famines or Mozambique's recent floods) leads to poverty and mass migrations that disrupt trade relations and absorb large amounts of Western aid. The implication of both these analyses is that Western states must start to categorize foreign countries according to the type and degree of environmental threat they pose, while devising appropriate preventative and curative strategies (through a mixture of trade, aid, diplomacy, military action, etc.).[18]

More generally, environmental in/security relates to both transnational and "leaky local" environmental problems. In this case, the former often correspond to the "threats without enemies" scenario mentioned earlier since no single nation-state is culpable. In the case of "leaky local" environmental problems, though, the "culprits" are often more readily identifiable. For instance, when Iraq invaded Kuwait in 1991, its control of a set of "local resources" – its own large oil fields and those of Kuwait – posed a security threat to the oil-dependent G7 nations. Less confrontationally, the successive attempts since Rio to negotiate a binding Convention on Biodiversity have seen the USA, as the world's largest supplier of pharmaceuticals, resist attempts by biodiverse states to retain legal control of their plant, animal, and insect resources. By insisting that biodiversity is the "common heritage of human kind" and linking TRIPs (that is, the Trade Related Aspects of Intellectual Property agreement) to the current world trade regime, the USA has sought to gain free access to foreign bioresources while retaining economic control of any commercial products arising from the development of those resources by US biotech companies (like Monsanto).[19]

Together, the language and practices of global environmental management and environmental in/security have, over the last decade or so, fed into the development of an array of new geopolitical regimes. In these regimes, several nation-states cooperate over an actual or potential environmental problem. They do so within a common discursive frame (a specific global environmental discourse) which engenders a set of shared global environmental practices, often ratified in international protocols and conventions. What's so remarkable about these regimes is that they have tackled the hoary problem of how to get multiple different states to cooperate in the absence of a central, supra-state body with the power to enforce compliance (what Rosenau and Cziempel, 1992, call "governance without government"). These regimes include the new Convention on Climate Change (CCC), the partly successful protocols on chloroflurocarbons and sulfur/nitrous oxides, the International Whaling Commission (IWC) led regulation of commercial whaling, the Basel Convention on Transboundary Hazardous Wastes and the Convention on Biological Diversity.[20] True, not all of these regimes have held together particularly well. Donnelly (1986) usefully distinguishes declaratory regimes (the need for a regime is announced by one or more states), promotional regimes (one or more state actively pushes for regime creation), implementation regimes (specific global environmental discourses and practices are developed) and enforcement regimes (the regime comes into practical existence), where the first is merely verbal and the last fully-fledged and properly policed. Although there are currently few enforcement regimes in existence, it is still the case that there are far more promotional and implementation regimes on environmental matters than ever before in world political history.

How, then, are we to evaluate these several geopolitical environmental regimes? How, in other words, do we judge the new environmental geopolitics of global environmental management and environmental in/security? Since I've been arguing that the answer to questions like these depends very much on who or what is doing the answering, I propose to offer three different responses. These responses correspond to the three main traditions or paradigms of geopolitical analysis to be found in political geography and the discipline of international relations. They are, in other words, part of the formal geopolitics expounded by academics and think-tanks outside the government apparatus. But, as we'll see, they can and do affect practical environmental geopolitics by either bolstering or criticizing current discourses and practices. In other words, each tradition is itself a global environmental discourse which may (or may not) engender certain global environmental practices. In each case, I consider different transnational environmental problems and leaky local environmental problems in order to give a range of examples. And in each case I show how quite different causes of and solutions to these problems are posited.[21]

The realist response

Realism is an old and once-dominant approach to understanding interstate relations, stretching back to the hard-headed analyses of figures like Mackinder and Hans Morgenthau and represented more recently in Hedley Bull's (1977) *The Anarchical Society*.[22] It sees states as the dominant actors in world politics, notwithstanding the undoubted power of non-state actors like multinational companies and international NGOs. The reason for this is that realists attach great importance to sovereignty. Even with globalization and the rise of supranational organizations, realists see states as the ultimate legal–political authorities: there is no power above or below nation-states with comparable control over peoples, territories, and environments.

For realists, world politics is essentially anarchic, a ruthless and competitive process of dog-eat-dog. As MacGrew (1992, p. 19) puts it, "With no overarching authority in the global system, a situation . . . exists in which states must acquire power to defend and protect their vital interests." Interstate relations is thus about power politics, with states selfishly pursuing their own interests by whatever means possible. Since this "classical realist" view is rather extreme, a neo-realist perspective has recently emerged which emphasizes the ways "hegemonic powers" prevent all-out anarchy by cementing state relations in particular ways (the "hegemonic stability thesis;" see Vogler, 1992, pp. 128–30).

For realists and neo-realists, most transnational and "leaky local" environmental problems are caused by one or more states acting "rationally" according to their own needs. For instance, many transnational environmental problems relate to so-called "common resources," that is, resources which are not owned or controlled by any one nation-state but which are free to be used by several nation-states. Take greenhouse warming. The atmosphere is a "global commons" because it is a trans-boundary system that escapes sovereign control. Accordingly, states have treated it historically as a "free good." There is, in short, no incentive for states to spend money on reducing carbon dioxide pollution at home because other states may simply go on emitting CO_2 and other greenhouse gases indefinitely.[23] Similary, states that control "strategic resources" of value to the wider world community are seen as having a vested interest in maximizing any advantages to be gained from such control (think, for example, how Middle Eastern states raised world oil prices by deliberately withholding their oil supplies in 2000).

Why, then, have states in fact cooperated in trying to solve transnational and "leaky local" environmental problems in recent years? And why have environmental regimes not arisen where, arguably, they should have? Realists and neo-realists (like Kaplan and Homer-Dixon) argue that regimes will only arise where cooperation is in the national interest of one or more key states or where a key state has nothing to lose but something to gain from joining a regime. As Vogler (1992, p. 129) puts it, "A realist answer to our . . . question regarding the strength and weaknesses of regimes will thus be simple and robust. Regime arrangements will reflect bargains struck between the national interests of those possessing power in a particular issue area and will be sustained by just those power capabilities." In other words, states will adopt the global environmental discourses and practices of a particular regime if some or all of the following conditions apply: (i) mutual advantage (i.e. several states see the benefits of cooperation); (ii) credible threat (i.e. states see that non-compliance is not in their national interest); and (iii) credible penalty (i.e. states see that non-compliances will incur penalties – legal, economic, military, etc. – from other states).

The realist/neo-realist view is seemingly supported by the evidence of several recent environmental regimes (successful and not successful). For instance, as more scientific evidence has become available to show that greenhouse warming is a real socio-ecological threat, more states have come to realize the future economic, environmental, and human costs of dealing with changed weather patterns, sea-level rises and the like. Importantly, key states like Britain, France, and Germany have been at the forefront of CCC debates over "managing" the greenhouse effect, but the reluctance of the USA to sign-up to the CCC also seems indicative of the neo-/realist position: for the USA alone currently emits around 20% of world "greenhouse gases" and has thus required extra persuasion to spend the huge sums necessary to reduce its atmospheric pollution. By contrast, the USA *has* been a hegemonic presence in the conservationist whaling regime brokered by the IWC since the mid-1980s. The virtual cessation of commercial whaling for over a decade is, it appears, an achievement wrought not out of US altruism but out of the fact that the USA has no indigenous whaling industry to imperil through reducing whale kills. In a climate where the publics and governments of most Western nations regard whaling as cruel, the USA has sought to appear "environmentally friendly" by using trade sanctions and diplomatic pressure to persuade whaling nations, such as Japan, to allow endangered whale populations to recover their numbers (see Stoett, 1993).[24] In contrast to this hegemon-led regime, Israel, Jordan, and Syria have failed to agree to an integrated water management regime when, arguably, they should have. With all three states fast developing in what is a semi-arid area, water shortages have become acute. So why have they failed to agree on a fair way to access the transboundary river and groundwater resources they share? The neo-/realist answer would point to the severe religious divides between Israel (a Jewish state) and Syria and Jordan (Muslim states) and Israel's superior military capacity to gain access to land adjacent to the River Jordan (the West Bank, a territory Jordan has tried to reclaim for over two decades; see Gleick, 1993). Finally, the success of the USA and other G7 countries in leading a concerted military campaign against Iraq in 1991 seems all too redolent of the neo-/realist power-political view of how vulnerable states might deal with an environmental security threat (see Barber and Dickson, 1995).

The liberal-pluralist response

Plausible though the neo-/realist perspective may appear, it does not tell the whole story about environmental geopolitics today. For liberal-pluralists, like Keohane (1984), supranational problems do arise in large part because states (and their home populations) pursue goals related to their domestic needs. However, when thinking about solutions to these problems, liberal-pluralists emphasize that states are not the only or most important actors in world politics, that supra-state organizations have relative autonomy from the states they represent, and that states are not primarily selfish and individualistic in their actions. In McGrew's words, "liberal-pluralists... describe the global system as a polyarchical, mixed-actor or complex conglomerate system to denote the incredible variety of actors... Order is thus achieved... through a complex web of criss-crossing governing arrangements which bind states... together" (1992, p. 20). In this view, then, (i) nonstate organizations like Fortune 500 companies (e.g. Shell, DuPont, Ford), international NGOs (e.g. the International Union for the Conservation of Nature) and the media are very influential in interstate negotiations; (ii) supra-state organizations like the UN, International Monetary Fund, and EU matter immensely; and (iii) states are capable of acting towards common goals where appropriate. Arguably, the global managerialism of someone like Al Gore is liberal-pluralist in that it envisages states and nonstate actors coming together consensually to mitigate commonly shared environmental problems.

One of several environmental regimes that appears to support the liberal-pluralist case is that relating to atmospheric ozone destruction. The Montreal Protocol and its successor agreements have succeeded in virtually phasing out commercial production of the most harmful chlorofluorocarbons (CFCs). This seems odd at first sight, because the big CFC producers – such as the USA – faced enormous expense in finding commercially viable replacements. So why did countries like the USA sign-up to a binding agreement on CFCs from 1987, even as they were under pressure from chemical giants like ICI and Dupont not to do so? From a liberal-pluralist perspective the answer is three-fold. First, scientists from several countries offered strong proof early on that ozone thinning was a *direct* result of CFC emissions and pushed hard – through the UN – for governments to respond to this fact. Second, states quickly realized that ozone thinning was a *mutual threat*, leading, for example, to possible skin cancers in their domestic populations. Finally, because only a few Western countries (China excepted) were "big players" in CFC production, companies like ICI could be assured *that regime rules would apply across the board*, so that few or no competitors could undercut them through continued CFC manufacture (see Vogler, 1992, p. 132).

The critical response

As with the neo-/realist perspective, the liberal-pluralist position appears to have much to recommend it as an explanation of contemporary geopolitical realities, namely the environment. However, both responses look less convincing when seen from a more radical perspective. Following Cox's (1984) celebrated distinction, neo-/realism and liberal-pluralism can be seen as "problem-solving" approaches to geopolitics that can be opposed to "critical" approaches. Problem-solving approaches take geopolitical "realities" at face value and try to establish how transnational and "leaky local" environmental problems are dealt with in the corridors of power. They accept, rather than question, the broad structure of world politics and claim to be analytical rather than prescriptive. From a critical perspective, though, realist and liberal-pluralist perspectives are superficial and normative, both concealing and implicitly supporting "business-as-usual" in world politics. In other words, critical geopolitical commentators see realism and liberal-pluralism as global environmental discourses in their own right which are helping to consolidate questionable global environmental discourses and practices in the world at large. They are part of the problem not the solution.

The critical perspective is by no means unified. It can be seen as a hybrid of "geopolitical economy" (after, e.g., Agnew and Corbridge, 1995) and the "critical geopolitics" developed by geographers such as Simon Dalby and Gearoid Ó Tuathail.[25] The former is a Marxist-inspired approach which sees the words and deeds of states in the political system as being decisively affected by the structure of the world economic system. Geopolitical economy agrees with neo-/realists that states can and do aggressively pursue their self-interests, and it agrees with liberal-pluralists that there are other key actors in world politics. But the point is that state actions are enormously influenced by a state's particular location in the world economic system.[26] National sovereignty means that the world economy is split up not just into different political units, but also into territorial parcels of different economic size and importance. Accordingly, geopolitical economy sees interstate relations as part-determined by relative degrees of economic power.[27]

Drawing on the ideas of Dalby, Ó Tuathail and other political geographers, we can complement geopolitical economy with the insights of "critical geopolitics." The global environmental discourses expounded by a given state will, as noted, be conditioned by that state's economic interests. But critical geopolitics urges us to see these discourses as both materially effective in their own right (forms of "power-knowledge") and as relatively autonomous from economic influences. Given the multiple actors in world politics, Dalby et al. enjoin us to see a set of *competing* global environmental discourses in any one issue area, many of which seek to resist (often unsuccessfully) the economically self-interested discourses and practices advocated by powerful states. Together, geopolitical economy and critical geopolitics seek to *contest* existing geopolitical arrangements, uncovering the power relations inherent in them and the possibilities for more just interstate relations.[28]

To my mind, a critical perspective offers us the most powerful and accurate way to understand which global environmental discourses and practices win-out on any given environmental issue at any given time. Let's take the already mentioned example of global warming.[29] In what sense is global warming "global"? It clearly affects the whole planet, but there are in fact complex asymmetries in who is responsible for it, who should pay for it and who will win/lose as atmospheric temperatures rise during the decades ahead. To begin with, the massive rise in greenhouse gas levels over the last two centuries is largely the result of Western industrialization. The unchecked emission of carbon dioxide and other pollutants allowed Western economies to develop without paying for the resulting environmental damage – damage with which the world community as a whole has had to live. Secondly, this fact indicates that the G7 countries should currently bear the cost of reducing greenhouse gas emissions *worldwide*. But it has become clear since Rio that the global environmental discourses and the fast-solidifying practices surrounding the Climate Change Convention will brook no such radical solution. Instead, the greenhouse effect is being framed as a "technical" issue within a "management" discourse that looks to a limited form of financial redistribution to developing countries from the G7 (through the GEF and the carbon tax regime now in force). Yet these countries are effectively being asked to forego the economic benefits Western countries have historically enjoyed through burning fossil fuels, yet without any of the financial help needed to find and fund "clean power sources" (Sachs, 1993). Symptomatically, the USA, the one country which has most to lose economically from signing-up to the CCC, has delayed for more than a decade, bowing to pressure from its domestic power companies and other Western multinationals dependent on fossil fuels. With a former oil-tycoon, George W. Bush, now occupying the US Presidency, there is little chance that the USA's foot-dragging will change.

Of course, poor countries who have most to lose from global warming (like Bangladesh, a third of which is likely to be inundated if sea levels rise by 1 meter), have sought to put forward alternative global environmental discourses recommending more radical and just global environmental practices. Likewise, Greenpeace and other environmental NGOs have pressed for revolutionary changes in the global economic and political system, wherein *all* fossil fuel burning is phased out and a more ecologically sane way of living is adopted by peoples worldwide. The "eco-centric" or "green" critique made by these organizations is that the "growthmania" endemic to capitalism is fundamentally anti-ecological. The problem in getting these more radical global environmental discourses and practices on the interstate agenda, though, is that Western countries hold too many of the world's economic cards. With most developing countries dependent on the West and its supra-national organizations (the IMF and World Bank) for a mixture of trade, loans, aid and military assistance, they have little power to enforce a more economically just and ecologically sound environmental geopolitics.

Beyond Environmental Geopolitics

In my view, the current trend in interstate relations is to manage environmental issues in ways that fundamentally challenge neither current inequities in world economic power, nor the ecologically destructive practices on which those inequities are based. Nevertheless, there are some positive trends as well. Both globally and regionally, nation-states are more alive than ever before to the urgent need to conserve resources and protect the environment. The result has been a set of precedent-setting accords, agreements and protocols that have replaced a Cold War politics of antagonism with a less aggressive politics of mutual self-interest (however much skewed by power relations). Although, even from a neo-/realist and liberal-pluralist view, there is far more to be done to make these supranational arrangements work more effectively, states are certainly a long way down the road to ecological enlightenment. In the long run, however, the more critical perspective offered here suggests the need for all of us to develop discourses and practices which can lead us into a future based on *qualitatively different* geopolitical and environmental criteria (see also Laferriere and Stoett, 1999). It is, alas, an economically and ecologically just future unlikely to arrive any time soon. The challenge therefore – and one being taken up by socialists, anarchists and environmentalists worldwide – is not to lie down and accept the global environment as it seems to be but to fight for the environment we all deserve.

Acknowledgments

Thanks to Katharyne Mitchell for helpful editorial suggestions and to Nick Scarle for drawing the figure.

ENDNOTES

1. See, for instance, Herod et al. (1998, chapter 1).
2. For example, during the 1910s the relations between Canada, the USA, Japan, and Russia were heavily influenced by their mutual involvement in hunting the north Pacific fur seal to the point of extinction. See Castree (1997).
3. These problems may relate to environmental "threats" and "opportunities." An example of the former is where a country controlling a resource valued by other countries acts

unilaterally to cut off supplies. An example of the latter is where one country seeks to develop the resources of another country (e.g. its copper resources) and must overcome opposition within that country to get its way.

4. These two broad classes of environmental problem do not exhaust those currently of relevance to interstate relations. For a fuller inventory of supranational environmental problems see Braden and Shelley (2000, pp. 104–12).
5. Though not complex *enough*, because in the short space of a chapter like this I can only gesture towards some of the depth and sophistication needed to deal adequately with a topic as large and important as environmental geopolitics.
6. I will say more about Gore later. For now it's enough to note that he's one of the very few leading Western politicians who's explicitly interested in environmental issues.
7. These two terms will be defined later. For now, it's sufficient to note that they refer to critical perspectives on the nature and implications of current geopolitical arrangements worldwide.
8. See Dodds (2000) for a useful introduction to geopolitics.
9. All three men became leading geopolitical theoreticians and strategists in their home countries.
10. For a detailed account of successive geopolitical orders see Agnew and Corbridge (1995) and Taylor and Flint (2000, chapter 2).
11. A good example is the dominant discourse around global warming. Many environmental organizations – like Friends of the Earth – want to see a major shift away from technologies based on burning fossil fuels. However, most Western governments, in particular, deem this far too radical. Consequently, much of the intergovernmental debate on what to do about global warming has focused on ways to reduce, rather than eradicate, pollution resulting from burning coal, oil, and petroleum. The whole notion of eliminating greenhouse gas emissions is thus effectively deemed an illegitimate subject of interstate debate.
12. Until recently, an example was the international whaling regime, led by the USA, which used the principle of wildlife conservation to place a ban on commercial whaling in the face of opposition from whaling nations like Japan.
13. My use of the term regime should not be confused with the "environmental regimes" discussed by "realist" and "liberal-pluralist" analysts of world politics (I discuss these two perspectives later in the chapter). An introduction to the literature on environmental regimes can be found in Vogler (1992).
14. This is not to discount the importance of how the public understands environmental problems and interstate actions in relations to those problems. However, for reasons of space I want to focus on what are undoubtedly the "big players" in environmental geopolitics today: that is, states, other large institutions in world politics (like the UN) and the words and deeds of those analysts who formally study – and make recommendations about – geopolitical realities.
15. See Dodds (2000, pp. 110–18) and Connelly and Smith (1999, chapter 7) for a short history of intergovernmental environmental action pre- and post-Rio.
16. For a summary and critique of Gore's arguments see Anderson (1996, pp. 93–107).
17. For more on global warming and world politics, see Paterson (1996).
18. Dalby (1998b) offers a fuller consideration of environmental in/security.
19. See part 4 of Ó Tuathail et al. (1998) for key readings on contemporary environmental geopolitics, including those by Kaplan and Homer-Dixon.
20. Braden and Shelley (op. cit., pp. 115–7) offer a full list of major international environmental agreements. Of course, the geopolitics of environment does not always involve multiple states simultaneously. On a sub-continental scale, there may be bi- and trilateral environmental issues (as in the ongoing US–Canadian rifts over Atlantic and Pacific fisheries).

21. For reasons of space, I can only offer a skeletal account of the three approaches to geopolitics here. For more detail see McGrew (1992), Vogler (1992), and Vogler and Imber (1996).
22. Mackinder was a British geographer and Morgenthau a German-American political scientist, both writing in the early twentieth century. Bull is a late twentieth century political scientist, who worked in Britain.
23. For more on environmental commons see Vogler (1996, chapter 1).
24. Regrettably, Japan has recently broken the ban on large-scale whaling by claiming the need to harvest many whales for "scientific" purposes.
25. For a rather different take on the critical perspective see, for example, Laferriere and Stoett (1999, chapters 1 and 5).
26. Taylor and Flint's (2000) "world systems" approach is closely linked to Agnew and Corbridge's geopolitical economy, in spirit if not in detail.
27. For a geopolitical economy view of world environmental politics, see Lipschutz and Conca (1993). Their book grounds interstate actions over environmental issues in the realities of economic power differentials between countries.
28. For an example, see Warner's (2000) biting critique of the discourse and practices surrounding "global environmental in/security."
29. The following account applies equally well to other transnational environmental problems – see Litfin (1994) for example – and to a leaky local environmental problem like biodiversity conservation: see McAfee (1999).

BIBLIOGRAPHY

Agnew, J. 1998. *Geopolitics: Revisioning World Politics*. London: Routledge.

Agnew, J. and Corbridge, S. 1995. *Mastering Space*. London: Routledge.

Anderson, P. 1996. *The Global Politics of Power, Justice and Death*. London: Routledge.

Barber, J. and Dickson, A. 1995. Justice and order in international relations. In D. Cooper and J. Palmer (eds.) *Just Environments*. London: Routledge, 121–36.

Benedick, R. E. 1999. Tomorrow's environment is global. *Futures*, 31, 937–47.

Bernstein, S. 2000. Ideas, social structure and the compromise of liberal environmentalism. *European Journal of International Relations*, 6, 464–512.

Braden, K. and Shelley, F. 2000. *Engaging Geopolitics*. Harlow: Longman.

Bull, H. 1977. *The Anarchical Society*. New York: Columbia University Press.

Castree, N. 1997. Nature, economy and the cultural politics of theory: the "war against the seals" in the Bering Sea, 1870–1911. *Geoforum*, 28, 1–20.

Connelly, G. and Smith, G. 1999. *Politics and the Environment*. London: Routledge.

Cox, R. 1981. Social forces, states and world orders. *Millennium*, 10, 126–55.

Cox, R. 1984. The crisis in world order and the challenge to international cooperation. *Cooperation and Conflict*, 29, 99–113.

Dalby, S. 1998a. Introduction. In G. Ó Tuathail et al. (eds.) *The Geopolitics Reader*. London: Routledge, 179–87.

Dalby, S. 1998b. Geopolitics and global security. In G. Ó Tuathail and S. Dalby (eds.) *Rethinking Geopolitics*. London: Routledge, 295–313.

Dodds, K. 2000. *Geopolitics in a Changing World*. London: Routledge.

Donnelly, J. 1986. International human rights: a regime analysis. *International Organisation*, 40, 599–42.

Gleick, P. 1993. Water and conflict. *International Security*, 18, 79–112.

Gore, A. 1993. *The Earth in Balance*. New York: Plume.

Held, D. et al. 1999. *Global Transformations*. Oxford: Polity.
Herod, A. et al. (eds.). 1998 *An Unruly World?* London: Routledge.
Kaplan, A. 1994. The coming anarchy. *The Atlantic Monthly*, 273(2), 44–76.
Keohane, R. 1984. *After Hegemony*. Princeton, NJ: Princeton University Press.
Laferriere, E. and Stoett, P. 1999. *International Relations and Ecological Thought*. London: Routledge.
Litfin, K. 1994. *Ozone Discourses*. New York: Columbia University Press.
Lipschutz, R. and Conca, K. (eds.). 1993. *The State and Social Power in Global Environmental Politics*. New York: Columbia University Press.
McAfee, K. 1999. Selling nature to save it? *Society and Space*, 17, 133–54.
McGrew, A. (1992) Conceptualising global politics. In A. McGrew et al. (eds.) *Global Politics*. Oxford: Polity, 1–30.
Ó Tuathail, G. 1998. Postmodern geopolitics. In G. Ó Tuathail and S. Dalby (eds.) *Rethinking Geopolitics*. London: Routledge, 16–38.
Ó Tuathail, G. and Dalby, S. 1998. Rethinking geopolitics. In G. Ó Tuathail and S. Dalby (eds.) *Rethinking Geopolitics*. London: Routledge, 1–15.
Ó Tuathail, G., Dalby, S., and Routledge, S. (eds.). 1998. *The Geopolitics Reader*. London: Routledge.
Paterson, M. 1996. *Global Warming and Global Politics*. London: Routledge.
Prins, G. (ed.). 1993. *Threats Without Enemies*. London: Earthscan.
Rosenau, J. and Cziempel E. (eds.). 1992. *Governance without Government*. Cambridge: Cambridge University Press.
Sachs, W. (ed.). 1993. *Global Ecology: a new arena of political conflict*. London: Zed.
Stoett, P. 1993. International politics and the protection of great whales. *Environmental Politics*, 2, 277–302.
Taylor, P. and Flint, C. 2000. *Political Geography*. London: Prentice Hall.
Vogler, J. 1992. Regimes and the global commons. In A. McGrew et al. (eds.) *Global Politics*. Oxford: Polity, 118–37.
Vogler, J. 1996. *The Global Commons: a regime analysis*. Chichester: Wiley.
Vogler, J. and Imber, M. (eds.) 1996. *The Environment and International Relations*. London: Routledge.
Warner, J. 2000. Global environment insecurity: an emerging concept of control? In P. Stott and S. Sullivan (eds.) *Political Ecology*. London: Arnold, 247–66.

Chapter 28

Green Geopolitics

Simon Dalby

Geopolitics, Globalization, and Environment

As the twentieth century, the hundred years that Adam Hochschild so aptly calls "the century of famines and barbed wire" (1998, p. 231) came to an end, the themes of resources and boundaries, scarcity and violence, and nature and its destruction, were explicitly connected in discussions of world politics. As the twenty-first century begins, the legacy of these debates shape contemporary discussion. In some texts environmental degradation is viewed as a potential threat to the political stability of modernity; in other discussions the dominant theme is the need to enhance security by stabilizing states in the South and strengthening boundaries against the disruptive flows of migrants and refugees (Chase et al., 1999; Kaplan, 2000). Whatever the particular emphasis in these political discussions, the interconnections between humanity and nature, North and South, and resource production and consumption reflect the key themes of contemporary green thinking and challenge the "modern" assumptions of geopolitics.

Being modern these days is usually understood as living in a stable, affluent democratic state, one protected from threats from across borders, and where the good life is defined in terms of "our" ability to consume, without interruption, the ever-changing and improving products of scientific research and related industrial production. Modernity is a very ambiguous term indeed, but it usually refers to who we are as knowledgeable consumers in a way that directly links to our identities as citizens of Northern nation-states in the rich, stable and "successful" part of the global economy dedicated to the rational pursuits of our individual interests (Taylor, 1999). "We" are part of the modern geopolitical arrangements of mutually recognized sovereign national societies that now cover the whole globe, while also having a collective identity provided by participation in the consumer economy.

But simple geopolitical assumptions about modern states, economies, and societies are now very frequently challenged by discussions of globalization, with all the "postmodern" emphases on the connections and flows of ideas, cultures, and products between distant places, and the related matters of the changing identities of who "we" are as a consequence of these connections. Globalization is in part a challenge to simple assumptions that we modern people are safely "here" protected from all kinds of dangers by our states and our technologies (Held et al., 1999). Instead "we" are now part of an interconnected global economy, influenced by distant political crises or social changes precisely because of our connections to those places. Investors are encouraged to take advantage of these interconnections as mutual fund companies advertise their effectiveness in seeking investment in "emerging markets" in the South. Tourism promises ever more exotic destinations in the South as ever more landscapes become objects for "consumption" there too.

The sovereignty of states is in doubt now because of both economic interconnections and human migration. But there are many conceptual and political dangers in assuming that all the hype about the global reach of corporations and the flows of images, information, and commodities are either new or threatening. The history of the rise of European, and subsequently American, power was to a substantial extent based on the resources and wealth brought to Europe by the colonization of distant lands (Hochschild, 1998). Resources, most obviously now oil, flow from the relatively poor and weak states to the powerful affluent powers of the North fueling its industrial and military superiority over states in the South (Athanasiou, 1996). What now also flows, although much less freely, in the same direction are poor people in search of jobs and a better life.

Adam Hochschild's (1998) phrase concerning barbed wire and famines, that I have borrowed for the leitmotif of this chapter, refers more specifically to the legacy of starvation and illness that inmates of concentration and prison camps suffer, but the image catches the links between modern political power that divides the world into states surrounded by fences, and the deprivation and suffering of the poorest parts of humanity, who in many cases are starving in part because of the operation, and sometimes violent policing, of state boundaries (Falk, 1999). As United Nations reports on the state of the world's economy repeatedly remind us, the disparities between rich and poor are still growing. The number of people moving, either to escape their fates in the zones of "complex humanitarian emergencies" or to try to improve their economic lot, is increasing and in the process generating various moral panics and populist fears of migrants in the Northern zones of economic affluence (Black, 1999).

In these discussions the conventional political assumptions of modernity, of clearly delineated political spaces of the nation-state as the basic premise for discussion, analysis and policy action are both reasserted frequently in the face of flows of people and pollution, and simultaneously rendered inadequate precisely by these transgressions of state boundaries (Litfin, 1998). Invoking the term "globalization" as a catchall term has become the common strategy for defining and labeling all these matters (Held et al., 1999). Interconnections and flows are often constructed as threats to various identities and places, understood to be part of contemporary global transformations, but at the same time the flows also come from places also understood as economic opportunities for global corporations.

But the fact that specific places have been constructed by the prior intersections of previous flows is often difficult to incorporate into modes of reasoning that operate to fix the fluxes of time into the supposedly stable spaces of modern political life (Agnew, 1998). Historical connections and disruptions are obscured in the simple geopolitical assumptions of separate societies in autonomous states. Security is especially tied to assumptions of stability, to identities "in here" in the supposedly protected spaces of home, community, nation, and state. But the growing phenomenon of refugees and the difficulties of "managing displacement" require thinking carefully about the processes of political change (Hyndman, 2000). In particular, the simplistic Malthusian assumptions that Southern "overpopulation" is threatening the affluence of the North is an argument that is always a dangerous temptation when the world is viewed in these ways (Hartmann, 1999).

As discussions of global climate change in particular are making clear, the prevalent geopolitical assumptions of here and there, of protected local spaces and distant dangers are both a very tempting formulation within which to try to grapple with these issues and increasingly inadequate at the same time (Harris, 2000). Ecological thinking, emphasizing interconnections and complex interactions likewise suggests that geopolitical preoccupations with territorially bounded states are inappropriate tools for analysis or policy instruments (Dalby, 2000). Thinking in these terms also challenges the modern urban assumptions of nature as something "out there" external to humanity to be used as a source of resources for human consumption at precisely the moment that humanity is about to become, for the first time, a predominantly urban species.

These ironies and complexities are the context for thinking about green geopolitics. They go to the core of the geographic discipline by asking questions that connect concerns with spaces and places to the use of resources and the understanding of environment. The geopolitical assumption of environment as something "out there," separate from us "in here," maps onto popular fears of migrants, concerns with "overpopulation" and threats to Northern affluence (Chapman et al., 1997). It does so precisely as it operates to obscure the connections across these boundaries – "our" use of oil, rubber, timber, coffee, beans and vegetables, and numerous other things that come from the South and frequently disrupt the ecologies and societies there – while providing "us" in the North with the materials for our consumer lifestyle, all in the name of Southern "development" (Dalby, 1999).

One chapter cannot deal with all these issues in detail. Instead, the argument here focuses on a couple of current themes which invoke particular geopolitical specifications of the world in analysing contemporary environmental themes. The argument that follows shows that specific geopolitical assumptions are built into discussions of global environmental concerns, and frequently these modes of geopolitical reasoning are less than helpful for either analysing contemporary problems or suggesting practical policy initiatives. In juxtaposing "modern" and "postmodern" geopolitical modes of thinking the chapter suggests that the spatializations in contemporary environmental politics are a crucial consideration in both analysis and advocacy.

Modern and Postmodern Geopolitics

Geopolitics is about both the world and politics, but it is crucially a way of knowing the world constructed in visual terms. It is literally a modern way of seeing the world that has profound consequences, not only for the world but in terms of how we who see the world in these terms understand our place within it. Often the visual sense invokes a detachment, a supposedly objective view "from nowhere" of the whole planet, a way of seeing supported by the technological marvels of computer enhanced satellite imagery and the photographic efforts of a generation of astronauts (Cosgrove, 1994). But, as with so many human aspirations to transcendence, this "God trick" of total surveillance often tricks those who use it, in this case, into forgetting that it is but one way of viewing things and that it is linked directly to complex social arrangements of power and privilege.

Most explicitly, this geopolitical imagination is linked to the ability of the rulers of large and powerful states to construct the technologies that do such surveillance, and to the practices that specify where and what precisely is monitored because it may endanger the political order run by the rich and powerful (Nye and Owens, 1996). This is especially important because in the routine daily practice of politics, "[t]he world is actively 'spatialized,' divided up, labeled, sorted out into a hierarchy of places of greater or lesser 'importance' by political geographers, other academics and political leaders. This process provides the geographical framing within which political elites and mass publics act in the world in pursuit of their own identities and interests" (Agnew, 1998, p. 2).

The modern geopolitical imagination in Agnew's (1998) terms is both a Eurocentric world view and a global vision, the product of an historic process connected to the expansion of European power over the last half millennium. We see the world as a whole then divide it into a hierarchy of places, blocs, and states that have attributes of political importance. In the process we make a series of conceptual transformations of time into space: modernity is here, primitiveness there. Territorial state assumptions, of sovereignty, states as political containers, and a pervasive foreign/domestic dichotomy also play a prominent role in the modern geopolitical imagination. We make sense of the world by sorting out the messiness of complex geographies into simple classifications based on clean lines that draw distinctions between conveniently labeled places and peoples. Finally, Agnew (1998) argues, the geopolitical vision is also one that assumes a struggle between states is the normal condition of politics. Security is supposedly provided by seeking primacy in a competitive arena of great powers. "In sum modern geopolitics is a condensation of Western epistemological and ontological hubris – an imagining of the world from an imperial point of view" (Ó Tuathail, 1997, p. 42).

Near the end of the twentieth century it appeared that such concerns were in some ways being left behind. As the Berlin Wall fell in 1989, the end of geopolitics was declared by numerous politicians. With the world no longer starkly divided between East and West, communist and capitalist, great power rivalries could be portrayed as a thing of the past. Global media images on CNN and satellite TV suggested that space was apparently giving way to time, territory to cyberspace, location to mobility, states to a global economy, and military aggression to a "new world order" in American president George Bush's phrase. And yet new dangers appeared in Washington, in particular, and elsewhere in the capitals and think-tanks of the Western world (Prins, 1993). New geographies of danger, of rogue states, failed states, global environmental threats, and related phenomena inscribed the world in terms of tame zones and wild zones, stable centers and threatening peripheries (Kaplan, 2000). These new specifications of danger are suggesting that threats are in some senses global while also being about "borderlessness, state failure and de-territorialization" (Ó Tuathail, 1999, p. 18).

But if we understand geopolitics in terms of an "envisioning, strategizing and disciplining of global space," a sense of postmodern geopolitics is possible in which the new modes of thinking about government stretch the categories of the modern geopolitical imagination but nonetheless continue to link space and power in politically efficacious ways (Ó Tuathail, 1997, p. 46). The extension of the geopolitical imagination to include these "postmodern" themes of borderless dangers suggests that the term geopolitics itself is best understood not as a stable signifier, but instead as a shifting congealment of knowledge practices, power, and spatiality. Modes of power and surveillance shift in changing circumstances. Postmodern geopolitics is thus not a replacement of an earlier practice but an upgrading and extension of earlier themes (Ó Tuathail, 1998). In particular, the extension of the discussion of global threats suggests the importance of thinking carefully about the geopolitical premises in contemporary political discourse, precisely where the spatial specifications are apparently taken-for-granted common sense.

Greening Geopolitics

Part of the debate about the end of geopolitics has been caught up in the discussion of global environmental change and the potential dangers of resource shortages, scarcity-induced displacements, and the degradation of renewable resources (Suliman, 1999). When linked to a concern with national security on the part of the United States (Allenby, 2000), or the states making up the North Atlantic Treaty Organization (NATO) more generally, the implications of the postmodernization of geopolitics become especially clear. Environmental degradation is a problem frequently attributed to the global "South" where it is also frequently linked to Malthusian assumptions of overpopulation to construct an argument that the South is the source of global environmental danger (Dalby, 1999). Once again, the geopolitical impulse to divide the world into blocs with specific attributes and assign them relative importance in a global order comes into play in such formulations. But when we consider it more carefully, the environmental crisis can be used as a means of thinking carefully about the way geopolitics works as knowledge production and policy framework (Dalby, 2000). Environment is both a new threat in the security discourses of NATO and specifically in Washington, and also a phenomena that often works to reveal the limits of both modern and postmodern variants of geopolitical thinking.

One productive way to consider these matters is to follow Gearóid Ó Tuathail's (1999) discussion of Ulrich Beck's theory of risk society. Beck (1992) contrasts industrial society of the first half of the twentieth century, where conflicts were primarily over the distribution of wealth in a context of scarcity, with that of the latter half of the twentieth century when conflicts were more frequently about the distribution of risks and hazards in a techno-scientific society of relative plenty. The success of industrial production in the latter part of the twentieth century "solved" matters of poverty for the majority of the population living in Western states. Beck suggests that this technological and scientific production system brings with it numerous side effects and hazards which are contested through technological arguments. Planning disputes, environmental impact assessments, and arguments over "safe" levels of toxins in food, air and drinking water engage science directly in politics. This frequently pits "experts" against each other in direct argument during public discussion. In the process, this ensures that science can no longer be seen as a neutral arbiter; expertise is not outside politics and technical controversy. It is now part and parcel of political struggle.

But this is a "reflexive" modernity in the sense that it is dealing with problems literally of its own making, rather than with matters outside the modern economy. The technological controversies about water quality, environmental change, and food safety are the result of industrial production. Beck (1996) also points out that the side effects of scientific production are no respecters of state boundaries; these "fabricated uncertainties" of risk society can endanger people in many apparently distant places. Whether it is radioactive fallout from Chernobyl or the international health hazards of "mad cow" disease, global interconnections are supposedly reducing the importance of state boundaries to manage risk. Ozone holes, biodiversity loss, climate change, and a variety of diseases with the potential to cause major medical problems are symptomatic of risk society and a threat to "global security."

A moment's reflection on the 1918 influenza pandemic, or earlier plague and cholera epidemics confirms that these hazards and interconnections are probably only new in so far as they can be designated as more artificial than earlier disasters. But even that point needs careful qualification because a number of major human diseases such as smallpox, that have had a dramatic effect on our ecological history, and played a significant role in the European colonization of substantial parts of the world, are diseases derived from the human domestication of animals (Diamond, 1997). Nonetheless, with the end of the Cold War the interconnections between places are perhaps more evident, certainly the artificial production of risks from complex chemicals in the food chain or radioactive particles in the atmosphere suggest that humanity is dangerously altering its environment.

Ó Tuathail (1999) uses this argument to make the point that while modernity endangers itself in many ways, these dangers are still frequently not understood in terms of risk society but are still blamed on external causes. When such things as global environmental threats appear in contemporary states' national security policies, they usually do so in ways that suggest that at least most of the threats have their origins outside the boundaries of the state in question. Posing environmental matters as a concern for states and national security, arguing that it is at this scale that policies and priorities can be formulated to protect societies from environmentally induced disruptions, is a tempting policy option for many analysts (Allenby, 2000). But the geography of specifying the environmental threat as somehow external obscures the fact that these threats might better be understood in terms of the unintended long-term and long-distance consequences of our own actions (Dalby, 1999).

In the case of global environmental disruption, the overall impacts of North American and European consumption are large in comparison to other parts of the world. Michael Redclift (1999) extends this point to argue that competition for resources is the root of many environmental changes and consequently the source of many related conflicts. In seeking ever more sources of resources, disruptions and change may have unforeseen distant and long-term consequences. This perspective suggests, in a manner loosely consistent with risk society theory, that environmental change is better understood directly as a manner of changes and disruptions wrought by the processes of modernity. Such thinking, a necessary prerequisite for taking matters of sustainability seriously, requires both an understanding of interconnections, and a willingness to critically reassess modernity and its spatial assumptions of both autonomous states and individuals (Bulkeley, 1997). In risk society, "we" are no longer "here" separate from dangers from "there."

The overall policy implications of presenting global change as an intrinsic part of our modernity are very different from a modern version of security that constructs environmental change as an external threat from "there" to an insecure modernity "here." But the danger of understanding security as a matter of boundaries and external threats looms over all this, in the process threatening military responses to political instability in a crisis, while the underlying connections and the modern causes of these disruptions are ignored in the interests of short term order and the protection of modernity (Deudney, 1999). While this is obviously a matter of ignoring many different views of environment and development in the North and South, and of failing to understand the different experiences and popular understandings of widely disparate human conditions of living (Chapman et al., 1997), it also forgets the complex histories obscured by the geopolitical shorthands of environmental threats originating in the South.

Forgotten Histories

Concerns about environmental dangers, in particular, and global security, more generally, frequently pose their arguments in terms of looming catastrophes – threats to the common fate of humanity which require policy innovation in the North to save the South, and by extension the North too, from various perils (Barnett, 2001). Formulated in these terms the necessity to intervene in the South to "save" either the Earth or humanity appears an obvious moral duty. But in making such claims the long history of prior interconnections between North and South are forgotten. The long-term historical changes wrought by the expansion of European colonization and the transformation of rural ecologies of the South to supply urban markets are the precursors of many contemporary concerns with resource shortages and degradation.

Discussing these matters only in terms of states, and their policy priorities, forgets the long-term impact of European and subsequent American and Japanese involvement in the South. In Richard Grove's (1997, p. 183) succinct summary:

> Colonial ecological interventions, especially in deforestation and subsequently in forest conservation, irrigation and soil "protection", exercised a far more profound influence over most people than the more conspicuous and dramatic aspects of colonial rule that have traditionally preoccupied historians. Over the period 1670 to 1950, very approximately, a pattern of ecological power relations emerged in which the expanding European states acquired a global reach over natural resources in terms of consumption and then too, in terms of political and ecological control.

Development in the last few decades has frequently been about extending this control further in Southern states, by extracting more resources from the rural areas to feed the growing demands in both the cities in the South and the export markets supplying food and other commodities to the distant metropoles of the North (Peet and Watts, 1996). Resistance to these changes has occurred frequently too, although only sometimes does opposition take the high profile forms that attract the attention of the North's media. Chiapas is perhaps emblematic of rural resistance to encroachment on common property use and the further alienation of agricultural resources by the commercial sector (Esteva and Prakesh, 1998). The complexity of political actions in resisting dam building, oil extraction, mining, and agricultural expansion provides a much more nuanced understanding of political violence related to environmental change than the simple assumptions in environmental security discussions that environmental changes cause potentially violent conflict (Kaplan, 2000).

The history of violent resource extraction and the coercion used by Europeans to collect and produce the commodities is not that far back in history but it is frequently forgotten. The violent colonization of what became the Congo, in search of ivory and subsequently rubber, may have claimed as many as ten million lives towards the end of the nineteenth century and the beginning of the twentieth century (Hochschild, 1998). And yet in discussing contemporary developments in the region and the possibilities of commercial activities to improve the lot of Africans, popular media representations of the region frequently simply omit the history of prior violent usurpation in constructing the area as untouched and in need of modernization and development (see, for instance, Goldberg, 1997). Again modern geopolitical specifications of the world into territorial states and autonomous spaces works to occlude the depredations of the past and obscure the processes currently in motion. While it might be argued that it is not of great importance that such presentations of the world in leading American media outlets get Africa so wrong, it is clear that these failures to contextualize the African experience, and the persistent representation of it in neo-Malthusian terms, also occurs within the halls of international financial institutions, where decisions are made concerning the appropriate allocation of development funding and debt relief on precisely such misunderstandings (Williams, 1995).

Environmental Connections

While such concerns are the stuff of geopolitical discourse and scholarly analysis, the practical implications of both risk society theorizing and critical geopolitical analysis can be helpfully illustrated by focusing on the practicalities of everyday life in the North. One small episode in this author's life, a visit to a friend's house in Vancouver during the northern hemisphere summer of the year 2000, both provides part of the logic for this chapter and makes the point about the connections that are frequently obscured in the conventional discussions of global security. Surveying the suburban lawn of her house in the aftermath of her daughter Olivia's fourth birthday party, and observing the collection of plastic toys strewn in all directions, my friend Karin Hall exclaimed "Middle class North American children are an environmental disaster!" She went on to catalog the various "essential large plastic things" that were then *de rigeur* in every child's life at the time in the city, expressing mild exasperation at how she too, despite her own environmentalist inclinations, and her marriage to someone with a geography degree, had come to have a lawn covered in plastic.

"Essential plastic things" have come to be both status symbols and part of everyday life, and yet they present substantial environmental problems of waste disposal, whether buried in a landfill, incinerated or processed in some recycling program. Plastic has in many ways eclipsed the earlier importance of rubber, a commodity that has, as Hochschild (1998) documents, such a bloody history in Africa. But the consequences of the extraction of plastic's raw materials, frequently from rural areas in the South, have important political parallels that point to the importance of how environmental questions are framed in geopolitical language. Whether in the case of the U'wa people in rural Columbia (del Pilar Uribe Marin, 1999), or the more high profile case of the problems in Ogoniland in Nigeria (Watts 1998), peoples distant from the suburban lawns of North America have their lives often violently disrupted by the activities of oil companies. The petroleum that is the raw material for the plastic in Olivia's toys may not have come from oil wells in either of these troubled places, but the pattern of resource extraction, and the environmental consequences of Northern consumption of resources from the South, once again repeats the long history of distant colonial appropriations that have so transformed ecologies in numerous "Southern" places.

Some weeks after Olivia's birthday party I found myself in another neighborhood of Vancouver in a furniture store talking to a shopkeeper about the "Chinese water buckets" on display in the window. These were being sold to middle-class consumers in Vancouver to use as stylish magazine holders because, it was explained to me, rural Chinese people have abandoned these traditional heavy, ornate and cumbersome wooden items for lighter, more practical plastic ones. Thus, peasant vernacular technology in one part of the globe is commodified and becomes a consumer item due to its exotic *cachet* on the opposite side of the Pacific Ocean. I wondered if the use of plastic buckets had increased water use in rural China. Fear of water shortages, driven in part by the rapid urbanization of the coastal regions of China, and the increasingly meat-rich diet as a consequence of increasing affluence, which increases demand for feed grains for livestock, and hence demands for irrigation water, is part of the discussion of potential Chinese dangers to the international political order due to environmentally induced conflicts, potential disruptions of international agricultural trade, and even civil war between parts of China resulting from potential shortages in the future (Boland, 2000).

Focusing on such matters in terms of globalization has the advantage of emphasizing the connections between Northern affluence and Southern commodity provision, but insofar as globalization suggests either novelty, or economic and cultural convergence, it is not helpful in tracking the specific connections that matter. Tracing matters in terms of commodity chains which focus on the specific trading links between North and South provides a better understanding of the geographical connections. Food production for export in the South is often effectively controlled by large-scale retailers in the North. This is not by direct control but, as an examination of Kenyan vegetable exports to England makes clear, by a complex system of contracts and, in particular, by the quality control requirements of Northern corporations, which effectively limit what gets put on airplanes in Kenya (Barrett et al., 1999).

The connections between the Shell Corporation in Europe, where it sells petrol, and its involvement in Nigeria, where the oil it sells in Europe comes from, has made the corporation itself the focus of human rights activists in the North concerned to expose the violence, political corruption, and environmental destruction in Ogoniland (Human Rights Watch, 1999). This has also driven a sophisticated public relations response from Shell Corporation in which it tries to distance itself from any complicity in the political violence in Nigeria in the region where its oil supplies originate (de Larrinaga, 2000).

Corporate Disconnects

In this Shell is not alone – other corporations have entered the public debate about environmental matters but frequently in a way that suggests that their products are the solution to various environmental problems. Corporations and political leaders have been engaging in various attempts to either co-opt or marginalize the more radical environmental critiques (Beder, 1997). This is sometimes done by rendering political arguments into matters of aesthetics in terms of wilderness preservation and the protection of beautiful places. The consequences of consumption are frequently ignored in a discourse that reduces environment to pretty parks, recreational opportunities and national pride in "our" part of the world (Benton and Short, 1999).

In the case of pollution problems, complex political matters are reduced to technical questions for managerial solutions, rather than discussed in frameworks that challenge contemporary political arrangements, although as risk society theory suggests this frequently results in technical disputes that become part of political debate. As part of an increasingly prevalent discourse of "ecological modernization," which is a complex corporate and official state approach to these matters which suggests that technological innovation and sophisticated management is the answer to all environmental problems, corporations finesse environmentalist critiques by repackaging them as technical concerns (Hajer, 1995).

The larger environmentalist critiques of modern consumer identities are also challenged by corporate arguments that one can be an environmentally responsible consumer if one purchases the particular products produced by a specific corporation. From technical specifications of poverty in the South, definitions of the "problem" of food shortages in ways that affirm technological innovation by seed companies as the "solution" to world hunger, to the opportunist advertising campaign by British Petroleum to reinvent itself as an energy company interested in moving "beyond petroleum" in 2000, just when petroleum product prices caused protests in many Northern countries, corporations have moved to reinvent themselves and reinscribe contemporary problems in modernist economic terms (Beder, 1997).

Precisely these practices of modernity with their assumptions of individual autonomy and rights to consumption, are what more critical brands of environmentalism challenge. Consequently there have been plenty of political reactions, by those writers who understand modernity as the ultimate value, to denigrate and marginalize environmentalists, or dismiss their arguments altogether as simply wrong-headed. Corporate spokespeople sometimes present environmentalists as neo-Luddite types with an agenda to prevent progress (Rowell, 1997). Environmentalists are also sometimes portrayed as deranged terrorists planning to disrupt the functioning of "normal" societies. Notably Tom Clancy (1999), the best-selling American author of geopolitical "techno-thrillers," has joined this chorus linking the mad scientist genre to deep ecology in his novel *Rainbow Six*. In this book, a genetically enhanced ebola virus designated "Shiva" is designed, and plans made to deploy it, to kill off most of *Homo sapiens* and hence to "save" the world. While the dramatic changes wrought by human activities are recognized in places in this novel, how they might be tackled is not considered. Rather, the assumption that they need to be is dismissed by linking profound concerns with environmental destruction to eco-terrorism. By such marginalizations and the reassertion of a benign American identity, the larger questions of global climate change, biodiversity loss, and related matters are obscured.

The political struggle over popular representations of global environmental concern is thus related to the question of corporate and government technological mastery, a geopolitical vision that in its more extreme advertising tropes allows for the appropriate management of all problems. Global security is now frequently understood as requiring global surveillance. In one prominent rendition, American superiority in information technology is argued to be a competitive advantage in assessing and dealing with new threats to the global order (Nye and Owens, 1996). Telephone companies and Internet providers now frequently use symbols of the globe and images of satellites to ensure that one can remain in touch no matter where you are on the planet. Corporations have appropriated the symbol of the whole planet in such advertising (McHaffie, 1997).

But the assumption that technology is the answer to all manner of difficulties often also occludes the crucial questions about the causes of insecurity. Many of these might be better understood as consequences of the globalization of modernity and the disruptions implicit in its dynamics. Risk society suggests that technological interventions are themselves risky endeavors, the assumption of control frequently obscures side effects and the consequences for poor and marginal populations. Corporate claims of technological expertise operate to assure that even the most arcane risks of postmodern economies are at least ostensibly susceptible to their control. Despite all the problems around the world, further economic growth is still presented as the only possibility. Now, of course, it is dressed up in the technological and corporate rhetoric of sustainable development, but while development it most assuredly still is, the claims to sustainability are much less clear.

In all these discussions geopolitics has been turned into technical controversy. But as risk society arguments emphasize, the modern faith in science as the neutral arbiter is undermined precisely by the ways in which corporations and governments try to reassure worried publics about environmental risks. Whether it is mad cows, radioactive rain or how best to dispose of oil drilling platforms, frequently people simply don't believe the corporate and government experts and protest accordingly.

Sustainable Security?

In much of the contemporary literature on environment and security, as well as on development and sustainability, there is a growing recognition of the importance of matters of practical security as the condition for sustainable development, and likewise the need for various forms of development to enhance security in many ways (Suliman, 1999). But how these interconnections are to be thought through, and in what geopolitical framework any of this can be made meaningful, remain the big political issues for attempts to think intelligently about how to "green" any understanding of geopolitics.

Most obviously the question of geographic scale looms over these considerations. Can a town in England be considered sustainable if it is dependent on oil supplies from Nigeria and fresh vegetables grown in Kenya and flown in daily to Heathrow? Could it be considered so if it included in its boundaries woodland containing enough trees to remove an equivalent quantity of carbon dioxide from the atmosphere to that added by its population by driving cars and heating their homes? Should the fuel, from Nigerian oil wells possibly, that is used to fly the vegetables from Kenya be counted in too? What if the town was to buy some land in Kenya or Nigeria to grow trees there to absorb the carbon dioxide instead, because it's cheaper to buy land and pay Southern wage levels to workers there to look after the trees? But what then of the local people who may need the trees for firewood or shelter, or who may be trying to use the land to grow crops for export to improve their economic condition?

These are the kinds of questions that have to be asked if the simplistic geopolitical assumptions of states as the containers of political communities that decide these matters are to be removed and more complicated ecological arrangements considered. They are precisely the questions already being asked by numerous "green" campaigners for Southern debt relief, human rights, and ecological reform in the North. They are being asked by ecologists trying to find innovative strategies for Northern communities concerned to reduce the environmental impact in the North without precluding economic opportunities for the poor in the South (Sachs et al., 1998). However, apart from questions of "emissions trading" and "carbon sinks" in greenhouse gas negotiations, they frequently are not the kinds of questions being asked (yet?) by politicians and corporate executives as they survey the global scene in search of political dangers or business opportunities.

Scholars preoccupied with security still frequently focus on states and their stability. Development and the possibilities for peaceful cooperation are understood in terms of the operation of formal commodity markets and the frequently ethnocentric assumptions of Western-style democratic institutions as the only option. But as scholars familiar with anthropology are increasingly pointing out, this preoccupation with states and security also obscures the more general patterns of the insecurity of numerous marginal peoples (Weldes et al., 1999). Environmental insecurity is not new in many of the places that detailed anthropological studies have documented. The historical struggles over access to rural land and resources in particular places offer correctives to simple Malthusian assumptions that overpopulation is a problem, and link up with the larger literature in political ecology that is charting the critical connections between North and South (Peet and Watts, 1996).

Understanding the historical interconnections suggests rather different policy options and shows the importance of geopolitical concepts in formulating contemporary security thinking. If security is extended to consider people, not just states as has traditionally been done, then there ought to be possibilities for thinking about human security in geographic terms, but in ways that focus explicitly on these interconnections. Thus, the consequences of climate change on marginal peoples in the South are understood as partly caused by suburban driving patterns in North America; struggles over land use in Kenya are seen in conjunction with the vegetable purchasing patterns of suburbanites in London; plastics are understood as both a problem of disposal and related to the expropriation of land and resources from distant rural communities.

Green Geopolitics

This focus on interconnections, rather than on the recent artificial constructions of modern territorial state boundaries, is appropriate for considerations of the ecological dimensions of politics at the biggest scale. But in the context of global environmental change it is important to remember, as Beck's theory of risk society outlines, that all human activities now in some fashion happen in artificial environments. The scale of human actions has changed, even if only slightly, the air that we all breathe. Our uncertainties are increasingly fabricated, the distinction between natural and artificial is collapsing. This is not to suggest that we are conquering nature, in the sense of subduing it to our collective will. Diseases are showing great resilience in the face of antibiotics and climate changes are likely to produce many phenomena that may endanger the marginal and vulnerable. But our practical decisions as to which species and ecosystems survive or are eliminated are now changing the global ecological situation in unpredictable ways and in turn changing the context within which such decisions are made.

The inadequacy of territorially based national security strategies to deal with the complex problems of postmodern geopolitics show that we need to get our political theories past the conceptual roadblocks of modern spatial assumptions. Then, hopefully, we will see more clearly how interconnectedness is the postmodern condition and that the processes in motion can only be partly controlled by state boundaries. The political slogan "think globally, act locally" has long played with this theme, its users well aware of the local consequences of global actions, but the theories of politics that struggle to deal with globalization frequently remain stuck focused mainly on states. Neither should we prematurely announce the end of the nation-state or the failure of many of its functions, although only some states are capable of carrying out those things that Northern consumers take for granted as the backdrop to their lives. Nonetheless the analysis in this chapter does suggest clearly that Northern states have to evolve rapidly in the face of both global connections and the potential for disruptions. Innovative changes are frequently happening outside state control and leadership in such things as the growth of organic farming, the spread of urban recycling systems, and global protests against specific corporations.

Political leaders will no doubt be especially reluctant to let go of some of the functions of the nation-state, not least because the strategy of blaming external forces for policy failures remains a useful ploy for staying in power in many places. But as the more innovative political thinking and most vocal critics of globalization understand, state leaders are increasingly incapable of addressing the political necessities of developing the much more sustainable modes of living that contemporary risk society clearly needs. Greening geopolitics is about changing the ways we think politically, asking questions about the practical impacts of Northern lifestyles and always remembering the frequently forgotten connections across boundaries.

BIBLIOGRAPHY

Agnew, J. 1998. *Geopolitics: Revisioning World Politics*. London: Routledge.
Allenby, B. R. 2000. Environmental security: Concept and implementation. *International Political Science Review*, 21(1), 5–21.
Athanasiou, T. 1996. *Divided Planet: The Ecology of Rich and Poor*. Boston: Little Brown.
Barnett, J. 2001. *The Meaning of Environmental Security*. London: Zed.
Barrett, H. R., Ilbery, B. W., Browne, A. W., and Binns, T., 1999. Globalization and the changing networks of food supply: the importation of fresh horticultural produce from Kenya into the UK. *Transactions of the Institute of British Geographers, N.S.*, 24, 159–74.
Beck, U. 1992. *Risk Society: Towards a New Modernity*. London: Sage.
Beck, U. 1996. World risk society as cosmopolitan society? Ecological questions in a framework of manufactured uncertainties. *Theory, Culture and Society*, 13, 1–32.
Beder, S. 1997. *Global Spin: The Corporate Assault on Environmentalism*. Totnes, Devon: Green Books.
Benton, L. M. and Short, J. R. 1999. *Environmental Discourse and Practice*. Oxford: Blackwell.
Black, J. K. 1999. *Inequity in the Global Village: Recycled Rhetoric and Disposable People*. West Hartford, CT: Kumarian Press.
Boland, A. 2000. Feeding fears: Competing discourses of interdependency, sovereignty and China's food security. *Political Geography*, 19(1), 55–76.
Bulkeley, H. 1997. Global risk, local values? Risk society and the greenhouse issue in Newcastle Australia. *Local Environment*, 2(3), 261–74.
Chapman, G., Kumar, K., Fraser, C., and Gaber, I. 1997. *Environmentalism and the Mass Media: The North-South Divide*. London: Routledge.
Chase, R., Hill, E., and Kennedy, P. (eds.). 1999. *The Pivotal States: A New Framework for U.S. Policy in the Developing World*. New York: Norton.
Clancy, T. 1999. *Rainbow Six*. New York: Berkley.
Cosgrove, D. 1994. Contested global visions: one-world, whole earth, and the Apollo space photographs. *Annals of the Association of American Geographers*, 84(2), 270–94.
Dalby, S. 1999. Threats from the South? Geopolitics, equity and environmental security. In D. Deudney and R. Matthew (eds.) *Contested Grounds: Security and Conflict in the New Environmental Politics*. Albany, NY: State University of New York Press, 155–85.
Dalby, S. 2000. Geopolitics and ecology: Rethinking the contexts of environmental security. In M. Lowi and B. Shaw (eds.) *Environment and Security: Discourses and Practices*. London: Macmillan, 84–100.
de Larrinaga, M. 2000. (Re)Politicizing the discourse: Globalization is a S(h)ell game. *Alternatives*, 25(2), 145–82.
del Pilar Uribe Marin, M. 1999. Where development will lead to mass suicide. *The Ecologist*, 29(1), 42–6.
Deudney, D. 1999. Environmental security: A critique. In D. Deudney and R. Matthew (eds.) *Contested Grounds: Security and Conflict in the New Environmental Politics*. Albany, NY: State University of New York Press, 187–219.
Diamond, J. 1997. *Guns, Germs and Steel: The Fates of Human Societies*. New York: Norton.
Esteva, G. and Prakesh, S. 1998. *Grassroots Post-Modernism: Remaking the Soil of Cultures*. London: Zed.
Falk, R. 1999. *Predatory Globalization: A Critique*. Cambridge: Polity.
Goldberg, J. 1997. Their Africa problem and ours. *The New York Times Magazine*, 03.02.97, 33–9, 59–62, and 75–7.

Grove, R. 1997. *Ecology, Climate and Empire: Colonialism and Global Environmental History, 1400–1940*. Cambridge: White Horse.

Hajer, M. A. 1995. *The Politics of Environmental Discourse: Ecological Modernization and the Policy Process*. Oxford: Oxford University Press.

Harris, P. G. (ed.). 2000. *Climate Change and American Foreign Policy*. New York: St. Martin's Press.

Hartmann, B. 1999. Population, environment and security: A new trinity. In J. M. Silliman and Y. King (eds.) *Dangerous Intersections: Feminist Perspectives on Population, Environment and Development*. Boston: South End, 1–23.

Held, D., McGrew, A., Goldblatt, D., and Perraton, J. 1999. *Global Transformations: Politics, Economics and Culture*. Stanford, CA: Stanford University Press.

Hochschild, A. 1998. *King Leopold's Ghost: A Story of Greed, Terror, and Heroism in Colonial Africa*. Boston: Houghton Mifflin.

Human Rights Watch. 1999. *The Price of Oil: Corporate Responsibility and Human Rights Violations in Nigeria's Oil Producing Communities*. New York: Human Rights Watch.

Hyndman, J. 2000. *Managing Displacement: Refugees and the Politics of Humanitarianism*. Minneapolis, MN: University of Minnesota Press.

Kaplan R. D. 2000. *The Coming Anarchy: Shattering the Dreams of the Post Cold War*. New York: Random House.

Litfin, K. (ed.). 1998. *The Greening of Sovereignty in World Politics*. Cambridge, MA: MIT Press.

McHaffie, P. 1997. Decoding the globe: Globalism, advertising and corporate practice. *Environment and Planning D: Society and Space*, 15(1), 73–86.

Nye, J. S. and Owens, W. A. 1996. America's information edge. *Foreign Affairs*, 75(2), 20–36.

Ó Tuathail, G. 1997. At the end of geopolitics? Reflections on a plural problematic at century's end. *Alternatives*, 22(1), 35–55.

Ó Tuathail, G. 1998. Postmodern geopolitics? The modern geopolitical imagination and beyond. In G. Ó Tuathail and S. Dalby (eds.) *Rethinking Geopolitics*. London: Routledge, 16–38.

Ó Tuathail, G. 1999. De-territorialized threats and global dangers: Geopolitics and risk society. In D. Newman (ed) *Boundaries, Territory and Postmodernity*. London: Frank Cass, 17–31.

Peet, R. and Watts, M. (eds.). 1996. *Liberation Ecologies: Environment, Development, Social Movements*. New York: Routledge.

Prins, G. (ed.). 1993. *Threats without Enemies: Facing Environmental Insecurity*. London: Earthscan.

Redclift, M. 1999. Environmental security and competition for the environment. In S.C. Lonergan (ed.) *Environmental Change, Adaptation and Security*. Dordrecht: Kluwer, 3–16.

Rowell, A. 1996. *Green Backlash: Global Subversion of the Environmental Movement*. London: Routledge.

Sachs, W., Loske, R., and Linz, M. 1998. *Greening the North: A Post-Industrial Blueprint for Ecology and Equity*. London: Zed.

Suliman, M. (ed.). 1999. *Ecology, Politics and Violent Conflict*. London: Zed.

Taylor, P. J. 1999. *Modernities: A Geohistorical Interpretation*. Minneapolis, MN: University of Minnesota Press.

Watts, M. 1998. Nature as artifice and artifact. In B. Braun and N. Castree (eds.) *Remaking Reality*. London: Routledge, 243–68.

Weldes, J., Laffey, M., Gusterson, H., and Duvall, R. (eds.). 1999. *Cultures of Insecurity: States, Communities and the Production of Danger*. Minneapolis, MN: University of Minnesota Press.

Williams, G. 1995. Modernizing Malthus: The World Bank, population control and the African environment. In J. Crush (ed.) *Power of Development*. London: Routledge, 158–75.

Chapter 29

Environmental Justice

Brendan Gleeson and Nicholas Low

Introduction

The *Dictionary of Human Geography* defines "environmental justice" as a "socio-political movement that seeks to articulate environmental issues from a social justice perspective" (Johnston et al., 2000, p. 2). This description only partly captures the diverse contemporary phenomena that evoke environmental justice, including social movements, legislation, institutional frameworks, and theoretical debate, at a variety of political scales. The environmental justice rubric was born in grassroots community struggles and has in time acquired theoretical and juridical definition and a broader political currency (Harvey, 1996; Low and Gleeson, 1998). Wenz (1988) was the first to tackle the theoretical problems posed by environmental justice.

Gathering momentum in the 1980s, a movement for environmental justice grew in the United States out of civil rights struggles over the siting of toxic and hazardous industries (Towers, 2000). In that country charges of "environmental racism" resonated with a history of racial and ethnic politics and a historic concern with the ethic of rights (Alston, 1990; Sarokin and Schulkin, 1994). In the early 1990s, a series of legislative and institutional initiatives, especially by the US federal government, sought to address the problem of environmental injustice.

Contemporaneously there developed a lively – and sometimes intersecting – international academic discourse of environmental justice that explores the human consequences of environmental ethics. The justice of environmentalism was brought to the foreground of the international political agenda at the United Nations Conference on Environment and Development in 1992, the "Earth Summit." Sustainable development (q.v.) is both about justice between the present and future generations, and justice between rich and poor nations. Although environmental justice mostly refers to the distributional fairness of the human impact upon local environments, the most difficult philosophical question concerns the ethical relationship between humans and the rest of the natural world.

This chapter is divided into three main parts. In the first, we trace the origins of the environmental justice rubric and its early translation to political and policy spheres. After this, the discussion turns to the longer term consequences of environmental justice debates for political theory and action. The final section maps the spatiality of environmental justice, tracing its widening political and geographic resonances.

The Grass Roots Movement for Environmental Justice

In the USA, the debate about environmental justice had its origins in the grassroots struggles of local communities during the 1970s against "environmental racism" (Alston, 1990; Sarokin and Schulkin, 1994). These struggles, involving both local communities of color and a range of progressive groupings (notably churches and civil rights organizations), sought to oppose the racially discriminatory distribution of hazardous wastes and polluting industries in the USA. A range of minority groups were involved in these campaigns, including urban African American and Latino communities and Native American peoples residing on traditional lands (many of which had been poisoned by military and industrial uses). Importantly, this grassroots campaign emerged outside, even at times in opposition to, the mainstream of the environmental movement in the USA (Hofrichter, 1993). Activists pointed out that the environmental movement had concentrated on the ecological concerns of white, middle class Americans, and had failed to identify and oppose the disproportionate burden of toxic contamination on minority communities. Hofrichter (1993, p. 2) attributes the toxic burden on communities of color to "the unregulated, often racist, activities of major corporations who target them for high technology industries, incinerators and waste."

A defining moment in the environmental racism campaign was provided in 1982 by the vigorous protests against the siting of a PCB landfill in a black community within Warren County, North Carolina (Cutter, 1995; Mohai and Bryant, 1992). The Warren County action saw prominent national civil rights leaders uniting with the local community in a campaign of civil disobedience (resulting in 500 arrests) reminiscent of the racial justice struggles of the 1960s (Goldman, 1996; Heiman, 1996). Shortly afterwards, a federal government study found evidence of racial discrimination in the location of commercial toxic waste landfills in one region of the USA (USGAO, 1983). Following this in 1987, the influential United Church of Christ (UCC) Commission for Racial Justice report on toxic waste patterns demonstrated that race was the central determining factor in the distribution of chemical hazard exposure in the USA (UCC, 1987). The broad findings of the landmark UCC report were confirmed by later social scientific studies (e.g. Bullard, 1990a, 1990b, 1992; Bullard and Wright, 1990; Mohai and Bryant, 1992), although in recent years a considerable number of analyses (e.g. Been, 1993, 1994; Boerner and Lambert, 1995) have also sought to "debunk" the environmental racism thesis on methodological grounds. However, both Goldman (1996) and Heiman (1996) point out that many of these sceptical studies have been funded by risk producing and waste management industries.

By the early 1990s, several thousand groups had emerged to oppose inequitable distributions of land uses which threatened the environmental health of local communities (Bullard, 1993a, b and c). In many instances, community action was successful in either preventing the establishment of polluting facilities or ameliorating their effects through both voluntary and enforced agreements on site conditions. Also, the 1990s saw the environmental racism movement refocus its politico-ethical ideals around the broader notion of "environmental justice" (Cutter, 1995). In 1991, more than 650 activists from over 300 local grassroots groups attended the First National People of Color Environmental Leadership Summit in Washington, DC (Goldman, 1996). The summit adopted 17 principles of environmental justice which extend the movement's focus on race to include other concerns, such as class and nonhuman species. Cutter (1995, p. 113) argues that the movement has now transcended, without abandoning, its concern with communities of color to "include others (regardless of race or ethnicity) who are deprived of their environmental rights, such as women, children and the poor" – a definition endorsed by Hofrichter (1993).

This broadening of political purpose described above has also extended the social and institutional reach of the environmental justice movement, which has "moved from street-level protests to federal commissions, corporate strategies, and academic conferences" (Goldman, 1996, p. 131). Not surprisingly then, the emergent environmental justice movement attracted regional and national political attention during the late 1980s and early 1990s. Goldman reported that by 1996: "Numerous federal, state, and local bills have been introduced to address various aspects of environmental injustice, addressing fair siting, citizen participation, compensation, and health research" (1996, p. 127).

In 1992, the US Environmental Protection Agency established an Office of Environmental Equity and published a report on the national distribution of ecological risks (USEPA, 1992). The movement's official recognition reached its zenith in February, 1994, when President Clinton signed Executive Order 12898, which required that every federal agency consider the effects of its own policies and program on the health and environmental well-being of minority communities (Cutter, 1995; Goldman, 1996). To date, the achievements of the federal environmental justice program have been modest, but certainly worthwhile, including remediation works at a number of contaminated sites, the improvement of some community health services, the funding of education and training campaigns in hazard awareness and monitoring, and the targeting of minority business enterprises in the awarding of EPA contracts (USEPA, 1995).

Despite its institutional successes, there are growing indications that the environmental justice movement may have reached its political high-tide mark, at least for the foreseeable future. Earlier, Goldman (1996, p. 126) argued that the movement "may be entering the most difficult phase of its early history" as political opposition to environmentalism strengthens in the federal and state legislatures. In the 1990s, the so-called "Wise Use" campaign has united many industry and resource user groups who argue that US environmental standards and regulations are too stringent and represent an "unjust" circumscription of private property rights (Helvarg, 1994). The Wise Use lobbies were able to influence the environmental and resource policies of the Republican Party which controlled Capitol Hill and many state legislatures for much of the 1990s (Helvarg, 1995). Conservative political forces during this period called for reductions to both funding and programmatic support for a range of state functions (see, e.g., Gillespie and Schellhas, 1994). Critically, these targeted functions included affirmative action programs, redistributive social policies, and environmental regulation, all key public policy elements for the environmental justice movement (Goldman, 1996; Heiman, 1996; Moore and Head, 1993; Sarokin and Schulkin, 1994). During 1994–5, the reality of the Wise Use threat to environmental justice was underscored when the Republican Party proposed radical funding cuts to the federal EPA. Green lobbies and most Democrats countered that these cuts, if realized, would undermine the principal federal institutional base of the environmental justice movement.

In 2000, an EPA Commissioner, Carol Browner, accused the Republican-dominated Congress of weakening environmental policy frames (Katz, 2000, p. 1). Early indications suggested that the new Bush administration would water down many of the environmental policy initiatives undertaken during the Clinton Presidency. Amongst other changes undertaken in the early phases of the Bush Presidency, the EPA's enforcement budget was reduced, thus possibly inhibiting the agency's ability to implement and monitor the Clinton Executive Order on environmental justice. In April 2001, a coalition of environmental groups protested at the White House over a perceived "rollback" of environmental policy and program funding under the Bush Presidency (Greenpeace, 2001).

A further threat emerged during the 1990s in the form of industry-sponsored research and legal maneuvers which have sought to oppose the claims and activities of the environmental justice movement. As Goldman (1996, p. 132) notes, polluting industries and their allies in waste management have engaged a range of "expert" commentators in order to deflect the political arguments of environmental justice activists with legal and scientific complexities: "Now the academic guns have been loaded to defend the turf of expertise, raise the threshold of entry into the debate, and ensure that the burden of proof remains squarely on the backs of the victims of pollution." Here we have a struggle to contain the environmental problematic within the administrative state and its professional ancillaries.

The environmental justice movement also faces a number of internally generated challenges and threats. There is a need for the movement to better define its politico-ethical purpose: many commentators and activists now argue that its established focus on the distribution of environmental well-being has actually entrenched the political power of polluting industry and waste management corporations. As both Heiman (1996) and Cutter (1995) have noted, the movement has tended to pursue "environmental equity," meaning the equitable distribution of negative externalities. Heiman argues that the realization of this political goal will hardly trouble polluting industry, which, after adjusting its locational prerogatives, will resume its risk generating production, only with the EPA assurance of fair equality of opportunity to pollute and be polluted (Heiman, 1996, p. 114).

The environmental justice movement gave new life to the idea of justice as a social ethic. The argument is that it is unfair for the risks and costs of production to be loaded on to the living environments of some people, while others enjoy the benefits but avoid the costs. The linking of "social justice" and "environment" opened environmentalism to increased reflection on the social consequences and conditions of its own ethics. At the same time, it made possible a powerful political and discursive coalition between two congeries of politically mobilized groups: social reformers and environmentalists. The broad mobilization of social justice causes under the environmental justice banner that was achieved during the 1990s may not be sustained in the new millennium. Pulido (2000), for example, has attempted to reinstate race and racism as key politico-theoretical imperatives for environmental justice debates.

The practices and concepts of the environmental justice movement made waves of critical theory in politics, and in fact helped shape a new branch of politics: ecological politics. In the first wave, the experience of the movement as a form of political life – conjoined with new social critiques and ideas about participation – generated new reflections on the meaning of pluralism and liberal democracy. In the second wave, the concept of environmental justice fired the imagination of political theorists worldwide as a potent ground for conceptualizing the environmental crisis and the relationship between social justice and environmentalism. This second wave both internationalized environmental justice and used the concept to address global problems of "sustainability." In the next section we discuss the first of these waves.

The Environmental Justice Movement as a New Paradigm of Politics

At first political commentators did not see much that was new in the environmental justice (EJ) movement. Some sought critically to interpret the movement within established analytical frameworks: pluralism, elitism, populism, political economy (e.g. Boerner and Lambert, 1995; Comacho, 1998; Heiman, 1996; Szasz, 1994). Comacho (1998), for instance, conceptualizes power as the ability of A to prevail over B, "insiders" and "outsiders" in relation to the policy system, questions of incorporation of insurgents, links between racial and class politics, and the success or failure of the environmental justice movement in altering the political agenda.

Foreman (1998) agrees with the conclusions of Boerner and Lambert that only tenuous support, at best, can be found for racial inequity in the siting of hazardous processes or exposure to risk. EJ rhetoric, he argues, discourages citizens from "thinking in terms of risk priorities" (ibid., p. 117), so larger risks may be taken by retaining the *status quo* than by allowing the local siting of plant to deal with toxic hazards. Moreover, he asserts, "The movement is too weak, has too few resources, and has too strong a local orientation to be a significant presence on such national and international matters as global warming, acid rain, airborne particulates, or the future of the electric car" (ibid., p. 122). Szasz (1994, p. 164) observed that the hazardous waste and toxics movement (sic) merely rediscovered "a quite traditional form of popular grass-roots organizing. The only thing that was new about it was the issue that sparked it off." From his analysis of "ecopopulism," he draws the lesson that movements become interesting and influential when they transcend the boundaries of a particular zone of political action: "the challenge and conceptual richness lie in identifying the conditions that allow action to jump from one zone to another and in describing the synergies that then arise" (ibid., p. 163).

Two major efforts to reconceptualize environmental justice in terms of the problem of siting of hazardous facilities came from Davy (1997) and Schlosberg (1999). Davy points to the ineluctable difference between environmental efficiency and environmental compassion. The critics of the EJ movement, such as Foreman, argue justice from the viewpoint of efficiency – within an implicit utilitarian framework: it is unjust for a majority to suffer disbenefit from the activities of a minority even if that disbenefit is severe and cumulative. The environmental justice movement argues justice on the basis of the rights of all people everywhere to a clean and safe environment, an argument that can be traced to the uniform vulnerability of the person and thus to common compassion. Davy (1997, p. 36) points out that standard siting procedures in the USA could have been much improved if the recommendations of the National Governors Association (NGA) in 1981 had been adhered to, the first of which was "to forge a broad consensus on a siting process which is understandable, fair, and leads to a timely, legitimate decision" (NGA, 1981). Such a process would require "early and thorough public participation and education."

The law on hazardous facility siting in the USA and Germany has, in Davy's analysis, created an elaborate language game and a sort of procedural dance around the concept of risk distinguished from the ordinary "wear and tear" of life. The decision of the administrative authority governing siting is deeply rooted in the assessment of experts, depends heavily on technology, and the decision is frequently taken out of the hands of elected local authorities (Davy, 1997, p. 88). Once a permit is issued it creates a legal shield against hostile legal action. This game was put to the test and found extremely deficient in the case, minutely examined by Davy, of the waste incinerator proposal for East Liverpool, Ohio. The proposal was the subject of a long struggle by local people and eventually challenged in the courts by Greenpeace. Although the appellants were ultimately unsuccessful, the case led the Clinton administration to issue an executive order mandating each federal agency to make environmental justice part of its mission "by identifying and addressing, as appropriate, disproportionately high and adverse human health or environmental effects of its programs, policies and activities on minority populations and low-income populations in the United States" (Executive Order of the President of The United States No. 12898, February 11, 1994).

The key to understanding the failure of the legal system is, in Davy's view, the "monolithic miscommunication" which it encourages between proponents and opponents of a hazardous facility. First, the system tends rapidly to divide the multiple stakeholders into just two cases and two camps: "for" and "against." Then (in his case study) "like monoliths, the two 'cases' stood apart, with each side trying to overpower, dominate, and outrun the other" (Davy, 1997, p. 130). Typically a "dialogue" takes place over many years in which statements are made but entirely without communication because neither side hears or responds to the other. As the dialogue proceeds it generates a "rhetoric of distrust" and a strategy on both sides of "whatever it takes" to win.

Injustice is inevitable from someone's point of view, Davy argues, at least when justice is viewed as an "essential" norm: one single, all-embracing concept to determine all situations from all social perspectives. There is no way of resolving three versions (at least) of justice: the elitist version in which developers are regarded as noble competitors in a free market, the social justice version in which the recipients of development are wronged, even when all are wronged equally, and the utilitarian version in which a public authority seeks to uphold the common good. Where any two sides insist on their own version of justice, violence inevitably ensues, and continues until one or other party is utterly defeated. The answer is not justifiable at all but political. Davy's message is this: if only the various parties to a dispute would listen to and validate each other's stories and arguments, they might find that they had much more of a common purpose than they thought, and they might find compromises that did not result in violence or victory/defeat in court.

Davy's prescription, drawn from the negative, in some ways mirrors that of Schlosberg drawn from the positive aspect of the politics of environmental justice. Schlosberg (1999) views the practices of the EJ movement through the lens of a new "critical" pluralism. Traditional pluralism has welcomed difference and "the many." The "old" pluralist theory as it developed in the USA (e.g. Bentley, 1935 [1908], Dahl, 1961; Truman 1965, [1951]) regarded the interest group as the principal unit of political analysis, acknowledged the centrality for democracy of tolerance of difference among interest groups, analysed policy as the outcome of variable "pressure" on the political system on the part of groups with varying resources of power, and held that the rules of liberal democracy, as long as they favored competition, were such as to permit the balance of these pressures to find expression in public policy.

Although Schlosberg finds in Lindblom's work and in that of antecedents such as J. S. Mill an interest in norms of political participation, this aspect of democracy receded in the mainstream of American pluralism, and participation was either taken for granted or its importance downplayed. The practices of the environmental justice movement threw new light on participation, approaching the question: how to be politically effective while at the same time acknowledging, validating, and nurturing cultural difference, and bridging between political concerns. Environmental justice provided not only a unifying discourse for culturally diverse activist groups, but also the way this diversity was handled in practice to make effective decisions enacted a new paradigm of politics.

Central to this new politics are the principles of "agonistic respect," "intersubjectivity," and collaboration rather than competition. Agonistic respect moves from neutral tolerance to positive recognition of the other. Schlosberg draws on the work of the work of James (1979 [1896]) and Connolly (1991, 1995), as well as critics such as Iris Young (1990, 1996) in applauding the discursive ideal of "a diverse banquet, where all the qualities of being respect one another's personal sacredness, yet sit at the common table of space and time" (James, 1979 [1896], p. 201). Reciprocity, as Schlosberg remarks, is the key: "we need not suffer fools, fascists nor those who refuse to grant us recognition, gladly" (Schlosberg, 1999, p. 75). Such a statement abandons relativism and seeks a new political norm.

Intersubjectivity is the process of asserting oneself while recognizing the other; more than this it means accepting the achievement of the life goals of the other into the set of one's own goals – and only then setting about reconciliation of conflicts within that whole set of goals. Finally, critical pluralism throws into stark contrast the norm of competition, making us aware of how far the institution of the competitive market has penetrated liberal political thought. Critical pluralism explicitly rejects the view that politics is about "what qualities add up to victory in group competition." Critical pluralism is about argument, respectful discourse and "mutual adjustment" (Charles Lindblom's term). Agreement on action is viewed as open to continual negotiation: "a critical pluralism demands a conception of unity which does not deny the basis of 'mobile arrangements' or alliances among the differing experiences of those that construct them. Difference, and the proliferation and juxtaposition of diversity, may spawn a plethora of possible unities" (Schlosberg, 1999, p. 101).

The environmental justice movement in America gained its strength from practices of networking and alliance building among grassroots groups and individuals. These practices responded both to the alienation bred by the centralized organization of the mainstream environmentalist groups, and to the community practices familiar to activist groups based on everyday relationships within localities. "Networking," Schlosberg notes, "goes beyond the organizational form. Networks and alliances validate multiplicity and create political relations of solidarity" (Schlosberg, 1999, p. 144).

Critical pluralism calls for "communicative designs." The participatory public inquiries conducted by Thomas Berger on pipeline construction in the Canadian Arctic and the condition of indigenous people in Alaska are cited as precedents (Berger, 1977, 1985; Schlosberg, 1999, p. 147). The inquiries were characterized by open access to both the process of decision and information. Traditionally underrepresented groups (indigenous peoples, rural municipalities, environmental groups, and small businesses) were granted funds to participate on a more equal basis with the large companies. All parties were required to share relevant information. The process allowed for stark value differences to be "aired in a public arena where intersubjective understanding was a goal and agonistic respect the means of achieving it" (Schlosberg, 1999, p. 148).

The substantive meaning of the environmental justice framework is supplied by what it includes, not by what it excludes. Meaning is constructed by the multiple practices and values it embraces rather than by formulaic statement. Nevertheless the First National People of Color Environmental Leadership Summit (FNPCELS) did conclude with a now famous statement about environmental justice. The process through which this statement was shaped was characterized by pluralist, communicative practice. The process, indeed, was held to be as important as the outcome:

> The process by which the principles came into being "illustrated the extraordinary degree of openness, patience, dedication, and sacrifice required to foster sufficient trust across immense cultural barriers to create a powerful multiracial, multinational people's movement" (Madison et al. 1992: 49)....The entire process was an example of the construction of the communicative tools essential to collaboration across differences: respect, openness, intersubjectivity, and solidarity (Schlosberg, 1999, pp. 157–8).

The examples Schlosberg examines are not discussed in sufficient detail to provide convincing evidence of critical pluralism in action. In fact, he cautions that the practices of the environmental justice movement do *not* always follow the ideal model of critical pluralism. Rather, "critical pluralism" is an idea shaped by reflection on the experience of the movement: "much of the effort of the environmental justice movement has been in attempts to create less distorted, and more open, participatory modes of communication both within the movement and in the companies, communities, and governmental agencies it addresses" (ibid., p. 179).

The discussion of environmental justice has turned from principles of justice to principles of politics. There is an immanent critique of the all-pervasive norm of market society, and a return from geographical economics to geographical politics (see Harvey, 1996; Smith, 1994). Szasz (1994) appreciated the revolutionary potential of environmental justice for politically problematizing environmental issues: a potential not restricted to localities or national boundaries. True enough that the environmental justice movement in America had predominantly local battles to fight, but the idea of environmental justice came to have much wider scope and application. Writers now began to focus attention on a multiplicity of struggles worldwide against environmental depredation (e.g. Shiva et al., 1995), to re-examine the conceptual foundations of justice (Low and Gleeson, 1999), and to link environmental justice with environmental governance in the global political arena. We now consider this turn of events.

Internationalizing and Globalizing Environmental Justice

Urban planning, at least since World War II, relied on an implicit conception of social justice. By the 1990s, the social justice agenda had begun to change. On the one hand, the dominance of social justice in political thought had been replaced by the utilitarianism and liberalism of the market, leading to dilemmas for planning (see Gleeson and Low, 2000; Low, 1994). On the other, environmental questions about intergenerational, international, and interspecies justice became mainstream debates (Cooper and Palmer, 1995). Planning practice was acquiring a new agenda: the impact of urban life on the wider natural environment. So it seemed necessary to bring social justice and environmental responsibility together under the rubric of "environmental justice" (Low and Gleeson, 1997). A fuller exploration of the terms of the debate led Low and Gleeson (1998) to conclusions similar to those of Schlosberg and Davy: there could be no single, all-embracing conception of justice, but that did not imply value relativism. Rather, justice must be viewed as a dialectical concept continuously stimulating new political ideas and practices and, importantly, revealing new truths about the relations between human individuals, their societies and nature (Low and Gleeson, 1998, p. 195). There is a universal ethic recognizing the capacity for human beings for doubt, thought and debate based on mutual recognition of equal personal sovereignty in a community of equals.

Out of the debate on justice new principles emerged which define humanity by its relationships: compassion (Davy, 1997), care (Warren, 1999), enlarged thought (Benhabib, 1992), agonistic respect and intersubjectivity (Schlosberg, 1999), inclusion (Young, 1990), recognition (Plumwood, 1999), mutual dependence (FNPCELS), autonomy (Held, 1995), and consent (Tully, 1995). These principles lay foundations for a new post-liberal politics (Gleeson and Low, 2001). Rather than the diminished and furiously competing "self" of utilitarian market ethics, this new politics views the human "self" as defined by its connection with others, the boundaries of "the other" extending in some conceptions to the whole of the natural world (Low and Gleeson, 1998, pp. 67–8).

The new environmental politics is a politics of space and place, but also a global politics. The ideas emerging from the EJ movement in the USA could not be contained within the political context of a racially infused struggle over rights – as already recognized by the movement itself in the first principle of the People of Color Environmental Leadership Summit: "Environmental Justice affirms the sacredness of Mother Earth, ecological unity and the interdependence of all species, and the right to be free from ecological destruction." From a European vantage point, Dobson (1998) explored the relationship between social justice and environmental sustainabilities – for like justice there is no single valid version of environmental sustainability. He concluded, however, that "no theory of justice can henceforth be regarded as complete if it does not take into account the possibility of extending the community of justice beyond the realm of present generation human beings" (1998, p. 244).

A conference was held in Melbourne, Australia, in 1997 to mark these expansions in space, time, and species. The conference brought together people thinking about environmental ethics, scholars of the EJ movement in the USA, and activists and environmental philosophers from around the world. Opening the conference, the Norwegian environmentalist Arne Naess remarked that the frontier of debate is long: "there is an immense need patiently to disseminate information, to dwell repeatedly on the concrete cases of injustice and on the concrete cases of ecological unsustainability" (Naess, 1999, p. 28). Robert Bullard and Vandana Shiva spoke of distributional injustices in America and India, Kristin Shrader Frechette and Clive Hamilton of global threats from Chernobyl and the enhanced greenhouse effect. David Harvey noted: "Of course a universal environmental ethic is impossible – and of course it is desirable!" Other speakers debated key issues of principle: postcapitalist economy, animal rights, the ecological knowledge of indigenous peoples, and global political justice (see Low, 1999 for keynote papers).

The focus of environmental justice shifted to global governance. Just as certain problems of the environment are perceived to be global in scope – such as global warming and the depletion of biodiversity – so global governance comes to be seen as the necessary scale of the solution. Mason (1999) from a North American and British perspective concludes his discussion of "environmental democracy" with a plea for transnational liability for environmental harm and global environmental citizenship. Low and Gleeson (1998) go further, arguing that global economic governance already exists, and global environmental governance must follow. A world constitution for environmental and ecological justice is needed, equipped with institutions such as a global environmental organization with powers comparable to those of the World Trade Organization, a directly elected World Environment Council to provide a forum for public debate, and an International Court of the Environment (see also Postiglione, 2001).

Questions of global environmental justice, however, are problematic. Dryzek, (1999) points to the potential danger of a strong global state arguing instead for a different discursive "software" of politics, embodying reflexive modernization rather than any more global institutional "hardware." Archibugi (2001) agrees, calling for more attention to the task of creating a democratic international politics rather than designing new, comprehensive global institutions, whatever the democratic claims of the latter. Semmens (2001) raises a further difficulty, pointing to the ambiguity of political space in the contemporary global political order and the consequences of this for environmental refugees and their politico-economic oppressors. Both in a sense "suffer" from a deeper deterritorialization inflicted on the world's peoples by the disembedded global economy and its institutional handmaidens.

The ethic preferred by the rich nations for the solution to global warming – at least those nations that want a global agreement at all – is market justice. Market justice rewards the rich. In deciding who pays for global warming, market justice would enable the rich to buy the right to pollute the planet more than the poor. This is the implicit basis of the Kyoto agreement. Even though the developing nations are not yet to be bound by it, the assumption is that at some point, when it has become widely accepted, they will join up. The basis of at least fifty per cent of warming mitigation is to be achieved through the trading of carbon credits in a carbon market.

By contrast, it has been argued by a group of American scientists in an article published in the prestigious American journal *Science* that the basis of the Kyoto protocol is unjust (Baer et al., 2000). Any future agreement should be based not on market justice but on the *equal* rights of all persons rich and poor. Global carbon emissions today average about 1 ton per year per person. The *per capita* emissions of the USA average 5 tons per person per year. In comparison, in developing countries including China and India the average is about 0.6 tons per person per year. To stabilize the climate, average world emissions must be stabilized at about 0.3 tons per person per year. A preferred regime for climate equity based on equal rights would allocate the right to pollute on an equal per capita basis.

Conclusion

Environmental justice was born of grassroots struggle and in many minds it remains firmly rooted in the thickets of everyday community struggles for equity and amenity. This "grounded" interpretation remains an indisputable "core reality" of environmental justice and its increasingly complex and wide ranging theoretical, political, and spatial resonances. However, as our account has attempted to show, the rubric has escaped its original emplacement in community politics to assume a higher profile in national and international discussions about the distribution of environmental quality. Environmental justice put the political spotlight firmly on the *environment* of justice – the ecological dimensions of civil rights. Increasingly, the environment signaled by the rubric is a multi-scalar reality that embraces every political and geographic level at which environmental differentiation is evident, from the neighborhood to the globe. Finally, through its intersection with mainstream green debates, the notion has acquired a temporality, such that we may speak of environmental injustices wreaked on past and future generations of living beings.

What must not be overlooked in this linear tale of widening reach is the more subtle, reflexive impact of environmental justice debates and struggles. We draw attention here to the ways in which environmental justice has worked back on the larger concepts that form its apparent politico-theoretical compass. Seen this way, environmental justice is not simply a practical amalgam of larger separate debates centering on environmental quality and social justice. As our review has shown, the rubric and its analysis have, in turn informed, challenged, and broadened these larger, more familiar socio-political values and the debates that surround them. For one, we have seen increasing awareness within environmental action and theorization of the centrality of social justice and spatial equity for any viable solution to ecological problems. In turn, this has generated new understandings of the wider political constitution of environmental struggle and the implications this raises for a "reflexive modernity" (Beck) that is coming to terms with the ecological and social failures of many modernist institutions and values.

The concept of social justice has not survived the advent of environmental justice unchanged. As part of this, the "social" has broken up; giving way to, and making room for, nature and its human constitution in the form of built and occupied "environments." And in truth the struggle to protect the social and environmental integuments of society against a predatory market capitalism has always been a single struggle even though its social movements have diverged (Low, 2002).

Environmental justice was born in places, pin pricks on the global map. But in time the rubric has acquired multiple and expanding resonances in larger theoretical debates and political spheres. Perhaps the rubric's central and continuing value lies in its ability to inspire both grassroots action in places and a rethinking of politics and concepts across space and time.

BIBLIOGRAPHY

Agnew, J. 1998. *Geopolitics: Revisioning World Politics*. London: Routledge.

Alston, D. (ed.). 1990. *We Speak for Ourselves: Social Justice, Race and the Environment*. Washington, DC: Panos Institute.

Archibugi, D. 2001. The politics of cosmopolitical democracy. In B. J. Gleeson and N. P. Low (eds.) *Governing for the Environment, Global Problems, Ethics and Democracy*. Basingstoke, UK: Palgrave, 196–210.

Baer, P., Harte, J., Haya, B., et al. 2000. Equity and greenhouse gas responsibility. *Science*, 289, 22.

Been, V. 1993. What's fairness got to do with it? Environmental justice and the siting of locally undesirable land uses. *Cornell Law Review*, 78, 1001–85.

Been, V. 1994. Locally undesirable land uses in minority neighborhoods: disproportionate siting or market dynamics? *Yale Law Review*, 103, 1,383–422.

Benhabib, S. 1992. *Situating the Self: Gender, Community and Postmodernism in Contemporary Ethics*. Cambridge, MA: Polity.

Bentley, A. 1935 [1908]. *The Process of Government*. Illinois: Principia.

Berger, T. R. 1977. *Northern Frontier, Northern Homeland: Report of the MacKenzie Valley Pipeline Inquiry*. Toronto: James Lorimer.

Berger, T. R. 1985. *Village Journey: The Report of the Alaska Native Review Commission*. New York: Hill and Wang.

Boerner, C. and Lambert, T. 1995. Environmental injustice. *Public Interest*, Winter 95, 61–82.

Bullard, R. 1990a. *Dumping in Dixie*. Boulder, CO: Westview.

Bullard, R. 1990b. Ecological inequalities and the new South: black communities under siege. *The Journal of Ethnic Studies*, 17(4), 101–15.

Bullard, R. 1992. Environmental blackmail in minority communities. In B. Bryant and P. Mohai (eds.) *Race and the Incidence of Environmental Hazards: a Time for Discourse*. Boulder, CO: Westview, 82–95.

Bullard, R. 1993a. Anatomy of environmental racism and the Environmental Justice Movement. In R. Bullard (ed.) *Confronting Environmental Racism: Voices from the Grassroots*. Boston, MA: Southend.

Bullard, R. 1993b. Waste and racism, a stacked deck? *Forum for Applied Research and Public Policy*, Spring 1993, 29–35.

Bullard, R. 1993c. Anatomy of Environmental Racism. In R. Hofrichter (ed.) *Toxic Struggles: the Theory and Practice of Environmental Justice*. Philadelphia, PA: New Society, 25–35.

Bullard, R. and Wright, B. H. 1990. Toxic Waste and the African American Community. *The Urban League Review*, 13(1–2), 67–75.

Comacho, D. E. (ed.). 1998. *Environmental Injustices, Political Struggles, Race, Class and the Environment*. Durham, NC: Duke University Press.

Connolly, W. 1991. *Identity/Difference: Democratic Negotiations of Political Paradox*. Ithaca, NY: Cornell University Press.

Connolly, W. 1995. *The Ethos of Pluralization*. Minneapolis, MN: University of Minnesota Press.

Cooper, D. E. and Palmer, J. A. (eds.). 1995. *Just Environments, Intergenerational, International and Interspecies Issues*. London: Routledge.

Cutter, S. 1995. Race, Class and Environmental Justice. *Progress in Human Geography*, 19(1), 111–22.

Dahl, R. A. 1961. *Who Governs? Democracy and Power in an American City*. New Haven, CT: Yale University Press.

Davy, B. 1997. *Essential Injustice: When Legal Institutions Cannot Resolve Environmental and Land Disputes*. Vienna: Springer.

Dobson, A. 1998. *Justice and the Environment*. Oxford: Oxford University Press.

Dryzek, J. 1999. Global Ecological Democracy. In N. Low (ed.) *Global Ethics and Environment*. London: Routledge, 264–82.

Foreman, C. H. 1998. *The Promise and Peril of Environmental Justice*. Washington, DC: Brookings Institution Press.

Gillespie, E. and Schellhas, B. 1994. *Contract with America*. New York: Times.

Gleeson, B. J. and Low, N. P. 2000. *Australian Urban Planning, New Challenges, New Agendas*. Sydney: Allen and Unwin.

Gleeson, B. J. and Low, N. P. (eds.). 2001. *Governing for the Environment, Global Problems, Ethics and Democracy*. Basingstoke, UK: Palgrave.

Goldman, B. 1996. What is the future of Environmental Justice? *Antipode*, 28(2), 122–41.

Greenpeace. 2001. *Greenpeace rallies at White House to "take back the Earth" from Bush*. National press release, Greenpeace US, April 18.

Harvey, D. 1996. *Justice, Nature and the Geography of Difference*. Oxford: Blackwell.

Heiman, M. K. 1996. Race, Waste, and Class: New Perspectives on Environmental Justice. *Antipode*, 28(2), 111–21.

Held, D. 1995. *Democracy and the Global Order*. Stanford, CA: Stanford University Press.

Helvarg, D. 1994. *The War Against The Greens*. San Francisco: Sierra Club Books.

Helvarg, D. 1995. Legal assault on the environment. *The Nation*, 30.01.95, 126–30.

Hofrichter, R. 1993. Introduction. In R. Hofrichter (ed.) *Toxic Struggles: the Theory and Practice of Environmental Justice*. Philadelphia, PA: New Society, 1–10.

James, W. 1979 [1896]. *The Will to Believe and Other Essays in Popular Philosophy*. Cambridge, MA: Harvard University Press.

Johnston, R., Gregory, D., Pratt, G., and Watts, M. (eds.). 2000. *The Dictionary of Human Geography*, 4th edn. Oxford: Blackwell.

Katz, A. 2000. EPA chief raps GOP "bottom-line" attitude. *New Haven Register*, 04.27.2000, 1.

Low, N. P. 1994. Planning and Justice. In H. Thomas (ed.) *Values in Planning*. Aldershot, UK: Avebury.

Low, N. P. (ed.). 1999. *Global Ethics and Environment*. London: Routledge

Low, N. P. 2002. Ecosocialisation and Environmental Planning, A Polanyian Approach. *Environment and Planning A*, 34, 43–60.

Low N. P. and Gleeson, B. J. 1997. Justice in and to the environment, ethical uncertainties and political practices. *Environment and Planning A*, 29, 21–42.

Low, N. P. and Gleeson, B. J. 1998. *Justice, Society and Nature: an Exploration of Political Ecology*. London: Routledge.

Low, N. P. and Gleeson, B. J. 1999. Geography, justice and the limits of rights. In J. D. Proctor and D. M. Smith (eds.) *Geography and Ethics, Journeys in a Moral Terrain*. London: Routledge, 30–43.

Madison, I., Miller, V., and Lee, C. 1992. The principles of environmental justice: formation and meaning. In C. Lee (ed.) *Proceedings: The First National People of Color Environmental Leadership Summit*. New York: United Church of Christ Commission for Racial Justice.

Mason, M. 1999. *Environmental Democracy*. London: Earthscan.

Mohai, P. and Bryant, B. 1992. Environmental injustice: weighing race and class as factors in the distribution of environmental hazards. *University of Colorado Law Review*, 63, 921–32.

Moore, R. and Head, L. 1993. Acknowledging the past, confronting the future: environmental justice in the 1990s. In R. Hofrichter (ed.) *Toxic Struggles: The Theory and Practice of Environmental Justice*. Philadelphia: New Society.

Naess, A. 1999. An outline of the problems ahead. In N. Low (ed.) *Global Ethics and Environment*. London: Routledge.

National Governors Association (NGA). 1981. Siting hazardous waste facilities. *The Environmental professional*, 3, 133–42.

Plumwood, V. 1999. Ecological ethics from rights to recognition: multiple spheres of justice for humans, animals and nature. In N. Low (ed.) *Global Ethics and Environment*. London: Routledge.

Postiglioni, A. 2001. An International Court of the Environment. In B. J. Gleeson and N. P. Low (eds.) *Governing for the Environment, Global Problems, Ethics and Democracy*. Basingstoke: Palgrave, 211–20.

Pulido, L. 2000. Rethinking environmental racism: white privilege and urban development in Southern California. *Annals of the Association of American Geographers*, 90(1), 12–40.

Sarokin, D. J. and Schulkin, J. 1994. Environmental justice: co-evolution of environmental concerns and social justice. *The Environmentalist*, 14/2, 121–9.

Schlosberg, D. 1999. *Environmental Justice and the New Pluralism, The Challenge of Difference for Environmentalism*. New York: Oxford University Press.

Semmens, A. 2001. Maximizing justice for environmental refugees; a transnational institution on behalf of the deterritorialized. In B. J. Gleeson and N. P. Low (eds.) *Governing for the Environment, Global Problems, Ethics and Democracy*. Basingstoke: Palgrave, 72–87.

Shiva, V., Ramprasad, V., Hegde, P., Krishnan, O., and Holla-Bhar, R. 1995. *The Seed Keepers*. New Delhi: Research Foundation for Science, Technology and Ecology.

Smith, D. M. 1994. *Geography and Social Justice*. Oxford: Blackwell.

Szasz, A. 1994. *Ecopopulism, Toxic Waste and the Movement for Environmental Justice*. Minneapolis, MN: University of Minnesota Press.

Towers, G. 2000. Applying the political geography of scale: grassroots strategies and environmental justice. *The Professional Geographer*, 52(1), 23–36.

Truman, D. B. 1965 [1951]. *The Governmental Process*. New York: Knopf.

Tully, J. 1995. *Strange Multiplicity: Constitutionalism in an Age of Diversity*. Cambridge: Cambridge University Press.

United Church of Christ. 1987. *Toxic Wastes and Race in the United States: a National Report on the Racial and Socioeconomic Characteristics of Communities with Hazardous Waste Sites*. New York: United Church of Christ Commission for Racial Justice.

United States Environmental Protection Agency (USEPA). 1992. *Environmental Equity: Reducing Risk for all Communities*. Washington, DC: Government Printing Office.

United States Environmental Protection Agency (USEPA). 1995. *Waste Programs Environmental Justice Accomplishments Report – Factsheet*. Washington, DC: USEPA Office of Solid Waste and Emergency Response.

United States General Accounting Office (USGAO). 1983. *Hazardous and Nonhazardous Waste: Demographics of People Living near Waste Facilities*. Washington, DC: Government Printing Office.

Warren, K. 1999. Care-sensitive ethics and situated universalism. In N. Low (ed.) *Global Ethics and Environment*. London: Routledge, 131–45.

Wenz, P. S. 1988. *Environmental Justice*. New York: State University of New York Press.

Young, I. M. 1990. *Justice and the Politics of Difference*. Princeton, NJ: Princeton University Press.

Young, I. M. 1996. Communication and the Other: Beyond Deliberative Democracy. In S. Benhabib (ed.) *Democracy and Difference: Contesting the Boundaries of the Political*. Princeton, NJ: Princeton University Press.

Chapter 30

Planetary Politics

Karen T. Litfin

Globalization, the buzzword of the early twenty-first century, is generally understood in economic, political, and cultural terms. Although the global reach of multinational corporations, the mass media, and international organizations became evident in the last half of the twentieth century, globalization has been underway since the dawn of the colonial era. Similarly, transnational environmental problems have existed for many hundreds of years, since nations first shared waterways and occupied the territories of migratory species. But only in recent decades have environmental problems become *planetary* in scope. The emergence of planetary politics is a direct result of the fact that, by the late twentieth century, humanity had become *a geophysical force*, affecting not only local and regional ecosystems but *all* of Earth's systems: the atmosphere, the biosphere, the hydrosphere, etc. This chapter discusses planetary politics in terms of two key global issues: stratospheric ozone depletion and global climate change. While other issues could be considered, the global politics of the atmosphere are particularly illustrative of the new phenomenon we might call "planetary politics."[1]

Planetary politics entails a distinctive set of dynamics: complex linkages between the local and the global; the necessity and inherent difficulty of North–South cooperation; intergenerational time horizons which are typically articulated on the basis of scientific models; a strong tendency towards a holistic understanding of the Earth's systems; and an incremental institutionalization of a precautionary approach. The causes of global environmental problems span the local to the global. Climate change, for instance, is caused by the local actions of automobile drivers and rice farmers; it is also caused by a global economy that is almost entirely driven by fossil fuel consumption. The effects, however, are truly planetary, bearing little or no relation to their place of origin. The greatest levels of human-induced ozone depletion, for instance, are over Antarctica, despite the fact that almost all ozone-depleting chemicals are released in the northern hemisphere. Likewise, greenhouse gas emissions from Europe and North America threaten to submerge tiny island states in the South Pacific and the Caribbean. These far-flung causal connections can only be grasped with a holistic perspective of the Earth's atmosphere (in practical terms, using computer models), but they also render the assignment of culpability highly difficult. The long atmospheric lifetimes of ozone-depleting chemicals and greenhouse gases – over one hundred years for some – means that the effects of actions taken today will not be felt tangibly for decades. Given the scientific uncertainties on both issues, there is also some concern over the possibility of irreversibility. Finally, since the causes of global environmental problems are global, efforts to address them require global cooperation – particularly between North and South. Yet, given the enormous disparity of resources and given the fact that industrialized and developing countries are far from equally responsible for causing the problem, cooperation is typically a thorny process.

From a theoretical perspective, the emergence of planetary politics poses some overarching challenges to traditional modes of conceptualizing international relations and political geography. Sovereignty, the nation-state, and the pre-eminence of material power have been some of the bedrock principles of world politics during the modern era (Lipschutz and Conca, 1993; Litfin, 1998). Yet each of these is problematized by the emergence of planetary politics – called into question, to be sure, but not necessarily eliminated or altogether superceded by new political structures and practices. Sovereignty, or the notion of fixed, territorially defined and mutually exclusive domains of legitimate rule, is challenged by the holistic understanding of a spatially and temporally interdependent web of human communities. At a minimum, planetary politics requires that sovereignty's traditional emphasis on rights be supplemented by assigning some modicum of global and intergenerational responsibility to the nation-state. More radically, planetary politics may eventually necessitate new, postmodern orientations towards political authority and identity. For the time being, while nation-states remain key players, they are generally pushed and pulled by new actors who are increasingly empowered by the very nature of planetary politics: scientists and environmental non-governmental organizations (ENGOs).

Material power, still an important factor in planetary politics, is being substantially displaced by ideational forms of power, especially the knowledge-based power of science and the new cultural norms of ecological sensibility and planetary stewardship. Yet, while planetary politics is driven by science, science is far from determining; rather, scientific knowledge is framed and interpreted in light of contending material and ideal interests. Environmental problems, global or otherwise, are not simply physical events; they are discursive phenomena that can be studied as struggles among contested knowledge claims which become incorporated into divergent narratives about risk and responsibility. In the planetary politics of ozone and climate change, the proponents of these divergent discourses include a wide array of actors: ENGOs, scientists, businesses, international organizations, and governments.

The continuing importance of material power, however, is evinced in the ability of large firms to shape outcomes; the salience of scientific knowledge and environmental norms does not eliminate the power of big money. In the ozone issue, a handful of chemical companies persuaded the international community to move towards its favored substitutes for chlorofluorocarbons (CFCs). With respect to climate change, the automobile and fossil fuel industry has been successful in blocking meaningful action in the USA. Moreover, the traditional realist claim that the great powers ultimately determine the course of world politics is supported by the ability of the USA to shape the agenda and dominate the outcomes for both issues. Notions of realpolitik, however, offer little insight into planetary politics. The decisions of the great powers to support this or that policy cannot be grasped by conventional power-based explanations of world politics. And, of course, military power is essentially useless in addressing such problems as ozone depletion and climate change. Material power and state power still matter, but the ends to which they are deployed can only be understood with reference to ideational factors.

This chapter will first explore the emergence of planetary politics through efforts to address two primary global environmental problems: ozone depletion and climate change. In addition to offering a historical overview of regime negotiation for the issue, each section will explore the ways in which the issue at hand exemplifies the distinctive dynamics of planetary politics: the complexity of local–global linkages; the dilemmas of North–South cooperation; and the instantiation of intergenerational time horizons and holism. Moreover, each section will examine the ways in which the nation-state system, sovereignty, and the pre-eminence of material power are challenged by efforts to institutionalize global governance for the issue under consideration. Following the analysis of specific issues, the chapter will look more broadly at normative questions involving planetary management and planetary forms of identity and authority.

Stratospheric Ozone Depletion[2]

Planetary politics can be dated from the 1980s, with the first formal efforts to address a truly global environmental problem through an international treaty on the ozone layer. As usual, political action lagged behind the science. In 1974, F. S. Rowland and Mario Molina, two US chemists, hypothesized that, although CFCs were chemically inert at ground level, odd chlorine atoms could be photochemically released from CFCs in the stratosphere where they could initiate a catalytic chain reaction with ozone molecules. While a handful of countries, including the USA, took unilateral action in the late 1970s to eliminate CFCs in aerosols, there was no international consensus on either the gravity of the problem or on how to address it. It was not until 1987 that the first treaty regulating the use of CFCs, the Montreal Protocol, was signed.

CFCs, first discovered by Du Pont Corporation in 1930, were used widely during the late twentieth century by industrialized countries for refrigeration, air conditioning, foam blowing, aerosol propellants, and as a solvent in the electronics industry. These "miracle compounds" were nontoxic, nonflammable, and chemically inert at ground level. A related class of ozone-depleting chemicals, halons, contain bromine instead of chlorine and were used primarily by the US military for firefighting. Although halons were produced in small quantities, they were of special concern because of their extremely high ozone depletion potentials. By 1986, CFC sales in the USA were about $1 billion annually, representing about 30 percent of the world market. Because CFCs were used in expensive products like cars and refrigerators, the value of equipment relying upon CFCs was far higher, estimated at $135 billion in the USA (Alliance for a Responsible CFC Policy, 1987, p. V-1). A handful of US and European companies were responsible for virtually all CFC production.

Responding to citizen pressure following the Rowland-Molina discovery, the USA and a handful of other countries implemented domestic legislation in the late 1970s banning CFCs in aerosol propellants. Internationally, the European Community, resisting any reductions in CFC production, and the USA, pushing for a global aerosol ban, were at loggerheads. Only in 1985 did twenty states, including all of the major CFC producers, sign the Vienna Convention for the Protection of the Ozone Layer, which called for information sharing on ozone-related science and technology. While that treaty failed to mandate precautionary action to control CFCs, it did establish the basis for a normative shift by obligating states to refrain from activities *likely* to damage the ozone layer. The key question, then, became how to know which activities were likely to be harmful; the answer hinged upon the interpretation of the available scientific knowledge.

The first comprehensive international report on ozone (WMO/NASA, 1986) predicted global ozone losses of 5–9 percent by the end of the twenty-first century. From a political perspective, the report's conclusions offered something for everyone. Advocates of the *status quo* argued that the predictions were not dire and would not occur for a long time. Supporters of precautionary action pointed out that predicted ozone losses would increase skin cancer rates by 10–20 percent. More importantly, the discovery of the Antarctic ozone hole in 1985 lent credence to the precautionary position; the hole was not conclusively linked to CFCs until 1988. Because it was not predicted by the computer models, the hole's emergence revealed the inadequacy of the models and suggested that the consequences of under-reacting might be worse than the consequences of over-reacting.

Despite the weak environmental record of the Reagan administration, US negotiators were the key proponents of a precautionary approach. Three factors explain this apparent anomaly: (i) Lee Thomas, the new EPA Administrator, was deeply concerned about the Antarctic ozone hole, supported the precautionary discourse and took a personal interest in the issue. (ii) Because of the earlier aerosol ban, US industry had already invested substantially in research for CFC substitutes. (iii) The Natural Resources Defense Council, a US ENGO, threatened to sue the EPA under the Clean Air Act to reduce CFC emissions, a move that could have put US industry at a competitive disadvantage. Thus, calling for a 95 percent reduction of CFC emissions from 1985 by the year 2000, the USA took the lead in moving the world towards a global phaseout of CFCs.

The Montreal Protocol on Substances that Deplete the Ozone Layer, signed in 1987, required 50 percent reductions of domestic consumption of CFCs and halons. Production cuts could lag by 10 percent to supply importing countries and to allow the EC to rationalize production, while developing countries could delay implementation for ten years. Scientific reviews were mandated every four years; these generated a spate of treaty revisions during the 1990s. The treaty was heralded as an unprecedented instance of global precautionary action. Mostafa Tolba, head of the United Nations Environment Programme (UNEP), called it "the first truly global treaty that offers protection to every single human being on this planet, . . . unique because it seeks to anticipate and manage a world problem before it becomes an irreversible crisis" (Tolba, 1987).

No sooner was the treaty signed, than a scientific consensus emerged on three core issues: the Antarctic ozone hole was definitively linked to CFCs; global ozone losses were confirmed; and a significant risk of depletion over the Arctic was recognized. These conclusions induced a deepening sense of crisis among scientists, diplomats, and ENGOs. Given the generally glacial pace of international law, this sense of crisis led to a remarkably rapid series of treaty amendments over the next several years. Meeting in London in 1990, the parties agreed to a global ban on CFCs, halons, and two hitherto unconsidered ozone depleting chemicals (carbon tetrachloride and methyl chloroform) by the year 2000. Two years later, in Copenhagen, the parties advanced the phaseout dates for all of these chemicals to 1994–6, mandated a freeze on the production of methyl bromide (an agriculture chemical) by 1995, and banned a primary family of CFC substitutes (HCFCs) by 2030. (For a summary of the targets and timetables mandated by the Montreal Protocol and subsequent treaty revisions, see table 30.1.)

Developing countries, having been relatively silent on the ozone issue prior to Montreal, raised their voices loudly once a CFC ban was on the table. The ten-year grace period would become a moot point with a global ban. They argued that they should not have to forego necessities, like refrigeration, in order to solve a problem caused by the industrialized countries; nor should they be forced to pay higher prices for substitutes. For their part, the chemical companies would not compromise their intellectual property rights by simply giving over the new chemicals to developing countries. China and India submitted an innovative proposal, eventually adopted in 1990, for a multilateral ozone fund to finance the transfer of substitute technologies to developing countries. The USA only reluctantly backed the fund, and insisted upon a clause specifying that it would be established "without prejudice to any future arrangements" (London Revisions, Article 10, paragraph 10). This reference was a thinly veiled claim that the ozone fund should not be construed as a precedent for the climate issue. Nonetheless, the ozone fund established an important precedent in recognizing that global cooperation required institutional mechanisms to facilitate Third World participation, especially when their participation would entail economic sacrifice. In the case of the ozone fund, the amounts in questions were small – in the order of $100 million annually for several years. A comparable fund for climate change would be much larger, and hence far more difficult to institutionalize. Even for the relatively small ozone fund, contributions from industrialized countries to the multilateral ozone fund have lagged behind their pledges.

Table 30.1 Montreal Protocol Revisions: amounts and dates to final reductions for ozone-depleting chemicals

	Montreal (1987)	London (1990)	Copenhagen (1992)[a]
CFCs	50% by 2000*	100% by 2000	100% by 1996
Halons	Freeze by 1992	100% by 2000	100% by 1994
Methyl chloroform	–	100% by 2005	100% by 1996
Carbon tetrachloride	–	100% by 2000	100% by 1996
HCFCs	–	–	100% by 2030[b]
Methyl bromide	–	–	Freeze by 1995

[a]The base year used for calculating reductions for each agreement is the year preceding the agreement.
[b]Applies only to industrialized countries.

The decision to eliminate CFCs sparked a struggle over replacements. CFC producers quickly seized the opportunity to promote two sets of compounds: HCFCs, which disintegrate more rapidly in the lower atmosphere because they contain hydrogen; and HFCs, which are ozone-friendly but are extremely potent greenhouse gases. While HCFCs could only be used transitionally because they deplete ozone, the chemical companies argued that they could be used safely until at least 2030. The conflict over CFC substitutes pitted ENGOs against the chemical industry, which found an ally in the World Bank. For the time being, industry seems to have prevailed. At the insistence of the USA, the World Bank, with its long history of supporting capital-intensive development strategies that generate profits for multinational firms, obtained primary control of the Multilateral Fund.[3] Hence, most projects supported by the Fund have involved transfer of HCFC technologies to developing countries. In the atmosphere of crisis that emerged after 1988, eliminating the major ozone depleting chemicals became the primary task; how it was done was a secondary consideration.

The Montreal Protocol and its subsequent amendments are regarded as the world's strongest and most successful environmental regime. Within fifteen years of a framework convention, a whole family of chemicals was outlawed globally. Large ozone losses continue to occur annually over Antarctica and, to a lesser extent, over the Arctic, but this is to be expected given the long atmospheric lifetimes of CFCs. Nonetheless, there is a general consensus that the ozone layer will begin recovering early in the twenty-first century, with close to full recovery within a hundred years. Between the initial discovery in 1974 that CFCs could deplete stratospheric ozone and the global ban on those compounds in 1996, twenty-two years elapsed.[4] Given the creeping pace of international law, this represents rapid progress. There is also a general consensus that the ozone case was a relatively easy one for building global cooperation. The CFC industry was relatively small and self-contained, dominated by only a handful of companies, and substitute chemicals were available without significant economic sacrifice. Scientific uncertainty was low, particularly after the Antarctic ozone hole was linked to CFCs in 1988. Nonetheless, while all of these factors make the ozone issue an "easy case," the achievement of global cooperation should not be minimized.

Lessons for planetary politics

The key positive lesson to be drawn from the ozone case is that concerted global action in the face of environmental crisis is possible. Science played a crucial role in facilitating international cooperation, but its influence was not definitive.[5] Clearly, scientific knowledge was responsible for putting the issue on the agenda. Without the Rowland-Molina discovery and later research, without the discovery of the Antarctic ozone hole by ground-based scientists and its subsequent confirmation by satellite data, without the long-term predictions generated by sophisticated computer models, the ozone layer would not have become an issue of planetary politics. Prior to Montreal, scientific information did not provide any clear policy direction. Rather, science was framed and interpreted in light of contending discourses about risk and responsibility. After Montreal, the science supported a decision to ban CFCs and other halogens, but it gave no clear guidance on how to proceed with respect to chemical substitutes. Overall, science facilitated the incremental institutionalization of a global norm protecting the ozone layer.

Perhaps most significant in terms of its contribution to planetary politics, ozone science was fundamentally holistic and intergenerational in its orientation. The long atmospheric lifetimes of ozone-depleting chemicals meant that the scientific projections of ozone loss extended into future centuries. The inherently global character of the ozone layer was evident in the fact that the location of ozone losses was irrelevant to the location of CFC emissions. Moreover, the computer models used for predicting future ozone depletion were intrinsically holistic. The territorial boundaries entailed in the notion of sovereignty were thereby rendered meaningless as an institutional framework for addressing the problem. The problem required global cooperation, particularly in the face of developing countries' aspirations to increase their own CFC consumption.

The ozone case is also instructive with respect to North–South cooperation in planetary politics. Initially, developing countries were unconcerned about proposals to reduce CFC consumption globally because of exceptions that allowed them to continue increasing their own consumption. However, while the South's per capita CFC use was a tiny fraction of the North's, *projections* indicated that CFC consumption in the South would eventually outstrip that of the North. Therefore, industrialized countries recognized that a viable ozone regime would have to be global. In 1989, once a global ban on CFCs was under serious consideration, developing countries mobilized to promote the multilateral ozone fund. Despite US opposition, the fund established an important precedent for financial compensation and technology transfer to developing countries in exchange for their participation in the regime. In the eyes of its supporters, it represents the only fair way to gain Third World cooperation; in the eyes of its opponents, it represents a new form of environmental blackmail.

Environmental NGOs have been prominent players on the ozone issue, but to a lesser extent than on other issues. ENGOs induced the USA to ban CFCs in aerosols, but were less visible during the Montreal Protocol negotiations. The threat of a lawsuit by the Natural Resources Defense Council was only one factor among many in the EPA decision to promote strong precautionary action on CFCs prior to 1987. In the late 1980s and early 1990s, however, ENGOs became a powerful presence on a host of international environmental issues, including ozone. In the aftermath of Montreal, ENGOs launched a number of campaigns targeting municipal and industrial users of CFCs, persuading McDonald's to abandon its use of styrofoam containers made with CFCs, for instance. Greenpeace and Friends of the Earth worked hard to include HCFCs and methyl bromide in the list of phase-out chemicals. There is little doubt that the presence of ENGOs at post-Montreal negotiations, and as a constant background political presence affected the eventual outcomes.

In sum, the ozone case illustrates the key dynamics of planetary politics: the complexity of local–global linkages; the importance of science and global civil society; the necessity and inherent difficulty of North–South cooperation; intergenerational time horizons and a holistic perspective; and the problematic nature of sovereignty as a framework for addressing problems of global ecology. As the first clear instance of the formal institutionalization of planetary politics, the ozone case offers a sense of hope that global cooperation can be achieved in the face of ecological crisis.

Global Climate Change[6]

While the climate change issue illustrates all of the same dynamics, it offers less reason for optimism because the pace of change has been much slower. Unlike the ozone case, the climate change issue involves several trace gases, especially carbon dioxide, which are byproducts of virtually every aspect of industrial society. Nonetheless, like the ozone case, it also exemplifies the dynamics of planetary politics.

The "greenhouse effect," caused by the radiation-trapping properties of certain gases, is one of the most widely accepted theories in the atmospheric sciences, although climate change only recently became a policy problem. Until 1957, most scientists believed that the oceans would absorb virtually all anthropogenic CO_2; in that year, two scientists concluded that the oceans would absorb only half of it. In an oft-repeated phrase, they declared that "mankind is conducting a great one-time geophysical experiment" (Revelle and Suess, 1957, p. 27). This somber warning, however, attracted little policy attention until the discovery in the 1980s that other trace gases, including methane, CFCs, and nitrous oxides, would nearly double the warming trend expected from CO_2 alone.

The level of political involvement by climate change scientists is unprecedented in planetary politics. Beginning in 1985, scientific conferences generated a plethora of strong policy recommendations, calling for a rapid reduction fossil fuel dependence, increases in energy efficiency by 50 percent, and an end to deforestation, with the objective of limiting the rate of global warming to 0.1°C per decade (WMO, 1985). At a 1988 Toronto conference, over 300 scientists and policy makers called for reducing CO_2 emissions by 20 percent by 2005. That target, which scientists believed would fall far short of stabilizing the climate, was mostly a matter of political feasibility. The US EPA, for instance, estimated that stabilizing atmospheric concentrations of CO_2 at 1988 levels would require reducing greenhouse gas emissions by 50 to 80 percent, returning them to 1950s levels. The values implicit in scientists' policy recommendations reflect an underlying precautionary approach – a belief that it is unwise to alter the global climate system in ways that might induce conditions with which the human species has no experience. One important set of uncertainties, "climate change surprises," could include disturbances of the gulf-stream and possible triggering of an ice age in the northern hemisphere (Kerr, 1998).

Within months of the Toronto conference, the Intergovernmental Panel on Climate Change (IPCC) was established. The first IPCC assessment, drawing on the work of over three hundred scientists, only accepted as certain the greenhouse effect, that human activities were increasing atmospheric concentrations of greenhouse gases, and that global temperatures would rise 1.5 to 5.0°C by 2050. As for environmental effects, only sea-level rise was predictable with any certainty, and even then the range was sizable: between 20 and 140 centimeters. The observed warming was consistent with the modeled predictions, but it also fell within the range of natural variability (IPCC, 1990, pp. 2–5).

With no clear "signal" that global warming was underway, opponents of regulatory action were able to ensure that the Framework Convention on Climate Change, signed at the United Nations Conference on Environment and Development (UNCED) in 1992, contained no binding measures to reduce greenhouse gas emissions. In the evolution of planetary politics, UNCED is noteworthy because of the unprecedented number of NGOs in attendance. The Climate Action Network, an umbrella group of NGOs around the world working on the climate change issue, was particularly prominent at UNCED and since. Despite the disappointment of NGOs, the treaty committed its parties to "protect the climate system for the benefit of present and future generations of humankind, on the basis of equity and in accordance with their common but differentiated responsibilities and respective capabilities" (Framework Convention on Climate Change, Article 3). Throughout the international negotiations, the notion of "common but differentiated responsibilities" has framed North–South relations. The normative commitments entailed in the FCCC helped NGOs to later press governments to act more decisively.

Unlike the ozone issue, conflict between North and South characterized the climate negotiations from the start. Because industrialization has traditionally been premised upon fossil fuel consumption, many developing countries feared that efforts to address climate change would hinder their own economic development. Developing countries were quick to point out the inherent injustice in equating the "luxury emissions" of the rich with "survival emissions" of the poor (Agarwal and Narain, 1991), The South, however, has not been unified on this issue. Dozens of Third World countries, the small-island and low-lying countries, are most at risk and have promoted radical measures to reduce greenhouse gas emissions. Their umbrella group, the Alliance of Small Island States, has worked closely with ENGOs. At the other extreme, OPEC member states have been joined by the fossil fuel industry in attempting to block progress in the negotiations. As in the ozone issue, there is an international consensus in support of a technology transfer fund for developing countries, at least in principle, but little headway has been made towards formal implementation.

The IPCC released its second assessment report late in 1995, concluding that "the balance of evidence . . . suggests a discernible human influence on global climate" (IPCC, 1996, p. 5). Once climate change was scientifically documented as empirical reality rather than modeled prediction, the stage was set for a legally binding regulatory agreement. The Kyoto Protocol, signed in December 1997, mandates reductions in overall greenhouse gas emissions by thirty-nine industrialized countries by 5.2 percent of 1990 levels by the year 2012. The targeted reductions are not the same for all, with most European countries bound by 8 percent targets, the US by 7 percent, and Australia permitted to increase emissions by 8 percent. The Kyoto Protocol also authorizes tradeable permits, enabling countries to buy "credits" from other countries whose emissions levels are below the 1990 level. Countries can count forests as carbon "sinks," so that their actual reductions can be lower than the mandated targets. Developing countries are not bound by any formal regulations, but do commit themselves to a "clean development mechanism" which would permit industrialized countries to offset their own emissions reductions by financing projects in developing countries. Many observers have been critical of such "loopholes" as carbon sinks, emissions trading, and the clean development mechanism for permitting affluent countries to avoid reducing their own greenhouse gas emissions.

At first glance, the Kyoto Protocol's requirements seem minimalist. Yet this would be a serious misreading of the treaty's practical and symbolic significance. Given that emissions were projected to rise by 30 percent, the protocol actually requires meaningful reductions from where they would have been (Bolin, 1998). Moreover, Kyoto should not be viewed as the last word, but rather as an important first step. From toxic waste trade to ozone depletion, planetary politics has proceeded on an incremental basis. Thus, Kyoto sets the course for future developments, but its fate is tied to the willingness of governments to implement it.

The USA, accounting for about 25 percent of the world's greenhouse gas emissions, is a crucial player. Yet, as home to five of the seven major oil companies and the center of global automobile production, it has a powerful incentive to oppose international regulatory action. The Global Climate Coalition, the largest industry group involved in the climate change issue and closely tied to the fossil fuel industry, has been especially influential in the USA, where its member companies are among the largest contributors to congressional campaigns (Levy and Egan, 1998). In opposing a regulatory protocol, the fossil fuel industry has made tremendous efforts to contest the scientific authority of the IPCC, sponsoring its own studies that dispute the IPCC findings and going directly to the mass media.

Yet industry had been more influential in the USA than internationally. In comparison to governments, international institutions are relatively more insulated from the pressures of capital and popular concerns about jobs and fuel prices (Levy and Egan, 1998). In other words, because their political authority is not rooted in popular consent, international institutions need not be so preoccupied with economic performance. Thus, industry groups like the GCC enjoy their greatest influence at the national level and only in certain countries. Indeed, NGOs appear to have the upper hand in the international arena. Most noteworthy is the transnational Climate Action Network, which has distributed daily newsletters to the delegates at every Conference of the Parties since Rio. Nonetheless, industry groups have so far been successful in preventing ratification of the Kyoto Protocol by the US Senate, a development that could seriously hamper implementation elsewhere.

Unlike other transnational environmental problems, climate change uniquely touches the heart of industrial society. Thus, we may anticipate that states would strongly resist any proposals to reduce greenhouse gas emissions. Furthermore, given the significant uncertainties in both the models and the data, we might expect some strong challenges to the authority of climate science – not just from interest groups like the automobile, coal, and petroleum industries, but also from states themselves. Yet an international political consensus has emerged which recognizes the authority of climate science. Whatever its tangible results in terms of the Earth's climate, the Kyoto Protocol provides evidence for a growing commitment on the part of political leaders to the authority of climate science, even if that authority might encroach upon a central dimension of their political authority – their ability to promote economic growth. Thus, the climate regime may point to the emergence of a new global norm of intergenerational responsibility, a norm that could have momentous implications for planetary politics in the twenty-first century.

Lessons for Planetary Politics

Although the pace of treaty-making has been much slower than the ozone case, international action on climate change exhibits many of the same dynamics: the incremental institutionalization of environmental norms based upon changing scientific knowledge and a holistic approach; the difficulty and necessity of North–South cooperation; complex linkages between the local and the global; the inadequacy of sovereignty as an institutional framework for addressing the problem; and a growing recognition of the need for new norms of intergenerational responsibility.

Science plays a crucial, yet somewhat ambiguous, role in planetary politics, setting the agenda but never determining the outcomes. Science both renders the invisible visible and extends the temporal horizons of policy actors. As one long-time analyst observes, "Science makes the environment speak. Without science, trees have no legal standing, ecosystems degrade unrecognized, and species are lost without our knowing" (Moltke, 1997, p. 265). Neither ozone nor climate change would exist as political issues without the computer-based predictions of scientists. Science also interjects an intergenerational time frame into planetary politics, but the decision to take precautionary action is not scientifically mandated. Rather, it reflects a moral and political commitment to act, even in the face of scientific uncertainty, to avert or lessen environmental risk.

Although holism and norms of intergenerational responsibility are integral to planetary politics, the climate case interjects a cautionary note. The Earth may be a unified whole, but global politics does not yet reflect that reality. Politics, even planetary politics, remains a game played by self-interested, and often short-sighted, actors. The threat of catastrophic climate change may simply not be enough to compel governments, industries, and citizens to wean themselves from their fossil-fuel addiction. Similarly, although the institution of sovereignty may be inadequate to the tasks of planetary politics, this does not mean that it will automatically disappear. The ability of the USA to block implementation of the Kyoto Protocol, at least for the time being, demonstrates the enduring influence of both sovereignty and material power.

Paradoxically, however, the inherent difficulties in addressing global climate change could spur a broader and deeper institutionalization of planetary politics. For some observers, this means the growth of global civil society, which consists of "self-conscious constructions of networks of knowledge and action by decentred, local actors that cross reified boundaries of space as if they were not there" (Lipschutz, 1992). Other observers see the North–South divide as providing a potential basis for this shift, since justice (to say nothing of Third World compliance) will eventually require a global transfer of resources to help developing countries cope with the problems of global warming and switch to alternative energy resources (Hertsgaard, 1998). A large-scale transfer of wealth, it is argued, would require a global administrative body with more power and legitimacy than the UN currently enjoys. Thus, the climate change issue could provide the rationale for institutionalizing some form of world government (Low and Gleeson, 1998). This possibility may seem remote from today's vantage point, but it is nonetheless a distinct possibility.

While the pace of change may seem slow, we should remind ourselves that planetary politics is a very new phenomenon, emerging only in the past two decades. Some of the questions that it raises are just appearing on the horizon. Does a planetary politics entail formal institutionalization on a global scale, or can it be approached in a more decentralized manner? To what extent are the norms and practices of sovereignty consistent with planetary politics? Does a planetary politics entail planetary forms of social identity, and to what extent is it realistic to presume that these will subvert existing forms of national, racial, gender, and class identity? These are only some of the far-reaching questions raised by the emergence of planetary politics, but given the scope and pace of environmental degradation, we can only expect these questions to become more salient in the coming century. Other problems, such as biodiversity loss or deforestation, are sometimes called global, but are less clearly so because their loci of cause and effect are more geographically circumscribed. While both occur across the planet, and the global economy is a primary cause of both, they are comparatively more amenable to national solutions because the resources in question fall under the sovereign jurisdiction of various states. Other atmospheric issues, like acid rain or urban smog, are clearly local and regional, rather than global, in scope. The Earth's ozone layer and climate system, in contrast, are characterized by truly planetary relations of causality that can only be understood and addressed holistically.

ENDNOTES

1. International "regimes" covering a wide range of environmental issues could be addressed. But none is as truly planetary as ozone depletion or global climate change.
2. Most of the information for this section is drawn from Litfin (1994) and Benedick (1999).
3. Since the early 1980s, the World Bank has been heavily criticized by the NGO community for its poor environmental and human rights record. In response, the Bank has implemented certain procedural and program reforms, the adequacy of which is sharply debated (see Reed, 1997).
4. The subtitle of a book on the "ozone crisis" published in 1989 is "the fifteen-year evolution of a sudden global emergency" (Roan, 1989).
5. The question of the extent to which science itself is socially constructed is hotly debated among historians of science and sociologists of knowledge (Latour and Woolgar, 1979).
6. The material for this section is drawn primarily from Houghton (1994) and Litfin (2000).

BIBLIOGRAPHY

Agarwal, A. and Narain, S. 1991. Global warming in an unequal world: a case of environmental colonialism. *Earth Island Journal*, Spring 91, 39–40.

Alliance for a Responsible CFC Policy. 1987. *Montreal: A Briefing Book*. Rosslyn, VA: ARCFCP.

Benedick, R. 1999. *Ozone Diplomacy: Safeguarding the Planet*, 2nd edn. Cambridge, MA: Harvard University Press.

Bolin, B. 1998. The Kyoto Negotiations on Climate Change: a science perspective. *Science*, 279(16), 328–32.

Hertsgaard, M. 1998. *Earth Odyssey: Around the World in Search of Our Environmental Future*. New York: Broadway.

Houghton, J. 1994. *Global Warming: The Complete Briefing*. Cambridge: Cambridge University Press.

Intergovernmental Panel on Climate Change (IPCC). 1990. *IPCC First Assessment Report*. Geneva: WMO/UNEP.

Intergovernmental Panel on Climate Change (IPCC). 1996. *IPCC Second Assessment Report*. Geneva: WMO/UNEP.

Kerr, R. 1998. Warming's Unpleasant Surprises. *Science*, 281, 156–8.

Latour, B. and Woolgar, S. 1979. *Laboratory Life: The Social Construction of Scientific Facts*. Beverly Hills: Sage.

Levy, D. L. and Egan, D. 1998. Capital Contests: National and Transnational Channels of Corporate Influence on the Climate Change Negotiations. *Politics and Society*, 26(3), 337–61.

Lipschutz, R. 1992. Reconstructing World Politics: The Emergence of Global Civil Society. *Millennium*, 21, 389–420.

Lipschutz, R. and Conca, K. (eds.). 1993. *The State and Social Power in Global Environmental Politics*. New York: Columbia University Press.

Litfin, K. T. 1994. *Ozone Discourses: Science and Politics in Global Environmenal Cooperation*. New York: Columbia University Press.

Litfin, K. T. 1997. Sovereignty in World Ecopolitics. *Mershon International Studies Review*, 41(2), 167–204.

Litfin, K. T. (ed.). 1998. *The Greening of Sovereignty in World Politics*. Cambridge, MA: MIT Press.

Litfin, K. T. 2000. Environment, Wealth, and Authority: Global Climate Change and Emerging Modes of Legitimation. *International Studies Review*, Fall 2000, 119–48.

Low, N. and Gleeson, B. 1998. *Justice, Society and Nature: An Exploration of Political Ecology*. London: Routledge.

von Moltke, K. 1997. Institutional Interactions: The Structure of Regimes for Trade and the Environment. In O. Young (ed.) *Global Governance: Drawing Insights from the Environmental Experience*. Cambridge: MIT Press, 247–72.

Reed, D. 1997. The Environmental Legacy of Bretton Woods: The World Bank. In O. Young (ed.) *Global Governance: Drawing Insights from the Environmental Experience*. Cambridge: MIT Press, 227–46.

Revelle, R. and Suess, H. E. 1975. Carbon dioxide exchange between atmosphere and ocean and the question of an increase of atmospheric CO2 during the past decades. *Tellus*, 9, 18–27.

Roan, S. 1989. *Ozone Crisis: The Fifteen-Year Evolution of a Sudden Global Emergency*. New York: Wiley.

Tolba, M. 1987. Press Statement (September 22, 1987). Nairobi: United Nations Environment Programme.

World Meteorological Organization (WMO). 1985. *Report of the International Conference on the Assessment of the Role of Carbon Dioxide and Other Greenhouse Gases in Climate Variations*. Geneva: WMO-661.

World Meteorological Organization/National Aeronautics and Space Administration (WMO/NASA. 1986. *Atmospheric Ozone, 1985*. Washington, DC: WMO/NASA.

Index